The field of plant breeding covers a broad range of different species and categories of plants. While there are many techniques and issues that are similar across these species, many more are unique to each genus.

The *Handbook of Plant Breeding* is organized by major crop categories and includes the most up-to-date molecular techniques being used. It will serve as a resource for plant breeding laboratories in both the university and industrial setting.

More information about this series at http://www.springer.com/series/7290

Antonio M. De Ron
Editor

Grain Legumes

Springer

Editor
Antonio M. De Ron
Misión Biológica de Galicia (MBG)
Spanish National Research Council (CSIC)
Pontevedra
Spain

ISSN 2363-8478 ISSN 2363-8486 (electronic)
Handbook of Plant Breeding
ISBN 978-1-4939-2796-8 ISBN 978-1-4939-2797-5 (eBook)
DOI 10.1007/978-1-4939-2797-5

Library of Congress Control Number: 2015942916

Springer New York Heidelberg Dordrecht London

Printed on acid-free paper

Springer Science+Business Media LLC New York is part of Springer Science+Business Media (www.springer.com)

Preface

Legume species belong to the Fabaceae family and are characterized by their fruit, usually called pods. Several species of this family were domesticated by humans, such as soybean, beans, faba bean, pea, chickpea, lentil, peanut, lupine, pigeon pea, mung bean, peanut, or cowpea and many of them are of great relevance as human food and animal feed. Food legumes are typically consumed as dry seeds, which have high protein content, and in some cases as immature seeds or pods.

Members of the legume family, the Fabaceae or Leguminosae, fill critical niches in most terrestrial biomes. This is one of the few plant families whose species are capable of "fixing" nitrogen from the air, through association with specialized soil bacteria, for use as a natural fertilizer, thus reducing fertilizer requirements. The family has traditionally been divided into three subfamilies: Caesalpinioideae, Mimosoideae and Papilionoideae, this latter subfamily contains most of the major food and feed legumes.

Several grain legume crops are crucial elements of global agriculture and nutrition, both as food and feed since they are major sources of plant protein. Legumes contribute to the sustainable improvement of the environment when grown in agricultural rotations due to their ability of biological nitrogen fixation and their effects on the soil, and yield of the next crop, and the services given to other components of agroecosystems such as pollinators. Legumes play a key role in the diversification and sustainable intensification of agriculture, particularly in light of new and urgent challenges such as climate change. The overall objective is to increase the sustainability of the food and feed chain at all its steps, meet the requirements of citizens for safe, healthy and affordable food via the nutritional prevention of diet-related diseases and assure food quality and authenticity. Reducing energy and water consumption and optimizing process control contribute to making food processing and distribution more sustainable and the food sector more competitive.

The demand for plant proteins for human nutrition has increased over the past few decades in many countries due to: (i) demographic growth and urbanization, (ii) the limited land areas which can be used for production of food crops while farming systems are changing towards specialized cereal and oilseed production, (iii) a decrease in animal protein production due to shortage of irrigation and/

or rainfall especially, and (iv) deliberate reduction in red meat consumption for health reasons. Because of the high protein content of their seeds, grain legumes are attractive candidates to overcome the deficiency in plant protein production. However, in comparison to cereals, limited improvement in farming practices has been achieved over the past few decades to enhance the production of important grain legumes. A number of limiting factors affect legume yield, with water deficiency in quantity or quality being among the key ones, to obtain more stable and more reliable production. Even though these constraints have become structural in many agrosystems, very limited research and development efforts have been devoted to strategies to improve grain legume production under stress conditions to contribute to the development of sustainable agriculture worldwide.

Further, the decrease in legume cropping is linked to a heavier use of chemical fertilizers, pesticides and herbicides than in the past and an overall simplification of agricultural systems. This has reduced the level of above-and below-ground biodiversity in terms of macro- and microorganisms living in the agroecosystem and has caused an increased pollution of the environment, impairing the beneficial effects biodiversity has on crop productivity and the maintenance of agroecosystem services for future generations. In addition, the decrease in legume cropping in some agricultural areas urgently needs to be reversed as nitrogen fertilizers costs are increasing with rising energy costs, leading to high production costs for farmers, and substantial greenhouse gas emissions linked to the use of nitrogen fertilizers.

Also social and scientific issues should be considered. Interest in legumes has been decreasing among many farmers, breeders, processing sector entrepreneurs and scientists. Most worrying is the fact that knowledge on grain legumes with regard to growing legumes in rotations, appropriate harvesting, storage and preparation of the seed for further reproduction or processing have progressively been lost. In addition, the use of legumes in human diet is decreasing in many developed countries and knowledge on how to use legumes in food preparations is being lost, despite continued calls by the medical professions to include a wider range of plant proteins in the diet. To reverse these current trends, actions must be taken, to promote wider use of legumes in crop production that will enable significant benefits in economic, environmental and climate change spheres.

Approaches aimed at the improvement and exploitation of legume nutritional and technological qualities are needed and can be expected to drive consumers and farmers towards new, diverse, healthier and more sustainable choices. To contribute to the development of sustainable agriculture, special attention has to be paid to the factors limiting legume yield to obtain more consistent production and to fill the knowledge and development gap on strategies to improve grain legume production under stress conditions.

The decrease in manufacture of inorganic N fertilizers will result in reducing the emission of greenhouse gas. Nitrous oxide (N_2O) is produced naturally in the soil during the microbial processes of nitrification and denitrification; considered over a 100-year period, N_2O is a greenhouse gas with tremendous global warming potential (GWP) when compared to carbon dioxide (CO_2) since it has 310 times the ability per molecule of that gas to trap heat in the atmosphere. The decline of soil fertility with loss of organic matter, the excessive use of chemical fertilizers, the inappropriate use of the scarce water resources and the increase in soil acidity and salinity, particularly in dry regions, all pose real threats to economic, social and environmental sustainability. Agricultural systems involving legumes represent a cheaper and more sustainable alternative to conventional practices by symbiotically capturing atmospheric N_2, thus reducing the use of industrially produced nitrogen in the production of field crops. Improved N management is needed not only to optimize economic returns to farmers but also to minimize environmental concerns associated with N use, namely leaching problems and water pollution.

Intercropping or crop rotation including legumes is a promising strategy for more sustainable crop production in many agricultural systems through the N transfer and N release from legume residue. In crop rotation, legume crops can be used in between of cereals or other cash crops (e.g., vegetables). The final contribution of fixed N_2 to the soil depends upon the legume species N balance, environmental conditions and agricultural practices.

Globally, grain legumes are the most relevant source of plant protein, especially in many countries in Asia, Africa, and Latin America, but there are some constraints in their production, such as poor adaptation, pests and diseases, and unstable yield. Current research trends in legumes are focused on new methodologies involving genetic and -omic studies, as well as new approaches to the genetic improvement of these species, including the relationships with their symbiotic rhizobia.

The book on grain legumes includes two parts. The first one consists of eight crop-specific chapters devoted to the most produced and consumed worldwide grain legume crops covering the whole range of topics related to breeding: origin and evolution, genetic resources, breeding achievements, specific goals and techniques, including the potential and actual integration of new technologies. The second part includes five cross chapters covering topics that relate to the different crops of the general chapters. All the chapters have been written by outstanding breeders and scientists with wide experience in their crops and topics. This handbook contains all the basic and updated information on the state of the art of breeding grain legumes. The vast amount of knowledge collected in this volume should not only serve breeders but also researchers, students and academicians. It may be regarded as a scientific knowledge platform that provides practical plant breeders with new scientific information, but also to make molecular biologists more familiar with the peculiarities of breeding of the main grain legume species.

Pontevedra, Spain Antonio M. De Ron

Acknowledgement

The editor acknowledges the excellent contributions of all the authors, as well as the support by the Springer Editorial team: Kenneth K. Teng, Hanna Smith, Elizabeth Orthmann and Megha Koirala. Special thanks to the Editors-in-Chief of this series Handbook of Plant Breeding, Jaime Prohens, Fernando Nuez and Marcelo Carena, for giving me the great opportunity to edit this Handbook on Grain Legumes.

Contents

Contributors

Michael Abberton International Institute of Tropical Agriculture, Ibadan, Nigeria

Jelena M. Aleksić Laboratory for Plant Molecular Biology, Institute of Molecular Genetics and Genetic Engineering (IMGGE), University of Belgrade, Belgrade, Serbia

Nora Alkama Department des Sciences Agronomiques, Universite de Mouloud Mammeri, Tizi Ouzou, Algeria

Nuno Felipe Almeida Instituto de Tecnologia Química e Biológica António Xavier (ITQB), Oeiras, Portugal

Moez Amri Regional Field Crop Research Center of Beja (CRRGC), Beja, Tunisia

Adnane Bargaz Biosystems and Technology, Swedish University of Agricultural Sciences, Alnarp, Sweden

James Beaver Department of Crop and Agro-Environmental Science, University of Puerto Rico, Mayaguez, PR, USA

Penelope J. Bebeli Department of Crop Science, Laboratory of Plant Breeding and Biometry, Agricultural University of Athens, Athens, Greece

Deng-Jin Bing Agriculture and Agri-Food Canada, Lacombe, AB, Canada

Elena Bitocchi Dipartimento di Scienze Agrarie, Food and Environmental Sciences, Università Politecnica delle Marche, Ancona, Italy

Roberto Bollini CNR Institute of Agricultural Biology and Biotechnology, Milan, Italy

Lech Boros Department of Seed Science and Technology, Institute of Plant Breeding and Acclimatization—National Research Institute, Radzików, Poland

Ousmane Boukar International Institute of Tropical Agriculture, Cowpea Breeding Unit, Kano, Kano, Nigeria

Mark A. Brick Department of Soil and Crop Sciences, Colorado State University, Fort Collins, CO, USA

Judith Burstin INRA, UMR1347 Agroecology, Dijon, France

Pedro A. Casquero Engineering and Agricultural Sciences, University of León, León, Spain

Patricia Castro Genetic Improvement of Fruits and Vegetables Lab, Beltsville, MD, USA

Timothy J. Close Department of Botany and Plant Sciences, University of California, Riverside, CA, USA

Eleonora Cominelli CNR Institute of Agricultural Biology and Biotechnology, Milan, Italy

Clarice J. Coyne USDA, Washington State University, Pullman, WA, USA

José I. Cubero Departmento de Genetica, Universidad de Córdoba, Córdoba, Spain

Antonio M. De Ron Biology of Agrosystems, Misión Biológica de Galicia (MBG), Spanish National Research Council (CSIC), Pontevedra, Spain

Daniel G. Debouck Genetic Resources Program, International Center for Tropical Agriculture (CIAT), Cali, Valle del Cauca, Colombia

Girish Prasad Dixit Indian Institute of Pulses Research, Kanpur, Uttar Pradesh, India

Claire Domoney Department of Metabolic Biology, John Innes Centre, Norwich Research Park, Norwich, UK

Thomas F. Döring Department of Agronomy and Crop Science, Humboldt University Berlin, Berlin, Germany

Jean-Jacques Drevon Eco&Sols, INRA, Montpellier, France

Gérard Duc UMR1347 Agroecologie, INRA, Dijon, France

Jeffrey D. Ehlers Department of Agricultural Development, Bill and Melinda Gates Foundation, Seattle, WA, USA

Thomas Henry Noel Ellis CGIAR Research Program on Grain Legumes, Patancheru, Telangana, India

Aaron Fait Blaustein Institutes for Desert Research, Ben Gurion University, Midreshet Ben-Gurion, Israel

Christian A. Fatokun International Institute ot Tropical Agriculture (IITA), Ibadan, Oyo State, Nigeria

Deidré Fourie Plant Breeding, ARC-Grain Crops Institute, Potchefstroom, South Africa

Valérie Geffroy Université Paris Sud, Orsay, France

Ana M. González Biology of Agrosystems, Misión Biológica de Galicia (MBG), Spanish National Research Council (CSIC), Pontevedra, Spain

Thomas J.V. Higgins CSIRO Agriculture Flagship, Canberra, ACT, Australia

Bao Lam Huynh Department of Nematology, University of California—Riverside, Riverside, CA, USA

Katarzyna Anna Kamel Department of Genomics, Institute of Plant Genetics, Polish Academy of Sciences, Poznan, Poland

Jaime Kigel Hebrew University of Jerusalem, Rehovot, Israel

Magdalena Kroc Department of Genomics, Institute of Plant Genetics, Polish Academy of Sciences, Poznan, Poland

Stephen Kyei-Boahen International Institute of Tropical Agriculture (IITA), Nampula, Mozambique

Antonio Leonforte Nuseed Innovation Centre, Horsham, VC, Australia

Rafael Lozano Department of Biology (Genetics), Research Center on Agricultural and Food Biotechnology (BITAL), University of Almería, Almería, Spain

Eva Madrid Institute for Sustainable Agriculture, CSIC, Córdoba, Spain

Pascal Marget UMR1347 Agroecologie, INRA, Dijon, France

Frédéric Marsolais Agriculture and Agri-Food Canada, Genomics and Biotechnology, Southern Crop Protection and Food Research Centre, London, ON, Canada

Phillip McClean Department of Plant Science, North Dakota State University, Fargo, ND, USA

Rebecca J. McGee USDA-ARS, Grain Legume Genetics and Physiology Research Unit, Washington State University, Pullman, WA, USA

Kevin E. McPhee Department of Plant Sciences, North Dakota State University, Fargo, ND, USA

Aleksandar Mikić Forage Crops Department, Institute of Field and Vegetable Crops, Novi Sad, Serbia

Teresa Millán Department of Genetics, Campus de Rabanales, Córdoba, Spain

Reid G. Palmer Department of Agronomy, Iowa State University, Ames, IA, USA

Roberto Papa Dipartimento di Scienze Agrarie, Food and Environmental Sciences, Università Politecnica delle Marche, Ancona, Italy

Jeffrey Paull School of Agriculture, Food and Wine, The University of Adelaide, Glen Osmond, SA, Australia

Robert J. Redden Agriculture Productivity. Department of Economic Development, Jobs, Transport & Resources, Australian Temperate Field Crops Collection Horsham, VIC, Australia

Philip A. Roberts Department of Nematology, University of California—Riverside, Riverside, CA, USA

A. Paula Rodiño Biology of Agrosystems, Misión Biológica de Galicia (MBG), Spanish National Research Council (CSIC), Pontevedra, Spain

Leah Rosental French Associates Institute for Agriculture & Biotechnology of Drylands, Jacob Blaustein Institutes for Desert Research, Ben-Gurion University of the Negev, Beer Sheva, Israel

Diego Rubiales Institute for Sustainable Agriculture, CSIC, Córdoba, Spain

Josefa Rubio Area de Mejora y Biotecnologia, IFAPA Centro "Alameda del Obispo", Córdoba, Spain

Marta Santalla Biology of Agrosystems, Misión Biológica de Galicia (MBG), Spanish National Research Council (CSIC), Pontevedra, Spain

Olaf Sass Norddeutsche Pflanzenzucht Hans-Georg Lembke KG, Holtsee, Germany

Petr Smýkal Department of Botany, Palacky University in Olomouc, Olomouc, Czech Republic

Francesca Sparvoli CNR Institute of Agricultural Biology and Biotechnology, Milan, Italy

Thomas R. Stefaniak North Central Research Extension Center, North Dakota State University, Minot, ND, USA

Frederick L. Stoddard Department of Agricultural Sciences, University of Helsinki, Helsinki, Finland

Kiriya Sungthongwises Faculty of Agriculture, Khon Kaen University, Khon Kaen, Thailand

María José Suso Plant Breeding. Instituto de Agricultura Sostenible (CSIC), Córdoba, Spain

Wojciech Święcicki Department of Genomics, Institute of Plant Genetics, Polish Academy of Sciences, Poznan, Poland

Ana M. Torres Area de Mejora y Biotecnología, IFAPA Centro Alameda del Obispo, Córdoba, Spain

Albert Vandenberg Department of Plant Sciences/Crop Development Centre, University of Saskatchewan, Saskatoon, SK, Canada

Maria Carlota Vaz Patto Instituto de Tecnologia Química e Biológica António Xavier (ITQB), Oeiras, Portugal

Margarita Vishnyakova Genetic Resources of Grain Legumes, Vavilov Institute of Plant Industry, Saint-Petersburg, Russian Federation

Thomas D. Warkentin Crop Development Centre, University of Saskatchewan, Saskatoon, SK, Canada

Norman Weeden Department of Plant Sciences and Plant Pathology, Montana State University, Bozeman, MT, USA

Zong Xuxiao Chinese Academy of Agricultural Sciences (CAAS), Institute of Crop Science (ICS), Beijing, Beijing, China

Fernando Juan Yuste-Lisbona Department of Biology (Genetics), Research Center on Agricultural and Food Biotechnology (BITAL), University of Almería, Almería, Spain

Mainassara Zaman-Allah CIMMYT, Southern Africa Regional Office, Harare, Zimbabwe

About the Author

Antonio M. De Ron is Research Professor at the Misión Biológica de Galicia (MBG), Spanish National Research Council (CSIC) in Pontevedra, Spain. He received both his graduate and doctorate in Biology from the University of Santiago de Compostela (USC, Spain). Initially he worked as a postgraduate student in plant protection at the National Institute for Agricultural Research (INIA, Spain). In 1988, Prof De Ron gained a staff position as scientist at the MBG working on legume genetics, germplasm and breeding and he is currently the leader of the Biology of Agrosystems Research Group. He was also part-time lecturer of genetics and breeding at the USC. Currently he is also serving as Leader of the Protein Crops Working Group of the European Association for Research on Plant Breeding (EUCARPIA, The Netherlands). He has published many scientific articles, books and monographs, as well as several educational publications. Prof De Ron has won awards for his work and is well recognized in the international scientific community for his achievements.

Chapter 1
Common Bean

Antonio M. De Ron, Roberto Papa, Elena Bitocchi, Ana M. González, Daniel G. Debouck, Mark A. Brick, Deidré Fourie, Frédéric Marsolais, James Beaver, Valérie Geffroy, Phillip McClean, Marta Santalla, Rafael Lozano, Fernando Juan Yuste-Lisbona and Pedro A. Casquero

1 Introduction

The common bean (*Phaseolus vulgaris* L.) is a diploid annual species and is predominantly self-pollinating. Common bean consists of two major gene pools, Mesoamerican and Andean, characterized by partial reproductive isolation, that include wild populations and cultivated varieties. The common bean is the third most important food legume crop worldwide, surpassed only by soybean (*Glycine max* (L.) Merr.) and peanut (*Arachis hypogea* L.). Among the main food crops, the common bean shows the greatest variation in growth habit, seed characteristics (size, shape and colour) and maturation time. This variability enables its production in a wide range of cropping systems and environments as diverse as the Americas, Africa, the Middle East, China and Europe (Blair et al. 2010). Despite being cultivated for its fresh pods and grains, beans are produced and consumed mainly as dry grain.

A. M. De Ron (✉) · A. M. González · M. Santalla
Biology of Agrosystems, Misión Biológica de Galicia (MBG), Spanish National Research Council (CSIC), El Palacio-Salcedo, 36143 Pontevedra, Spain
e-mail: amderon@mbg.csic.es

A. M. González
e-mail: amgonzalez@mbg.csic.es

M. Santalla
e-mail: msantalla@mbg.csic.es

R. Papa · E. Bitocchi
Dipartimento di Scienze Agrarie, Food and Environmental Sciences, Università Politecnica delle Marche, Via Brecce Bianche, 60131 Ancona, Italy
e-mail: r.papa@univpm.it

E. Bitocchi
e-mail: e.bitocchi@univpm.it

A. M. De Ron (ed.), *Grain Legumes*, Handbook of Plant Breeding 10,
DOI 10.1007/978-1-4939-2797-5_1

The common bean has lately gained attention as a functional food due to its health benefits and human disease prevention. In fact, its inclusion in diets is linked to reduce risk of obesity, diabetes, cardiovascular diseases, and colon, prostate and breast cancer (Correa 1981; Hangen and Bennink 2003; Thompson et al. 2009). These health benefits may be attributed to its important fibre and starch content, ability to regulate glycaemia and gastrointestinal function, as well as to its antioxidant properties provided by the presence of phenolic compounds and proteins.

D. G. Debouck
Genetic Resources Program, International Center for Tropical Agriculture (CIAT), Km 17 recta a Palmira, Cali, Valle del Cauca AA 6713, Colombia
e-mail: d.debouck@cgiar.org

M. A. Brick
Department of Soil and Crop Sciences, Colorado State University, 307 University Avenue, Fort Collins, CO 80532, USA
e-mail: Mark.Brick@Colostate.edu

D. Fourie
Plant Breeding, ARC-Grain Crops Institute, 114 Chris Hani Street, Private Bag X 1251, Potchefstroom, South Africa
e-mail: FourieD@arc.agric.za

F. Marsolais
Agriculture and Agri-Food Canada, Genomics and Biotechnology, Southern Crop Protection and Food Research Centre, 1391 Sandford St., London, ON N5V 4T3, Canada
e-mail: Frederic.Marsolais@agr.gc.ca

J. Beaver
Department of Crop and Agro-Environmental Science, University of Puerto Rico, Call Box 9000, Mayaguez, PR 00681-9000, USA
e-mail: James.Beaver@upr.edu

V. Geffroy
Batiment 630, Université Paris Sud, Orsay, France
e-mail: Valerie.geffroy@u-psud.fr

P. McClean
Department of Plant Science, North Dakota State University, Fargo, ND, USA
e-mail: phillip.mcclean@ndsu.edu

R. Lozano
Department of Biology (Genetics), Research Center on Agricultural and Food Biotechnology (BITAL), University of Almería, Almería, Spain
e-mail: rlozano@ual.es

F. J. Yuste-Lisbona
Department of Biology (Genetics), Research Center on Agricultural and Food Biotechnology (BITAL), University of Almería, Almería, Spain
e-mail: fyuste@ual.es

P. A. Casquero
Engineering and Agricultural Sciences, University of León, León, Spain
e-mail: pedro-casquero@unileon.es

For centuries, farmers have maintained their heirloom varieties and have exchanged their seeds with surrounding areas, mainly in local markets. It is not always easy to know the use given by farmers to their old landraces, and it must be assumed that snap and dry beans have probably been selected under dissimilar criteria and pressure. This results in a very different set of characteristics for size, shape, tenderness and cooking quality of the edible parts of plant. Therefore, the traditional varieties are a valuable source of well-adapted germplasm of common bean. The current common bean germplasm collections show a wide variation of phenotypes, although in many developed countries where landraces are being replaced by elite cultivars the genetic erosion is affecting the species. Also the traditional intercropping with maize in many countries is into abeyance, and sole cropping of bean may become unsustainable in some environments as the soil is eroded and the pressure of pests and diseases builds up (Davis and Woolley 1993).

The current integration of genomic data into gene bank documentation systems and its combination with genetic, taxonomic, agronomic, phenotypic and ecological data will usher in a new era for the valorization of the common bean genetic resources.

2 Origin and Systematics

2.1 Phylogeny

Most of the *Phaseolus* species are geographically distributed in Mesoamerica, and for this reason the genus is considered to have originated in Mesoamerica (Freytag and Debouck 2002; Delgado-Salinas et al. 2006) between 6 and 4 million years (Ma) ago (Delgado-Salinas et al. 2006). This indicates that the *Phaseolus* genus originated after the late Miocene (ca. 7 Ma ago, Coates et al. 2004) when the closure of the Isthmus of Panama allowed the connection of Mesoamerica and South America through a land bridge. Eight well-supported crown clades characterize the *Phaseolus* genus, with an average age of ca. 2 Ma, thus indicating that most of the diversity came into existence after the formation of the actual geographical and geological form of Mexico (ca. 5 Ma ago; Delgado-Salinas et al. 2006). Among the eight *Phaseolus* clades, the *vulgaris* group is the oldest, at ~4 Ma. Along with *P. vulgaris,* there are three other domesticated *Phaseolus* species that belong to this group *(P. dumosus, P. coccineus, P. acutifolius)*, with the most closely related species to *P. vulgaris* being *P. coccineus* and *P. dumosus*. Gepts et al. (1999) suggested that *P. vulgaris* diverged from *P. dumosus* and *P. coccineus* some 2 Ma ago, through an analysis of the sequence data of the α-amylase inhibitor gene. The other domesticated species, *P. lunatus,* is most distantly related to *P. vulgaris* (Delgado-Salinas et al. 2006).

2.2 Origin

Wild forms of *P. vulgaris* occur from northern Mexico to northwestern Argentina, and they are characterized by three distinct gene pools (Fig. 1.1): Mesoamerica, the Andes and northern Peru–Ecuador (Debouck et al. 1993; Kami et al. 1995). The Mesoamerican and Andean are the two main gene pools, where the geographical structure is evident also for the domesticated forms, as it has been demonstrated

Fig. 1.1 Common bean gene pools

through studies based on *P. vulgaris* morphology, seed proteins, allozymes, multilocus molecular markers and nucleotide data (Bellucci et al. 2014a). The third gene pool is constituted by wild populations that grow in a small geographical area on the western slopes of the Andes, the distinctiveness of which is the specific phaseolin (main seed-storage protein), type I ('Inca', Kami et al. 1995). This phaseolin type has not been found in the other two gene pools.

Until recently, the most credited origin of the species was the northern Peru–Ecuador hypothesis, as suggested by Kami et al. (1995) who sequenced a portion of the gene coding for phaseolin and reported that the phaseolin type I gene does not have the tandem direct repeats that are, instead, characteristic of the Mesoamerican and Andean phaseolin types. Considering that duplications, which generate tandem directs repeats, are more likely to occur than deletions, which specifically eliminate a member of a tandem direct repeat, Kami et al. (1995) suggested that *P. vulgaris* originated from the wild populations of northern Peru and Ecuador and subsequently spread northwards (from Colombia to northern Mexico) and southwards (from southern Peru to Argentina).

The alternative hypothesis describes a Mesoamerican origin for *P. vulgaris*. Bitocchi et al. (2012) investigated the nucleotide diversity at five gene fragments across a wide sample of wild *P. vulgaris* accessions that were representative of the entire geographical distribution of the species. In particular, three main outcomes supported a Mesoamerican origin of the common bean. (i) A strong reduction in the genetic diversity (90 %) of the Andean compared to Mesoamerican wild forms, indicating the occurrence of a bottleneck in the Andean gene pool that predates its domestication. (ii) A clear population structure is highlighted in Mesoamerica, with four different genetic groups (B1, B2, B3 and B4) that characterize the accessions from this gene pool. The B1 group included accessions distributed across all the Mesoamerica, while the other three groups were characterized by only Mexican accessions; in particular, the B2 group spread from central to southern Mexico, and the B3 and B4 being characteristic of a wide area of central Mexico. Such a population structure had not been identified before in previous studies, the main reason for which was probably related to the nature of the markers used; indeed, compared with multilocus molecular markers, sequence data are less prone to homoplasy (e.g. Wright et al. 2005; Morrell and Clegg 2007), and the assumption of no recombination is less likely to be violated, and thus, these sequence data were very useful to address evolutionary issues (Bitocchi et al. 2012, 2013). (iii) There is no clear distinction between the Mesoamerican and Andean wild gene pools, which was indicated by the phylogenetic relationships between the four different Mesoamerican genetic groups with the South American gene pools.

Considering all of these data, Bitocchi et al. (2012) suggested the Mesoamerican origin of *P. vulgaris,* with Mexico being the more likely cradle of diversity of this species, where all of the four different genetic groups are present. Moreover, they suggested that the wild common bean that grows in northern Peru and Ecuador represents a relict population that only includes a fraction of the genetic diversity of the ancestral population, with phaseolin type I appearing to be extinct in Mesoamerica. This result was recently confirmed by the resequencing of 60 wild *P. vulgaris* genotypes (Schmutz et al. 2014).

2.3 Domestication

The domesticated forms of *P. vulgaris* have important traits that distinguish them from the wild forms, such as reduced and loss of the dissemination mechanisms, loss of seed dormancy and photoperiod insensitivity, greater seed size and determinate growth habit. The main effect of domestication was a reduction in the genetic diversity in the domesticated forms that was imposed by founder effect (i.e. genetic drift) and selection at loci controlling domestication traits. This reduction has been clearly identified in the Mesoamerican domesticated gene pool in several studies (Papa et al. 2005; Papa et al. 2007; Rossi et al. 2009; Kwak and Gepts 2009; Nanni et al. 2011; Bitocchi et al. 2013). The same studies have shown that, in contrast, in the Andean gene pool, the bottleneck of domestication was less evident; in particular, Bitocchi et al. (2013) showed a reduction in the diversity that was threefold greater in Mesoamerica as compared with the Andes.

Bellucci et al. (2014b) applied next-generation sequencing technology (RNA-Seq) to investigate, not only at nucleotide but also at transcriptome level, the domestication process in Mesoamerica. They analysed nucleotide polymorphism and gene differential expression in wild and domesticated forms at 27,243 contigs, each representing a putative single gene. Their results showed that domestication not only led to a drastic reduction of diversity (~60%) but also decreased diversity of gene expression (~18%). Another important outcome of this study was the detection of ~9% of contigs being affected by selection during domestication (directly targets of selection or physically linked to the selected genes). The findings indicated that positive selection was the rule, even if, in a few cases, selection increased the nucleotide diversity in the domesticated forms at target loci associated with abiotic stress responses, flowering time and morphology.

A still open debate concerns the occurrence of single or multiple domestications within the two main gene pools, with studies suggesting both single (Papa and Gepts 2003; Kwak and Gepts 2009; Kwak et al. 2009; Rossi et al. 2009) and multiple (Singh et al. 1991a, b, c; Chacón et al. 2005) events. However, the most recent studies support a single domestication, in both Mesoamerica and the Andes (Bitocchi et al. 2013).

Mamidi et al. (2011) analysed sequence data from 13 loci and dated the domestication bottlenecks to ca. 8000 and ca. 7000 years before the present for the Mesoamerican and Andean gene pools, respectively. In Mesoamerica, two different domestication geographical areas have been suggested recently: Rio Lerma–Rio Grande de Santiago basin in west-central Mexico (Kwak et al. 2009) and in Oaxaca Valley (Bitocchi et al. 2013). Similarly, in the Andes, Chacón et al. (2005) indicated central-southern Peru as the geographical area where *P. vulgaris* was domesticated; in contrast, other studies have suggested Bolivia and northern Argentina (Beebe et al. 2001; Bitocchi et al. 2013).

2.4 Diffusion and Evolution Out of the Americas

The diffusion of *P. vulgaris* out of the American domestication centres appears to have been very complex and to have involved numerous introductions into different continents and countries. Several of these have been proposed as secondary centres of diversification, such as Europe (Santalla et al. 2002; Angioi et al. 2010; Gioia et al. 2013), central-eastern and southern Africa, Brazil and China (Bellucci et al. 2014a). In particular, *P. vulgaris* from Europe is characterized by a higher frequency of the Andean (ca. 70%) as compared to Mesoamerican types (Gepts and Bliss 1988; Gil and De Ron 1992; Logozzo et al. 2007; Angioi et al. 2010). In Brazil, Burle et al. (2010) reported that the Mesoamerican types are fourfold more frequent than the Andean. In Africa, there is an equal frequency of the two types (Gepts and Bliss 1988; Asfaw et al. 2009; Blair et al. 2010), while China shows a predominance of the Mesoamerican types (Zhang et al. 2008).

Moreover, once out of the Americas, the spatial isolation between the Mesoamerican and Andean gene pools was not maintained, which provided increased potential for their hybridization and introgression. In Europe, this aspect is very important for breeding; indeed, their hybridization has led to the recombination of the Mesoamerican and Andean traits that has resulted in the production of novel and useful genotypes and phenotypes (i.e. resistance to biotic and abiotic stress; Rodiño et al. 2006; Angioi et al. 2010; Blair et al. 2010; Santalla et al. 2010). However, various studies suggest that in other continents, the introgression between these gene pools appears not to be as relevant as it has been in Europe.

3 Genetic Resources and Utilization

Somewhere in Central America during the Pliocene and for 4 Ma (Delgado-Salinas et al. 2006), a group of legumes evolved in what is today the section Phaseoli of the *Phaseolus* genus (Freytag and Debouck 2002). One of them, *P. vulgaris* L., migrated northwards and to the Andes and has survived as wild in montane forests to this date. When humans crossing Beringia during the last Ice Age colonized the Americas, they found common beans growing wild from Mexico down to Argentina. Genetic studies with the help of molecular markers have shown these beans to be diverse though grouped in 2–3 pools (Tohme et al. 1996). For reasons possibly linked to food shortages, about 8000 years ago (Mamidi et al. 2011), Amerindians started planting beans, that is, initiated a domestication process. This happened independently in western Mexico (Kwak et al. 2009) and in the central Andes (Chacón et al. 2005), possibly at the same time or slightly earlier in the Andes. Beans planted with corn were a basic staple for all New World civilizations from the Carolinas (USA) down to Jujuy (Argentina). In 1493, the Spanish galleons brought common beans to the Old World where new processes of selection and recombination resumed. Not surprisingly, new landraces and some recombinants occurred in these new lands of adoption such as Spain, Italy, eastern Africa and China. If time correlates with the piling up of genetic diversity, useful sources are clearly in the secondary gene pool of Phaseoli and in the wild forms (Porch et al. 2013).

Bean breeding has often focused first on transfer of resistances to diseases and pests because of the imperative to secure the 'meat of the poor' throughout Latin America and Africa (where the highest consumption per capita is registered). Yield per se, tolerance to drought, adaptation to low phosphorus soils and nutritional quality are priorities of bean breeders since the 1990s (Broughton et al. 2003). Although not the entire germplasm has been collected nor evaluated, many interesting traits have been disclosed in ex situ collections (Table 1.1) and have been used to get yield gain close to 20% over the past 50 years (Singh et al. 2007). While many landraces were topping at 400 kg/ha, yields of 2900 kg/ha are no longer the exception. Growth habit from a vine liana has been 'domesticated' too, namely with the selection of type II, for mechanical harvesting, and changing the original poor root system is coming into the horizon, by using the secondary gene pool (Porch et al. 2013). Although current ex situ collections harbour diversity (Table 1.2, where the top five gene banks have 34% of total accessions worldwide), wild species and secondary gene pools are not yet fully represented nor evaluated, an obvious and timely priority.

Table 1.1 Some bean germplasm used to overcome limiting factors in bean production

Trait looked for	Material used[a]
Abiotic stresses	
Aluminium toxicity	G35346 (*P. coccineus*, from Oaxaca)
Drought	Common red Mexican G11212; G21212 landrace from Colombia
Low phosphorus	G19227A; Chaucha Chuga G19833
Diseases	
Angular leaf spot	Interspecific hybrids with *P. coccineus*; Bolivian G8719; Mexican G2726
Anthracnose	Aliya G02333; Kaboon G1588; Cornell 49–242 G5694
Anthracnose	Interspecific hybrids with *P. coccineus*
Ascochyta blight	*P. dumosus* G35182 from Guatemala
BGYMV	Royal Red G04450; *coccineus* G35172 from Rwanda
BCMV	Porillo Sintético G4495, Royal Red G04450
Beet curly top virus	California Pink G06222, Red Mexican G05507
Beet curly top virus	Porillo Sintético G04495, Burtner, Tio Canela 75
Common bacterial blight	Interspecific hybrids with *P. acutifolius* VAX4, MBE7
Common bacterial blight	Montana No. 5; PI 207262
Halo blight	Montcalm G06416, ICA Tundama G14016
Halo blight	Pinto US 14 G18105
Halo blight	Wis HBR 72 G03954
Fusarium root rot	Porillo Sintético G04495; wild *P. vulgaris* G12947
Pythium root rot	PI 311987 G02323

Table 1.1 (continued)

Trait looked for	Material used[a]
Rhizoctonia solani Rot	N203 G00881
Rust	Compuesto Negro Chimaltenango G05711
Rust	Redlands Pioneer G05747
Rust	PI 260418
Web blight	BAT 93; Flor de Mayo G14241
White mould	Interspecific hybrids with *P. coccineus* G35172
White mould	Interspecific hybrids with *P. costaricensis* G40604
Pests	
Acanthoscelides weevil	Wild *P. vulgaris* from western Mexico G12952, G2771
Apion godmani pod weevil	Amarillo 154 G03982; G03578
Empoasca leafhoppers	California dark red kidney, from USA G17638
Ophiomyia bean fly	*P. coccineus* G35023 and G35075, and interspecific hybrids
Whiteflies Aleyrodidae	DOR 303
Zabrotes weevil	Wild *P. vulgaris* from Chiapas, Mexico G24582
Nitrogen fixation	
N_2 fixation under low P	Bituyano from Cajamarca, Peru, G19348
Yield	
Favourable QTLs	Wild *P. vulgaris* from Colombia G24423
Favourable QTLs	Wild *P. vulgaris* from Colombia G24404
Nutritional traits	
Seed protein quantity	PI 229815
High zinc content	Peruvian landrace G23823
High iron content	Peruvian landrace G23823
Polyphenols	Wild *P. vulgaris* from Mexico G11025

[a] G numbers refer to the International Center for Tropical Agriculture (CIAT) genebank, while PI numbers refer to the Western Regional Plant Introduction Station at Pullman, Washington, USA
BCMV bean common mosaic virus, *BGYMV* bean golden yellow mosaic virus, *QTLs* quantitative trait loci

Table 1.2 Major germplasm collections of *Phaseolus* beans, and type of accessions. FAO (2010)

Gene bank	Accessions (%)	Landraces (%)	Wild species (%)
CIAT, Colombia	35,891 (14)	30,507 (85)	2153 (6)
USDA, USA	14,674 (6)	9832 (67)	880 (6)
Embrapa, Brazil	14,460 (6)	5784 (40)	–
INIFAP, Mexico	12,752 (5)	7014 (55)	2168 (17)
IPK, Germany	8680 (3)	5729 (66)	87 (1)

4 Varietal Groups: Market Classes

Bean consumers of different countries and regions show specific preferences for various combinations of seed size, shape, colour, cooking time, broth appearance and storability (De Ron et al. 2000). Therefore, a classification often used for common bean is the one into commercial types, which is based predominantly on characteristics of grain colour and size, and is related to market preferences. The wide range of seed characteristics has been formalized in the bean world into distinct commercial or market classes. Among the bean varieties grown in the world, 62 dry bean market classes are recognized (Santalla et al. 2001; FAO 2002) according to consumer preferences, production and market price (Fig. 1.2). Dry bean market classes are produced under recommended agronomic practices and traded according to the defined class attributes. Thus, classes must be segregated throughout production and distribution.

Increased diversity of commercial market classes has been achieved to meet market and consumer interests. Among the Durango beans, the most important market classes are 'great northern' and 'pinto'. The most abundant market classes that represent race 'Nueva Granada' are 'dark red kidney', 'white kidney', 'calima' and 'large cranberry' beans. Regarding the race Mesoamerica, the most popular bean market classes are 'navy', 'small white', 'mulatinho', 'carioca' and the classes of small black seed. The Chilean market classes most accepted and consumed are 'tórtola' and 'coscorrón'. In addition, other minor market classes, such as 'manteca', 'sapito' and 'cuyano', are also consumed in more specific areas. Race Peru is char-

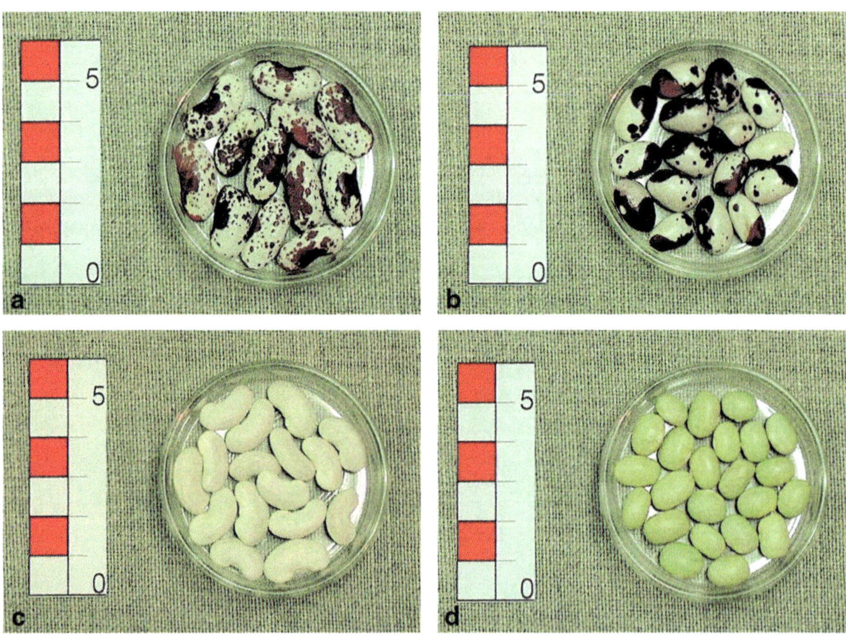

Fig. 1.2 Common bean international market classes. **a** Favada Pinto (race Nueva Granada). **b** Red Caparron (race Peru). **c** Hook (race Durango). **d** Small Yellow (race Mesoamerica)

acterized by large seeds which are often round or oval but can also be elongated. Its most popular types are 'yellow canario' and 'azufrado' beans.

Market classes usually include improved germplasm and thus tend to show a low level of variability. However, the range of commercially available bean cultivars and varieties in different market classes is constantly changing. New cultivars are released for their increased yield potential, pest and disease resistance, full-season and early double-cropped growth potential and improved market quality. Public and private plant breeders develop new varieties by adding desirable features to old cultivars or create new and better cultivars by recombining the best traits from available germplasm.

The polymorphism of common bean is so great that, in each region, and even in each locality, different varieties with similar characteristics correspond to different names. There are several ethnic varieties or 'heirloom' varieties, which are characteristic of an area or region, and they can be designated with different names. These landraces evolved from ancient types by conscious or unconscious selection and are currently well adapted to the agroecological conditions under which they have been grown for centuries. In Europe, the high appreciation by consumers of these 'heirloom' varieties is recognized by the attribution of the protected geographical indication (PGI), one of the European Union marks attributable to traditional foods. With the increased interest in 'heirloom' varieties (seeds passed down from generation to generation), many fine old-fashioned varieties have been reintroduced recently by various seed companies.

5 Major Breeding Achievements and Specific Goals in Current Breeding

5.1 Achievements in Dry Bean Breeding in the USA

Along with corn (*Zea mays* L.) and squash (*Cucurbita* spp.), dry bean was among the earliest crops domesticated in the Americas (Kaplan 1956). Native Americans commonly grew beans as a companion crop with corn and squash in what is termed the 'three sisters' or milpa method that originated in Mesoamerica and spread northward into Mexico and the southwestern USA. Some of the old landraces were eventually selected and produced by the New World settlers for local consumption. Beans were also introduced into Europe from the New World as early as AD 1500 by the early explorers (Zeven 1997). Subsequently, they were reintroduced into the eastern USA by Europeans that migrated from Europe to the USA. The first large-scale commercial production of dry edible beans in the USA occurred in Orleans County, New York in 1839. New York became one of the first important producers of dry beans and maintained its dominance until the early 1900s when Michigan became the leading producer.

A significant change in dry bean production occurred in 1917, when seed production began shifting from the central and the eastern USA to the semiarid western USA, where today most commercial bean seed is produced (Brick and Lowry

2000). This shift initially occurred because seed-borne pathogens, such as anthracnose (ANT; *Colletotrichum lindemuthianum* (Sacc *et* Magn.) Scrib.) and common bacterial blight (*Xanthomonas axonopodis* pv. *phaseoli* (Smith) Dye), became serious problems in commercial production fields (Adams 1996). Idaho was among the first states to produce large quantities of commercial dry bean seed and still produces more certified bean seed than any other state.

5.2 Genetic Improvement

Several books have been published that address dry bean improvement, production challenges and genetic resources in the USA and the Europe: *Genetic Resources of Phaseolus Beans* by P. Gepts (ed) in 1988; *Common Bean Production Problems in the Tropics* by H. F. Schwartz and T. Pastor-Corrales (eds) in 1989; Common Beans: *Common Beans: Research for Crop Improvement* by A. van Schoonhoven and O. Voysest (eds) in 1991; *Phaseolus spp: Bean Science* by R. Maiti (ed) in 1997; *Common Bean Improvement in the Twenty-First Century* by S. P. Singh (ed) in 1999; *Catalogue of Bean Genetic Resources* by J. M. Amurrio, M. Santalla and A. M. De Ron (eds) in 2001; *Handbook on Evaluation of Phaseolus Germplasm* by C. De la Cuadra, A. M. De Ron and R. Schachl (eds) in 2001; and *Compendium of Bean Diseases* (2nd edn) by H. F. Schwartz, J. R. Steadman, R. Hall and R. L. Forster (eds) in 2005.

Early breeding efforts primarily focused on improved disease resistance and adaptation to local environments, later efforts also focused on improved seed quality, improved plant architecture and breeding for yield. Among the early bean researchers, R. A. Emerson, renowned for his research on maize genetics, worked on beans at the University of Nebraska from 1898 until 1912. The Michigan Agricultural College (currently Michigan State University) was among the first institutions in the USA to employ a full-time dry bean breeder in 1906 followed by the University of Idaho in 1925 (Singh et al. 2007). Michigan State University released the first USA navy bean cultivar 'Robust' in 1915 as a selection from locally grown landraces. In the early twentieth century, breeding programmes at Cornell University and Michigan Agricultural College focused on disease resistance, primarily resistance to ANT (Burkholder 1930) and common bacterial blight (Adams 1996). Additional research in the western USA focused on developing resistance to a range of pathogens, including rust (*Uromyces appendiculatus* Pers: Unger.), white mould (caused by *Sclerotinia sclerotiorum* (Lib.) DeBary), bacterial blights, viruses, root pathogens and *beet curly top virus* (BCTV) transmitted by the beet leafhopper (*Circulifer tenellus* (Baker)).

5.3 Seed Yield

Many review papers and chapters have been published that summarize breeding strategies to increase yield in dry bean (Beaver 1999; Brick and Grafton 1999; Singh 1999a,

1999b; Urrea and Singh 1994; Kelly 2004; Kelly and Cichy 2012; Vandemark et al. 2014). Some strategies employed by dry bean breeders to improve yield include interracial and interspecific crosses, gamete selection, early generation testing, recurrent selection, ideotype breeding and selection for physiological efficiency.

To ensure that breeding programmes have optimum genetic diversity for yield improvement, Kelly et al. (1998) suggested a 'three-tiered' pyramidal breeding strategy to manage germplasm in a breeding programme. The three tiers were composed of three levels of germplasm improvement/advancement in the breeding programme and included types of crossing protocols to use in each tier. The apex of the pyramid consisted of elite, agronomically acceptable germplasm within the target market class and the use of single-seed descent to advanced lines and testing of advanced lines. Germplasm in this tier would be used to develop cultivars that are commercially acceptable to the industry and have high yield. The intermediate tier of the pyramid has diverse germplasm outside of the market class and includes the use of interracial material, and pedigree and inbred backcross breeding methods. The base tier places no restrictions on germplasm, including interspecific and interracial material, and no restriction on breeding methods employed including gamete selection, congruity backcrosses and conical crossing. This system would advance germplasm up the tiers or maintain them as they became more adapted to optimize improvement at each tier of the breeding pyramid.

Improvements in yield have also been achieved in some cases by selection for yield components. However, because seed size is a descriptor of market class, only the yield components pod number and seed number can be exploited to increase yield. Selection of hybrid populations was especially relevant to crosses between small-seeded Mesoamerican and large-seeded Andean germplasm because it prevented breeders from combining the high pod load potential of small-seeded navy beans with very large seed size of a kidney bean (White and Gonzales 1990), even though maximum genetic diversity could be attained by crosses between the Middle American and the Andean gene pools (Becerra-Velásquez and Gepts 1994). Studies with interracial crosses have shown mixed results to improve yield (Singh and Urrea 1994; Singh et al. 2002; González et al. 2009). Interracial hybridization between beans from races Durango and Mesoamerica has been used to improve pinto, great northern, small red and pink beans (Singh et al. 1993). Urrea and Singh (1994) compared breeding methods in interracial crosses for beans and suggested that early generation testing and selection should be used to more efficiently manage populations from interracial crosses. Singh and Urrea (1994) made crosses between races of Andean and Middle American origin and found that on average mean yield was higher in the interracial crosses that within race crosses. It is known that epistasis can play a role in the performance of progeny that result from interracial crosses (Johnson and Gepts 2002; Moreto et al. 2012).

During the early development of some market classes, yield gains were achieved by selection for a more vigorous vine that produced higher biomass than traditional landraces. However, cultivars developed by selection for more vigorous vine growth had increased risk of white mould disease due to denser plant canopies that retained canopy humidity. Subsequently, breeders developed cultivars with semi-vine habit to enhance disease avoidance mechanisms, allow the plant canopy to dry faster and allow the plant material to dry in the field after undercutting during harvest. Yield gains that have occurred over the past 30 years appear to be linear over time and should continue increasing due to improved plant architecture, disease resistance and avoidance, drought and water-use efficiency and other traits. Today, approaches using molecular markers have enhanced the process of breeding for yield; however, yield potential has not been reached in most market classes evidenced by the linear relationship between seed yield over time, with no indication of a yield plateau. Vandemark et al. (2014) reported that genetic gain in dry bean during the past 30 years based on common trials was 13.9 kg ha^{-1} $year^{-1}$ (0.77% $year^{-1}$) and 17.4 kg ha^{-1} $year^{-1}$ (0.85% $year^{-1}$) for navy and pinto bean cultivars, respectively. Vandemark et al. (2014) concluded that continued introgression of germplasm from other races of common bean should provide new sources of genetic diversity to enhance yield in the future.

5.4 Plant Architecture

The concept of an ideal plant type or 'ideotype' was first proposed in cereal crops and later suggested for common bean by Adams (1982). Adams proposed the strategy of ideotype breeding to improve yield potential and stability of the crop; today, that concept is highly associated with upright or type II architecture. Based on the success of ideotype breeding in small-seeded 'navy' and 'black' beans, Kelly et al. (1990) used recurrent selection to develop and released the first upright type II pinto cultivar 'Sierra'. Breeding for upright type II plant architecture has continued to be an important component of breeding some bean market classes in the western USA and Canada since 1990. Upright architecture also provides a level of disease avoidance for some fungal pathogens such as white mould, enables the crop to be direct harvested thus reducing seed loss during cutting and field curing and facilitates the use of furrow irrigation.

5.5 Disease Resistance

Diseases are one of the most important factors limiting bean production globally. Recent reviews summarizing the current state of breeding for disease resistance in dry bean were reported by Schwartz et al. (2005), Miklas et al. (2006), Terán et al. (2009), Schwartz and Singh (2013) and Tryphone et al. (2013).

Common bacterial blight (*CBB*; Fig. 1.3), caused by *Xanthomonas axonopodis* pv. *phaseoli* (Xap) (Smith) Vauterin, Hoste, Kosters and Swings and its fuscans variant, *Xanthomonas axonopodis* pv. *phaseoli* var. *Fuscans,* is considered one of the most important production constraints worldwide. Singh and Muñoz (1999) and Yu et al. (2000) recently published reviews on breeding for CBB resistance. The identification of 22 quantitative trait locus (QTL) across all 11 chromosomes il-

Fig. 1.3 Common bacterial blight in common bean pod and seeds. (Courtesy HF Schwartz)

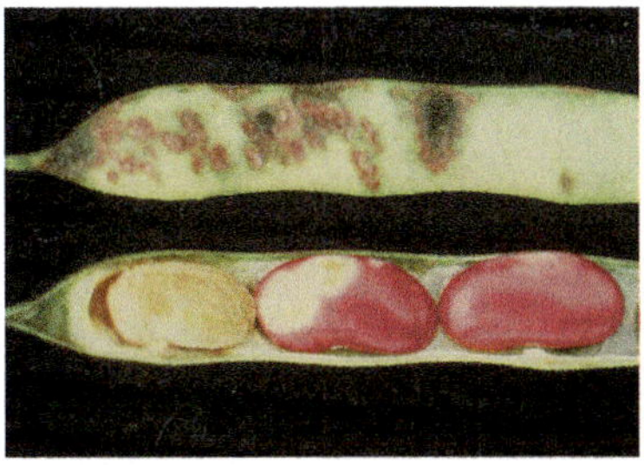

lustrates the complexity of breeding for resistance to this disease. Resistance is quantitatively inherited with largely additive effects, and the heritability is low to moderately high. The combination of resistance sources from the primary, secondary and tertiary gene pools has resulted in the development of improved CBB resistant lines such as XAN 159, OAC 88-1, Wilk 2 and VAX 1 through VAX 6 (Miklas et al. 2006; Singh and Schwartz 2010; Viteri et al. 2013).

Halo blight (HB), caused by the bacterium *Pseudomonas savastanoi* pv. *phaseolicola* (Burkh.) Gardan et al., is a major disease of dry beans produced in moderate to cool production areas. Pathogenic variation within the pathogen population exists with nine races reported worldwide (Taylor et al. 1996a). Five putative R genes (R1, R2, R3, R4 and R5) were tentatively identified by Teverson (1991) and Taylor et al. (1996b). The resistant alleles were renamed *Pse-1* to *Pse-5* to standardized naming by the Bean Improvement Cooperative Genetics Committee (Bassett and Myers 1999). Races 7 and 8 were found to be most prevalent in Spain and South Africa, respectively (Rico et al. 2003; Fourie 1998), and race 6 was the most widely distributed in East Africa and the USA (Lamppa et al. 2002; Taylor et al. 1996a). A new gene *Pse-6* was found located on chromosome Pv04 within a well-described R-gene cluster that conditions resistance to ANT and rust that conditions resistance to races 1, 5, 7 and 9 (Miklas et al. 2014).

Liebenberg and Pretorius (2010) recently published a comprehensive review on the pathology and control of *common bean rust,* caused by *Uromyces appendiculatus* (Pers.Pers.) Unger. High variability exists within the rust pathogen which coevolved with Andean and Middle American beans. High levels of resistance have been identified from the Andean and Middle American gene pools, and nine major resistance genes, *Ur-3, Ur-4, Ur-5, Ur6, Ur-7, Ur-9, Ur-11, Ur-12* and *Ur-13*, have been identified, tagged and mapped by several authors as reviewed by Singh and Schwartz (2010) and Miklas et al. (2006). Except for *Ur-12,* that conditions adult plant resistance, all these genes are race specific (Miklas et al. 2006). Designation of a new gene, *Ur-14,* characterized from the cultivar 'Ouro Negro' was recently proposed by de Souza et al. (2011). PI 310762 was reported to have high levels of resistance and is considered an important resistance source to use in developing cultivars with broad resistance (Pastor-Corrales et al. 2012).

Angular leaf spot (ALS), caused by *Pseudocercospora griseola* (Sacc.) Crous and Braun, is a serious disease in South America, Central America and eastern Africa. The pathogen is highly variable, and distinct Andean and Middle American

Fig. 1.4 Anthracnose on bean leaves

races exist as a result of coevolution with the bean host (Singh and Schwartz 2010). Despite this variability, high levels of resistance have been identified by a number of authors in both gene pools (Singh and Schwartz 2010; Miklas et al. 2006). Resistance has been reported from the secondary gene pool (*P. coccineus* and *P. dumosus*) by Mahuku et al. (2003). ALS resistance is controlled by single dominant as well as recessive genes, and a number of sequence characterized amplified regions *(SCAR)* markers linked to these resistance genes have been developed (Singh and Schwartz 2010; Miklas et al. 2006).

ANT (Fig. 1.4), caused by *Collectotrichum lindemuthianum* ((Sacc. and Magnus) Briosi and Cavara), is a widespread fungal disease of common bean that causes significant yield losses worldwide. The pathogen is highly variable, and a number of Andean and Middle American races have been identified (Kelly and Vallejo 2004; Singh and Schwartz 2010). High levels of resistance in both the primary and secondary gene pools have been reported globally, including interspecific breeding lines derived between the two gene pools (Singh and Schwartz 2010). Resistance has been reported to be controlled by both single and multiple gene models (Singh and Schwartz 2010) with at least 15 named *Co*-genes identified. Designation of a new gene, *Co*-15, characterized from the cultivar Corinthiano was recently proposed. Most of the resistance genes have been mapped to the integrated bean linkage map, and molecular markers have been developed for use in marker-assisted selection (MAS; Miklas et al. 2006; Singh and Schwartz 2010).

In 2013, Schwartz and Singh (2013) published a comprehensive review on breeding for resistance to *white mould* disease, caused by *Sclerotinia sclerotiorum* ((Lib.) deBary), in common bean. White mould is a problem in many dry bean production areas worldwide (Steadman 1983; Schwartz and Steadman 1989). Plant architecture and genetic resistance have been identified as mechanisms for control of white mould. Germplasm from race Durango does not possess adequate levels of resistance to white mould; consequently, breeders have incorporated upright plant architecture from race Mesoamerican to provide avoidance to the disease (Kelly et al. 1990; Miklas et al. 2013; Kolkman and Kelly 2003; Osorno et al. 2010; Brick et al. 2011). In addition, resistant QTLs from Andean germplasm have been introgressed into breeding material (Singh et al. 2007, 2014; Griffiths 2009; Soule et al. 2011). Resistant QTL from the secondary gene pool has also been identified (Miklas et al. 1998; Singh et al. 2009a, b, 2013).

Root rots, caused by *Fusarium, Rhizoctonia* and *Pythium* species, are common in most bean production areas. A primary source of resistance to root pathogens is the plant introduction from Mexico, 'N203' (PI 203958) collected by Oliver Norvell (Wallace and Wilkinson 1965). Resistance genes acquired from N203 were introgressed into many market classes of bean. Despite extensive information on Middle American *Fusarium* resistance sources, the transfer of resistance into Andean bean cultivars has been limited (Román-Avilés and Kelly 2005).

The development of germplasm and cultivars with multiple disease and pest resistance has become a common achievement in bean breeding programmes worldwide. The releases BelDakMik-RMR 14 to BelDakMik-RMR 23 carry pyramided resistance to all known races of rust in the USA and resistance to all strains of *bean common mosaic virus* (BCMV) and *bean common mosaic necrotic virus* (BCMNV; Terán et al. 2009). It is common to combine the recessive *bc* alleles with the dominant *I* gene, termed 'protected*I*gene', to provide resistance to all known strains of BCMV or BCMNV (Brick and Grafton 1999). Markers linked to most resistant alleles have been published (Haley et al. 1994; Kelly et al. 1995; Miklas et al. 1996; Johnson et al. 1997).

5.6 *Protein Quality and Metabolites*

Protein quality in common bean is suboptimal, like in other grain legumes, and limited by the low concentration of sulphur amino acids, methionine and cysteine. The sum of methionine and cysteine is considered as a nutritionally relevant parameter when assessing protein quality (FAO 2013). Protein quality improvement was a major focus of breeding research in the 1970s and 1980s. The goals were both to increase seed protein concentration and to balance the composition of essential amino acids (Bliss and Brown 1983). This work relied on a microbiological assay for bioavailable methionine (Kelly and Bliss 1975). There was good correlation between bioavailable methionine measured with this assay and total methionine, as well as the sum of total methionine and cystine.

The 7S globulin phaseolin is abundant in seed, accounting for up to 50% of total protein in cultivated varieties (Vitale and Bollini 1995). Lectins account for 5–10% of total protein, and there are only low levels of the 11S globulin legumin (Mühling et al. 1997). Efforts to improve protein quality relied on the availability of alleles conferring a deficiency in phaseolin and erythroagglutinating phytohemagglutinin, from *P. coccineus* Mexican Red Runner, and Great Northern 1140 or Pinto 111, respectively (Gepts and Bliss 1984; Osborn and Bliss 1985; Voelker et al. 1986). From the similar pattern of DNA hybridization observed between a group of accessions of different geographical origin, Bollini et al. (1985) concluded that there could be a unique genetic source for phytohemagglutinin deficiency. While the 7S globulins from legume crops are relatively poor in sulphur amino acids, it was found that methionine is actually positively correlated with phaseolin levels (Gepts and Bliss 1984). This is in agreement with biochemical data on protein fractions indicating that most of the methionine is associated with phaseolin, whereas cysteine is found in non-phaseolin proteins (Chagas and Santoro 1997). Removing phytohemagglutinin resulted in increased phaseolin concentration, probably as a compensatory mechanism to maintain a stable seed protein concentration (Osborn and Bliss 1985). By contrast, introducing arcelin-1 from a wild accession makes this lectin the most abundant seed protein and lowers phaseolin concentration by approximately threefold (Romero Andreas et al. 1986). Given the positive relationship between phaseolin and total available methionine, research was conducted to increase the percentage of phaseolin (Delaney and Bliss 1991a, b). A secondary aspect of this research concerns the removal of anti-nutritional proteins. Since erythroagglutinating phytohemagglutinin is toxic when consumed in raw bean, research has been performed, leading to the release of a phytohemagglutinin-deficient cultivar (Campion et al. 2009a).

More recently, research on nutritional improvement has shifted to micronutrients and the removal of anti-nutritional factors. Recognizing that deficiencies in zinc and iron are associated with stunted growth, decreased immune function and anaemia, the Consultative Group on International Agricultural Research (CGIAR) has undertaken a major initiative towards the bio-fortification of common bean with high concentration of these minerals, under the Harvest Plus programme. This work has been covered in several recent reviews (Beebe 2012; Blair 2013; Kelly and Cichy 2012). Screening germplasm identified a range of concentrations, with some intergene pool landraces of the northern Andes and *Phaseolus dumosus* accessions having over twofold the average mineral concentrations. The high concentrations in these intergene pool landraces are likely a result of transgressive segregation, and this has been confirmed in recombinant inbred populations (Blair et al. 2009). The enhanced materials have been used to breed cultivars with intermediate to high concentrations for Latin America and eastern Africa, including the so-called gorilla beans, with further breeding on the way.

Phytic acid is negatively associated with mineral bioavailability, due to the formation of insoluble complexes. An ethyl methanesulphonate (EMS)-induced *lpa* mutant has been isolated, having low phytic acid concentration (Campion et al. 2009b). This mutant is affected in an ABC transporter gene, *Pvmrp1,* required for phytic acid and raffinose oligosaccharide accumulation (Panzeri et al. 2011). Genotypes having low phytic acid concentration have also been isolated by conventional breeding. A QTL responsible for population variation in phytic acid concentration has been associated with a member of the gene family of D-myo-inositol-3-phosphate synthases, involved in the first committed step of myo-inositol biosynthesis, located on linkage group (LG)b01 (Blair et al. 2012). Recently, it was shown that the *lpa* mutation enhances the bioavailability of iron in women (Petry et al. 2013), which indicates that reduction of phytic acid is a promising approach for biofortification.

A trait widely disseminated in recent years is slow darkening in pinto beans (Junk-Knievel et al. 2008; Sanchez-Valdez et al. 2004; Singh et al. 2006). Slow darkening is controlled by a single, recessive gene, *sd,* distinct from *j,* associated with non-darkening (Elsadr et al. 2011). Although the *sd* gene has been mapped (Felicetti et al. 2012), its nature and function are currently unknown. It is thought that the slow darkening trait is related to the biosynthesis of tannins, which are associated with reduced bioavailability of micronutrients. It will be interesting to understand the molecular and biochemical phenotype of this trait and its possible relationship with nutritional quality.

The work involving modification of protein composition has been continued, taking advantage of genetically related lines integrating the same alleles discussed above, conferring a deficiency in phaseolin and phytohemagglutinin (Osborn et al. 2003). The parental background is the Sanilac cultivar, and the wild-type line, SARC1, integrates the lectin arcelin-1 as its major seed protein. As compared with SARC1, the absence of phaseolin and phytohemagglutinin in SMARC1N-PN1 is associated with a 10 % increase in methionine and 70 % increase in cysteine, mostly at the expense of the abundant non-protein amino acid, *S*-methylcysteine (Taylor et al. 2008). The combined concentration of methionine and cysteine, expressed in mg per g protein, is increased by approximately 40 %, slightly exceeding the Food and Agriculture Organization (FAO) guidelines. In this genotype, the concentration of another essential amino acid, leucine, may be slightly limiting (House and Marsolais unpublished results). These changes are associated with an increased concentration of legumin and several other sulphur-rich proteins, including albumin-2, defensin D1, Bowman–Birk-type proteinase inhibitor 2 and albumin-1A and albumin-1B (Marsolais et al. 2010; Yin et al. 2011). Two proteins of 56 and 54 kDa had been shown by Burow et al. (1993) to be elevated following suppression of phaseolin and lectins. They likely correspond to group 3 late embryogenesis abundant protein and the α-subunit of legumin, respectively (Yin et al. 2011). Profiling in developing seeds using a high-density Combimatrix 90K microarray identified transcripts coding for additional sulphur-rich protein types likely to be elevated in the absence of phaseolin and lectins, including the basic 7S globulin and Kunitz trypsin protease inhibitor (Liao et al. 2012). Genetic research is currently underway to evaluate whether the increased cysteine concentration observed in SMARC1N-PN1 could be used for protein quality improvement. In a subset of lines derived from a cross between the cultivar Morden003 and SMARC1N-PN1, the combined molar concentration of methionine and cysteine could be improved by 18–30 % as compared with parental values (Hou et al. 2014).

Since dry bean is primarily a source of protein and fibre in nutrition, and that there are significant health benefits from dietary fibre intake, there is significant interest in breeding for improved dietary fibre concentration, while reducing raffinose oligosaccharides, responsible for flatulence, an important trait related to consumer preference. From an initial assessment of genetic variability, there are excellent prospects for breeding genetic materials with improved dietary fibre and reduced raffinose oligosaccharide concentrations (Brick et al. 2014).

6 Breeding Methods and Specific Techniques

Although common beans generally have a low outcrossing proportion (Brunner and Beaver 1989), environmental factors may affect the level of outcrossing (Ibarra-Pérez et al. 1997). The breeding objectives should identify the set of traits that improved cultivars must possess for the target environment and identify other traits that would be desirable for improved cultivars to possess. Kelly et al. (1998) noted that the most appropriate breeding method may vary within an integrated genetic improvement programme, depending on breeding objectives. In developing countries, the lack of a formal system for the release, multiplication and dissemination of seed of improved bean cultivars may require the use of more participatory, community-based breeding methods.

The pedigree method is the most common procedure used to improve dry beans (Kelly and Cichy 2012). Pedigree selection may be more useful in populations derived from crosses between elite breeding lines. These populations are more likely to produce progeny in early generations that possess the desired combination of traits. MAS with codominant markers could be used in combination with pedigree selection in earlier generation to identify bean breeding lines that possess desired alleles of traits. Pedigree selection can be accelerated with the use of off-season nurseries and the simultaneous screening of breeding lines for specific traits such as disease resistance in greenhouse trials or with MAS (Osorno et al. 2010).

If the objective is rapid generation advance, single-seed or single-pod selection can be employed in greenhouses or winter nurseries, which are not necessarily representative of the target environment. Urrea and Singh (1994) proposed the modified pedigree (single-seed descent) method as a means to maintain genetic variability and rapidly advance lines to more advanced generations where greater progress can be made in the selection of quantitative traits such as seed yield. The multiple-seed procedure or harvesting pods rather than single seed from individual plants should generate sufficient genetic variability to select advanced lines for quantitatively inherited traits such as seed yield. Backcrossing is useful for the introgression of simply inherited traits from different races or gene pools of the common bean into cultivars or elite bean breeding lines.

Gamete selection may be an effective breeding method for the selection of multiple traits in populations derived from crosses involving more than two parents (Singh 1994, 1999a, b). Gamete selection proved to be successful in the development of breeding lines with multiple disease resistance, in part, because dominant genes conferred resistance to several of the diseases under selection (Terán et al. 2009). The availability of codominant markers for recessive traits such as the *bc-3* allele for BCMNV resistance (Drijfhout 1978) and the *bgm* gene for bean golden yellow mosaic virus (BGYMV) resistance (Velez et al. 1998) would enhance the effectiveness of gamete selection.

The combination of all favourable alleles for a quantitative trait into a single genotype is highly unlikely. Therefore, multiple cycles of selection and recombination may be needed to improve quantitative traits such as seed yield and resistance to abiotic stress.

In many ways, Andean beans are often more difficult to breed than smaller-seeded beans from the Middle American gene pool. Andean beans have less genetic variability that can be exploited by plant breeders (Beebe et al. 2001). Movement of genes between gene pools is impeded by hybrid dwarfism in F_1 plants caused by the complementary dominant genes Dl_1 (Middle American) and Dl_2 (Andean; Singh and Gutiérrez 1984). Nevertheless, bean breeders have been successful in transferring between gene pools specific genes for disease resistance such as BGYMV, ANT and rust (Pastor-Corrales et al. 2007; Pastor-Corrales 2003). Due to the co-evolution of beans and pathogens within the Andean and Middle American gene pools, plant breeders and pathologists have found pyramiding resistance genes from both gene pools can result in more durable resistance to diseases such as rust, ANT and ALS (Kelly and Miklas 1998).

An important challenge facing contemporary plant breeders is the need to be proficient in an ever-widening range of knowledge and techniques. How can a bean breeder use these tools in an efficient and cost-effective manner? The cost of conducting molecular tests from large commercial laboratories may continue to decline due to gains in efficiency from economy of scale. Outsourcing these services may be more cost-effective, reliable and more rapid than trying to maintain a state of the art molecular breeding laboratory.

7 Integration of New Biotechnologies in Breeding Programmes

7.1 Plant Breeding, Molecular Markers and Genetic Maps

The goal of plant breeding is to increase the number of favourable alleles needed for a specific production condition. The quicker these alleles can be introduced and stabilized in a breeding programme, the faster the rate of gain. A new technology, MAS, emerged in the late 1990s and early 2000s that aided breeding programmes. Molecular markers became available that tagged specific genes such as *Ur-3* and *Co-2* (Geffroy et al. 1998), and programmes were able to track those genes in their programmes which lessened, but did not eliminate the need for phenotypic testing. Yet, the development of the molecular markers, while not labour intensive, was time-consuming. New technologies built around the recently released common bean genome sequence (Schmutz et al. 2014) are now becoming available.

Genetic linkage maps are highly valuable tools for the identification of genomic regions carrying major genes and QTLs controlling agronomical traits as well as for comparative genome analyses. In common bean, the first molecular marker-based genetic maps were developed 20 years ago and were built mainly with restriction fragment length polymorphisms (RFLPs) and random amplified polymorphic DNAs (RAPDs) and contained only about hundreds of markers (Vallejos et al. 1992; Freyre et al. 1998).

In recent years, the establishment of genetic maps has benefited from the development of new types of molecular markers which take advantage of automated sequencing technologies, in particular the single-nucleotide polymorphisms (SNPs) markers. The existence of high-throughput methods for assaying SNP is continually reducing the cost of genotyping, and millions of SNP markers are now available in common bean (Hyten et al. 2010; Felicetti et al. 2012; Souza et al. 2012; Blair et al. 2013; Goretti et al. 2014; Zou et al. 2014). The resulting dense genetic maps are very useful for precisely localizing major genes and QTL involved in agronomic traits. These dense maps are also very useful tools to assist sequence assembly in whole-genome sequencing projects. For example, a genetic map based on 7015 SNP markers was used to assemble the common bean reference genome sequence (see after). These 7015 SNP markers were developed based on next-generation sequencing data for 14 genotypes from the major gene pools and market classes of common bean. In addition, by integrating genetic map data with genotyping data generated from collections of genotypes, linkage disequilibrium (LD) patterns across the genome can be investigated. This is a prerequisite for precise 'genome-wide association studies' (GWAS) or association mapping. GWAS is an alternative strategy to the classical biparental QTL mapping strategy to identify the genetic basis of quantitative traits (Rafalski 2010). Since the diversity of markers and the extent of LD may vary depending on the history of the collections, they should be investigated prior to GWAS design (Shi et al. 2011; Galeano et al. 2012). Ultimately, because of the low sequencing costs, high-density marker genotyping could be replace by 'genotyping by sequencing' (GBS) based on a robust, cost-effective, highly multiplexed approach (Elshire et al. 2011).

7.2 Sequencing the Genome: An Ancient Orphan Crop Joins Modern Era

The genome of an Andean genotype from CIAT, G19833, was sequenced (Schmutz et al. 2014) and recently released to the public (http://www.phytozome.org). The total genome size is 521 Mb and represents 89% of the 587-Mb genome (http://data.kew.org/cvalues/). The assembled genome sequence was annotated using transcriptome data and *ab initio* approaches, and 27,197 gene models and 4441 alternately spliced transcripts identified a total of 31,632 protein-coding sequences. In addition, it was determined that the transposable elements represent ~45% of genome, and ~40% of the genome is retrotransposon. The vast majority of these were inserted <2 MA. A first draft of the entire common bean genome sequence of a Mesoamerican genotype (BAT93) was also developed under the framework of the PhasIbeAm consortium (Vlasovab et al. 2014).

The genome sequence has an immediate application by providing a reference from which new markers can be developed. The synteny with other legume species is also a relevant genetic tool. One example is the large collection of robust indel markers discovered by low pass sequencing 14 diverse genotypes (Mafi Moghaddam et al. 2014) and mapping the reads to the reference genome to discover insertion and deletion events. Scaffolds developed during the sequencing project were used to discover a large number of single-nucleotide polymorphisms (SNPs). Careful filtering of the SNPs leads to the development of Illumina Infinium SNP chip (BARCBean6K3) containing ~6000 SNPs. The careful design provided a platform that can be used for any market class and many markers that are suitable for bi-parental QTL mapping as well as GWAS.

7.3 Current Challenge in the Post-Genomic Area for Breeding Programmes

An important long-term challenge is the discovery of the gene(s) that control important production traits. This will need to be a cooperative worldwide effort that involves breeders, geneticists and genomic and bioinformatics experts. Breeders provide the essential skills of phenotyping and the identification and development of genetic populations. Connecting phenotyping with the functional gene requires the skills of pathologists, physiologists, and those with a deep knowledge of plant anatomy. Those skilled with genomics and bioinformatics provide the expertise to link the phenotypic and genotypic data with candidate genes. Once a candidate gene is defined and the causative mutation is discovered, breeders will then have access to best possible marker, one that is in the gene controlling the important phenotype.

7.4 Induced Mutagenesis as a Genetic Tool for Improvement

In common bean, both physical and chemical mutagens have been used to induce mutations. Andersen and Down (1956) were the first to report an induced mutant in this species using X-rays to modify growth habit, which resulted in the development of the determinate navy bean cultivar Sanilac. Gamma radiation has also been used to study and improve important traits affecting plant morphology (Frazier and Davis 1966a; Nagata and Bassett 1984; Tulmann-Neto and Sabino 1994; Avinash and More 2010), yield (Sarafi 1973), seed quality (Frazier and Davis 1966b; Hussein and Disouki 1976; Wyatt and Dukes 1980; Allavena 1989), or resistance to pests (Mohan et al. 1980) and diseases (Tulmann-Neto and Ando 1976; Zogorcheva and Poriazov 1983; Allavena 1989).

Among chemical mutagens, EMS is most widely used in plants for the development of large mutant populations. Porch et al. (2009) generated a common bean population of 3000 M2 mutant lines using EMS, which was organized and classified based on mutated tissue, including root, stem, leaf, seed and whole plant traits. EMS has also been used to generate mutants affected in plant architecture (Motto et al. 1975), flower and seed coat colour (Moh 1971; Avinash and More 2010), seed development (Silue et al. 2006), biological nitrogen fixation (Davis et al. 1988; Park and Buttery 1989; Gautam et al. 1998) and phytic acid biosynthesis (Campion et al. 2009b). In addition, other chemical mutagens have been used for the induction of common bean mutants, such as sodium azide (NaN_3; Cary 1982; Jeng et al. 2010) and *N*-ethyl-*N*-nitrosourea (ENU; Svetleva 2004).

Mutations can also be induced using molecular tags or DNA insertions. If the tags insert into a gene coding or regulation region, they will disrupt the gene function. The most commonly used random insertion mutagens are the retrotransposon *Tos17,* T-DNA, and the *Ac/Ds* and *En/Spm* transposon systems. Of these, T-DNA insertional mutagenesis is the favoured approach due to its stability through generations of insertions and the low copy number per mutated genome, whereas mutations promoted by transposable element are often unstable (Delseny et al. 2001). As the sequence of the inserted element is known, the genomic region flanking the insertion can be easily identified using standard cloning and polmerase chain reaction (PCR)-based strategies, such as plasmid rescue (Behringer and Medford 1992), inverse PCR (IPCR; Ponce et al. 1998) and thermal asymmetric interlaced PCR (TAIL PCR; Liu et al. 1995). Thereby, an insertional mutant collection constitutes a useful tool for forward and reverse genetic approaches to identify new genes and analyse their functions. Currently, there are several publicly available mutant databases, mainly in the species *Arabidopsis thaliana, Oryza sativa* and *Zea mays.* However, in species such as *P. vulgaris,* the lack of efficient transformation systems is hindering the development of insertional mutagenesis resources.

Methods currently available for common bean transformation include indirect gene transfer using *Agrobacterium tumefaciens* or *A. rhizogenes* and direct gene transfer techniques, mainly particle bombardment or electroporation (Dillen et al. 1995; Kim and Minamikawa 1996; Aragão et al. 2002; Rech et al. 2008; Amugune et al. 2011). Common bean, like other legumes, is generally considered recalcitrant to *Agrobacterium*-mediated transformation due to poor regeneration in tissue culture (Svetleva et al. 2003; Colpaert et al. 2008; Arellano et al. 2009). However, high transformation efficiency rates (75–90 % frequency) have been achieved by means of *A. rhizogenes*-mediated root transformation (Estrada-Navarrete et al. 2006). Recently, Aragão et al. (2013) has carried out the molecular characterization of the first commercial transgenic common bean immune to BGMV. Molecular analyses showed that the transgenes were structurally stable for eight self-pollinated generations and after backcrosses with a non-transgenic commercial variety. In addition, the levels of small interfering RNA (siRNA) were analysed in seeds cooked for 10 min, demonstrating that transgenic beans are free of siRNA signals after cooking and therefore suitable for human consumption.

Genetic transformation causes some public concern, especially in Europe. In contrast, novel lines obtained by mutagens are much more acceptable to consumers, breeders and governments. In this context, Targeting Induced Local Lesions in Genome (TILLING) technology has been developed as an alternative to insertional mutagenesis. TILLING is a non-transgenic method that uses gene-specific primers for the identification of mutants of a gene of interest from a large mutagenesis population (McCallum et al. 2000). TILLING has gained popularity as a reverse genetic approach because it can produce an allelic series of mutants, including knockouts, and it does not rely on the transformation method for gene discovery and verification. Significant advances have been made in the development of a TILLING platform in common bean because of the lack of insertional mutagenesis resources.

A TILLING consortium for tool development has been created, including the University of Geneva (Geneva, Switzerland), USDA/ARS/TARS (Mayagüez, Puerto Rico) and CIAT (Cali, Colombia). To date, a population of 3000 M2 mutant lines for TILLING has been generated by Porch et al. (2009) from the genotype BAT93, a representative of the Mesoamerican gene pool. However, it is necessary to increase this population, because a population of over 5000 mutant lines is required for adequate genome coverage and an effective TILLING approach (Porch et al. 2009). Moreover, the TILLING protocol for common bean has yet to be optimized. Once the BAT93 TILLING project is completed, it will provide a source of characterized genes for their application in molecular breeding of traits of interest.

Developing varieties with improved agronomic traits is a primary goal of the common bean breeding programmes. In this context, induced mutation breeding has become an effective method to increase common bean genetic variability available to the plant breeder. Furthermore, renewed interest is being generated in induced mutations since, as mentioned before, the sequencing of the common bean genome is currently in its final stages, and its availability will significantly broaden opportunities for functional genomics research. Consequently, induced mutagenesis will soon become a powerful tool for the isolation and functional characterization of agronomically interesting genes, which can be used in common bean improvement.

8 Seed Production

8.1 Introduction

Successful crop production depends initially on the availability of high-quality seed; therefore, production of high-quality seed has a major impact on growing beans. Low-quality bean seed can cause a poor and an uneven stand resulting in incidence of diseases, uneven maturity, harvesting problems and yield losses. High-quality seed will reduce the incidence of seed-borne bacterial and fungal diseases and has to guarantee the genetic purity of the genotype, and standards are established by national seed laws and seed certification agencies to assure bean growers that the seed they buy is accurately labelled with the correct variety.

8.2 Flowering and Pollination

Nowadays, almost all varieties of common bean grown at high latitudes are day-neutral in their flowering response and plants flower as soon as they are physiologically ready. Bean flowers are self-pollinated as the anther sacs are borne directly adjacent to the stigma and the pollen is released the day before the flower opens. Normally, the stigma is self-pollinated before the flower opens and is accessible to pollinating insects. However, a considerable amount of out crossing can occur under certain conditions. High temperatures induce or increase the physiological deterioration of seeds (Hampton et al. 2013). There is genetic variation for this condition among bean varieties, and the pollen of one variety may lose viability while another does not under similar conditions. If multiple varieties flower during a hot period, it is possible that insects may move pollen from a fertile variety to another cultivar lacking fertile pollen. Others factors that affect to outcrossing are the presence of barrier plants and isolation distance between plots.

8.3 Isolation Distance

Gene flow is a common phenomenon even in self-pollinated plant species. Faria et al. (2010) evaluated the frequency of gene flow using transgenic bean cultivars resistant to the herbicide glufosinate ammonium and their conventional counterparts as recipients of the transgene. The outcross occurred at a rate of 0.0074% observing that the frequency of gene flow was cultivar dependent and most of the observed outcross was within 2.5 m from the edge of the pollen source. Ferreira et al. (2007) observed that the highest frequency of natural hybrids, 0.136%, occurred at a distance of 0.5 m between the bean cultivars; the natural outcrossing rate was practically zero beyond a distance of 3.25 m. Isolation requirements vary depending on flower characteristic and biodiversity in the environment. For bean commercial production, an isolation distance of 15 m is recommended with a taller crop barrier (several rows of corn) to prevent pollinator movement. For stockseed where genetic purity is essential, this distance should be increased to 50 m with a tall barrier crop as well.

8.4 Genetic Maintenance

Population size depends on the variation within the population of beans to be grown for seed. In general, commercial varieties are highly uniform and have little heterozygosity; these varieties are mostly derived from a single elite plant. A bean variety derived in this way has very little inherent genetic diversity, and therefore, 20 plants saved for seed should be enough to preserve its genetic integrity.

Selection, or rouging, refers to removing unwanted off-types in the population that is mainly practiced by growers. This selection is critical to maintain desirable variety characteristics. Rouging must be done throughout the growing season inspecting the entire plant. During crop, it is needed to pay attention to earliness, foliage colour, leaf shape, flower colour, growth form, trueness-to-type, vigour, and disease and insect resistance. Rouging has to be done again on the seed. If there are off-types at this stage to evaluate the purity of seed, crop is needed. It is critical to select against disease in bean crops in order to produce quality seed, minimizing risk of seed-transmitted diseases.

8.5 *Seed-Transmitted Diseases*

The most important diseases for bean seed production are the seed-borne diseases that may transmit the pathogen via the seed to the next generation. Common beans would have to be monitored for most important seed-transmitted diseases.

Viral Pathogens BCMV is the most common and widespread virus of common bean because it is transmitted by seed and by aphids. BCMNV is considered to be endemic to Africa, and it has been spread throughout the world in infected seeds (Davis et al. 2004).

Bacterial Pathogens Halo blight *(Pseudomonas syringae* pv. *phaseolicola)* occurs worldwide and can cause extensive losses under moderate temperatures and humid moist conditions. Bacterial brown spot *(P. syringae* pv. *syringae)* grow in wet and cool conditions. Bacterial blight *(Xanthomonas campestris* pv. *phaseoli)* is favoured by conditions of high moisture and humidity. Bacterial wilt *(Curtobacterium flaccumfaciens* pv. *flaccumfaciens)* and fuscous blight *(Xanthomonas fuscans)* are becoming a problem in bean seed production.

Fungal Pathogens ANT *(Colletotrichum lindemuthianum)* develops in regions with wet periods, high relative humidity and moderate temperatures. Young plants are infected from spores carried on seed or spores splashed from debris or nearby infected plants. Fusarium *(Fusarium oxysporum* f. sp. *phaseoli)* can survive in soil for long periods, and it also has been reported to be an external contaminant of seed.

Preventative measures for management of diseases of beans must include at least a 3-year crop rotation, and all diseased plants have to be removed from the field to prevent spread of the disease and also to avoid mechanical damage when the crop is wet and clean equipment to prevent spread from one field to another.

8.6 *Harvesting*

Bean seed production is better done in the dry regions where pods can be left on the plant to dry until harvest, without fear of disease. Timing of harvest is important in order to produce high-quality bean seed that is mature, has an optimum germination percentage and has high storage potential. Each variety has its own specific harvest timing; the initial signal that the crop is ready to harvest is the relative maturity of the pods and their yellow colour. Pods should generally be yellow coloured at harvest in order to mature properly in the field, but the exact colour is specific of the variety.

Acknowledgments A. M. De Ron thanks the INIA Project RFP2013-00001 from the Spanish Government. R. Lozano and F. Yuste-Lisbona thank Junta de Andalucía grant to the PAIDI Research Group AGR176 and Excellence Programme Project P10-AGR-06931 and Campus de Excelencia Internacional Agroalimentario-CeiA3. P. A. Casquero thanks INIA project RTA 2011-00076-C02-02. The authors thank Marcial (Talo) Pastor-Corrales for scientific assistance and Howard F. Schwartz for providing pictures of bean diseases.

References

Adams MW (1982) Plant architecture and yield breeding. Iowa State J Res 56:225–54

Adams MW (1996) An historical perspective on significant accomplishments in dry bean research. Annu Rpt Bean Improv Coop 39:33–45

Allavena A (1989) Modification of the seed coat color associated to the *I*gene conferring resistance to BCMV. Annu Rpt Bean Improv Coop 32:90–91

Amugune NO, Anyango B, Mukiama TK (2011) *Agrobacterium*-mediated transformation of common bean. Afr Crop Sci J 19:137–147

Andersen AL, Down EE (1956) Agronomic use of an x-ray induced mutant. Science 124:223–224

Angioi SA, Rau D, Attene G et al (2010) Beans in Europe: origin and structure of the European landraces of *Phaseolus vulgaris* L. Theor Appl Genet 121:829–843

Aragão FJL, Vianna GR, Albino MMC et al (2002) Transgenic dry bean tolerant to the herbicide glufosinate ammonium. Crop Sci 42:1298–1302

Aragão FJL, Nogueira EOPL, Tinoco MLP et al (2013) Molecular characterization of the first commercial transgenic common bean immune to the *Bean golden mosaic virus*. J Biotechnol 166:42–50

Arellano J, Fuentes SI, Castillo-España P et al (2009) Regeneration of different cultivars of common bean (*Phaseolus vulgaris* L.) via indirect organogenesis. Plant Cell Tiss Organ Cult 96:11–18

Asfaw A, Blair MW, Almekinders C (2009) Genetic diversity and population structure of common bean (*Phaseolus vulgaris* L.) landraces from the East African highlands. Theor Appl Genet 120:1–12

Avinash TB, More AD (2010) Induced flower colour mutations in *Phaseolus vulgaris* Linn through physical and chemical mutagens. Adv Biores 1:22–28

Bassett MJ, Myers JR (1999) Report of BIC genetic committee. Annu Rpt Bean Improv Coop 42:vi.2001

Beaver J (1999) Improvement of large-seeded race Nueva Granada cultivars. In: Singh SP (ed) Common bean improvement in the twenty-first century. Kluwer, Dordrecht, pp 275–288

Becerra-Velásquez VL, Gepts P (1994) RFLP diversity in common bean (*Phaseolus vulgaris* L.). Genome 37:256–263

Beebe S (2012) Common bean breeding in the tropics. In: Janick J (ed) Plant breeding reviews, vol 36. Wiley, New York, pp 357–426

Beebe S, Rengifo J, Gaitan E et al (2001) Diversity and origin of Andean landraces of common bean. Crop Sci 41:854–862

Behringer FJ, Medford JI (1992) A plasmid rescue technique for the recovery of plant DNA disrupted by T-DNA insertion. Plant Mol Biol Rept 10:190–198

Bellucci E, Bitocchi E, Rau D et al (2014a) Genomics of origin, domestication and evolution of *Phaseolus vulgaris*. In: Tuberosa R, Graner A, Frison E (eds) Genomics of plant genetic resources. Springer, The Netherlands, pp 483–507

Bellucci E, Bitocchi E, Ferrarini A et al (2014b) Decreased nucleotide and expression diversity and modified coexpression patterns characterize domestication in the common bean. Plant Cell. doi:10.1105/tpc.114.124040

Bitocchi E, Nanni L, Bellucci E et al (2012) Mesoamerican origin of the common bean (*Phaseolus vulgaris* L.) is revealed by sequence data. Proc Natl Acad Sci U S A 109:E788–E796

Bitocchi E, Bellucci E, Giardini A et al (2013) Molecular analysis of the parallel domestication of the common bean in Mesoamerica and the Andes. New Phytol 197:300–313

Blair MW (2013) Mineral biofortification strategies for food staples: the example of common bean. J Agr Food Chem 61:8287–8294

Blair MW, Astudillo C, Grusak MA et al (2009) Inheritance of seed iron and zinc concentrations in common bean (*Phaseolus vulgaris* L.). Mol Breeding 23:197–207

Blair MW, González LF, Kimani PM et al (2010) Genetic diversity, inter-gene pool introgression and nutritional quality of common beans (*Phaseolus vulgaris* L.) from Central Africa. Theor Appl Genet 121:237–248

Blair MW, Herrera A, Sandoval T et al (2012) Inheritance of seed phytate and phosphorus levels in common bean (*Phaseolus vulgaris* L.) and association with newly-mapped candidate genes. Mol Breeding 30:1265–1277

Blair MW, Cortes AJ, Penmetsa RV et al (2013) A high-throughput SNP marker system for parental polymorphism screening, and diversity analysis in common bean (*Phaseolus vulgaris* L.). Theor Appl Genet 126:535–548

Bliss FA, Brown JWS (1983) Breeding common bean for improved quantity and quality of seed protein. In: Janick J (ed) Plant breeding reviews, vol 1. Wiley, New York, pp 59–102

Bollini R, Allavena A, Vitale A (1985) Genomic analysis of phytohemagglutinin-deficient *Phaseolus vulgaris* cultivars. Annu Rpt Bean Improv Coop 28:82

Brick MA, Grafton KF (1999) Improvement of medium seeded race Durango cultivars. In: Singh SP (ed) Common bean improvement in the twenty-first century. Kluwer, Dordrecht, pp 223–253

Brick MA, Lowry G (2000) Bean seed production in the western United States. In Singh SP (ed) Bean research, production & utilization. Proc of the Idaho Bean Workshop celebrating 75 years of bean research & development and 50 years of the cooperative dry bean nursery. Univ of Idaho, Twin Falls, pp 153–164

Brick MA, Ogg JB, Schwartz HF et al (2011). Registration of Croissant pinto bean. J Plant Regist 5:1–5

Brick MA, Echevarria D, Kleintop A et al (2014) Dietary fiber and oligosaccharide content of dry bean varieties. Annu Rpt Bean Improv Coop 57:195–196

Broughton WJ, Hernández G, Blair MW et al (2003) Beans (*Phaseolus* spp.) model food legumes. Plant Soil 252:55–128

Brunner BR, Beaver JS (1989) Estimation of outcrossing of the common bean in Puerto Rico. HortScience 24 (4):669–671

Burkholder WH (1930) The bacterial diseases of the bean: a comparative study. Cornell Univ Agric Exp St Mem 127:1–88

Burle ML, Fonseca JR, Kami JA et al (2010) Microsatellite diversity and genetic structure among common bean (*Phaseolus vulgaris* L.) landraces in Brazil, a secondary center of diversity. Theor Appl Genet 121:801–813

Burow MD, Ludden PW, Bliss FA (1993) Suppression of phaseolin and lectin in seeds of common bean, *Phaseolus vulgaris* L.: increased accumulation of 54 kDa polypeptides is not associated with higher seed methionine concentrations. Mol Gen Genet 241:431–439

Campion B, Perrone D, Galasso I et al (2009a) Common bean (*Phaseolus vulgaris* L.) lines devoid of major lectin proteins. Plant Breeding 128:199–204

Campion B, Sparvoli F, Doria E et al (2009b) Isolation and characterisation of an *lpa* (low phytic acid) mutant in common bean (*Phaseolus vulgaris* L.). Theor Appl Genet 118:1211–1221

Cary JW (1982) Bean mutations from azide: an iron chlorotic strain, but no induced cold hardiness. Annl Rpt Bean Improv Coop 25:21–22

Chacón SMI, Pickersgill B, Debouck DG (2005) Domestication patterns in common bean (*Phaseolus vulgaris* L.) and the origin of the Mesoamerican and Andean cultivated races. Theor Appl Genet 110:432–444

Chagas EP, Santoro LG (1997) Globulin and albumin proteins in dehulled seeds of three *Phaseolus vulgaris* cultivars. Plant Foods Hum Nutr 51:17–26

Coates AG, Collins LS, Aubry MP et al (2004) The geology of the Darien, Panama, and the late Miocene-Pliocene collision of the Panama arc with north-western South America. Geol Soc Am Bull 116:1327–1344

Colpaert N, Tilleman S, Van Montagu M et al (2008) Composite *Phaseoulus vulgaris* plants with transgenic roots as research tool. Afr J Biotechnol 7:404–408

Correa P (1981) Epidemiological correlations between diet and cancer frequency. Cancer Res 41:3685–3690

Davis JHC, Woolley JN (1993) Genotypic requirement for intercropping. Field Crops Res 34:407–430

Davis JHC, Giller KE, Kipe-Nolt J et al (1988) Non-nodulating mutants in common bean. Crop Sci 28:859–860

Davis RM, Hall AE, Gilbertson R (2004) UC IPM Pest Management Guidelines: dry beans. Statewide IPM Program, University of California Agriculture and Natural Resources. ANR Publication 3446. http://www.ipm.ucdavis.edu/PMG/r52101611.html. Accessed 20 Nov 2013
De la Cuadra C, De Ron AM, Schachl R (eds.) (2001) Handbook on evaluation of *Phaseolus* germplasm. PHASELIEU—FAIR3463/ MBG-CSIC. PHASELIEU, Pontevedra
De Ron AM, Jacobsen HJ, Santalla M (2000) Production and consumption of major Market classes of dry bean (*Phaseolus vulgaris*) in Europe. In: 75 years of Bean research and development and 50 years of cooperative dry bean nursery. Workshop and Field Day, University of Idaho, Kimberley, USA, 3–4 August 2000
De Souza TLPO, Dessaune SN, Sanglard DA et al (2011) Characterization of the rust resistance gene present in the common bean cultivar Ouro Negro, the main rust resistance source used in Brazil. Plant Pathol 60:839–845
Debouck DG, Toro O, Paredes OM et al (1993) Genetic diversity and ecological distribution of *Phaseolus vulgaris* in northwestern South America. Econ Bot 47:408–423
Delaney DE, Bliss FA (1991a) Selection for increased percentage phaseolin in common bean. 1. Comparison of selection for seed protein alleles and S1 family recurrent selection. Theor Appl Genet 81:301–305
Delaney DE, Bliss FA (1991b) Selection for increased percentage phaseolin in common bean. 2. Changes in frequency of seed protein alleles with S1 family recurrent selection. Theor Appl Genet 81:306–311
Delgado-Salinas A, Bibler R, Lavin M (2006) Phylogeny of the genus *Phaseolus* (Leguminosae): a recent diversification in an ancient landscape. System Bot 31:779–791
Delseny M, Salses J, Cooke R et al (2001) Rice genomics: present and future. Plant Physiol Biochem 39:323–334
Dillen W, Engler G, Montagu M et al (1995) Electroporation-mediated DNA delivery to seedling tissues of *Phaseolus vulgaris* L. (common bean). Plant Cell Rept 15:119–124
Drijfhout E (1978) Genetic interaction between *Phaseolus vulgaris* and bean common mosaic virus with implications for strain identification and breeding resistance. Agr Res Rpt 872:1–98
Elsadr HT, Wright LC, Pauls KP et al (2011) Characterization of seed coat post harvest darkening in common bean (*Phaseolus vulgaris* L.). Theor Appl Genet 123:1467–1472
Elshire RJ, Glaubitz JC, Sun Q et al (2011). A robust, simple genotyping-by-sequencing (GBS) approach for high diversity species. Plos ONE. doi:10.1371/journal.pone.0019379
Estrada-Navarrete G, Alvarado-Affantranger X, Olivares JE et al (2006) *Agrobacterium rhizogenes*-transformation of the *Phaseolus* spp.: a tool for functional genomics. Mol Plant Micr Inter 19:1385–1393
FAO (2002) FAO. Vol 56. Production yearbook. Rome, Italy
FAO (2010) The second report on the state of the world's plant genetic resources for food and agriculture. FAO, Rome
FAO (2013) Dietary protein quality evaluation in human nutrition: report of an FAO expert consultation. FAO Food and Nutrition Paper 92. FAO, Rome
Faria JC, Carneiro GES, Aragao FJL (2010) Gene flow from transgenic common beans expressing the bar gene. GM Crops 1:94–98
Felicetti E, Song Q, Jia G et al (2012) Simple sequence repeats linked with slow darkening trait in Pinto bean discovered by single nucleotide polymorphism assay and whole genome sequencing. Crop Sci 52:1600–1608
Ferreira JL, Carneiro JED, Teixeira, AL et al (2007) Gene flow in common bean (*Phaseolus vulgaris* L.). Euphytica 153:165–170
Fourie D (1998) Characterization of halo blight races on dry beans in South Africa. Plant Dis 82:307–310
Frazier WA, Davis DW (1966a) Inheritance of dark green savoy mutant. Annu Rpt Bean Improv Coop 9:22
Frazier WA, Davis DW (1966b) Inheritance of silver mutant. Annu Rpt Bean Improv Coop 9:22–23
Freyre R, Skroch P, Geffroy V et al (1998). Towards an integrated linkage map of common bean. 4. Development of a core linkage map and alignment of RFLP maps. Theor Appl Genet 97:847–856

Freytag GF, Debouck DG (2002) Taxonomy, distribution, and ecology of the genus *Phaseolus* (Leguminosae-Papilionoideae) in North America, Mexico and Central America. SIDA Bot Misc 23:1–300

Galeano CH, Cortes AJ, Fernandez AC et al (2012). Gene-based single nucleotide polymorphism markers for genetic and association mapping in common bean. BMC Genet 13:48

Gautam AS, Sood KC, Mittal RK (1998) Mutagenic effectiveness and efficiency of gamma-rays and ethyl methane sulphonate in rajma (*Phaseolus vulgaris* L.). Legume Res 21:217–220

Geffroy V, Creusot F, Falquet J et al (1998). A family of LRR sequences in the vicinity of the Co-2 locus for anthracnose resistance in *Phaseolus vulgaris* and its potential use in marker-assisted selection. Theor Appl Genet 96:494–502

Gepts P, Bliss FA (1984) Enhanced available methionine concentration associated with higher phaseolin levels in common bean seeds. Theor Appl Genet 69:7–53

Gepts P, Bliss FA (1988) Dissemination pathways of common bean (*Phaseolus vulgaris*, Fabaceae) deduced from phaseolin electrophoretic variability. II Europe and Africa. Econ Bot 42:86–104

Gepts P, Papa R, Coulibaly S et al (1999) Wild legume diversity and domestication—insights from molecular methods. In: Vaughan D (ed) Wild legumes, Proc 7th MAFF International Workshop on Genetic Resources. National Institute of Agrobiological Resources, Tsukuba, pp 19–31

Gil J, De Ron AM (1992) Variation in *Phaseolus vulgaris* in the northwest of the Iberian Peninsula. Plant Breeding 109:313–319

Gioia T, Logozzo G, Attene G et al (2013) Evidence for introduction bottleneck and extensive inter-gene pool (Mesoamerica × Andes) hybridization in the European common bean (*Phaseolus vulgaris* L.) germplasm. PLoS One 8:e75974

González AM, Rodiño AP, Santalla M et al (2009) Genetics of intra-gene pool and inter-gene pool hybridization for seed traits in common bean (*Phaseolus vulgaris* L.) germplasm from Europe. Field Crops Res 112:66–76

Goretti D, Bitocchi E, Bellucci E et al (2014) Development of single nucleotide polymorphisms in *Phaseolus vulgaris* and related *Phaseolus* spp. Mol Breed 33:531–544

Griffiths PD (2009) Release of Cornell 601–606: common bean breeding lines with resistance to white mold. HortSci 44:463–465

Haley SD, Afanador L, Kelly JD (1994) Identification and application of a random amplified polymorphic DNA marker for the *I* gene (potyvirus resistance) in common bean. Phytopathol 84:157–160

Hampton, JG, Boelt, B, Rolston, MP, et al (2013) Effects of elevated CO_2 and temperature on seed quality. J Agr Sci 151:154–162

Hangen L, Bennink MR (2003) Consumption of black beans and navy beans (*Phaseolus vulgaris*) resduced azoxymethane-induced colon cancer in rats. Nutr Cancer 44:60–65

Hou A, Marsolais F, Pajak A et al (2014). Genetic improvement of protein quality in edible beans with adaptation to Manitoba. Annu Rpt Bean Improv Coop 57:129–130

Hussein HAS, Disouki IAM (1976) Mutation breeding experiments in *Phaseolus vulgaris* (L.). EMS and gamma-ray-induced seed coat colour mutants. Zeitschrift Pflanzenzüchtung 76:190–199

Hyten DL, Song Q, Fickus EW et al (2010) High through-put SNP discovery and assay development in common bean. BMC Genomics 11:475–483

Ibarra-Pérez FJ, Ehdaie B, Waines JG (1997) Estimation of outcrossing rate in common bean. Crop Sci 37:60–65

Jeng LT, Shih YJ, Lai CC et al (2010) Anti-oxidative characterisation of NaN_3-induced common bean mutants. Food Chem 119:1006–1011

Johnson W, Gepts P (2002) The role of epistasis in controlling seed yield and other agronomic traits in an Andean × Mesoamerican cross of common bean (*Phaseolus vulgaris.* L.). Euphytica 125:69–79

Johnson WC, Guzman P, Mandala D et al (1997) Molecular tagging of the bc-3 gene for introgression into and common bean. Crop Sci 37:248–254

Junk-Knievel DC, Vandenberg A, Bett KE (2008) Slow darkening in pinto bean (*Phaseolus vulgaris* L.) seed coats is controlled by a single major gene. Crop Sci 48:189–193

Kami J, Becerra-Velásquez V, Debouck DG et al (1995) Identification of presumed ancestral DNA sequences of phaseolin in *Phaseolus vulgaris*. Proc Natl Acad Sci USA 92:1101–1104
Kaplan, L (1956) The cultivated beans of the prehistoric southwest. Ann Missouri Bot Garden 43:189–251
Kelly JD (2004) Advances in common bean improvement: some case histories with broader applications. Acta Hort 637:99–122
Kelly JD, Bliss FA (1975) Quality factors affecting the nutritive value of bean seed protein. Crop Sci 15:757–760
Kelly JD, Cichy KA (2012) Dry bean breeding and production technologies. In: Siddiq M, Uebersax MA (eds) Dry beans and pulses production, processing and nutrition. Wiley-Blackwell, Hoboken, pp 23–54
Kelly JD, Miklas PN (1998) The role of RAPD markers in breeding for disease resistance in common bean. Mol Breed 4:1–11
Kelly JD, Vallejo VA (2004) A comprehensive review of the major genes conditioning resistance to Anthracnose in common bean. HortSci 39:1196–1207
Kelly JD, Adams MW, Saettler AW et al (1990) Registration of 'Sierra' pinto bean. Crop Sci 30:745746
Kelly JD, Afanador L, Haley SD (1995) Pyramiding genes for resistance to bean common mosaic virus Euphytica 82:207–212
Kelly JD, Kolkman JM, Schneider K (1998) Breeding for yield in dry bean (*Phaseolus vulgaris* L.) Euphytica 102:343–356
Kim JW, Minamikawa T (1996) Transformation and regeneration of French bean plants by particle bombardment process. Plant Sci Limerick 117:131–138
Kolkman JM, Kelly JD (2003) QTL conferring resistance and avoidance to white mold in common bean. Crop Sci 43:539–548
Kwak M, Gepts P (2009) Structure of genetic diversity in the two major gene pools of common bean (*Phaseolus vulgaris* L., Fabaceae). Theor Appl Genet 118:979–992
Kwak M, Kami J, Gepts P (2009) The putative Mesoamerican domestication center of *Phaseolus vulgaris* is located in the Lerma-Santiago basin of Mexico. Crop Sci 49:554–563
Lamppa RS, Gross PL, Del Rio LE (2002) Races of *Pseudomonas syringae* pv. *phaeolicola* in North Dakota. Annu Rpt Bean Improv Coop 45:104–105
Liao D, Pajak A, Karcz SR et al (2012) Transcripts of sulphur metabolic genes are co-ordinately regulated in developing seeds of common bean lacking phaseolin and major lectins. J Expt Bot 63:6283–6295
Liebenberg MM, Pretorius ZA (2010) Common bean rust: pathology and control. Hort Rev 37:1–99
Liu YG, Mitsukawa N, Oosumi T et al (1995) Efficient isolation and mapping of *Arabidopsis thaliana* T-DNA insert junctions by thermal asymmetric interlaced PCR. Plant J 8:457–463
Logozzo G, Donnoli R, Macaluso L et al (2007) Analysis of the contribution of Mesoamerican and Andean gene pools to European common bean (*Phaseolus vulgaris* L.) germplasm and strategies to establish a core collection. Genet Resour Crop Evol 54:1763–1779
Mafi Moghaddam S, Song Q, Mamidi S et al (2014). Developing market class specific InDel markers from next generation sequence data in *Phaseolus vulgaris* L. Frontiers Plant Sci 4 (in press)
Mahuku G, Cajiao CJC, Beebe S (2003) Sources of resistance to angular leaf spot (*Phaeoisariopsis griseola*) in common bean core collection, wild *Phaseolus vulgaris* and secondary gene pool. Euphytica 130:303–313
Mamidi S, Rossi M, Annam D et al (2011) Investigation of the domestication of common bean (*Phaseolus vulgaris*) using multilocus sequence data. Funct Plant Biol 38:953–967
Marsolais F, Pajak A, Yin F et al (2010) Proteomic analysis of common bean seed with storage protein deficiency reveals up-regulation of sulfur-rich proteins and starch and raffinose metabolic enzymes, and down-regulation of the secretory pathway. J Proteomics 73:1587–1600
McCallum CM, Comai L, Greene EA et al (2000) Targeting induced local lesions in genomes (TILLING) for plant functional genomics. Plant Physiol 123:439–442

Miklas PN, Johnson E, Stone V et al (1996) Selective mapping of QTL conditioning disease resistance in common bean. Crop Sci 36:1344–1351

Miklas PN, Grafton KF, Kelly JD et al (1998) Registration of four white mold resistant dry bean germplasm lines: I9365–3, I9365–5, I9365–31, and 92BG-7. Crop Sci 38:1728

Miklas PN, Kelly JD, Beebe SE et al (2006) Common bean breeding for resistance against biotic and abiotic stresses: from classical to MAS breeding. Euphytica 147:105–131

Miklas PN, Porter LD, Kelly JD et al (2013) Characterization of white mold disease avoidance in common bean. Eur J Plant Pathol 135:525–543

Miklas PN, Fourie D, Trapp J et al (2014) A new locus Pse-6 conferring resistance to halo bacterial blight in common bean. Crop Sci 54:1598–1608

Moh CC (1971) Mutation breeding in seed-coat colors of bean (*Phaseolus vulgaris* L.). Euphytica 20:119–125

Mohan DP, Benepal PS, Sheikh AQ et al (1980) Determination of optimal mutagenic dose of ethylmethane sulfonate, diethyl sulfate and ethidium bromide for beans (*Phaseolus vulgaris* L.). Annu Rpt Bean Improv Coop 23:115–117

Moreto AL, Ramalho MAP, Bruzi AT (2012) Epistasis in an Andean Mesoamerican cross of common bean. Euphytica 186:755–760

Morrell PL, Clegg MT (2007) Genetic evidence for a second domestication of barley (*Hordeum vulgare*) east of the fertile crescent. Proc Natl Acad Sci U S A 104:3289–3294

Motto M, Soressi GP, Salamini F (1975) Growth analysis in a reduced leaf mutant of common bean (*Phaseolus vulgaris* L.). Euphytica 28:593–600

Mühling M, Gilroy J, Croy RRD (1997) Legumin proteins from seeds of *Phaseolus vulgaris* L. J Plant Physiol 150:489–492

Nagata RT, Bassett MJ (1984) Characterization and inheritance of gamma ray induced mutations in common bean. J Am Soc Hortic Sci 109:513–516

Nanni L, Bitocchi E, Bellucci E et al (2011) Nucleotide diversity of a genomic sequence similar to SHATTERPROOF (PvSHP1) in domesticated and wild common bean (*Phaseolus vulgaris* L.). Theor Appl Genet 123:1341–1357

Osborn TC, Bliss FA (1985) Effects of genetically removing lectin seed protein on horticultural and seed characteristics of common bean. J Am Soc Hortic Sci 110:484–488

Osborn TC, Hartweck LM, Harmsen RH et al (2003) Registration of *Phaseolus vulgaris* genetic stocks with altered seed protein compositions. Crop Sci 43:1570–1571

Osorno JM, Grafton KM, Rojas-Cifuentes GA et al (2010) Registration of 'Lariat' and 'Stampede' pinto beans. J Plant Regist 4:5–11

Panzeri D, Cassani E, Doria E et al (2011) A defective ABC transporter of the MRP family, responsible for the bean lpa1 mutation, affects the regulation of the phytic acid pathway, reduces seed myo-inositol and alters ABA sensitivity. New Phytol 191:70–83

Papa R, Gepts P (2003) Asymmetry of gene flow and differential geographical structure of molecular diversity in wild and domesticated common bean (*Phaseolus vulgaris* L.) from Mesoamerica. Theor Appl Genet 106:239–250

Papa R, Acosta J, Delgado-Salinas A, Gepts P (2005) A genome-wide analysis of differentiation between wild and domesticated *Phaseolus vulgaris* from Mesoamerica. Theor Appl Genet 111:1147–1158

Papa R, Bellucci E, Rossi M et al (2007) Tagging the signatures of domestication in common bean (*Phaseolus vulgaris*) by means of pooled DNA samples. Ann Bot 100:1039–1051

Park SJ, Buttery BR (1989) Inheritance of nitrate-tolerant supernodulation in EMS induced mutants of common bean (*Phaseolus vulgaris* L.). J Hered 80:486–488

Pastor-Corrales MA (2003) Sources, genes for resistance, and pedigrees of 52 rust and mosaic resistant dry bean germplasm lines released by the USDA Beltsville bean Project in collaboration with the Michigan, Nebraska and North Dakota Agricultural experiment stations. Annu Rept Bean Improv Coop 46:235–241

Pastor-Corrales MA, Kelly JD, Steadman JR et al (2007) Registration of six great northern bean germplasm lines with enhanced resistance to rust and bean common mosaic and necrosis potyviruses. J Plant Regist 1:77–79

Pastor-Corrales MA, Shin S, Wolf J (2012) Exceptional rust resistance in Mesoamerican common bean accession PI 310762. Annu Rpt Bean Improv Coop 55:147–148
Petry N, Egli I, Campion B et al (2013) Genetic reduction of phytate in common bean (*Phaseolus vulgaris* L.) seeds increases iron absorption in young women. J Nutr 143:1219–1224
Ponce MR, Quesada V, Micol JL (1998) Rapid discrimination of sequences flanking and within T-DNA insertions in the Arabidopsis genome. Plant J 14:497–501
Porch TG, Blair MW, Lariguet P et al (2009) Generation of a mutant population for TILLING common bean genotype BAT 93. J Am Soc Hort Sci 134:348–355
Porch TG, Beaver JS, Debouck DG et al (2013) Use of wild relatives and closely related species to adapt common bean to climate change. Agron 3:433–461
Rafalski JA (2010) Association genetics in crop improvement. Curr Opin Plant Biol 13:174–180
Rech EL, Vianna RG, Aragão FJL (2008) High efficiency transformation by biolistics of soybean, common bean and cotton transgenic plants. Nat Protoc 3:410–418
Rico A, López R, Asensio C et al (2003) Nontoxigenic Strains of *Pseudomonas syringae* pv. *phaseolicola* are a main cause of halo blight of beans in Spain and escape current detection methods. Phytopathol 93:1553–1559
Rodiño AP, González AM, Santalla M et al (2006) Novel genetic variation in common bean from the Iberian Peninsula. Crop Sci 46:2540–2546
Román-Avilés B, Kelly JD (2005) Identification of quantitative trait loci conditioning resistance to Fusarium root rot in common bean. Crop Sci 45:1881–1890
Romero Andreas J, Yandell BS, Bliss FA (1986) Bean arcelin. 1. Inheritance of a novel seed protein of *Phaseolus vulgaris* L. and its effect on seed composition. Theor Appl Genet 72:123–128
Rossi M, Bitocchi E, Bellucci E et al (2009) Linkage disequilibrium and population structure in wild and domesticated populations of *Phaseolus vulgaris* L. Evol Appl 2:504–522
Sanchez-Valdez I, Acosta-Gallegos JA, Ibarra-Perez FJ et al (2004) Registration of 'Pinto Saltillo' common bean. Crop Sci 44:1865–1866
Santalla M, De Ron AM, Voysest O (2001) European bean market classes. In: Amurrio M, Santalla M, De Ron AM (eds) Catalogue of bean genetic resources. Fundación Pedro Barrié de la Maza/ PHASELIEU-FAIR3463/ MBG-CSIC, Fundación, Pontevedra, pp 77–94
Santalla M, Rodiño AP, De Ron AM (2002) Allozyme evidence supporting southwester Europe as a secondary center of genetic diversity for common bean. Theor Appl Genet 104:934–944
Santalla M, De Ron AM, De La Fuente M (2010) Integration of genome and phenotype scanning gives evidence of genetic structure in Mesoamerican common bean (*Phaseolus vulgaris* L.) landraces from the southwest of Europe. Theor Appl Genet 120:1635–1651
Sarafi R (1973) Utilisation de rayons ionisants dans l'amélioration du haricot (*Phaseolus vulgaris* L.). Annales Amélioration Plantes 23:77–81
Schmutz J, McClean PE, Mamidi S et al (2014) A reference genome for common bean and genome-wide analysis of dual domestications. Nat Genet. doi:10.1038/ng.3008
Schwartz HF, Singh SP (2013) Breeding common bean for resistance to white mold: a review. Crop Sci 53:1832–1844
Schwartz HF, Steadman JR (1989) White mold. In: Schwartz HF, Pastor Corrales MA (eds) Bean production problems in the tropics, 2nd edn. CIAT, Cali, pp 211–230
Schwartz HF, Steadman JR, Hall R et al (eds) (2005) Compendium of bean diseases, 2nd edn. APS, St. Paul
Shi C, Navabi A, Yu K (2011) Association mapping of common bacterial blight resistance QTL in Ontario bean breeding populations. BMC Plant Biol 11:52
Silue S, Lariguet P, Pankhurst C et al (2006) Screening *Phaseolus vulgaris* L. EMS mutants to isolate plants failing in seed development and to study genetics of embryogenesis. Annu Rpt Bean Improv Coop 49:149–150
Singh SP (1994) Gamete selection for simultaneous improvement of multiple traits in common bean. Crop Sci 34:352–355
Singh SP (ed) (1999a) Common bean improvement in the Twenty-First Century. Kluwer, Dordrecht
Singh SP (1999b) Improvement of large-seeded race Nueva Granada cultivar. In: Singh SP (ed) Common bean improvement in the twenty-first century. Kluwer, Dordrecht, pp 275–288

Singh SP, Gutiérrez A (1984) Geographical distribution of the DL1 and DL2 genes causing hybrid dwarfism in *Phaseolus vulgaris* L., their association with seed size, and their significance to breeding. Euphytica 33:337–345
Singh SP, Muñoz CG (1999) Resistance to common bacterial blight among *Phaseolus* species and common bean improvement. Crop Sci 39:80–89
Singh SP, Schwartz HF (2010) Breeding common bean for resistance to diseases: a review. Crop Sci 50:2199–2223
Singh SP, Urrea CA (1994) Selection for seed yield and other traits among early generations of intra-and interracial populations of the common bean. Rev Brazil Genet 3:299–303
Singh SP, Gepts P, Debouck DG (1991a) Races of common bean (*Phaseolus vulgaris* L., Fabaceae). Econ Bot 45:379–396
Singh SP, Gutiérrez JA, Molina A et al (1991b) Genetic diversity in cultivated common bean. II. Marker–based analysis of morphological and agronomic traits. Crop Sci 31:23–29
Singh SP, Nodari R, Gepts P (1991c) Genetic diversity in cultivated common bean. I. Allozymes. Crop Sci 31:19–23
Singh SP, Molina A, Urrea CA et al (1993). Use of interracial hybridization in breeding the race Durango common bean. Can J Plant Sci 73:785–793
Singh SP, Terán H, Muñoz CG et al (2002) Selection for seed yield in Andean intra-gene pool and Andean × Middle American inter-gene pool populations of common bean. Euphytica 127:437–444
Singh SP, Terán H, Lema M et al (2006) Registration of slow darkening pinto bean germplasm line SDIP-1. Crop Sci 46:2726–2727
Singh SP, Terán H, Lema M et al (2007) Seventy-five years of breeding dry bean of the western USA. Crop Sci 47:981–989
Singh SP, Terán H, Schwartz HF et al (2009a) Introgressing white mold resistance from *Phaseolus* species of the secondary gene pool into common bean. Crop Sci 49:1629–1637
Singh SP, Terán H, Schwartz HF et al (2009b) Development of white mold resistant interspecific common bean germplasm lines VCW 54 and VCW 55. J Plant Regist 3:191–197
Singh SP, Terán H, Schwartz HF et al (2013) White mold-resistant, interspecific common bean breeding line VRW32 derived from *Phaseolus costaricensis*. J Plant Registr 7(1):95–99
Soule M, Porter L, Medina J et al (2011) Comparative QTL map for white mold resistance in common bean, and characterization of partial resistance in dry bean lines VA19 and 19365–31. Crop Sci 51:123–139
Souza T, De Barros EG, Bellato, CM et al (2012) Single nucleotide polymorphism discovery in common bean. Mol Breeding 30:419–428
Steadman JR (1983) White mold—a serious yield limiting disease of bean. Plant Dis 67:346–347
Svetleva D (2004) EMS and NEU mutagenic efficiency and effectiveness in induction of morphological mutations in *Phaseolus vulgaris* L. Annu Rpt Bean Improv Coop 47:177–178
Svetleva D, Velcheva M, Bhowmik G (2003) Biotechnology as a useful tool in common bean (*Phaseolus vulgaris* L.) improvement. Euphytica 131:189–200
Taylor JD, Teverson DM, Allen MA et al (1996a) Identification and origin of races of *Pseudomonas syringae* pv. *phaseolicola* from Africa and other bean growing areas. Plant Pathol 45:469–478
Taylor JD, Teverson DM, Davis JHC (1996b) Sources of resistance to *Pseudomonas syringae* pv. *phaseolicola* races in *Phaseolus vulgaris*. Plant Pathol 45:479–485
Taylor M, Chapman R, Beyaert R et al (2008) Seed storage protein deficiency improves sulfur amino acid content in common bean (*Phaseolus vul*garis L.): redirection of sulfur from gamma-glutamyl-S-methyl-cysteine. J Agric Food Chem 56:5647–5654
Terán H, Lema M, Webster D et al (2009) Seventy five years of breeding pinto bean for resistance to diseases in the United States. Euphytica 167:341–351
Teverson DM (1991) Genetics of pathogenicity and resistance in the halo-blight disease of beans in Africa. PhD Dissertation. Univ of Birmingham, United Kingdom
Tohme J, González DO, Beebe S et al (1996) AFLP analysis of gene pools of a wild bean core collection. Crop Sci 36:1375–1384
Thompson MD, Brick MA, McGinley JN et al (2009) Chemical composition and mammary cancer inhibitory activity of dry bean. Crop Sci 49:179–186

Tryphone GM, Chilagane LA, Protas D et al (2013) Marker assisted selection for common bean diseases improvements in Tanzania: prospects and future needs. In: Andersen SB (ed) Plant breeding from laboratories to field. InTech. doi:10.5772/52823. http://www.intechopen.com/books/plant-breeding-from-laboratories-to-fields/marker-assisted-selection-for-common-bean-diseases-improvements-in-tanzania-prospects-and-future-need
Tulmann-Neto A, Ando A (1976) Induced mutation in beans (*Phaseolus vulgaris*) to obtain varieties resistant to golden mosaic virus. Annu Rpt Bean Improv Coop 19:86
Tulmann-Neto AT, Sabino JC (1994) Induction and use of early and determinate growth habit mutant in common bean (*Phaseolus-vulgaris* L). Rev Bras Genet 17:425–430
Urrea CA, Singh SP (1994) Comparison of mass, F_2-derived family, and single-seed descent selection methods in an interracial population of common bean. Can J Plant Sci 74:461–464
Vallejos CE, Sakiyama NS, Chase CD (1992) A molecular marker-based linkage map of *Phaseolus vulgaris* L. Genetics 131:733–740
Vandemark GJ, Brick MA, Osorno JM et al (2014) Yield gains in edible grain legumes. In: Specht J, Diers B, Carver B et al (eds) Genetic gains of major US field crops. CSSA Special Publication 33 ASA, CSSA, SSSA Press, Madison, pp 87–123
Velez JJ, Bassett MJ, Beaver JS et al (1998) Inheritance of resistance to bean Golden mosaic virus in common bean. J Am Soc Hort Sci 123:628–631
Vitale A, Bollini R (1995) Legume storage proteins. In: Kigel J, Galili G (eds) Seed development and germination. Marcel Dekker, New York, pp 73–102
Viteri DM, Cregan PB, Trapp JJ et al (2013) A new common bacterial blight resistance QTL in VAX 1 common bean and interaction with the new QTL, SAP6 and SU91 with bacterial strains. Crop Sci 54:1598–1608
Vlasovab A, Rendón-Anaya M, Capella-Gutiérrez S et al (2014) Analysis of the Mesoamerican common bean genome reveals correlations of non-coding RNAs with domestication genes. Nat Commun (submitted)
Voelker TA, Staswick P, Chrispeels MJ (1986) Molecular analysis of two phytohemagglutinin genes and their expression in *Phaseolus vulgaris* cv. Pinto, a lectin-deficient cultivar of the bean. EMBO J 5:3075–3082
Wallace DH., Wilkinson RE (1965) Breeding for Fusarium root rot resistance in beans. Phytopathol 55:1227–1231
White J, Gonzales A (1990) Characterization of the negative association between seed yield and seed size among genotypes of common bean. Field Crops Res 23:159–175
Wright SI, Bi IV, Schroeder SG et al (2005) The effects of artificial selection on the maize genome. Sci 308:1310–1314
Wyatt JE, Dukes PD (1980) A seed coat color mutation in beans induced by gamma radiation. Annu Rpt Bean Improv Coop 23:34
Yin F, Pajak A, Chapman R et al (2011) Analysis of common bean expressed sequence tags identifies sulfur metabolic pathways active in seed and sulfur-rich proteins highly expressed in the absence of phaseolin and major lectins. BMC Genomics 12:268
Yu K, Park SJ, Poysa V (2000) Marker-assisted selection of common beans for resistance to common bacterial blight: efficacy and economics. Plant Breeding 119:411–415
Zeven AC (1997) The introduction of the common bean (*Phaseolus vulgaris* L.) into Western Europe and the phenotypic variation of dry beans collected in The Netherlands in 1946. Euphytica 94:319–328
Zhang X, Blair MW, Wang S (2008) Genetic diversity of Chinese common bean (*Phaseolus vulgaris* L.) landraces assessed with simple sequence repeats markers. Theor Appl Genet 117:629–640
Zogorcheva L, Poriazov I (1983) Induction of mutation in green beans by gamma rays. Annu Rpt Bean Improv Coop 26:89–90
Zou XL, Shi C, Austin RS et al (2014) Genome-wide single nucleotide polymorphism and insertion-deletion discovery through next-generation sequencing of reduced representation libraries in common bean. Mol Breeding 33:769–778

Chapter 2
Pea

Thomas D. Warkentin, Petr Smýkal, Clarice J. Coyne, Norman Weeden, Claire Domoney, Deng-Jin Bing, Antonio Leonforte, Zong Xuxiao, Girish Prasad Dixit, Lech Boros, Kevin E. McPhee, Rebecca J. McGee, Judith Burstin and Thomas Henry Noel Ellis

1 Introduction

Pea (*Pisum sativum* L.) is one of the first domesticated crops and is currently grown in most temperate regions of the world. Pea belongs to the Leguminosae family and as such is capable of fixing atmospheric nitrogen, thereby greatly reducing the requirement for petrochemical-based inputs. World production of dry pea ranged from $9.4–11.3 \times 10^6$ t to $6.0–6.6 \times 10^6$ ha between 2000 and 2012 (FAOSTAT 2013). These totals have been relatively steady over the past 50 years; however, the key producing areas have shifted over that time. Eastern Europe was

T. D. Warkentin (✉)
Crop Development Centre, University of Saskatchewan, 51 Campus Drive,
Saskatoon, SK S7N 5A8, Canada
e-mail: tom.warkentin@usask.ca

P. Smýkal
Department of Botany, Palacky University in Olomouc, Slechtitelu 11,
Olomouc 783 71, Czech Republic
e-mail: petr.smykal@upol.cz

C. J. Coyne
USDA, Washington State University, 59 Johnson Hall, Pullman, WA 99164-6402, USA
e-mail: coynec@wsu.edu

N. Weeden
Department of Plant Sciences and Plant Pathology, Montana State University, Bozeman,
MT, USA
e-mail: nweeden@montana.edu

C. Domoney
Department of Metabolic Biology, John Innes Centre, Norwich Research Park, Norwich, UK
e-mail: claire.domoney@jic.ac.uk

D.-J. Bing
Agriculture and Agri-Food Canada, 6000 C and E Trail, Lacombe, AB T4L 1W1, Canada
e-mail: dengjin.bing@agr.gc.ca

A. M. De Ron (ed.), *Grain Legumes*, Handbook of Plant Breeding 10,
DOI 10.1007/978-1-4939-2797-5_2

the major producer from the 1960s to 1980s, then Western Europe from the 1980s to 1990s, and since then North America, primarily Canada. China and India have had relatively stable production of 2–3 × 10^6 t/year over the past 50 years. In terms of world production of dry legume crops, dry pea trails only common bean which had annual production of 17.6–23.3 × 10^6 t between 2000 and 2012, and the oilseed legumes soya bean (161.3–265 × 10^6 t) and groundnut (33.1–42.1 × 10^6 t) during the same period (FAOSTAT 2013). World production of vegetable pea ranged from 12.0–17.4 × 10^6 t to 1.6–2.2 × 10^6 ha between 2000 and 2012 (FAOSTAT 2013). Vegetable pea production has been rising steadily over the past 50 years with China and India being the major producers.

In order to expand world production of pea, breeders, agronomists, end users and producers face several challenges. Grain yield gains must continue for pea to remain an attractive option in crop rotations. This will require a concerted effort from pea breeders internationally. In Western Europe, pea production has declined in the past two decades as producers have focused on high-yielding winter wheat and winter canola crops. This has led to a decline in pea-breeding activity. Pea

---

A. Leonforte
Nuseed Innovation Centre, 5 Ballinger Street, Horsham, VC 3400, Australia
e-mail: tony.leonforte@au.nuseed.com

Z. Xuxiao
Chinese Academy of Agricultural Sciences (CAAS), Institute of Crop Science (ICS), No.12, Zhong Guan Cun South Street, Haidian District, Beijing 100081, Beijing, China
e-mail: zongxuxiao@caas.cn

G. P. Dixit
Indian Institute of Pulses Research, Near Kanpur University, Kanpur 208024, Uttar Pradesh, India
e-mail: gpdixit1966@rediffmail.com

L. Boros
Department of Seed Science and Technology, Institute of Plant Breeding and Acclimatization—National Research Institute, Radzików 05-870, Poland
e-mail: l.boros@ihar.edu.pl

K. E. McPhee
Department of Plant Sciences, North Dakota State University, 370G Loftsgard Hall, P.O. Box 6050, Fargo, ND 58108, USA
e-mail: kevin.mcphee@ndsu.edu

R. J. McGee
USDA-ARS, Grain Legume Genetics and Physiology Research Unit, Washington State University, 305 Johnson Hall, PO Box 646434, Pullman, WA 99164, USA
e-mail: rjmcgee@wsu.edu

J. Burstin
INRA, UMR1347 Agroecology, Bat. Mendel, 17 rue de Sully, Dijon 21065, France
e-mail: judith.burstin@dijon.inra.fr

T. H. N. Ellis
CGIAR Research Program on Grain Legumes, ICRISAT, Patancheru 502324, Telangana, India
e-mail: n.ellis@cgiar.org

production has increased in North America and Australia over the past two decades and similarly, pea-breeding efforts have increased.

In order to achieve yield gains in pea, many biotic and abiotic stresses must be addressed through breeding. These stresses are specific to each region; however, in general, fungal diseases are the key biotic stress in most pea-growing regions, followed by various insects and viruses. Heat stress at flowering is the key abiotic stress in many pea-growing regions, followed by early-season flooding. Addressing these stresses are key breeding objectives for pea breeders in their attempts to increase and stabilize grain yields.

Greater international exchange of germplasm and increased use of diverse *Pisum* accessions may aid in achieving new yield gains. Use of genomic tools should enhance breeders' ability to substantially enrich their breeding populations with desired alleles, prior to the expensive exercise of yield testing in field trials.

Greater market diversification for pea will create more demand and expand production. Dry pea has typically been used as dhal in Asian markets. A major new use for dry pea is the Chinese vermicelli market which utilizes pea starch which is effective because of its high amylose content. This market has expanded from zero to more than 700,000 t/year over the past two decades (FAOSTAT 2013). Further use of pea and pea fractions (protein, starch and fibre) in diverse food products could promote expansion of the crop. Pea has good potential in new food applications due to its moderate protein concentration, slowly digestible starch and high levels of soluble and insoluble fibre, all of which are attractive for addressing type 2 diabetes and obesity. In addition, pea has low allergenicity and to date is a non-genetically modified organism (GMO), both factors making it attractive compared to soya bean in some markets.

2 Origin and Systematics

2.1 Phylogeny and Taxonomy

Pea belongs to the Leguminosae plant family, the third largest flowering plant family with 800 genera and more than 18,000 species (Lewis et al. 2005). The Papilionoideae is the largest subfamily, with 476 genera and about 14,000 species, which shared a common ancestor around 50 MA (Doyle et al. 1997; Lavin et al. 2005). The largest group of papilionoids, Hologalegina, with nearly 4000 species in 75 genera, includes the large galegoid tribes (including Galegeae, Fabeae, Trifolieae), united by the loss of one copy of the chloroplast inverted repeat. Tribe Fabeae Rchb. (not Vicieae (Bronn) DC., nom. illeg.) currently consists of five genera: *Lathyrus* (grass pea/sweet pea, about 160 species); *Lens* (lentils, 4 species); *Pisum* (peas, 3 species); *Vicia* (vetches, about 160–250 species) and the monotypic genus *Vavilovia formosa* (Mikič et al. 2013; Smýkal et al. 2011; Schaefer et al. 2012). Tribe Fabeae is considered one of the youngest groups in the legumes (Kupicha 1981; Steele and Wojciechowski 2003), and Bayesian molecular clock and ancestral range analysis suggest a crown age of 23–16 MA, in the mid-Miocene (Lavin et al. 2005; Schaefer et al. 2012). The centre of diversity and postulated area of origin of the Fabeae is in the eastern Mediterranean (Kupicha 1981; Schaefer et al. 2012) with a minimum of three dispersal events to the middle Atlantic islands and seven to the Americas. The tribe is considered monophyletic, nested within the Trifolieae. The crown age of the *Pisum* clade is estimated to 2.3–0.8 MA, while the divergence between *Pisum* and *Vavilovia* dates back to 9.8–4.8 MA (Schaefer et al. 2012).

The genus *Pisum* L., originally described to be distinct from *Lathyrus* L. (Linnaeus 1753), has recently been shown to be included in the *Lathyrus*/*Vicia* complex (Schaefer et al. 2012). Interestingly, Lamarck (1778), who was certainly aware of Linné's description, designated pea as *Lathyrus oleraceus*. Depending on how the *Lathyrus*/*Vicia* complex is treated, the genus *Pisum* may be incorporated into a larger *Lathyrus* genus to achieve monophyly. Thus, the taxonomic nomenclature used here will undoubtedly be revised. The classification of taxa within *Pisum* L. based on morphology and karyology has changed over time from being considered a genus with five species (Govorov 1937) to the currently widely accepted version with two species, *P. fulvum* and *P. sativum,* recognized (Kupicha 1981; Davis 1970). Numerous names have been proposed for wild representatives of *P. sativum.* In the review of Yarnell (1962), *P. humile (P. syriacum, P. sativum* subsp. *sativum* var. *pumilio), P. elatius*, *P. abyssinicum* and *P. sativum* were considered conspecific, even though they often differ by inversions and translocations. In this chapter, we will refer to the following taxonomic definitions of *Pisum: P. sativum* L. with subsp. *sativum* (includes var. *sativum* and var. *arvense*), subsp. *elatius* (Bieb.) Aschers. and Graebn (includes var. *elatius,* var. *brevipedunculatum* and var. *pumilio*), and subsp. *abyssinicum* (A. Braun) Govorov and *P. fulvum* Sibth. and Sm.

P. abyssinicum, has been resurrected as a third species by some recent authors (Maxted and Ambrose 2001; Vershinin et al. 2003; Jing et al. 2007), but for reasons detailed below, this taxon will be maintained as a subspecies of *P. sativum* in this chapter. Other 'species' such as *P. jomardi, P. transcaucasicum* and *P. arvense* have also been included within *P. sativum* by most recent treatments (Jing et al. 2007; Zaytseva et al. 2012). The most appropriate status for *P. sativum* subsp. *abyssinicum* is still under debate (Maxted and Ambrose 2001; Zaytseva et al. 2012). This taxon is native to Ethiopia and Yemen and has very low genetic diversity as demonstrated by morphological, allozyme (Weeden and Wolko 2001) and DNA analyses (Pearce et al. 2000; Vershinin et al. 2003; Jing et al. 2005, 2010). It possesses a distinct phenotype (early flowering and strongly serrate leaflets) as well as unique alleles at particular loci. Similar to most *P. s.* subsp *elatius* accessions, this taxon differs from the standard *P. sativum* subsp. *sativum* karyotype by at least a reciprocal translocation (Ben-Ze'ev and Zohary 1973). Hence, it qualifies for species status on the basis of phenotype and biological isolation. However, recent DNA sequence comparisons have shown this taxon to fall within the *humile/elatius/sativum* cluster or between it and *P. fulvum,* depending on the sequence being analysed (Jing et al. 2007, 2010; Ellis 2011; Smýkal et al. 2011; Vershinin et al. 2003; Zaytseva et al. 2012). The taxon has been used as a bridge between *P. fulvum* and *P. sativum* because it crosses reasonably well with both. Many crosses have been attempted with abyssinicum lines, and the most fertile crosses were to *P. sativum* subsp. *sativum* germplasm rather than to subsp. *elatius* accessions, although the presence of the reciprocal translocation definitely leads to reduced fertility in the F1 and F2 generations (Weeden, personal communication). Thus, if the abyssinicum variation is to be given specific status, it appears appropriate for consistency sake to also raise at least a portion of the elatius accessions to species status. From a practical viewpoint, the current authors do not see much advantage to splitting the abyssinicum/elatius/sativum germplasm into three or four species at the present time. Another taxon that has recently been suggested to be included in *Pisum* (Maxted and Ambrose 2001), *V. formosa,* we retain as a distinct genus (Smýkal et al. 2013; Mikič et al. 2013).

The centre of pea genetic diversity is the broad area of the Fertile Crescent through Turkey, Syria, Iraq, Israel and Lebanon. It extends further east to Central Asia (Iran, Afghanistan, Pakistan and Turkmenistan; Smýkal et al. 2011). Ethiopia has been postulated as a secondary centre of diversity (van der Maesen 1998). Vavilov (1950) considered Ethiopia together with the Mediterranean and Central Asia as primary centres, and Near East as secondary. *Pisum sativum* subsp. *elatius* and subsp. *sativum* are found naturally in Europe, northwestern Asia and extend south to temperate Africa, while *P. fulvum* is restricted to the Middle East.

2.2 *Origin and Domestication*

Pea is one of the world's oldest domesticated crops. Archaeological evidence dates the existence of pea back to 10,000 BC in the Near East (Baldev 1988; Zohary and Hopf 2000) and Central Asia (Riehl et al. 2013). Pea, among other grain legumes, accompanied cereals and formed important dietary components of early civilizations in the Middle East and Mediterranean. In Europe, it has been cultivated since the Stone and Bronze Ages and in India from 200 BC (De Candolle 2007). The Near East and Mediterranean regions are also the area of origin and initial domestication. Cultivation of pea spread from the Fertile Crescent to today's Russia, and westwards through the Danube valley into Europe and/or to ancient Greece and Rome which further facilitated its spread to Northern and Western Europe. In parallel, pea was moved eastward to Persia, India and China (Makasheva 1979; Chimwamurombe and Khulbe 2011).

Phylogenetically, there are two wild populations variously described as subspecies of *P. sativum* or as species, *P. sativum* subsp. *elatius* Bieb. and *P. sativum* subsp. *sativum* (= *P. humile* Boiss and Noe (syn. *P. syriacum* (Berger) Lehm.; Ben-Ze'ev and Zohary 1973; Smýkal et al. 2011). These two wild groups are morphologically, ecologically and also genetically distinct (Ben-Ze'ev and Zohary 1973; Abbo et al. 2013). The domestication of cultivated pea from northern populations of '*humile*' was proposed by Ben-Ze'ev and Zohary (1973), but the source could equally be the 'northern elatius' (Kosterin et al. 2010; Smýkal et al. 2011). Recently, *P. humile* was included into so-called lost crops, for example, additional taxa that were at certain points in time and in certain locations' genuine crops, but were later abandoned

(Abbo et al. 2013). It is notable that despite its wild-type seed dispersal mode and wild-type seed dormancy, these southern *P. humile* are currently found only in secondary habitats and never invade adjacent less disturbed habitats, in contrast to *P. fulvum* and *P. elatius*. Cytogenetic differences and analyses of genetic diversity support the view that the majority of cultivated peas originated from a distinct gene pool within var. *humile* (Zohary and Hopf 2000), although recent molecular studies also highlight the likely genomic contribution from other wild forms and emphasize the importance of introgression and recombination within the complex (Jing et al. 2010).

The domestication of pea has been experimentally tested, both in order to determine the genetic basis which led to the cultivated crop from the wild plant (Weeden 2007), as well as wild pea harvesting (Abbo et al. 2010). The so-called domestication syndrome in the case of pulses applies to increases in seed size, reduction or elimination of pod shattering, and loss of germination inhibition, shoot basal branching and seed toxins and antimetabolites (Smartt 1990; Zohary and Hopf 2000; Weeden 2007). Altogether, at least 11 loci involved in domestication traits have been identified (Weeden 2007). In pea, explosive pod indehiscence and seed dormancy (hard seededness) were probably the greatest barriers to domestication (Smartt 1990). Pod dehiscence is primarily influenced by the *Dpo* gene (Lamprecht 1957a), although other genes also affect this trait (Weeden et al. 2002). The genetic basis of hard seededness has yet to be fully elucidated, although it is clear that the *a* mutation (lack of anthocyanin production) reduces testa thickness, thereby affecting dormancy. Other traits selected during domestication and development of modern cultivated forms, include several morphological characters that are determined by one or a few genes. These genes include *le* (semidwarf growth habit), *r* (wrinkled seed in garden types), *af* (conversion of leaflets to tendrils), and *p* and *v* (absence of sclerenchymatic tissue in pods). Both *a* and *r* improved seed palatability, *le* and *af* increased the efficiency of mechanical harvesting, and *p* and *v* lead to the development of edible-podded types.

Monogenic inheritance is also known for several physiological traits that have been altered during domestication. Wild *Pisum* in its native range displays a typical winter habit in which plants germinate in autumn, overwinter in the vegetative state and flower in response to increasing day length in spring (Weller et al. 2009; Abbo et al. 2013). The obligate or near-obligate requirement for long days suits pea to a winter cropping cycle and has been retained in some forage cultivars. However, most of the cultivated pea accessions from higher latitudes have a quantitative long-day response and are grown as a spring crop (Weller et al. 2009). Some pea varieties are very early flowering and not photoperiod sensitive. The genes controlling flowering in pea include *Lf, Sn, Hr* and *E* (Murfet 1973; Weller et al. 2009). The obligate long-day (wild type) genotype is *Lf, Sn, Hr,* with *E* or *e*. The quantitative long-day phenotype of many cultivars has the genotype *Lf, Sn, hr*. Day-neutral cultivars are *Lf, sn* (*Hr* is not strongly active in *sn/sn* plants), and the very early flowering types are *lf, sn*. Hence, *lf, sn* and *hr* could all be considered 'domestication' alleles. Domestication has also resulted in increased seed and pod size in pea, although not as markedly as in other crops, with a correlated increase in leaf size and stem strength (Swiecicki and Timmerman-Vaughan 2005; Weeden 2007). The genetic basis of seed size appears to be quantitative (Timmerman-Vaughan et al. 1996), and there are several genes known to influence pod and leaf size (Lamprecht 1953, 1954, 1957b, 1960, 1963). At present, no single or small set of these genes can be identified as crucial for domestication.

Based on these morphological and genetic studies, *P. humile/syriacum, P. elatius* and *P. fulvum* were identified as 'wild' germplasm in that they display traits such as dehiscent pods and seed dormancy (thick testa), that are necessary for survival in the wild and undesirable in a domesticated annual crop. In contrast, *P. sativum* including var. *arvense, transcaucasicum* and *asiaticum* generally display indehiscent pods and little seed dormancy, and could be considered domesticated. *P. abyssinicum* is early flowering, with indehiscent pods, moderately large seeds and lacks seed dormancy. Based on this phenotype, it has been identified as partially domesticated.

One interesting feature regarding the domestication of pea is that not all changes have been unambiguous improvements. Rather many are trade-offs sacrificing certain adaptations for other advantages. For instance, incorporation of the *a* allele into cultivars improved the taste of the seed but also made the plant more susceptible to pathogens such as *Pythium* and *Fusarium*. Elimination of the *Np* gene increases seed size but also increases susceptibility to bruchid attack (Berdnikov et al. 1992). Incorporation of the 'wrinkled seed' *(r)* and 'afila' *(af)* alleles leads to a reduction in yield in some environments. The sugar snap pea, with its combination of four or five mutations (*a, r, p* or *v, n* and *sin*) is notoriously susceptible to soil pathogens. Once genes controlling main domestication traits are identified, along with the full pea genome sequence, we might expect and look forward to comparative evolutionary studies across independently domesticated legumes.

2.3 Pea Genus Genetic Diversity

Based on morphology, *Pisum* sp. is one of the most diverse crop species known, comparable to *Zea mays, Cucurbita pepo* and *Brassica oleracea* (Hancock 2012). There are several user-defined classifications of cultivated pea diversity. Four simply inherited characters determine the main use types of peas within subsp. *sativum:* the presence or absence of pod parchment, flower anthocyanin, leaflets occurrence and whether the starch grains in the dry seed are simple or compound (Green 2008). This classification is similar to that proposed by Lehmann (1954) except for the *afila* type which was unknown at that time. Early data from electrophoretic patterns of major seed proteins: albumin and globulin (Waines 1975), allozymes (Hoey et al. 1996) and chloroplast DNA polymorphism (Palmer et al. 1985) separated *P. fulvum* as a distinct species and *P. sativum* as an aggregate of '*humile*', *P. sativum* subsp. *elatius* and *P. sativum*, in agreement with the current view of the genus. Chemosystematic studies using flavonoids (Harborne 1971) showed that *P. fulvum* contains quercetine 3-glucoside, primitive cultivars from Nepal and *P. abyssinicum* contain kaempferol and quercetine 3-sophoroside, while modern pea cultivars contain kaempferol and quercetine 3-(cumaroyl-sophorotrioside). Petals of wild peas contain delphinidin, petunidin and malvidin 3-rhamnoside-5-glucosides, while coloured petals of cultivated garden pea contain, in addition, pelargonidin, cyanidin and peonidin 3-rhamnoside-5-glucosides (Harborne 1971). Unfortunately, the yellow colour *P. fulvum* petals were not studied. Moreover, the *Pisum* genus contains the flavonoid phytoalexin pisatin, which is shared with the genus *Lathyrus* but not found in *Vicia* species (Bisby et al. 1994), which have wyerone instead. Serological reactions of *Pisum* taxa by Kloz and Turkova (1963) indicated a close relationship of all studied taxa, except of *P. fulvum* and *P. abyssinicum*. They were possibly the first to indicate that *P. abyssinicum* might have originated from hybridization between *P. sativum* subsp. *elatius* and *P. fulvum*. Hoey et al. (1996) using morphological, allozyme and random amplified polymorphic DNA (RAPD) characteristics on a set of Ben-Ze'ev and Zohary (1973) accessions showed a separation of *P. fulvum* and 'southern *humile*', while cultivated peas were among *P. sativum* subsp. *elatius* accessions. The position of 'northern *humile*' varied between a sister group to cultivated peas and *P. sativum* subsp. *elatius*. More recently, studies of internal transcribed spacer (ITS) sequence variation (Saar and Polans 2000) and histone H1 subtype five gene (Zaytseva et al. 2012) have supported this. Recent phylogenetic studies based on retrotransposon insertion markers support the model of *P. sativum* subsp. *elatius* as a paraphyletic group, within which all *P. sativum* are nested (Nasiri et al. 2010; Vershinin et al. 2003; Jing et al. 2005, 2010). Although pollination strategy is highly relevant for genetic diversity, it has not been properly studied in wild pea. Pea is considered a self-pollinating species; however, cross-pollination is likely to occur in wild pea populations. The reported cross-pollination rate in cultivation ranges from zero (White 1917) to 60 % (Harland 1948), depending on genotype and environment. The percentage reported for commercial cultivars is less than 1 % (Dostálová et al. 2005). In addition to biological consequences, self-pollination reinforced genetic barriers between wild and cultivated populations, facilitating fixation of the desired genotype (Zohary and Hopf 2000).

Molecular analysis of pea diversity preserved in germplasm collections was done using amplified fragment length polymorphism (AFLP; Ellis et al. 1998), its derived retrotransposon insertion-based marker method, sequence-specific amplification polymorphisms (SSAP; Pearce et al. 2000; Majeed et al. 2012; Vershinin et al. 2003) and gene sequences (Jing et al. 2007; Zaytseva et al. 2012). In all analyses, *P. fulvum* and *P. abyssinicum* formed neighbouring but separate branches, a subset of *P. sativum* subsp. *elatius* was positioned between *P. fulvum* and *P. abyssinicum,* and further branches were found within cultivated pea. The most recent studies of *P. abyssinicum* place it between *P. fulvum* and a subset of *P. sativum* subsp. *elatius* (Ellis 2011; Smýkal et al. 2011; Vershinin et al. 2003; Jing et al. 2010) and showed its very low genetic diversity, which could be explained by passage through a genetic bottleneck. Importantly, high conservation between retrotransposon sequence-specific amplification polymorphism (SSAP; Vershinin et al. 2003), retrotransposon insertions (Jing et al. 2005) and gene-based derived (Jing et al. 2007) trees was observed, in spite of the fact that they derive from different genomic compartments.

Another study on relationships among wild *Pisum*, using a combination of mitochondrial, chloroplast and nuclear genome markers (Kosterin and Bogdanova 2008; Kosterin et al. 2010), separated *P. fulvum* and *P. abyssinicum* accessions. Interestingly, Afghan types (originating from Afghanistan, Iran, Pakistan) are not nodulated by ordinary European and North American strains, but require specific *Rhizobium* strains for symbiosis (Young and Matthews 1982). Afghan types were clustered separately in diversity analysis based on retrotransposon insertions (Jing et al. 2010). Genetic discontinuity in root-nodulating bacteria of cultivated pea was shown in the trans-Himalayas between India and China (Rahi et al. 2012). The example of coevolution was found in a south Turkey pea line, which was found to form an effective symbiosis only with local *Rhizobium* strains but not with strains from other parts of Turkey (Lie et al. 1987). These authors suggested that the genetic uniformity of European *R. leguminosarum* strains is the result of selection and domestication of *Rhizobium* strains originally derived from the gene centres of pea.

Several studies of pea germplasm using morphological descriptors and agronomical traits and lately DNA markers have been published (see 'Genetic resources and utilization', Sect. 4). These gave a consistent view. In spite of being a rather small genus with two or three species, *Pisum* is very diverse and diversity is structured, showing a range of degrees of relatedness that reflect taxonomic identifiers, eco-geography and breeding gene pools (Ellis 2011; Smýkal et al. 2011; Jing et al. 2012).

In summary, pea belongs to the early domesticated legume crops accompanying cereals and formed an important dietary component of early civilizations. The Near East and Mediterranean regions are both the area of origin and initial domestication. In the process of domestication, two key traits have been modified, pod dehiscence and seed dormancy. Additional traits include seed size, flowering-time control and branching pattern. Pea belongs in the Fabeae tribe together with lentil, faba bean, common vetch and grass pea. Recent phylogenetic analysis has shown the *Pisum* L. genus to be positioned within the *Lathyrus/Vicia* complex to obtain monophyly. Despite high morphological variation and an extensive geographical range, two true species are recognized: *P. sativum/elatius* complex and the more distant *P. fulvum,* consisting of the secondary gene pool. These can be intercrossed and fertile progeny obtained, although there is some reduction in fertility due to chromosomal translocations and nucleo-cytoplasmic conflict. Phylogenetically related *V. formosa,* a perennial mountain species, consists of the tertiary gene pool.

3 Varietal Groups

Pea has a wide range of market classes and uses (See Fig. 2.1).

Field Pea Also known as dry pea and combining pea. The mature seed phenotype for field pea is round (genetically *RR*). Field pea includes yellow, green and red cotyledon varieties typically used in the dehulled/split form in foods such as dhal. New markets are emerging for pea flour in baked products, extruded snacks and

Fig. 2.1 Diversity of seed colour, shape and size in pea. Among 15 genotypes shown, testas have been removed from one seed of every line to show the cotyledon colour. The phenotypes reflect variation at several genetic loci, including *A*, *r*, *i* and *s*

noodles. Starch fractions, typically from yellow cotyledon pea, are used widely in China for production of vermicelli noodles. Protein and fibre fractions are also used in the food industry.

Smaller market classes include:

- Dun (pigmented seed coat) which is also used in the dehulled/split form for foods such as dhal
- Marrowfat (large seeds, blocky shape, green cotyledons, appealing flavour profile) for snacks and mushy pea
- Maple (mottled seed coat) for bird seed mixtures
- Forage (high biomass) cut prior to dry seed maturity for ruminant feed

- Sprouts, that is, germinated seedlings used as a vegetable
- Feed (can include any of the above types) for use in the compound feed industry typically for pigs and chickens

Vegetable Pea Also known as garden pea and vining pea. The mature seed phenotype for vegetable pea is wrinkled (genetically *rr*). Vegetable pea includes:

- Freezer pea
- Snow pea
- Snap pea

4 Genetic Resources and Utilization

Total pea genetic resources are extensive with ex situ germplasm holdings of 73,931 accessions in over 28 plus national and international collections with duplicate samples of 9670 accessions preserved at the Svalbard Global Seed Vault at Spitsbergen, Norway (Table 2.1).

The four largest active collections include the Australian Temperate Field Crop Collection, Horsham, of 7432 accessions, N.I.Vavilov Research Institute of Plant Industry (VIR), St. Petersburg, Russia, with 6790 accessions, the US Department of Agriculture (USDA) 6827 accessions, and the French National Institute for Agricultural Research (INRA) of 8839 accessions held at Dijon, France. The Vavilov Research Institute which originated in 1921 is the oldest and most famous of the large pea germplasm collections and was started with Vavilov's explorations. Very representative and arguably the best studied is the John Innes (JI), Norwich *Pisum* collection, containing 1200 *P. sativum* cultivars, 600 traditional landraces and 750 genetic stocks and reference lines together with wild *Pisum* samples. Australia has the least duplicative and most diverse ex situ collection to date for *Pisum*. Also, a large collection of 6105 is held by International Center for Agricultural Research in the Dry Areas (ICARDA) which could be accessed easily until recent times. In addition, there are other exciting national collections of pea germplasm. An example is in Israel, which hosts a collection of the crop's wild relatives *P. fulvum* and *P. sativum* subsp. *elatius* var. *pumilio* collected in the Middle East. An estimated 20 % of the world's ex situ pea germplasm is duplicative (Smýkal et al. 2013), which would leave 59,000 accessions as unique.

The highest traffic websites for ordering germplasm are the JI Centre (JIC; http://www.jic.ac.uk/germplasm/) and the USDA (http://www.ars-grin.gov/npgs/). These two sites have the highest turnover of international requests of readily available *Pisum* accessions. For example, the USDA sends 3140 pea accessions per year (5 year average) to researchers worldwide. The JIC Genetic Stocks (http://data.jic.ac.uk/cgi-bin/pgene/) accessed through the 'PGene Pisum Gene List' contains genes/traits recognized by the Pisum Genetics Association (http://hermes.bionet.nsc.ru/pg/) and the associated gene stock JI line. Duplicate genetic stocks are

Table 2.1 List of selected pea germplasm collections preserving significant *Pisum* diversity (data from Smýkal et al. 2013)

Country	Institute	Website	Cultivated *Pisum sativum*					Total
			Commercial varieties	Breeding lines	Land races	Mutant stock	Wild *Pisum* sp.	
Russia	N.I. Vavilov Research Institute of Plant Industry, St. Petersburg	http://www.vir.nw.ru	6653	–	–	–	–	6790
France	INRA CRG Légumineuse à grosses graines, Dijon	http://195.220.91.17/legumbase/	1315	496	174	4818	63	8839
Australia	Australian Temperate Field Crop Collection, Horsham	http://www2.dpi.qld.gov.au	1170	1420	4514	87	205	7432
USA	Plant Germplasm Introduction and Testing Research Station, Pullman	http://www.ars-grin.gov	1504	64	3935	37	84	6827
Syria	International Center for Agricultural Research in the Dry Areas, Aleppo	http://www.icarda.cgiar.org	1185	1305	3240	0	228	6105
Germany	Leibniz Institute of Plant Genetics and Crop Plant Research, Gaterleben	http://www.ipk-gatersleben.de/	3008	277	1553	71	47	5343
Italy	Istituto del Germoplasma, Bari	http://www.igv.cnr.it	0	0	4558	0	0	4558
China	Institute of Crop Science, Chinese Academy of Agricultural Sciences	http://icgr.caas.net.cn/cgris	532	45	3260	5 sets	0	3837
India	National Bureau of Plant Genetic Resources, New Delhi	http://www.nbpgr.ernet.in	210	1016	1207	8	0	3609
UK	John Innes Centre, Norwich	http://www.jic.ac.uk	1071	61	600	585	368	3567

Table 2.1 (continued)

Country	Institute	Website	Cultivated *Pisum sativum*					Total
			Commercial varieties	Breeding lines	Land races	Mutant stock	Wild *Pisum* sp.	
Poland	Plant Breeding and Acclimatization Institute Blonie, Radzikow	http://www.igr.poznan.pl	1254	638	21	297	75	2896
Sweden	NordGen, Nordic Genetic Resource Centre, Alnarp	http://www.nordgen.org/sesto	472	1528	471	0	32	2849
Bulgaria	Institute of Plant Introduction and Genetic Resources, Sadovo	http://www.genebank.hit.bg	750	500	196	150	3	2100
Brazil	National Center for Vegetable Crops Research (CNPH)/EMBRAPA	http://www.cnph.embrapa.br	N.A.	N.A.	N.A.	N.A.	–	1958
Spain	Junta de Castilla y León. Instituto Tecnológico Agrario de Castilla y León	http://www.itacyl.es	649	328	543	103	32	1772
Ethiopia		http://www.ibc.gov.et/	N.A.	N.A.	N.A.	N.A.	–	1768
Ukraine	Yurjev Institute of Plant Breeding, Kharkov	http://www.bionet.nsc.ru	N.A.	N.A.	N.A.	N.A.	–	1671
Spain	Institutot Nacional de Investigacion y Tecnologia Agraria y Alimantaria	http://www.inia.es	67	168	599	0	0	1648
Czech Republic	Centre for Applied Research of Vegetables and Special Crops, Olomouc	http://genbank.vurv.cz	1257	63	11	0	2	1414
Czech Republic	AGRITEC, Research, Breeding and Services Ltd., Sumperk	http://genbank.vurv.cz	972	257	54	0	9	1326

Table 2.1 (continued)

Country	Institute	Website	Cultivated *Pisum sativum*					Total
			Commercial varieties	Breeding lines	Land races	Mutant stock	Wild *Pisum* sp.	
Hungary	Institute for Agrobotany, Tapioszele	http://www.rcat.hu	933	0	79	0	4	1205
The Nether-lands	Centre for Genetic Resources, Wageningen	http://www.cgn.wur.nl/pgr/	504	86	345	0	0	1002
Serbia	IFVCNS, Novi Sad	http://www.nsseme.com/en/	665	127	74	42	55	991
Canada	Plant Gene Resources of Canada, Saskatchewan, Canada	http://www.agr.gc.ca/pgrc-rpc	449	76	55	0	11	616
Israel	Israel Plant Gene Bank, ARO Volcani Center	http://igb.agri.gov.il	0	0	0	0	333	343
Turkey	Aegean Agricultural Research Institute, Mene-men/ IZMIR	http://www.aari.gov.tr	0	0	236	0	0	236
Armenia	Institute of Botany NAS RA, Yerevan	http://www.sci.am/	4	0	0	0	0	19
Total			24,624	8455	25,725	6198	1551	73,931

N.A. not accounted

maintained also at the Nord Genebank (Blixt and Williams 1982). The JIC Pisum Genetic Stocks are partially duplicated at the USDA (Ambrose and Coyne 2009), and at Polish (297) and Bulgarian (150 accessions) genebanks. Also maintained by JIC is a fast neutron (FN) population started with 1400 seeds of JI 2822 subjected to 20 Grays FN radiation (Domoney et al. 2013; Hofer et al. 2009). A very fruitful targeting induced local lesions in genomes (TILLING) reference population from the pea cultivar 'Caméor' and database (UTILLdb) resource is available for forward and reverse genetics from INRA (*urgv.evry.inra.fr/UTILLdb*; Dalmais et al. 2008). The USDA G. A. Marx *Pisum* Genetic Stock collection was extracted from 80,000 lines Dr. Marx developed to study interactions between pea mutations (www.ars.usda.gov/). Murfet and Reid (1993) developed and maintain developmental mutants in Tasmania.

Only a small proportion (1876, approximately 2 %) of germplasm conserved accessions, represent wild collected pea. Of these, there are 706 of *P. fulvum,* 624 *P. s.* subsp. *elatius,* 1562 *P. s.* subsp. *sativum* (syn. *P. humile/syriacum*) and 540 *P. abyssinicum* accessions (Smýkal et al. 2013). Wild *Pisum* species and subspecies are expected to receive high interest as they are a fount of useful traits, for example, resistance to pea seed weevil (Clement et al. 2009; Byrne et al. 2008; Clement et al. 2002), rust (Barilli et al. 2010), a new resistance gene to powdery mildew (Fondevilla et al. 2007b) and yield components (Mikič et al. 2013). Less agronomically preferred germplasm (pigmented flower and pigmented seed coat) were proven excellent sources of resistance to Aphanomyces root rot (Hamon et al. 2011) and Fusarium root rots (Weeden and Porter 2007; Grunwald et al. 2003).

Pisum sativum subsp. *elatius* and subsp. *sativum* are found in Europe, northwestern Asia and temperate Africa, while *P. fulvum* is found in the Middle East (Abbo et al. 2008; Maxted and Ambrose 2001; Smýkal et al. 2011). A combined gap analysis was conducted for six legume genera using more than 2000 unique geo-referenced records. The resulting regression analysis illustrated that none of the countries rich in *Pisum* species can be considered over-sampled, with Turkey, Former Soviet Union (particularly the countries of the Caucasus), Syria and the Balkans warranting further ex situ collection, as there is potential for finding additional diversity (Maxted and Kell 2009). Passive in situ conservation of legume species, including pea, occurs in several protected landscape ecosystems in the Mediterranean and Near East regions, which are not aimed specifically to conserve wild crop relatives. Consequently, native legume populations are susceptible to genetic erosion or even extinction (Maxted and Bennett 2001). Three reserves were established within the Global Environment Facility project in Turkey (Kaya et al. 1998). While no in situ reserves are formally established, wild pea populations can still be found (Abbo et al. 2008, 2013; Mikič et al. 2012).

When available passport data on geographical origin are summarized, there is a large bias (17 %) towards Western and Central European accessions, as these regions represent modern pea-breeding activities. Substantially less well represented are Mediterranean (2.5 %) and Balkan (2 %) regions, and the Caucasus (0.8 %) and Central Asia (2 %) centres of pea crop domestication and diversity where higher variation can be anticipated (Smýkal et al. 2013). There are still important gaps in the ex situ collections, particularly of wild and locally adapted materials (Maxted and Kell 2009). An example of a significant gap was evident in a recent landrace collection study of pea from China (Zong et al. 2009). Li et al. (2013) used eco-geographical climatic characterization of 240 collection sites for 529 pea landraces in China to identify locations with long-term abiotic stresses, especially during the reproductive growth phase. This enabled 61 candidate accessions from these stress sites to be prioritized for phenotypic validation to confirm tolerances to frost, drought and heat. The Global Crop Diversity Trust (www.croptrust.org) has taken on a full assessment of the gaps in *Pisum* germplasm held ex situ and in situ.

There are several international collection databases, which possess information for pea, including: European Cooperative Program on Plant Genetic Resources (ECPGR), Genetic Resources Information Network (GRIN) and System-wide Information Network for Genetic Resources (SINGER) databases. Although these databases provide information on around two million crop accessions, this information is largely passport-based, and thus limited. The deposition and availability of molecular, agronomic and morphological trait data is a very critical issue. In the case of pea, this is largely handicapped by not being one of the CGIAR mandate crops, and consequently with no coordinated international funding support. Combining passport, morphological and genotypic data of many genebanks would both improve germplasm management and enable search/query data exploration for germplasm with multiple traits from a virtual world pea collection online (Furman et al. 2006; Smýkal et al. 2008a).

A critical question remains, how can we efficiently utilize the positive alleles found in the diversity of these *Pisum* germplasm resources? Climate change and population pressures dictate the necessity of solving this challenge soon (Wheeler and von Braun 2013). One proposal is to focus on reference sets for fine phenotyping and genotyping (Glaszmann et al. 2010). A recent meeting on crop wild relatives proposed a bolder three-step approach to mine biodiversity for crop improvement, all highly relevant to pea (McCouch et al. 2013). The first step is to sample sequence information (genotyping) of the genomes of all non-duplicate pea accessions in the world's gene banks. The second step is to analyse the phenotypes of these pea accessions to evaluate their traits and overall agronomic performance. The third step is to create an accessible bioinformatics infrastructure to catalogue the diversity held in the world's pea seed collections.

The following sections will summarize some of the progress to date in pea for these three steps in achieving efficient utilization of germplasm resources. For the analysis of pea diversity, simple sequence repeats (SSRs or microsatellites) have been popular because of their high polymorphism and information content, codominance and reproducibility (Baranger et al. 2004; Loridon et al. 2005; Smýkal et al. 2008b; Zong et al. 2009; Kwon et al. 2012; Table 2.2). More recently, expressed sequence tags (EST)-derived (eSSR) markers have become an important resource for gene discovery and comparative mapping studies (De Caire et al. 2012; Mishra et al. 2012). Alternately, retrotransposon repeats have been used to reveal diversity, first applied in fingerprinting format of SSAP; Ellis et al. 1998; Vershinin et al. 2003) and developed into a high-throughput locus-specific genotyping technology

Table 2.2 Summary of the genotyping of some of the national pea germplasm ex situ collections completed using DNA markers

Accessions	Marker class	Marker number (polymorphism)	Cluster estimate	Citation
4538[a]	RBIP, SSAP	27	3	Jing et al. (2012)
4429[a]	SSR, RBIP	21, 25	6, 8	Smýkal et al. (2011)
3020[a]	RBIP	45	7	Jing et al. (2010)
2120[a]	SSR	21(115)	3	Zong et al. (2009)
373	SNP	384(356)	–	Deulvot et al. (2010)
310[a]	SSR/RAPD	15/36(102)	3	Kwon et al. (2012)
299[a]	Isozymes	18	–	Swiecicki et al. (2000)
164	SSR, RBIP	10, 31	9	Smýkal et al. (2008b)
157	SSAP	129	4	Vershinin et al. (2003)
148[a]	SSR/RAPD/ protein	66(121)	8	Baranger et al. (2004)
122[a]	RBIP	18(56)	4	Martin-Sanz et al. (2011)
86	RAPD, SSR	24(95), 25(54)	–	Tar'an et al. (2005)
71[a]	SSAP	281	4	Ellis et al. (1998)
40	SRAP	7(162)	3	Esposito et al. (2007)
40	SSR	10(61)	2	Sarıkamış et al. (2010)
48[a]	Gene sequences	49	–	Jing et al. (2007)
35	SSR	15(41)	2	Ahmad et al. (2012)
21	RAPD, AFLP	20(175), 11(462)	2	Simioniuc et al. (2002)
20	SSR	14	2	Ford et al. (2002)

RBIP retrotransposon-based insertion polymorphism, *SSAP* sequence-specific amplification polymorphism, *SSR* simple sequence repeat, *SNP* single-nucleotide polymorphism, *RAPD* random amplified polymorphic DNA and ISSRs, *SRAP* sequence-related amplified polymorphism, *AFLP* amplified fragment length polymorphism

[a] Contain core collection(s)

based on insertion/deletion of Ty1-copia PDR1 element and used for phylogeny and genetic relationship studies in pea, providing a highly specific, reproducible and easily-scorable method (Jing et al. 2007, 2010, 2012; Smýkal et al. 2008b, 2011). Another class of Angela-family retrotransposon was identified and used for inter-retrotransposon amplified polymorphism (IRAP) fingerprinting (Smýkal 2006; Smýkal et al. 2009).

This knowledge has been used to characterize the distribution of genetic diversity in *Pisum* (Baranger et al. 2004; Ellis et al. 1998; Jing et al. 2005, 2007, 2010, 2012; Pearce et al. 2000, Martin-Sanz et al. 2011; Majeed et al. 2012; Smýkal et al. 2008b; Sarıkamış et al. 2010; Tar'an et al. 2005; Vershinin et al. 2003; Zong et al. 2008, 2009) and these studies provide a consistent view. In spite of being a rather small genus with two or three species, *Pisum* is very diverse and its diversity is structured, showing a range of degrees of relatedness that reflect taxonomic identifiers, eco-geography and breeding gene pools (Ellis 2011; Smýkal et al. 2011; Jing et al. 2012). Bayesian analysis of more than 4500 pea accessions from three large collections (JIC, ATFC, CzNPC) analysed by retrotransposon insertion markers, separated all wild peas (*P. fulvum, P. sativum* subsp. *elatius,* and *P. abyssinicum*) in one cluster, together with accessions of Afghan origin (Smýkal et al. 2011). Another cluster contained a large proportion of *P. sativum* subsp. *sativum* accessions of Ethiopian origin (Smýkal et al. 2011). One hundred and forty accessions of Chinese origin (Zong et al. 2009) were distributed more broadly into seven to eight clusters. It was proposed that the distinct differentiation of the Chinese *P. sativum* genotypes may in part reflect historic isolation of agriculture in eastern Asia from that in southern Asia, Europe and northern Africa (Zong et al. 2009). Three relatively distinct gene pools of Chinese pea landraces have been differentiated and formed under natural and artificial selections (Li et al. 2013).

Recently, the analysis was complemented with addition of 1518 further *Pisum* accessions selected from other major European collections leading to identification of further diversity and formulation of a core collection (Jing et al. 2012). These results showed that despite a wide diversity captured in historic cultivated germplasm, relatively few genotypes with a high degree of relatedness have been used as parents in modern pea-breeding programmes, leading to a narrow genetic base of cultivated germplasm (Ellis 2011; Jing et al. 2010, 2012; Smýkal et al. 2011); however, it is possible to broaden diversity using wild genotypes.

Although microsatellite and retrotransposon marker types are still used, their potential is at its limits. With advances in model legume sequencing and genomic knowledge, there is a switch to gene-based markers in pea (Aubert et al. 2006; Jing et al. 2007; Bordat et al. 2011). This trend can be expected to further proliferate in line with rapid advances in high-throughput single-nucleotide polymorphism (SNP) generation and detection assays (Deulvot et al. 2010; Bordat et al. 2011). Recently, a comprehensive transcriptome of pea was published (Franssen et al. 2011) and several high-throughput pea transcriptome sequencing projects are underway and should provide a complete set of pea genes. Much work remains to be done to achieve the first step of genotyping all non-duplicate pea accessions as suggested in the Feed the Future proposal. This is becoming increasingly feasible as the technologies for high-throughput sequencing continue to improve and costs are reduced (Varshney et al. 2009).

It is a daunting task to phenotype the non-duplicate pea germplasm accessions for the second step in the Feeding the Future proposal as it would involve tens of thousands of accessions. Several large studies were published primarily for quantitative disease reactions (Table 2.3). Germplasm data needs the same effort in curation as the seed banks themselves, as well as an efficient method to share phenotypic data. Generally, study authors are quite willing to share data, as indicated by Table 2.4.

The value of germplasm phenotyping is many fold. First is the value to all plant scientists looking to germplasm to study specific traits and genes. Second is the utilization by plant breeders for pea crop improvement. Once the variation is identified, breeders can test for heritability of a trait and move it into advanced breeding/elite materials. If no useful germplasm is identified (e.g. pea cyst nematode resistance), further screens can be conducted or alternative breeding methods pursued (e.g. mutation, genetic modification).

Table 2.3 Replicated trials of agronomic and disease/pest reactions of sets of pea germplasm

Traits	Accessions	Citation
100 seed weight[a]	6827[b]	Stout (personal communication)
Fusarium wilt, race 1[a]	3343[b]	Muehlbauer, McPhee, McGee (personal communication)
Fusarium root rot[a]	3080[b]	Grunwald et al. (2003)
Nematode resistance[a]	2253[b]	Tedford and Inglis (1999)
Aphanomyces root rot[a]	2195[b]	Malvick and Percich (1999)
Mycosphaerella blight[a]	1993[b]	Kraft et al. (1998)
Fusarium wilt, race 2[a]	1946[b]	McPhee et al. (1999)
Protein[a]	1355[b]	Jermyn and Slinkard (1977) and Coyne et al. (2005b)
Salinity, alkaline/acidity	780	Leonforte et al. (2013)
Mycosphaerella blight	558	Zhang et al. (2006b)
Eco-geographical climatic characterization	529	Li et al. (2013)
Sclerotinia white mould	497[b]	Porter et al. (2009)
Mycosphaerella blight	581	Priliouk et al. (1999)
Seed mineral nutrients[a]	458[b]	Grusak et al. (2004)
Agronomic, biomass[a]	392[b]	McPhee and Muehlbauer (2001)
Root traits[a]	330[b]	McPhee (2005)
Powdery mildew	317	Ali et al. (2007) and Azmat et al. (2012)
Pseudomonas syringae pv. pisi	242	Martín-Sanz et al. (2012)
Clover yellow vein virus[a]	215	Andrade et al. (2006)
Agronomic, mycosphaerella blight resistance, nutrition	169	Jha et al. (2013)
Pseudomonas bacterial blight	169[c]	Taylor et al. (1989)
Insects resistance *(Melanagromyza phaseoli, Chromatomyia horticola, Helicoverpa armigera)*	165	Mittal and Ujagir (2005)
Seed composition traits	157[c]	Matthews and Ambrose (1994)
Agronomic, powdery mildew	153	Ali et al. (2007)
Agronomic	105	Amurrio et al. (1995)
Perenospora viciae	88	Davidson et al. (2004)
Fusarium wilt race 2	80	Bani et al. (2012)
Mycosphaerella blight	78	Fondevilla et al. (2005)
Agronomic	76	Amurrio et al. (1992)
Rust resistance[a]	52	Barilli et al. (2009)
Adaptive traits	37	Annicchiarico and Iannucci (2008)
Pea weevil resistance[a]	31	Clement et al. (2002)

[a] Data available on Genetic Resources Information Network (GRIN)-Global www.ars-grin.gov/npgs
[b] USDA core collection included
[c] John Innes Centre core collection

Eight core collections have been developed for pea by the national pea germplasm programmes in Australia, China, Czech Republic, France, Poland, Spain, the UK and the USA. The first pea core was constructed at JIC following Brown (1989) recommendations and consists of 157 accessions with representatives from

Table 2.4 Pea genes influencing plant and seed traits that are important in pea-breeding programmes

Trait	locus	marker	number of programmes using MAS
Anthocyanin production	A	perfect	0
Enation virus resistance	En		2
Powdery mildew resistance	Er1	perfect	1
Fusarium wilt resistance (race 1)	Fw	several	0
Potyvirus resistance	sbm1, sbm2	perfect	0
Absence of leaflets	Af	several	0
Winter hardiness[a]	Hr	perfect	0
Ascochyta blight resistance[b]	QTL	several	>1
Frost tolerance[c]	QTL	several	>1
Photoperiod response	Sn	perfect	0
Photoperiod response	Lf	perfect	0
Aphanomyces tolerance[d]	QTL	several	>1
Lodging resistance[e]	QTL	A001, A004	1
Seed shape	R, Rb	perfect	0
Seed composition[f]	rugosus	perfect	3
Trypsin inhibitor[c]	Tri	perfect	>1
Pea Albumin 2	PA2	perfect	1
Protein content[g]	QTL	several	0
Protein content	Vc-2	perfect	0

[a] Lejeune-Hénault et al. (2008) and Weller et al. (2012)
[b] Hamon et al. (2011)
[c] Dumont et al. (2009)
[d] Pilet-Nayel et al. (2002, 2005)
[e] Zhang et al. 2006a
[f] *pgm* mutants entered a number of breeding programmes (UK, France, Denmark) but markers were not used for screening
[g] Burstin et al. (2007)
MAS marker-assisted selection, *QTL* quantitative trait loci

all the *Pisum* taxon phenotyped for nine seed quality characteristics (Matthews and Ambrose 1994). Simon and Hannan (1995) developed the USDA core collection of 504 lines of *P. sativum* with a few *P. sativum* subsp. *elatius* based on geography and flower colour.

The USDA core was refined to 310 accessions using 26 quantitative traits (Coyne et al. 2005a). The Spanish core is a landrace collection based on passport and quantitative data (Martin-Sanz et al. 2011). The Polish core of 266 accessions, selected for diversity of type, taxon and described genes, was studied using isozymes finding rare alleles in the landrace/primitive types (Swiecicki et al. 2000). Baranger et al. (2004) reported on the INRA core collection of 43 *Pisum* accessions selected based on protein and DNA marker diversity. With further use of molecular data, Zong et al. (2009) created a core of 146 Chinese landraces using molecular diversity data and then compared it to a core based on geography finding the molecular core more genetically diverse. Smýkal et al. (2011) proposed the application of a model-based method for the formation of a molecular/eco-geographically diverse international pea core collection for a practical start to phenotyping the non-duplicative pea germplasm. Core Hunter programme was developed and applied to pea germplasm, resulting in a representative set from the entire Czech National Pea Collection (CzNPC), Sumperk (1283 accessions), as well as six European composed collections (4429 accessions; Jing et al. 2012; De Beukelaer et al. 2012). International efforts are underway to phenotype and genotype various compositions of fluid cores based on volunteer entries and not a strategic core per se.

A major shift to allele mining (Reeves et al. 2012) and association mapping (Rafalski 2010) has occurred in pea germplasm collections. Rapidly, transcriptomes for marker (EST-SSRs and SNPs) development (Lucau-Danila et al. 2012; Kaur et al. 2012; Zhuang et al. 2012; McGee 2012a; Sanderson et al. 2011; Franssen et al. 2011; Deulvot et al. 2010; Duarte et al. 2014; Sindhu et al. 2014) are available, and we can imagine a complete gene list for pea in the near term. This shift to gene-based germplasm exploitation has necessitated additional approaches to pea genetic resource conservation. Traditionally, USDA pea accessions were maintained to preserve genetic diversity (e.g. landraces) within an accession. Pure lining accessions and creating a new pure-line accession has increased the utility of germplasm collections (Nelson 2011). Consequently, precision in phenotyping is achievable and high-throughput genotyping is feasible. Further, pure lining is essential for trait associations for genome-wide association studies (GWAS) using genotyping-by-sequencing and resequencing approaches (Brachi et al. 2011; Elshire et al. 2011). Genotyping-by-sequencing is very effective at generating 10,000 plus SNP genotypes in pea populations (Mazourek, personal communication). Larger GWAS studies on a global scale are in progress (Warkentin, personal communication). The choice of germplasm, extent of genome-wide linkage disequilibrium (LD), and relatedness within the population determine the mapping resolution, which together with marker density and statistical methods, are critical to the success of association analysis. Estimates of the rate of LD decay in pea within progressively more distantly related accessions tentatively suggest high LD among cultivars (Jing et al. 2007), comparable to rice and maize. This estimate should be considered preliminary, but would imply that a greater number of SNPs (hundred thousands) might be required for effective genome-wide association mapping.

A project aimed at sequencing the pea genome started in 2013 (McGee 2012b). Production of a draft pea genome will open opportunities for whole genome resequencing of pea germplasm resources for high-throughput SNP and gene identification similar to the case, for example, in rice (Xu et al. 2011) and chickpea (Varshney et al. 2013).

Databases have become essential to curate the growing genomic and agronomic information of pea and related legumes. Several of the pea transcriptomes are available online from Genbank. More useful are pea transcriptomes with user-friendly tools such as GBrowse available online from Legume Information System (www.comparative-legumes.org, Gonzales et al. 2005), KnowPulse (www.Knowpulse2.usak.ca, Sanderson et al. 2011) and Cool Season Food Legume Genome Database (www.gabcsfl.org, Main et al. 2013). Further, databases are needed to host increasingly larger phenotype and genotype data sets, such as GRIN-Global and Germinate 2.1 Pea (http://bioinf.scri.ac.uk/germinate_pea/app/index.pl; Lee et al. 2005). These databases use open source software so data can be extracted and imported into various software packages for analyses. A recent example of combining passport data with habitats for ecogeographic analysis identified potential sites for pea landraces with abiotic stress tolerances (Li et al. 2013).

5 Major Pea-Breeding Achievements

5.1 Canada

Over the past 20 years, steady progress has been made in improving the agronomic and quality characteristics of field pea as evidenced in the cultivars released. Yield gains in yellow cotyledon field pea cultivars, the most widely grown market class in western Canada, has been approximately 2 % per year over the past 20 years when comparing the relative yields of prominent varieties including Carneval, Grande, Alfetta and Eclipse (released between 1993 and 1999) versus Crop Development Centre (CDC) Golden, CDC Meadow, Agassiz and CDC Amarillo (released between 2002 and 2012), based on the post-registration Saskatchewan regional variety trials. Using the performance data of the second year entries in the Western Canada Field Pea Cooperative Variety Registration Tests in 2001–2012, the annual yield increase for yellow cotyledon pea entries arising from all breeding programmes represented was 122 kg ha^{-1} which is approximately a 3 % annual gain. Similar yield gains were achieved for green cotyledon entries over this time period.

The most important disease of field pea in western Canada is the Ascochyta blight complex (referred to as black spot in Australia), of which *Mycosphaerella pinodes* is the most important component. Twenty years ago, most cultivars were rated as having 'poor' resistance to mycosphaerella blight, while now most cultivars are rated as having 'fair' resistance, similar to that of the control cultivar. Radley. Kraft et al. (1998), in an extensive evaluation of pea germplasm under conditions of severe mycosphaerella blight, reported that no germplasm accessions had resistance superior to Radley. Powdery mildew has been considered the second most important disease of field pea in western Canada, as epidemics arise in most field pea regions in most years, typically late in the season. Twenty years ago, none of the pea cultivars recommended for western Canadian production was resistant to powdery mildew, while today, nearly all cultivars are resistant. Resistance in all cases is based on the single recessive gene *er-1*. *Aphanomyces euticheş* and various Fusarium diseases are becoming important in some regions of western Canada and are gaining increased attention from plant pathologists and breeders. Typically, these soil borne pathogens are associated with the wettest regions of western Canada, and those wet regions in which field pea has been grown extensively over the past 15 years or more.

After grain yield, typically the most desired trait of field pea growers is lodging resistance which facilitates harvest and reduces humidity in the canopy. Over the past 30 years, a nearly complete shift has been made in western Canada from 'leafy' cultivars to 'semileafless' cultivars. Semileafless cultivars are those carrying the *afila* gene that results in the replacement of leaflets by tendrils. Lodging ratings began in official trials in 1993 through the use of a 1–9 scale (1 = no lodging, 9 = completely lodged). The advantage of the semileafless trait for improved lodging resistance became apparent as varieties with this leaf type typically had lodging scores 2–3 points better than varieties with the leafy trait. The semileafless trait is a prerequisite for improved lodging resistance but, in addition, increased stem stiffness is required for good field resistance. Over the past 20 years, the typical lodging score of widely grown field pea cultivars has improved from scores of 5–6 to scores of 3–4 at present. Thus, lodging resistance remains a key breeding objective. Using the performance data of the second year entries in the Western Canada Field Pea Cooperative Variety Registration Tests in 2001–2012, lodging has been significantly reduced through breeding, with an annual reduction of 0.04 based on the 1–9 scale. It was interesting to note that, in general, high yield and lodging resistance were positively associated.

As yield increased, days to maturity also increased at a rate of 0.37 day per year (Western Canada Field Pea Cooperative Variety Registration Tests 2001–2012), indicating the challenge of developing high-yielding but early-maturing varieties. Under western Canadian conditions, field pea is typically the first crop harvested, thus development of earlier maturing varieties is not critical, except for the benefits of spreading out the harvest workload. The height (vine length) of breeding lines has been increasing at a rate of approximately 2 cm/year (2001–2012). Plant height was also positively related to yield; thus, it is likely that breeders have inadvertently selected taller varieties while selecting for higher yield.

Retention of green cotyledon colour, or bleaching resistance, is a critical quality trait in green field peas. Colour must be stable after harvest and after at least 1 year of storage. Steady improvement has been achieved over the past 15 years in green cotyledon bleaching resistance. Since 1994, green cotyledon varieties in official registration trials have been rated for bleaching resistance using a 1–5 scale, where 1 = no bleaching, 5 = completely bleached, and a key check cultivar arbitrarily assigned a score of 3 (or 2.5) in each trial. Most cultivars recommended for production in western Canada are now rated as having 'good' resistance to bleaching, while 15 years ago, most were rated as having only 'fair' resistance.

Marrowfat cultivars with large seed size (350–400 mg), fair–good green cotyledon bleaching resistance, powdery mildew resistance and yield 90 % of that of the yellow check cultivars are now available. Marrowfat cultivars with good lodging resistance are still lacking. Marrowfat peas are typically used in snack foods in Asia and as 'mushy peas' in the UK. Maple pea cultivars with powdery mildew resistance and yield 90 % of that of the yellow check cultivars are now available. Maple peas are typically used in seed mixtures for domestic and wild birds. One recently released dun pea cultivar (CDC Dakota) with powdery mildew resistance and yield greater than the yellow checks is now under pedigreed seed multiplication. Dun peas are typically dehulled and split then used in dhal applications, similar to regular yellow cotyledon peas. Forage pea cultivars which produce large biomass, small seed size, good lodging resistance, powdery mildew resistance and grain yield 90 % of that of the yellow check cultivars are now available. Forage peas are typically grown in mixtures with annual cereals such as oat, barley or triticale, harvested when the cereal reaches the soft dough stage and used as forage or silage for ruminants.

5.2 The USA

Pea-breeding achievements in the USA have been similar to those described above for Canada. Pea breeding has a longer history in the USA, particularly for the vegetable pea types. The US programmes have made excellent progress in green seed bleaching resistance, as well as in improving resistance to root rot pathogens and viruses.

5.3 Australia

Field pea production across the Mediterranean-type climates of Southern Australia was initially based on introduced and improved landrace plant types (i.e. dun types) that characteristically grew vigorously over winter and were tall (i.e. long internodes). The development of the modern semidwarf, semileafless varieties for Australia followed a breeding effort over more than 20 years to pyramid genes for adaptation and plant type. This process primarily involved complex crossing and recurrent selection programmes between introduced spring-type semileafless/semidwarf germplasm and Australian-adapted conventional cultivars. Specifically, the newly developed semidwarf germplasm was taller, more active in winter growth and showed reproductive development that commenced in mid-spring to avoid major frost risk (Leonforte and Brouwer 1999). A critical development for Australia was the trait pyramiding of lodging resistance (Leonforte et al. 2006) with reduced pod parchment (i.e. sugar pod trait) to reduce pod shattering at harvest. The unique combination of these traits led to the release of the first broadly adapted semidwarf cultivar Kaspa which has since rapidly dominated production across Southern Australia. The focus of recent breeding has been to optimize the phenology of the Kaspa plant type, which is late flowering and determinate for shorter season climates, particularly in Western Australia (WA). This has led to recent releases such as Pulse Breeding Australia (PBA) Gunyah and PBA Twilight that are early flowering, and display more stable yield than Kaspa in short season-climates. In Australia, the average rate of genetic gain for yield is estimated to be around 1–1.3 % per annum from field pea breeding over the past 20 years.

Field pea, like other pulses, is comparatively sensitive to a number of abiotic stress factors, particularly involving soil nutrition such as salinity and alkaline-induced boron toxicity, reproductive frost damage, heat stress (Dita et al. 2006) and specific herbicide actions. In Mediterranean-type climates, the main limitations to field pea production occur during reproductive development, as developing flowers, pods and seeds are highly sensitive to abortion and damage from increasingly high and low temperature extremes and water-limiting stress. A low-cost approach of long-term and strategic selection for grain yield in high-risk environments has, however, proven to be highly successful for genetically improving grain yield reliability of the pea crop for Mediterranean climates per se (McMurray et al. 2011). The component traits that contribute to higher water-limiting tolerance relate mostly to phenology and biomass production prior to flowering (Sadras et al. 2012). Higher stress tolerance has been identified in landrace accessions for toxicity to boron (Bagheri et al. 1994), salinity (Leonforte et al. 2013) and for iron deficiency (Kabir et al. 2012), and new seedling-screening techniques to improve tolerance levels are now being routinely used in Australia (Leonforte et al. 2009). There is also preliminary evidence for useful variation for heat stress tolerance during flowering and podding (Petkova et al. 2009), and some evidence for higher tolerance to reproductive frost damage based on the yield response of the Australian variety 'Sturt' (Hawthorne 2007) and pod damage (Shafiq et al. 2012).

Field pea is affected by many fungal and viral diseases, bacterial blight and pests, particularly in winter-sown cropping regions where there is a long vegetative growth phase. Black spot disease (Ascochyta blight), described earlier in this chapter, is a major limitation to field pea production in Southern Australia, and its management was recently reviewed by Khan et al. (2013). There is no robust source of resistance to this disease although the development of an erect plant type during vegetative growth has most likely improved overall resistance of the crop (Le May et al. 2005). Making large numbers of intercrosses between partially resistant lines and recurrent selection should help to develop more resistance (Fondevilla et al. 2007a). One Greek and one Ethiopian line have been used in a study involving two cycles of recurrent selection by Beeck et al. (2008), which showed significant simultaneous improvements in black spot resistance as well as stem strength.

Downy mildew caused by *Peronospora viciae* (Berk.) Caspary and powdery mildew caused by *Erysiphe pisi* DC. cause disease periodically in pea in Australia. Downy mildew is prevalent in cooler, wetter growing regions (i.e. southwestern Australia) and more typically affects seedling growth, but rarely causes systemic whole plant infection. Major genes for resistance have been identified and effective screening established (Davidson et al. 2004). However, rapid pathogen specialization is a widespread problem in many regions including Australia (Davidson et al. 2011). Powdery mildew can be a serious disease of field pea but more so in warmer and more humid Mediterranean climates at the start of flowering (i.e. South Australia, southern New South Wales). There are two major genes for resistance, *er1* and *er2,* conferring high and stable resistance to this disease (Katoch et al. 2010) which have been used extensively to develop resistant varieties globally. A third major gene *(er3)* conferring resistance has also been identified from *P. fulvum* Sibth. and Sm. (Fondevilla et al. 2008) but has not been exploited in breeding. Other more regionally important foliar fungal diseases include pea rust (*Uromyces pisi* (Pers.) Schrot. ; Barilli et al. 2009) and septoria blotch (*Septoria pisi* West). Both can cause severe damage in wet seasons; however, only moderate resistance has been found to rust (Rai et al. 2011) and low resistance to septoria blotch (Leonforte et al. 2004). Bacterial blight caused by *Pseudomonas syringae* pv. *pisi* Sackett and pv. *syringae* van Hall are other localized but potentially devastating diseases that can occur in cool temperate and Mediterranean-type climates. Breeding has mostly focused on pyramiding available major genes for race-specific resistance to pv. *pisi* (i.e. from seven races; Hollaway and Bretag 1995, Elvira-Recuenco et al. 2003).

Many aphid-transmitted viruses produce a range of disease symptoms individually or in combination in field pea. These include cucumber mosaic virus (CMV), pea early browning virus (PEBV), pea enation mosaic virus (PEMV), luteo viruses-pea leaf roll virus (PLRV) or bean leaf roll virus (BLRV), poty viruses-bean yellow mosaic virus (BYMV) and pea seed-borne mosaic virus (PSbMV), alfalfa mosaic virus (AMV), pea streak virus (PeSV) and red clover vein mosaic virus (RCVMV). Selection for major gene resistance to PSbMV and poty viruses is now incorporated into breeding strategies (van Leur et al. 2007).

Root rot diseases are less widespread in Mediterranean regions. They are generally caused by a combination of several common soil fungal pathogens: Aphanomyces root rot (*A. euteiches* Drechs.), Pythium tip blight *(Pythium ultimum),* Fusarium root rot (*Fusarium solani* f. sp. *pisi* (Jones) Snyder and Hansen), Rhizoctonia root rot (*Rhizoctonia solani* Khun) and Fusarium wilt (*Fusarium oxysporum* Schlecht.). Robust resistance is only found for Fusarium wilt; efforts are focused on improving resistance to Aphanomyces. Only partial resistance controlled by several quantitative trait loci (QTL) is available for this disease (Pilet-Nayel et al. 2002, 2005). In pea, useful pest resistance has only been identified for pea weevil (*Bruchus pisorum* L.) which is a widespread global problem. Resistance genes are in the secondary gene pool (*P. fulvum*; Clement et al. 2002). Transfer of genes for resistance from *P. fulvum* has now been completed by introgression from *P. fulvum* into cultivated field pea through backcrossing to produce several advanced pea weevil resistant lines (Clement et al. 2009; Aryamanesh et al. 2012).

5.4 Europe

A major achievement in modern pea breeding in Europe was the commercial deployment of the *afila* (semileafless) trait in the 1970s. The first successful cultivar was Solara from Cebeco Zaden (now Limagrain). Solara became a parent in numerous crosses by breeders throughout Europe over the next decade. Field pea-breeding activities expanded in Europe in the 1980s and 1990s as EU policy provided subsidies for the local production of high-protein crops to address the massive deficit in protein in the continent. Pea breeding in Europe is conducted primarily in the private sector with the public sector focused on upstream research. The companies having the most market share in the pea industry in Western Europe in the 1980s and 1990s were Cebeco Zaden (The Netherlands), Florimond Deprez (France), Serasem (France), Svalof Weibull (Sweden), Danisco Seeds (Denmark), DLF Trifolium (Denmark) and Sharpes (UK). In addition, many other small companies were actively involved. However, with changes in EU policy that were less favourable to protein crops, production of field pea and faba bean in Europe has declined over the past 15 years. This led to consolidation and reduction in pea-breeding activity. Currently, Limagrain, that consolidated the former Cebeco Zaden, DLF Trifolium and Sharpes programmes, is the largest pea-breeding programme in Europe, followed by Florimond Desprez and RAGT, that grouped breeding programmes from Serasem and GAE Recherche. In France, breeding programmes conducted during the past 20 years have mainly targeted yield, lodging and disease resistance. Post-registration trials conducted in 2012 and 2013 compared historical short semileafless varieties (Solara registered in 1987, Baccara registered in 1992) with the most recent cultivars (e.g., Arvalis registered in 2013). A yield increase of 1.1 t ha^{-1} has been achieved to reach 6.6 t ha^{-1}, accompanied by a much better lodging resistance (56 cm height at harvest for Kayanne, as compared to Solara, 26 cm, or Baccara, 22 cm).

Similar trends occurred in Eastern Europe in that pea-breeding expansion was led by the release of semileafless, semidwarf varieties such as Sum in Poland in the 1980s. Pea yields increased by 50 % over the period 1977–1995 in Poland primarily due to the cultivation of these new varieties (Swiecicki et al. 1997). Also in Poland, Prusinski (2007) estimated annual pea yield increases at 42–45 kg ha^{-1} from 1989 to 2006, and Boros estimated annual pea yield increases at 46 kg ha^{-1} for edible peas and 44 kg ha^{-1} for fodder peas for the period 2001–2010. Currently grown pea varieties have yield potential of 5–6 t ha^{-1}. The major pea-breeding companies centred in Eastern Europe are Poznań Plant Breeders Ltd., Plant Breeding Smolice Ltd., Danko Plant Breeders Ltd., and Selgen.

To supplement the benefits of cultivars with improved lodging resistance, use of fungicides for management of Ascochyta blight is most common in Europe, followed by North America and Australia. Breeding for powdery mildew resistance is common in Eastern European breeding programmes as the disease can cause yield losses of 10–65 % by significantly reducing seed weight (Ondrej et al. 2003). Improved resistance to downy mildew and Fusarium wilt races 1 and 2 are also important breeding achievements in Eastern Europe. In warm seasons, aphid dissemination of viral diseases is more prevalent. Their importance varies across regions depending on alternative hosts, seed transmission and aphid control. Screening methods were developed for virus diseases PEMV and PSbMV (Hochman and Dostalova 2010).

5.5 India

Field pea breeders have mainly concentrated on leaf type *(afila),* dwarf plant type and powdery mildew resistance over the past three decades. Recently, varieties have been developed which combine resistance to powdery mildew with dwarf stature and semileafless trait. Each of these characters has played a significant role in reverting the negative trend of area under the crop. The semileafless trait allows for penetration of sunlight to the lower canopy, and mechanical support to prevent lodging and reduce bird damage. The dwarfing gene has enhanced productivity through improved response to fertilizers, irrigation and dense plant population. Recently, several short duration varieties like Adarsh, Vikas and Prakash have been developed which escape terminal stress and have the yield potential of 2.5 t ha^{-1}. During the 1980s, a severe incidence of rust occurred in pea in northern and eastern regions of India and screening work initiated for resistance to this disease. In the past two decades, efforts have been made to identify slow rusting lines like FC1. A few rust tolerant varieties like HUDP 15, Prakash, Swati, Aman and Pant P 42 are now available.

5.6 China

Thirty years ago, local cultivars of field pea dominated the main dry pea production areas in rain-fed hilly fields of China. Since then, 45 cultivars were bred and released by pea breeders from nine public research institutions. Of these, 34 were for dry pea production, and 11 were for various types of vegetable production. Among the 34 dry pea cultivars, 26 have normal leaf type and 8 have semileafless type. Among the vegetable cultivars, six varieties are snow pea, three are snap pea, two are tendril-less for leaf vegetable production. All cultivars were bred through single plant or pure-line selection from segregating populations derived from crosses between cultivated genotypes or from natural mutations of local cultivars. Interspecies crosses and induced mutation methods were not applied.

While the majority of this chapter is focused on developments in dry pea breeding, many similar improvements in yield and disease resistance have been achieved in vegetable pea breeding over the past three decades. In North America and Europe, the majority of the vegetable pea-breeding activities have been conducted in a few private companies, with the leading contributors being General Mills, Inc. (Le Sueur, Minnesota), Syngenta (Nampa, Idaho), Crites Moscow (Moscow, Idaho), Pure-line Seeds (Moscow, Idaho), Brotherton Seed Company (Moses Lake, Washington) and Seminis (Filer, Idaho). These companies have released many varieties with key improvements including increased yield, resistance to soil-borne and foliar fungal pathogens, resistance to aphid-vectored virus diseases and resistance to abiotic stresses. In addition, they have commercialized snap pea and petit pois types, improved synchrony of harvest, and improved flavour and colour.

6 Specific Goals in Current Pea-Breeding Programmes

The objectives of the Canadian programmes (University of Saskatchewan at Saskatoon, Saskatchewan and Agriculture and Agri-Food Canada at Lacombe, Alberta) include the development of early maturing, high-yielding field pea cultivars with resistance to powdery mildew and improved resistance to mycosphaerella blight and lodging, with superior quality for export and domestic markets. Emphasis is placed on the development of cultivars for human consumption markets, with approximately 60 % of the activity focussed on the yellow cotyledon market class, 30 % of the activity on the green cotyledon market class and 10 % on speciality field pea markets including marrowfat, dun, maple and forage.

In the USA, there are two public pea-breeding programmes, that is, the USDA-Agricultural Research Service (ARS) programme is located at Washington State University in Pullman, Washington, and the North Dakota State University programme is based in Fargo, North Dakota. A private seed company, ProGene, LLC, is based in Othello, Washington. All three programmes breed peas for the target environments of the Pacific Northwest and Northern Plains regions of the country. Typically, the more arid regions favour green cotyledon varieties and yellow cotyledon varieties are grown in the more humid regions of the northern plains. Testing and evaluation of advanced lines is accomplished through a network of regional variety trials and state-wide variety trials conducted by agronomists at state universities and experimental stations. The primary breeding objectives of the programmes are similar to those of the western Canadian programmes. In addition, there is considerable breeding effort focused on the development of autumn-sown, food quality cultivars. Disease resistance breeding that is important in the USA includes resistance to PSbMV, BLRV, PEMV, Aphanomyces root rot, Fusarium root rot, Fusarium wilt (races 1 and 2) and powdery mildew. Recently, efforts have also been directed to screening for tolerance to low pH soils and aluminium toxicity.

In Australia, the national field pea-breeding programme is centred at Horsham, Victoria, with testing sites in key locations across the continent. The major objectives of the programme are similar to those described for western Canada, with specific emphasis on the dun seed type based on the cultivar Kaspa (light tan-red seed coat), resistance to bacterial blight, downy mildew, BLRV and PSbMV, tolerance to subsoil boron and NaCl toxicity, and reproductive frost damage.

In India, breeding efforts are focused on yellow cotyledon pea with non-pigmented seed coats, with smaller programmes on vegetable pea. Current breeding objectives include increasing harvest index, development of short duration (100–110 days) varieties, resistance to pea weevil, tolerance to terminal drought through early maturity and frost tolerance.

In China, semileafless leaf type is one of the specific breeding goals for dry pea production. Vegetable pea varieties are being developed with normal leaf type and short duration fitted to a two crop season, such as pea–maize, in areas of China at 40°N and above. Vegetable varieties of pea are special breeding goals for production around cities in southern China.

In Western Europe, high yield, lodging resistance and improved resistance to Aphanomyces root rot are the key objectives for spring field pea. Over the past four decades, the target market has primarily been the compound feed industry, and thus less emphasis is placed on seed visual quality than is the case in North America, but the seed protein content is critical for registration. The recent emergence of more frequent drought and heat waves during the reproductive period has led to consideration of tolerance to these traits in breeding programmes. In France, some breeding companies and the French national agronomic research institute INRA have undertaken to develop autumn sown field pea varieties, as a strategy to increase yield potential and stability through a longer crop cycle, higher biomass production and earlier maturity to avoid late-cycle drought and heat stress (Hanocq et al. 2009). The development of winter field pea varieties has added more breeding targets to those listed above for spring peas, particularly the improvement of winter hardiness, the fitting of flowering time to avoid flower initiation during frost at the end of winter, and also seed filling during drought and heat at the beginning of summer, and Ascochyta blight resistance. Following recent research, results on the control of flowering time and frost tolerance in pea (Lejeune-Hénaut et al. 2008), a new pea type registration has been launched: The winter peas responsive to photoperiod.

In Eastern Europe, breeding objectives for field pea include development of high-yielding varieties with improved resistance to lodging, multiple disease resistances (*E. pisi, Fusarium* subsp., *Ascochyta, Uromyces*), virus resistances (PEMV, PS-bMV), resistance to abiotic stress and improved grain quality. For green cotyledon peas, desirable traits include optimal size, shape and colour, high content of resistance starch, and high content of vitamins and carotenoids (luteins, β-carotene).

7 Breeding Methods and Specific Techniques

At the University of Saskatchewan, approximately 300 new crosses are generated each year during the course of three crossing cycles (summer, fall and winter). The F2-derived family method is used. Visual selection for seed quality traits is conducted in each generation, including size, shape, dimpling, cotyledon and seed coat colour and seed coat integrity. Selection for powdery mildew resistance is conducted in the F1 generation of complex crosses, or the F2 generation of single crosses based on natural epidemics late in the season. Approximately, 10,000 F2-derived F3 lines are evaluated in single replicate micro-plots. Approximately 15–25 % of these lines are advanced each generation, based on selection for lodging resistance, maturity, mycosphaerella blight resistance and seed quality, with the number of evaluation sites increasing through the generations. The Agriculture and Agri-Food Canada pea-breeding programme uses pedigree selection in combination with single-seed descent. The two-year Field Pea Co-operative (Co-op) Registration Test is grown at 10–12 locations in western Canada. Data summaries from the most promising varieties arising from the Co-op Test are presented to the Prairie Recommending Committee for Pulse and Special Crops. This group of experts votes on whether or not to recommend to the Canadian Food Inspection Agency that individual varieties be registered in Canada. Successful varieties are evaluated in provincial regional trials in western Canada to further evaluate performance in specific pea-growing regions. Data arising from these trials are summarized and published in provincial seed guides for use by seed growers, farmers and agronomists.

The USDA-ARS programme employs a pedigree breeding method. Approximately, 300 crosses are made each year. F1's are grown at the WSU Spillman Agronomy Farm (SAF). F2's are typically sent to a counter-season nursery and selected for simply inherited traits. The F3 through F6 generations are grown at SAF in paired rows and evaluated within and between families. Single plants are selected in F3–F5, and selected bulks are taken in F6. The selected F7's are evaluated for agronomic performance and produce sufficient seed for the preliminary yield trials (PYT) which are grown at one location. Breeding lines advanced from the PYT are entered into advanced yield trials (AYT) that are typically grown at four locations for several years. Lines selected from the AYT are entered into the appropriate state-wide and regional yield trials.

The field pea-breeding programme at Horsham, Australia, includes 450–550 crosses per year followed by pedigree selection. Individual plant or pod based selections (3000–5000) will be taken from F2 to F4 segregating populations in the field. Single-seed descent, recurrent and mass selection and summer seed multiplication are strategically employed. The introgression or maintenance of key major genes (e.g. *A, I, b, af, le, er 1, sbm-1, rpv, p, v*) is achieved through backcrossing, pedigree systems of selection and effective phenotyping (e.g. downy mildew, boron toxicity, PSbMV). For more complex traits (e.g. blackspot, bacterial blight, BLRV), segregating populations are maintained as bulk lines to increase frequency combinations of minor genes (e.g. improved lodging resistance) that will additively contribute to improved variation. Mass selection is used for eliminating genes in relation to poor grain quality traits (e.g. shape, size, dimples and colour). Progeny lines (i.e. 3000–5000) are initially tested in short paired rows. Preliminary testing occurs in short row plots at four to five regional locations with concurrent seed multiplication and yield testing in Victoria. Fixed line selections are grown over several years to permit observations of performance (e.g. grain yield) under different environmental conditions and enable the selection of lines that are more broadly adapted over years and environments.

Pea-breeding techniques in Europe are generally similar to those described above for North America and Australia. Promising lines are submitted by the breeding company to official testing for assessment of variety value for cultivation and use (VCU) and testing for distinctness, uniformity and stability (DUS). Positive results of the VCU and DUS assessments are a precondition to variety entry into a national list. DUS testing is required for granting of Plant Breeders' Rights and is conducted under The International Union for the Protection of New Varieties of Plants (UPOV) guidelines. The VCU process must be conducted in individual countries, while the DUS process need only be conducted once. Cultivars on the national list of one European country are eligible for production and marketing in the entire European Union territory after their admission to the Common Catalogue.

Several national and international organizations are actively engaged in field pea improvement in India, with the largest at the Indian Institute of Pulses Research (Kanpur), followed by GB Pant University of Agriculture and Technology (Pantnagar), CCS Haryana Agricultural University (Hisar), Punjab Agricultural University (Ludhiana) and Banaras Hindu University (Varanasi). In general, these field pea-breeding programmes include 250–300 crosses per year followed by pedigree selection. Single plant-based selections (2000–3000) are usually taken from F2 to F4 segregating populations in the field based on improved grain yield, dwarf plant type and powdery mildew resistance. The introgression of major genes is achieved through backcrossing, pedigree selection and phenotyping. Fixed line selections are grown over 3 years to permit observations of performance under different environmental conditions and enable the selection of lines that are more broadly adapted over years and environments.

Recently, EMS mutation methods have been introduced in pea variety improvement in China, by establishment of M_2 to M_3 populations of several green pea and dry pea varieties. This has led to the release of varieties including Zhong Wan No. 6.

8 Integration of New Biotechnologies in Breeding Programmes

Over the past three decades DNA markers, molecular techniques and genomic tools have been developed for pea (Smýkal et al. 2012), and it is of interest to ask how many of these are being used in public or private breeding programmes.

8.1 Gene Identification and Novel Allele Discovery

In pea, considerable effort has been directed towards gene identification. The genetic diversity within available germplasm collections is reasonably well characterized, having been analysed for variation in a wide range of molecular marker types (see 'Genetic Resources and Utilization' Sect. 4). Many polygenic traits important for pea breeding (flowering time, seed size, lodging susceptibility, resistance to many pests and pathogens—see sections above) have been subjected to genetic analyses, and at least specific regions of the genome influencing the respective trait have been determined. In a number of cases, particularly flowering time, plant height and branching and nodule formation, the actual genes have been identified and their interrelationships characterized. With the availability of many convenient SSR and SNP markers in pea, breeders are now routinely investigating traits important for specific habitats (boron susceptibility, bleaching susceptibility) or crop types (winter hardiness, amylose level). Mutation screening populations, and more recently TILLING populations, have been developed, greatly facilitating the recognition and confirmation of those genes conditioning specific phenotypes. For many crops, TILLING may be regarded as a useful and more practical alternative to GM technology for knocking out expression of target genes. TILLING in pea has been extremely successful (http://www-urgv.versailles.inra.fr/tilling/pea.htm, Dalmais et al. 2008), and a wide array of genes and alleles has been generated for the research community. TILLING targets have first been designed to prove or disprove the role of candidate genes in the processes involved, for example, internode length *(le)* and tendril-less *(tl)* mutations, where natural mutants have been available previously (Dalmais et al. 2008; Hofer et al. 2009).

Once the function of a gene is validated, reverse genetic methods allow the identification of novel alleles in mutant and/or genetic resource collections. TILLING or EcoTILLING screens can identify genotypes carrying different alleles of the genes of interest for their testing and subsequent use as donor progenitors in breeding programmes (e.g., Weller et al. 2012). It may be expected that novel mutants affecting flowering time, seed composition and yield are in the pipeline and will be in demand by breeders. Equally, the development of high-throughput and rapid screening methodologies for the detection of deletion mutants for seed quality targets (Domoney et al. 2013) may be expected to expand the potential for the identification and introduction of novelty by breeders.

8.2 Marker-Assisted Selection

In addition to the identification of genes responsible for or influencing important traits, the mapping of these genes on the pea genome and the understanding of the interactions of genes involved in the expression of quantitative traits have allowed pea breeders to organize their breeding activities and objectives into programmes that are more efficient and predictable. Genotypes with specific flowering times, pathogen responses, seed or pod qualities (see e.g., Fig. 2.2), or other traits can be selected before crosses are designed. Similarly, the steps necessary to obtain and the likelihood of recovering the desired genotypes can be determined with much greater precision. Linkage drag effects can be estimated, encouraging breeders to screen for specific recombinants before investing considerable efforts on genotypes with gene combinations difficult to break.

Marker-assisted selection (MAS) permits screening for genes before the character is expressed in the plant, by allowing the identification of heterozygous genotypes that carry the desired recessive allele, but will not express the trait, and by enabling the screening for individual genes controlling a polygenic trait so that those plants with more of the desirable alleles can be identified and used in future crosses. These more indirect uses of biotechnology have had a significant effect on the efficiency of pea-breeding programmes throughout the world.

Several markers have been associated to traits of interest in the past decade. These characters for which the genetic basis is at least partially known should be amenable to MAS. In cases where the gene(s) responsible for the trait have been identified, DNA sequence polymorphism can be used to generate a 'perfect' marker that correlates 100% with the presence or absence of the desired allele. This is particularly the case for mono- or oligogenic traits. For more complex traits, markers associated with QTL have been mapped. Table 2.4 lists a number of genes that are critical in pea breeding and for which 'perfect' markers or genetically linked markers have been identified.

Despite the availability of these useful markers, relatively few pea-breeding programmes are using MAS routinely. There are several reasons for this: In some cases (e.g. presence of anthocyanins, absence of leaflets and seed shape), the trait is easily scored directly, heterozygotes can be conveniently identified in the next generation, and often both parents were homozygous for the desired allele. Thus, these traits might not be expected to be prime candidates for MAS, despite the saving in space and time that can result from early selections based on markers particularly for

Fig. 2.2 A plant of JI 1194, a parent of one mapping population at JIC, Norwich

seed traits. Other genes that require phenotypic screening under specific conditions (e.g. presence of the pathogen, or under controlled photoperiods) are more suitable for MAS. However, for these traits with often complex heredity, the marker–trait linkage has to be tight enough for efficient MAS. Additionally, allele diversity is generally low in cultivated breeding pools. When broader parentage is used, there are greater incentives to screen for the retention of genes and markers representing the cultivated background, while introgressing the trait and region of the genome that contains a novel gene of interest.

There are however several instances where a MAS approach has been adopted thus far. For instance, MAS for enation mosaic virus is used in New Zealand because direct testing with the pathogen is not possible (there is no pea enation virus in, or allowed into, that country), and resistance to enation virus is highly desired in varieties released. MAS is also being used to introgress the pea seed albumin 2 mutation (Vigeolas et al. 2008) into a cultivated background for improved animal feed. The development of perfect markers for the locus *(Tri)* controlling the activity of trypsin inhibitors in pea seeds offered a facile alternative to cumbersome biochemical assays (Page et al. 2002) that is relevant to some winter pea-breeding programmes. Due to close linkage between *Tri* and *Vc-2* (a locus encoding a class of vicilin, a major seed storage protein), the marker for the latter (Chinoy et al. 2011) could be used to distinguish lines carrying particular combinations of alleles at the *Tri* locus and a pseudogene at *Vc-2,* where the latter allele may be associated with a reduced seed protein concentration. In France, MAS for pyramiding QTL for frost tolerance and *A. euteiches* disease resistance in elite genetic backgrounds is underway at INRA (Baranger et al. 2010).

The number of markers useful for selection has been relatively modest until recently (300 SSR markers were mapped in Loridon et al. 2005). Now high-throughput, cost-effective and easy to score marker technologies have been developed for pea in different programmes (e.g., Deulvot et al. 2010; Leonforte et al. 2013). These may enhance MAS in pea in the near future. This approach may be favoured by the generation of service genotyping platforms for these markers that will allow breeders who are not always familiar with and/or using marker technology to outsource this activity. Furthermore, as breeding programmes avail more of the wider pea germplasm and the ever-increasing diversity of traits offered by this resource, alongside the greater emphasis currently on translational research, it is likely that greater reliance will be placed on the use of markers. Finally, the availability of large SNP marker collections and the pea genome sequence (McGee 2012b) should allow genomic selection programmes to enhance breeding for complex traits.

8.3 Other Biotechnologies

As is the case for many pulses, pea is not particularly amenable to tissue culture. Although it can be done, growing plants from single cells or explants remains slow. Thus, recent advances are not groundbreaking but show promise. Regeneration protocols are available (Sanchez and Mosquera 2006; Rajput and Singh 2010), and Esposito et al. (2012) have developed a method for increasing perhaps tenfold the number of F1 plants produced from a cross. Both anther culture and double haploid production have been explored (Lulsdorf et al. 2011) but the approach has yet to be applied in breeding programmes.

The production of genetically modified pea plants and seeds expressing a foreign protein that conferred resistance to a major pest was a very early success story of legume crop transformation. In many parts of the world where pea is cultivated, the pea bruchid beetle *(B. pisorum)* prevails. Expression of the α-amylase inhibitor (α-AI) from French bean (*Phaseolus vulgaris* L.) in transgenic pea proved the utility of this inhibitor in protecting plants in the field and stored seeds from attack by the pea bruchid (Schroeder et al. 1995; Morton et al. 2000). However, the commercial development of these lines was abandoned when transgenic seeds appeared to invoke a T-cell response in mice which was not evident when either the non-transgenic pea or French bean seeds were tested. This observation and subsequent studies led to the conclusion that differential glycosylation of the α-amylase inhibitor expressed in the non-host plant had triggered a pre-immune response (Anonymous 2006); other minor changes to seed proteins were also documented for the transgenic seeds, likely reflecting pleiotropic effects on post-translational processing (Islam et al. 2009). The negative publicity that this research received is regrettable and, as noted by others (Anonymous 2006), has not served the best interests of scientists, breeders or consumers. In fact, the rigour applied to the testing of the modified seeds far surpassed any analysis that is performed on material from standard breeding programmes, including those involving wild relatives. It is difficult to assess to what extent the lack of acceptability of genetically modified foods in some parts of the world contributed to the negative publicity of this research and to the abandonment of other more applied research projects. Related research has produced additional transgenic pea lines, but none with such obvious agronomic benefit as the α-AI lines; pea lines having moderate reductions in seed trypsin inhibitor activity (Welham and Domoney 2000) and others with modified nodulation (Schneider et al. 1999) and resistance to PSbMV (Jones et al. 1998) contributed to fundamental research projects. These and others generated in the pursuit of academic research projects (Weigelt et al. 2009) have led to a fuller understanding of biological processes that may be investigated more thoroughly using mutagenesis. The latter programmes are more likely to lead to modified germplasm that will be accepted more readily into breeding programmes in the current climate. Transformation systems for pea, although not very efficient, are unlikely to represent a barrier and are also improving (Clemow et al. 2011). The exploitation of pea seeds as a vehicle for the production of valuable proteins, including pharmaceuticals, has also been demonstrated clearly (Mikschofsky and Broer 2012; Mikschofsky et al. 2009).

We may expect, with the current rate of development of new genomic tools (e.g., gene-editing techniques) coupled with high-throughput germplasm screens, the powers of association mapping and advanced phenotyping, plus advances in genome sequencing, that breeding programmes will move to a new level in the next 5–10 years. Programmes in which pyramiding of genes is currently very challenging (for instance, breeding for snap peas where at least three recessive genes must be combined to obtain the appropriate pod character, added to other desired disease resistance and flowering time traits) will become much more rapid and accessible.

9 Seed Production and Variety Commercialization

In the University of Saskatchewan programme, pre-breeder seed development is initiated concurrent with F6 yield testing. Single plants are selected from the best-performing F6 lines, then grown as F6-derived microplots concurrent with first-year Cooperative Registration Testing and as F6-derived long plots concurrent with second-year Cooperative Registration Testing. Breeders' seed of new varieties is typically bulked after long plot evaluation. Commercialization occurs through an agreement with Saskatchewan Pulse Growers (SPG), the organization which represents pulse growers. Variety release occurs on a royalty-free basis in exchange for SPG support of breeding. Breeders' seed is distributed by SPG to select-status seed growers who multiply the seed and ultimately sell certified-status seed to farmers.

In the Agriculture and Agri-Food Canada (AAFC) programme, selected F9 lines are grown in strips in the year prior to registration testing in the home location for seed multiplication and further purification, and for further observation and selection. Any off-type plants are removed, resulting in uniform and pure breeding lines. The commercialization of the AAFC's field pea varieties is based on a tendering mechanism. The breeder prepares the performance information on candidate varieties and provides it to the Office of Intellectual Property of Canada (OIPC). OIPC posts the information to the public as a proposal of commercialization. All proposals are reviewed by a committee, and a successful company is granted the right for commercialization of a specific variety. After the commercialization rights to a variety has been granted to a seed company, seed production is carried out at the AAFC Seed Unit, Indian Head, SK. Then the seed company multiplies the seed for future sales.

In the USDA-ARS breeding programme, lines identified for potential release are grown in strips 1–2 years before variety release is anticipated. Approximately 200 single plants are selected that conform to variety type. Seed from each plant is grown for 2 years in paired rows and selected for 'trueness to type'. After the second year, the remaining progeny are bulked and transferred to either the Washington State Crop Improvement Association or another licencing entity. That organization is responsible for the production of foundation and registered seed and marketing. The NDSU breeding programme utilizes a similar approach.

In Australia, lines identified for commercial release are included in both advanced breeding trials and the national variety testing (NVT) programme to maximize evaluation regionally. Multiplication of breeders' seed commences once lines are promoted to NVT testing. Commercialization occurs via an open tender process for either individual lines or on the basis of a short-term licence (5 years) for a variety pipeline. Pulse Breeding Australia (PBA) is an unincorporated joint venture between Australian state governments and GRDC, established in 2005 to manage the breeding and variety release strategy. PBA works closely with the seed licensee to develop a variety release strategy that maximizes uptake of new genetic advancements by industry. The seed licensee is responsible for commercial seed production and marketing and seed or endpoint royalty collection on behalf of stakeholders.

In European programmes, pre-breeder seed production typically begins once lines have entered multilocation yield trials, and it is carried out by individual breeding companies. Breeders' seed multiplication is accelerated after a variety succeeds in VCU and DUS evaluations.

The Indian Council of Agricultural Research (ICAR) organizes seed production of varieties of different crops through ICAR Institutes, State Agricultural Universities and organizations like the National Seed Corporation, State Farm Corporation of India. In China, public and private pea seed production systems are operating in northern China.

Acknowledgments Appreciation is expressed to Devini DeSilva for critically reviewing the references section of this chapter.

References

Abbo S, Zezak I, Schwartz E et al (2008) Experimental harvesting of wild peas in Israel: implications for the origins of Near East farming. J Archaeol Sci 5:922–929

Abbo S, Lev-Yadun S, Gopher A (2010) Agricultural origins: centers and noncenters; a Near Eastern reappraisal. Crit Rev Plant Sci 29:317–328

Abbo S, Lev-Yadun S, Heun M et al (2013) On the 'lost'crops of the neolithic Near East. J Expt Biol 64:815–822

Ahmad S, Singh M, Lamb-Palmer ND et al (2012) Assessment of genetic diversity in 35 *Pisum sativum* accessions using microsatellite markers. Can J Plant Sci 92:1075–1081

Ali Z, Qureshi AS, Ali W et al (2007) Evaluation of genetic diversity present in pea (*Pisum sativum* L.) germplasm based on morphological traits, resistance to powdery mildew and molecular characteristics. Pak J Bot 39:2739–2747

Ambrose MJ, Coyne CJ (2009) Formal collaboration between John Innes *Pisum* Collection and USDA-ARS Collection over *Pisum* genetic stocks. Pisum Genet 40:27

Amurrio JM, De Ron AM, Escribano MR (1992) Evaluation of *Pisum sativum* landraces from the northwest of the Iberian peninsula and their breeding value. Euphytica 66:1–10

Amurrio JM, De Ron AM, Zeven AC (1995) Numerical taxonomy of Iberian pea landraces based on quantitative and qualitative characters. Euphytica 82:195–205

Andrade M, Sato M, Uyeda I (2006) Two resistance modes to Clover yellow vein virus in pea characteried by a green fluorescent protein-tagged virus. Phytopathol 97:544–550

Annicchiarico P, Iannucci A (2008) Adaptation strategy, germplasm type and adaptive traits for field pea improvement in Italy based on variety responses across climatically contrasting environments. Field Crops Res 108:133–142

Anonymous (2006) Genetically modified mush (editorial). Nat Biotechnol 24:2

Aryamanesh N, Byrne O, Hardie DC et al (2012) Large-scale density-based screening for pea weevil resistance in advanced backcross lines derived from cultivated field pea (*Pisum sativum*) and *Pisum fulvum*. Crop Pasture Sci 63:612–618

Aubert, G, Morin J, Jacquin F et al (2006) Functional mapping in pea as an aid to the candidate gene selection and for investigating synteny with the model legume *Medicago truncatula*. Theor Appl Genet 112:1024–1041

Azmat MA, Ali Khan A, Asif S et al (2012) Screening pea germplasm against *Erysiphe polygoni* for disease severity and latent period. Int J Veg Sci 18:153–160

Bagheri A, Paull JG, Rathjen AJ (1994) The response of *Pisum sativum* L. germplasm to high concentrations of soil boron. Euphytica 75:9–17

Baldev B (1988) Origin, distribution, taxonomy, and morphology. In: Baldev B, Ramanujam S, Jain HK (eds) Pulse crops. Oxford & IBH Publishing, New Delhi pp 3–51

Bani M, Rubiales D, Rispail N (2012) A detailed evaluation method to identify sources of quantitative resistance to *Fusarium oxysporum* f. sp. *pisi* race 2 within a *Pisum* spp. germplasm collection. Plant Pathol 61:532–542

Baranger A, Aubert G, Arnau G et al (2004) Genetic diversity within *Pisum sativum* using protein- and PCR-based markers. Theor Appl Genet 108:1309–1321

Baranger A, Pilet-Nayel ML, Lecomte C et al (2010) Quel impact des innovations génétiques pour lever les facteurs limitant la production du prois protéagineux. Innovations Agron 11:59–78

Barilli E, Sillero JC, Moral A, Rubiales, D et al (2009) Characterization of resistance response of pea (*Pisum* spp.) against rust (*Uromyces pisi*). Plant Breeding 128:665–670

Barilli E, Satovic Z, Rubiales D et al (2010) Mapping of quantitative trait loci controlling partial resistance against rust incited by *Uromyces pisi* (Pers.) Wint. in a *Pisum fulvum* L. intraspecific cross. Euphytica 175:151–159

Beeck CP, Wroth JM, Falk DE et al (2008) Two cycles of recurrent selection lead to simultaneous improvement in black spot resistance and stem strength in field pea. Crop Sci 48:2235–2244

Ben-Ze'ev N, Zohary D (1973) Species relationships in the genus *Pisum* L. Israel J Bot 22:73–91

Berdnikov VA, Trusov YA, Bogdanova VS et al (1992) The neoplastic pod gene (Np) may be a factor for resistance to the pest *Bruchus pisorum* L. Pisum Genet 24:37–39

Bisby FA, Buckingham J, Harborne JB (eds) (1994) Phytochemical dictionary of the Leguminosae Vol. 1: plants and their constituents. Chapman and Hall, London

Blixt S, Williams JT (1982) The pea model. IBPGR, Rome

Bordat A, Savois V, Nicolas M et al (2011) Translational genomics in legumes allowed placing in silico 5460 unigenes on the pea functional map and identified candidate genes in *Pisum sativum* L. Genes, Genomes, Genetics 1:93–103

Brachi B, Morris GP, Borevitz JO (2011) Genome-wide association studies in plants: the missing heritability is in the field. Genome Biol 12:232

Burstin J, Marget P, Huart M et al (2007) Developmental genes have pleiotropic effects on plant morphology and source capacity, eventually impacting on seed protein content and productivity in pea. Plant Physiol 144:768–781

Byrne OM, Hardie DC, Khan TN et al (2008) Genetic analysis of pod and seed resistance to pea weevil in a *Pisum sativum* × *P. fulvum* interspecific cross. Crop Pasture Sci 59:854–862

Chinoy C, Welham T, Turner L et al (2011) The genetic control of seed quality traits: effects of allelic variation at the Tri and Vc-2 genetic loci in *Pisum sativum* L. Euphytica 180:107–122

Chimwamurombe PM, Khulbe RK (2011) Domestication. In: Pratap A, Kumar J (eds) Biology and breeding of food legumes. CABI, Massachusetts, pp 19–34

Clement SL, Hardie DC, Elberson LR (2002) Variation among accessions of *Pisum fulvum* for resistance to pea weevil. Crop Sci 42:2167–2173

Clement SL, McPhee KE, Elberson LR et al (2009) Pea weevil, *Bruchus pisorum* L. (Coleoptera: Bruchidae) resistance in *Pisum sativum* × *Pisum fulvum* interspecific crosses. Plant Breed 128:478–485

Clemow SR, Clairmont L, Madsen LH et al (2011) Reproducible hairy root transformation and spot-inoculation methods to study root symbioses of pea. Plant Methods 7:46

Coyne CJ, Razai L, Baik B-K et al (2005) Variation for pea seed protein concentration in USDA *Pisum* core collection. Pisum Genet 37:7–11

Dalmais M, Schmidt J, Le Signor C et al (2008) UTILLdb, a *Pisum sativum* in silico forward and reverse genetics tool. Genome Biol 9:R43

Davidson JA, Krysinska-Kaczmarek M, Kimber RBE et al (2004) Screening field pea germplasm for resistance to downy mildew (*Peronospora viciae*) and powdery mildew (*Erysiphe pisi*). Austral Plant Pathol 33:413–417

Davidson JA, Krysinska-Kaczmarek M, Leonforte A et al (2011) Resistance to downy mildew (*Peronospora viciae*) in Australian field pea germplasm (*Pisum sativum*). Austral Plant Pathol 40:575–582

Davis PH (1970) *Pisum* L. In: Davis PH (ed) Flora of Turkey and East Aegean Islands, vol 3. Edinburg University Press, Edinburgh, pp 370–373

De Beukelaer H, Smykal P, Davenport GF et al (2012) Core Hunter II: fast core subset selection based on multiple genetic diversity measures using Mixed Replica search. BMC Bioinformatics 13:312

De Caire J, Coyne CJ, Brumet S et al (2012) Additional pea EST-SSR markers for comparative mapping in pea (*Pisum sativum* L.). Plant Breed 131:222–226

De Candolle A (2007) Origin of cultivated plants. Kesinger Publishing, Montana

Deulvot CH, Charrel A, Marty F et al (2010) High-multiplexed SNP genotyping for genetic mapping and germplasm diversity studies in pea. BMC Genomics 11:468

Dita MA, Rispail N, Prats E et al (2006) Biotechnology approaches to overcome biotic and abiotic stress constraints in legumes. Euphytica 147:1–24

Domoney C, Knox M, Moreau C et al (2013) Exploiting a fast neutron mutant genetic resource in *Pisum sativum* L. (pea) for functional genomics. Funct Plant Biol 40:1261–1270

Dostálová R, Seidenglanz M, Griga M (2005) Simulation and assessment of possible environmental risks associated with release of genetically modified peas (*Pisum sativum* L.) into environment in Central Europe. Czech J Genet Plant Breed 41:51–63

Doyle JJ, Doyle JL, Ballenger JA et al (1997) A phylogeny of the chloroplast gene *RbcL* in the Leguminosae: taxonomic correlations and insights into the evolution of nodulation. Am J Bot 84:541–554

Duarte J, Rivière N, Baranger A et al (2014) Transcriptome sequencing for high throughput SNP development and genetic mapping in pea. BMC Genomics 15:126–141

Dumont E, Fontaine V, Vuylsteker C et al (2009) Association of sugar content QTL and PQL with physiological traits relevant to frost damage resistance in pea under field and controlled conditions. Theor Appl Genet 118:1561–1571

Ellis, THN (2011) *Pisum*. In: Kole C (ed) Wild crop relatives: genomic and breeding resources. Springer, Heidelberg, pp 237–248

Ellis THN, Poyser SJ, Knox MR et al (1998) Polymorphism of insertion sites of *Ty1-copia* class retrotransposons and its use for linkage and diversity analysis in pea. Mol Gen Genet 260:9–19

Elshire RJ, Glaubitz JC, Sun Q et al (2011) A robust, simple genotyping-by-sequencing (GBS) approach for high diversity species. PloS ONE 6:e19379

Elvira-Recuenco M, Bevan JR, Taylor JD (2003) Differential responses to pea bacterial blight in stems, leaves and pods under glasshouse and field conditions. Eur J Plant Pathol 109:555–564

Esposito MA, Martin EA, Cravero VP et al (2007) Characterization of pea accessions by SRAPs markers. Sci Hortic-Amst 113:329–335

Esposito MA, Almiron P, Gatti I et al (2012) A rapid method to increase the number of F-1 plants in pea (*Pisum sativum*) breeding programs. Genet Mol Res 11:2729–2732

FAOSTAT (2013) http://faostat3.fao.org/faostat-gateway/go/to/download/Q/*/E. Accessed 13 Dec 2013

Fondevilla S, Åvila CM, Cubero JI et al (2005) Response to *Mycosphaerella pinodes* in a germplasm collection *of Pisum* spp. Plant Breed 124:313–315

Fondevilla S, Cubero JI, Rubiales D (2007a) Inheritance of resistance *to Mycosphaerella pinodes* in two wild accessions of *Pisum*. Eur J Plant Pathol 119:53–538

Fondevilla S, Torres AM, Moreno MT et al (2007b) Identification of a new gene for resistance to powdery mildew in *Pisum fulvum*, a wild relative of pea. Breed Sci 57:181–184

Fondevilla S, Rubiales D, Moreno MT et al (2008) Identification and validation of RAPD and SCAR markers linked to the gene *Er3* conferring resistance to *Erysiphe pisi* DC in pea. Mol Breed 22:193–200

Ford R, Le Roux K, Itman C et al (2002) Diversity analysis and genotyping in *Pisum* with sequence tagged microsatellite site (STMS) primers. Euphytica 124:397–405

Franssen SU, Shrestha RP, Bräutigam A et al (2011) Comprehensive transcriptome analysis of the highly complex *Pisum sativum* genome using next generation sequencing. BMC Genomics 12:227

Furman BJ, Ambrose M, Coyne CJ et al (2006) Formation of Pea GRIC: an international consortium to co-ordinate and utilize the genetic diversity and agro ecological distribution of major collections of *Pisum*. Pisum Genet 38:32–34

Glaszmann JC, Kilian B, Upadhyaya HD et al (2010) Accessing genetic diversity for crop improvement. Curr Opin Plant Biol 13:167–173

Gonzales MD, Archuleta E, Farmer A et al (2005) The legume information system (LIS): an integrated information resource for comparative legume biology. Nucleic Acids Res 33 (Suppl 1):D660–D665

Govorov LI (1937) Pisum. In: Vavilov NI, Wulff EV (eds) Flora of cultivated plants IV: grain Leguminosae. State Agricultural Publishing Company, Moscow, pp 231–336

Green FN (2008) Classification of *Pisum sativum* subsp. *sativum* cultivars into groups and subgroups using simply inherited characters. Acta Hortic 799:155–62

Grunwald NJ, Coffman VA, Kraft JM (2003) Sources of partial resistance to Fusarium root rot in the *Pisum* core collection. Plant Dis 87:1197–1200

Grusak MA, Burgett CL, Knewtson SJ et al (2004) Novel approaches to improve legume seed mineral nutrition. Proceedings of the 5th AEP-2nd ICLGG conference, European Association for Grain Legume Research, Paris, France, 7–11 June 2004, pp 37–38

Hamon C, Baranger A, Coyne CJ et al (2011) New consistent QTL in pea associated with partial resistance to *Aphanomyces euteiches* in multiple field and controlled environments from France and the United States. Theor Appl Genet 123:261–281

Hancock JF (2012) Plant evolution and the origin of crop species. CABI, Wallingford

Hanocq E, Jeuffroy MH, Lejuene-Henault I et al (2009) Construire des idéotypes pour des systèmes de culture varies en pois d'iver. Innovations Agron 7:14–28
Harborne JB (1971) Distribution of flavonoids in the Leguminosae. In: Harborne JB, Boulter D, Turner BL (eds) Chemotaxonomy of the Leguminosae. Academic Press, London, pp 31–71
Harland, SC (1948) Inheritance of immunity to mildew in Peruvian forms of *Pisum sativum*. Hered 2:263–269
Hawthorne W (2007) A summary of pulse management options to minimize frost damage. Pulse Australia Bulletin 02
Hochman M, Dostalova R (2010) Pea breeding programs in the Czech Republic. Tag Ver Pflanzenzüchter Saatgutkaufleute Österr 61:95–96.
Hoey BK, Crowe KR, Jones VM et al (1996) A phylogenetic analysis of *Pisum* based on morphological characters, allozyme and RAPD markers. Theor Appl Genet 92:92–100
Hofer J, Turner L, Moreau C et al (2009) Tendril-less regulates tendril formation in pea leaves. The Plant Cell 21:420–428
Hollaway GJ, Bretag TW (1995) Occurrence and distribution of races of *Pseudomonas syringae* pv. *pisi* in Australia and their specificity towards various field pea (*Pisum sativum*) cultivars. Aust J Expt Agr 1995;35:629–32.
Islam N, Campbell PM, Higgins TJ et al (2009) Transgenic peas expressing an α-amylase inhibitor gene from beans show altered expression and modification of endogenous proteins. Electrophoresis 30:1863–1868
Jermyn WA, Slinkard AE (1977) Variability of percent protein and its relationship to seed yield and seed shape in peas. Legume Res 1:33–37
Jha AB, Arganosa G, Tar'an B et al (2013) Characterization of 169 diverse pea germplasm accessions for agronomic performance, *Mycosphaerella* blight resistance and nutritional profile. Genet Resour Crop Evol 60:747–761
Jing R, Knox MR, Lee JM et al (2005) Insertional polymorphism and antiquity of PDR1 retrotransposon insertions in a *Pisum* species. Genetics 171:741–752
Jing R, Johnson R, Seres A, Kiss G, Ambrose MJ, Knox MR, et al (2007) Gene-based sequence diversity analysis of field pea (*Pisum*). Genet 177:2263–2275
Jing R, Vershinin A, Grzebyta J et al (2010) The genetic diversity and evolution of field pea (*Pisum*) studied by high throughput retrotransposon based insertion polymorphism (RBIP) marker analysis. BMC Evol Biol 10:44
Jing R, Ambrose MA, Knox MR et al (2012) Genetic diversity in European *Pisum* germplasm collections. Theor Appl Genet 125:367–380
Jones AL, Johansen IE, Bean SJ et al (1998) Specificity of resistance to pea seed-borne mosaic potyvirus in transgenic peas expressing the viral replicase (NIb) gene. J Gen Virol 79:3129–3137
Kabir AH, Paltridge NG, Roessner U et al (2012) Mechanisms associated with Fe-deficiency tolerance and signaling in shoots of *Pisum sativum*. Physiol Plant 147:381–395
Katoch V, Sharma S, Pathania S et al (2010) Molecular mapping of pea powdery mildew resistance gene *er2* to pea linkage group III. Mol Breed 25:229–237
Kaya Z, Kun E, Guner A (1988) National plan for in situ conservation of plant genetic diversity in Turkey. In: Zencirci N, Kaya Z, Anikster Y et al (eds) Proceedings of international symposium on In Situ Conservation Of Plant Genetic Diversity. Central Research Institute for Field Crops, Ankara, Turkey, pp 33–47
Kaur S, Pembleton LW, Cogan NO et al (2012) Transcriptome sequencing of field pea and faba bean for discovery and validation of SSR genetic markers. BMC Genomics 13:104
Khan TN, Timmerman-Vaughan GM, Rubiales D et al (2013) *Didymella pinodes* and its management in field pea: challenges and opportunities. Field Crops Res 148:61–77
Kloz J, Turkova V (1963) Legumin, vicilin and proteins similar to them in the seeds of some species of the *Vicieae* tribe (a comparative serological study). Biol Plant 5:29–40
Kosterin OE, Bogdanova VS (2008) Relationship of wild and cultivated forms of *Pisum* L. as inferred from an analysis of three markers, of the plastid, mitochondrial and nuclear genomes. Genet Resour Crop Evol 55:735–755

Kosterin OE, Zaytseva OO, Bogdanova VS et al (2010) New data on three molecular markers from different cellular genomes in Mediterranean accessions reveal new insights into phylogeography of *Pisum sativum* L. sbsp. *elatius* (Bieb.) Schmalh. Genet Resour Crop Evol 57:733–739

Kraft JM, Dunne B, Goulden D et al (1998) A search for resistance in peas to *Mycosphaerella pinodes*. Plant Dis 82:251–253

Kupicha FK (1981). Vicieae (Adans.) DC. (1825) nom conserv prop. In: Polhill RM, Raven PH, (eds) Advances in legume systematics, vol 1. Royal Botanical Gardens, Kew, pp 377–381

Kwon S-J, Brown AF, Hu J et al (2012) Population genetic sub-structure within the USDA-ARS *Pisum* core collection and its potential as a platform for association mapping. Genes Genomics 34:305–320

Lamarck JB (1778) Flore françoise. Impr Royale, Paris

Lamprecht H (1953) Ein gen für schmale hülsen bei *Pisum* und seine koppelung. Agr Hort Genet 11:15–27

Lamprecht H (1954) Weitere studien über die vererbung der hülsenbreite von *Pisum*. Agr Hort Genet 12:202–210

Lamprecht H (1957a) The gene map of chromosome II of *Pisum*. Agr Hort Genet 15:12–47

Lamprecht H (1957b) Über die vererbung der hülsenbreite bei *Pisum*. Agr Hort Genet 15:105–114

Lamprecht H (1960). Weitere studien zur genenkarte von chromosom V von *Pisum*. Agr Hort Genet 18:23–56

Lamprecht H (1963) Zur vererbung der hülsenlänge bei *Pisum* sowie über ein neues gen für die reduktion der stipel. Agr Hort Genet 21:25–34

Lavin M, Herendeen PS, Wojciechowski M (2005) Evolutionary rates analysis of Leguminosae implicates a rapid diversification of lineages during the Tertiary. Syst Biol 54:575–594

Lee JM, Davenport GF, Marshall D et al (2005) GERMINATE. A generic database for integrating genotypic and phenotypic information for plant genetic resource collections. Plant Physiol 139:619–631

Lehmann C (1954). Das morphologische System der Saaterbsen (*Pisum sativum* L. sensu lato Gov. subsp. *sativum*). Der Züchter 24:316–337

Lejeune-Hénaut I, Hanocq E, Béthencourt L et al (2008). The flowering locus *Hr* colocalizes with a major QTL affecting winter frost tolerance in *Pisum sativum* L. Theor Appl Genet 116: 1105–1116

Le May C, Schoeny A, Tiroli B et al (2005) Improvement and validation of a pea crop growth model to simulate the growth of cultivars infected with Ascochyta blight. Eur J Plant Pathol 112:1–12

Leonforte A, Brouwer JB (1999) Expanding the adaptation of field pea. In: Langridge P, Barr A, Auricht G (eds) Proceedings of the 11th Australian plant breeding conference, CRC for Molecular Plant Breeding, Adelaide, Australia, pp 73–74

Leonforte A, Bretag T, Armstrong E et al (2004) Improving resistance to Septoria blotch (Septoria pisi) in field pea. Proceedings of the 5th European conference on Grain Legumes Research and 2nd International Conference on Legume Genetics and Genetics, Dijon, France, 7–11 June 2004, p 331

Leonforte T, Armstrong E, McMurray L et al (2006) Breeding reliable and lodging resistant semi-dwarf field peas for Australia. Proceedings of the 13th Australasian Plant Breeding conference. Christchurch, New Zealand, 18–21 April 2006

Leonforte A, Noy D, Redden R et al (2009) Improving boron and salinity tolerance in field pea (Pisum sativum L.). Proceedings of the 14th Australasian Plant Breeding Conference, Cairns, Australia

Leonforte A, Forster JW, Redden RJ et al (2013) Sources of high tolerance to salinity in pea (*Pisum sativum* L.). Euphytica 189:203–216

Lewis G, Schrirer B, Mackinder B et al (eds) (2005) Legumes of the world. Royal Botanical Gardens, Kew

Li L, Redden RJ, Zong X et al (2013) Ecogeographic analysis of pea collection sites from China to determine potential sites with abiotic stresses. Genet Resources Crop Evol 60:1801–1815

Lie TA, Goktan D, Engin M et al (1987) Co-evolution of the legume-*Rhizobium* association. Plant Soil 100:171–181

Linnaeus C (1753) Species plantarum, vol 2. Laurentius Salvius, Stokholm

Loridon K, McPhee KE, Morin J et al (2005) Microsatellite marker polymorphism and mapping in pea (*Pisum sativum* L.). Theor Appl Genet 111:1022–1031

Lucau-Danila A, Toitot C, Goulas E et al (2012) Transcriptome analysis in pea allows to distinguish chilling and acclimation mechanisms. Plant Physiol Bioch 58:236–244

Lulsdorf MM, Croser JS, Ochatt S (2011) Androgenesis and doubled-haploid production in food legumes. In: Pratap A, Kumar J (eds) Biology and breeding of food legumes. CABI International, Oxford, pp 159–177

Main D, Cheng C-H, Ficklin SP et al (2013) The cool season food legume database: an integrated resource for basic, translational and applied research. Proceedings of the Plant and Animal Genome XXI conference, Meeting Abstract, San Diego, USA, 12–16 January 2013, p W212

Majeed M, Safdar W, Ali B et al (2012) Genetic assessment of the genus *Pisum* L. based on sequence specific amplification polymorphism data. J Med Plants Res 6:959–967

Makasheva RK (1979). Gorokh (pea). In: Korovina ON (ed) Kulturnaya flora SSR. Kolos Publishers, Leningrad, pp 1–324

Malvick DK, Percich JA (1999) Identification of *Pisum sativum* germplasm with resistance to root rot caused by multiple strains of *Aphanomyces euteiches*. Plant Dis 83:51–54

Martin-Sanz A, Caminero C, Jing R et al (2011) Genetic diversity among Spanish pea (*Pisum sativum* L.) landraces, pea cultivars and the world *Pisum* sp. core collection assessed by retrotransposon-based insertion polymorphisms. Span J Agric Res 9:166–178

Martín-Sanz A, Pérez de la Vega M, Caminero C (2012) Resistance to *Pseudomonas syringae* in a collection of pea germplasm under field and controlled conditions. Plant Pathol 61:375–387

Matthews P, Ambrose MJ (1994) Development and use of a 'core' collection for the John Innes *Pisum* collection. In: Balfourier F, Perretant MR (eds) Evaluation and exploitation of genetic resources, Proceedings of the genetic resources section meeting of EUCARPIA. Clermont Ferrand, France, pp 99–107

Maxted N, Ambrose M (2001) Peas (*Pisum* L.). In: Maxted N, Bennett SJ (eds) Plant genetic resources of legumes in the Mediterranean. Kluwer, Dordrecht, pp 181–190

Maxted N, Bennett SJ (eds) (2001) Plant genetic resources of legumes in the Mediterranean. Springer Science+Business Media, Dordrecht

Maxted N, Kell SP (2009) Establishment of a global network for the in situ conservation of crop wild relatives: status and needs. FAO Consultancy Report, FAO, Rome, Italy, pp 1–265

McCouch S, Baute GJ, Bradeen J et al (2013) Agriculture: feeding the future. Nature 499:23–24

McGee RJ (2012a) Genomics-assisted breeding for cool season food legumes: from gene discovery to application. Proceedings of Plant and Animal Genome XX conference, San Diego, USA, 15–17 January 2012, p W205

McGee RJ (2012b) Mendel's legacy: an international pea sequencing project. Pisum Genet 44:13–14

McMurray LS, Davidson JA, Lines MD et al (2011) Combining management and breeding advances to improve field pea (*Pisum sativum* L.) grain yields under changing climatic conditions in south-eastern Australia. Euphytica 180:69–88

McPhee K (2005) Variation for seedling root architecture in the core collection of pea germplasm. Crop Sci 45:1758–1763

McPhee KE, Muehlbauer FJ (2001) Biomass production and related characters in the core collection of *Pisum* germplasm. Genet Resour Crop Evol 48:195–203

McPhee KE, Tullu A, Kraft JM et al (1999) Resistance to Fusarium wilt race 2 in the *Pisum* core collection. J Am Soc Hort Sci 124:28–31

Mikič A, Mihailovic V, Dimitrijevic M, et al (2012) Evaluation of seed yield and seed yield components in red–yellow (*Pisum fulvum*) and Ethiopian (*Pisum abyssinicum*) peas. Genet Resour Crop Evol 60:629–638

Mikič A, Smykal P. Kenicer GJ et al (2013) The bicentenary of the research on 'beautiful' vavilovia (*Vavilovia formosa*), a legume crop wild relative with taxonomic and agronomic potential. Bot J Linn Soc 172:524–531

Mikschofsky H, Broer I (2012) Feasibility of *Pisum sativum* as an expression system for pharmaceuticals. Transgenic Res 21:715–724

Mikschofsky H, Schirrmeier H, Keil GM et al (2009) Pea-derived vaccines demonstrate high immunogenicity and protection in rabbits against rabbit haemorrhagic disease virus. Plant Biotechnol J 7:537–549

Mishra RK, Gangadhar BH, Nookaraju A et al (2012) Development of EST-derived SSR markers in pea (*Pisum sativum*) and their potential utility for genetic mapping and transferability. Plant Breed 131:118–124

Mittal V, Ujagir U (2005) Field screening of pea, *Pisum sativum* L. germplasm for resistance against major insect pests. J Plant Protect Environ 2:50–58

Morton RL, Schroeder HE, Bateman KS et al (2000) Bean α-amylase inhibitor 1 in transgenic peas (*Pisum sativum*) provides complete protection from pea weevil (*Bruchus pisorum*) under field conditions. Proc Natl Acad Sci U S A 97: 3820–3825

Murfet IC (1973) Flowering in *Pisum*. *Hr*, a gene for high response to photoperiod. Heredity 31:157–164

Murfet IC, Reid JB (1993) Developmental mutants. In: Casey R, Davies DR (eds) Peas: genetics, molecular biology and biotechnology. CAB International, Wallingford, pp 165–216

Nasiri J, Haghnazari A, Saba J (2010) Genetic diversity among varieties and wild species accessions of pea (*Pisum sativum* L.) based on SSR markers. African J Biotechnol 15 3405–3417

Nelson RL (2011) Managing self-pollinated germplasm collections to maximize utilization. Plant Genet Resour 9:123–133

Ondrej M, Dostalova R, Hybl M et al (2003) Utilization of *afila* types of pea (*Pisum sativum* L.) resistant to powdery mildew (*Erysiphe pisi* DC.) in the breeding programs. Plant Soil Environ 49:481–485

Page D, Aubert G, Duc G et al (2002) Combinatorial variation in coding and promoter sequences of genes at the *Tri* locus in *Pisum sativum* accounts for variation in trypsin inhibitor activity in seeds. Mol Genet Genomics 267:359–369

Palmer JD, Jorgensen RA, Thompson WF (1985) Chloroplast DNA variation and evolution in *Pisum*: patterns of change and phylogenetic analysis. Genet 109:195–213

Pearce SR, Knox M, Ellis TH et al (2000) Pea Ty1-copia group retrotransposons: transpositional activity and use as markers to study genetic diversity in *Pisum*. Mol Gen Genet 263:898–907

Petkova C, Nikolova V, Kalapchieva SH et al (2009) Physiological response and pollen viability of *Pisum sativum* genotypes under higher temperature influence. Acta Hortic 830:665–672

Pilet-Nayel ML, Muehlbauer FJ, McGee RJ et al (2002) Quantitative trait loci for partial resistance to *Aphanomyces* root rot in pea. Theor Appl Genet 106:28–39

Pilet-Nayel ML, Muehlbauer FJ, McGee RJ et al (2005) Consistent quantitative trait loci in pea for partial resistance to *Aphanomyces euteiches* isolates from the United States and France. Phytopathol 95:1287–1293

Porter LD, Hoheisel G, Coffman VA (2009) Resistance of peas to *Sclerotinia sclerotiorum* in the *Pisum* core collection. Plant Pathol 58:52–60

Priliouk L, Vavilov NI, Robertson L et al (1999) Genetic diversity in pea germplasm from Vavilov Institute and ICARDA collections for black spot resistance and agronomic merit. Pisum Genet 31:53

Prusinski J (2007) Degree of success of legume cultivars registered by the Center for Cultivar Testing over the period of market economy. Acta Sci Pol Agric 6:3–16

Rafalski JA (2010) Association genetics in crop improvement. Curr Opin Plant Biol 13:174–180

Rahi P, Kapoor R, Young JP et al (2012) A genetic discontinuity in root-nodulating bacteria of cultivated pea in the Indian trans-Himalayas. Mol Ecol 21:145–159

Rai R, Singh AK, Singh BD et al (2011) Molecular mapping for resistance to pea rust caused by *Uromyces fabae* (Pers.) de-Bary. Theor Appl Genet 123:803–813

Rajput V, Singh NP (2010) Studies on *in vitro* regeneration and direct organogenesis in pea (*Pisum sativum* L.). Indian J Plant Physiol 15:246–249

Reeves PA, Panella LW, Richards CM (2012) Retention of agronomically important variation in germplasm core collections: implications for allele mining. Theor Appl Genet 124:1155–1171

Riehl S, Zeidi M, Conard NJ (2013) Emergence of agriculture in the foothills of the Zagros mountains of Iran. Science 341:65–67
Saar DE, Polans NO (2000) ITS sequence variation in selected taxa of *Pisum*. Pisum Genet 32: 42–45
Sadras VO, Lake L, Chenu K et al (2012) Water and thermal regimes for field pea in Australia and their implications for breeding. Crop Pasture Sci 63:33–44
Sanchez EA, Mosquera T (2006) Establishing a methodology for inducing the regeneration of pea (*Pisum sativum* L.) explants 'Santa Isabel' variety. Agron Colomb 24:17–27
Sanderson L, Krilow C, Vandenberg A et al (2011) KnowPulse: a breeder-focused web portal that integrates genetics and genomics of pulse crops with model genomes. Pisum Genet 43:46–47
Sarıkamış G, Yanmaz R, Ermis S et al (2010) Genetic characterization of pea (*Pisum sativum*) germplasm from Turkey using morphological and SSR markers. Genet Mol Res 9:591–600
Schaefer H, Hechenleitner P, Santos-Guerra A et al (2012) Systematics, biogeography, and character evolution of the legume tribe Fabeae with special focus on the middle-Atlantic island lineages. BMC Evol Biol 12:250
Schneider A, Walker SA, Poyser S et al (1999) Genetic mapping and functional analysis of a nodulation-defective mutant (sym19) of pea *(Pisum sativum* L.). Mol Gen Genet 262:1–11
Schroeder HM, Gollasch S, Moore A et al (1995) Bean α-amylase inhibitor confers resistance to the pea weevil (*Bruchus pisorum*) in transgenic peas (*Pisum sativum*). Plant Physiol 107:1233–1239
Shafiq S, Mather DE, Ahmad M et al (2012) Variation in tolerance to radiant frost at reproductive stages in field pea germplasm. Euphytica 186:831–845
Simioniuc D, Uptmoor R, Friedt W et al (2002) Genetic diversity and relationships among pea cultivars revealed by RAPDs and AFLPs. Plant Breeding 121:429–35
Sindhu A, Ramsay L, Sanderson LA et al (2014) Gene-based SNP discovery and genetic mapping in pea. Theor Appl Genet 127:2225–2241
Smartt J (1990) Grain legumes: evolution and genetic resources. Cambridge University Press, Cambridge
Smýkal P (2006) Development of an efficient retrotransposon-based fingerprinting method for rapid pea variety identification. J Appl Genet 47:221–230
Smýkal P, Coyne CJ, Ford R et al (2008a) Effort towards a world pea (*Pisum sativum* L.) germplasm core collection: the case for common markers and data compatibility. Pisum Genet 40:11–14
Smýkal P, Hybl M, Corander J et al (2008b) Genetic diversity and population structure of pea (*Pisum sativum* L.) varieties derived from combined retrotransposon, microsatellite and morphological marker analysis. Theor Appl Genet 117:413–424
Smýkal P, Kalendar R, Ford R et al (2009) Evolutionary conserved lineage of Angela-family retrotransposons as a genome-wide microsatellite repeat dispersal agent. Heredity 103:157–167
Smýkal P, Kenicer G, Flavell AJ et al (2011) Phylogeny, phylogeography and genetic diversity of the *Pisum* genus. Plant Genet Resources 9:4–18
Smýkal P, Aubert G, Burstin J et al (2012) Pea (*Pisum sativum* L.) in the genomic era. Agron J 2:74–115.
Smýkal P, Coyne CJ, Redden RJ et al (2013) Peas. In: Singh M, Bisht IS (eds) Genetic and genomic resources for grain legume improvement. Elsevier Insights, London, pp 41–80
Steele KP, Wojciechowski MF (2003) Phylogenetic analyses of tribes Trifoleae and Vicieae based on sequences of the plastid gene *mat*K (Papilionoideae: Leguminosae). In: Klitgaard BB, Bruneau A (eds) Advances in legume systematics, part 10. Royal Botanic Gardens, Kew, pp 355–370
Swiecicki WK, Timmerman-Vaughan G (2005) Localization of important traits: the example of pea. In: Nagata T, Lorz H, Widholm JM et al (eds) Molecular marker systems in plant breeding and crop improvement: biotechnology in agriculture and forestry, vol 55. Springer, Berlin, pp 155–169
Swiecicki W, Swiecicki WK, Wiatr K (1997) History and present achievements and prospectives of legume plants breeding in Poland. Zesz Probl Postęp Nauk Roliniczych 446:15–32

Swiecicki WK, Wolko B, Apisitwanich S et al (2000) An analysis of isozymic loci polymorphism in the core collection of the Polish *Pisum* genebank. Genet Resour Crop Evol 47:583–590
Tar'an B, Zhang C, Warkentin T, Tullu A et al (2005) Genetic diversity among varieties and wild species accessions of pea (*Pisum sativum* L.) based on molecular markers, and morphological and physiological characters. Genome 48:257–272
Taylor JD, Bevan JR, Crute IR et al (1989) Genetic relationship between races of *Pseudomonas syringae* pv. *pisi* and cultivars of *Pisum sativum*. Plant Pathol 38:364–375
Tedford EC, Inglis DA (1999) Evaluation of legumes common to the Pacific Northwest as hosts for the pea cyst nematode (*Heterodera goettingiana*). J Nematol 31:155–163
Timmerman-Vaughan GM, McCallum JA, Frew TJ et al (1996) Linkage mapping of quantitative trait loci controlling seed weight in pea (*Pisum sativum* L.). Theor Appl Genet 93:431–439
van der Maesen LJ (1998) Wild plants as genetic resources for crop improvement. In: Mathew P, Sivadasan M (eds) Diversity and taxonomy of tropical flowering plants. Mentor Books, Calicut, pp 93–112
van Leur JA, Aftab M, Leonforte A et al (2007) Control of pea seedborne mosaic virus in field pea through resistance breeding. Proceedings of the 16th biennial conference of the Australasian Plant Path Society, Adelaide, Australia, 24–27 September 2007, p 154
Varshney RK, Close TJ, Singh NK et al (2009) Orphan legume crops enter the genomics era! Curr Opin Plant Biol 12:202–210
Varshney RK, Song C, Saxena RK, Azam S et al (2013) Draft genome sequence of chickpea (*Cicer arietinum*) provides a resource for trait improvement. Nat Biotechnol 31:240–246
Vavilov NI (1950) The phytogeographic basis of plant breeding. In: Chester KS (translator) The origin, variation, immunity, and breeding of cultivated plants. Chronica Botanica, Massachusetts, pp 13–54
Vershinin AV, Allnutt TR, Knox MR et al (2003) Transposable elements reveal the impact of introgression, rather than transposition, in *Pisum* diversity, evolution, and domestication. Mol Biol Evol 20:2067–2075
Vigeolas H, Chinoy C, Zuther E et al (2008) Combined metabolomic and genetic approaches reveal a link between the polyamine pathway and albumin 2 in developing pea seeds. Plant Physiol 146:74–82
Waines JG (1975) The biosystematics and domestication of peas (*Pisum* L.). Bull Torrey Bot Club. 102:385–395
Weeden NF (2007) Genetic changes accompanying the domestication of *Pisum sativum*: is there a common genetic basis to the 'domestication syndrome' for legumes? *Ann Bot* 100:1017–1025
Weeden NF, Porter L (2007) The genetic basis of Fusarium root rot tolerance in the 'Afghanistan' pea. Pisum Genet 39:35–36
Weeden NF, Wolko B (2001) Allozyme analysis of *Pisum sativum* spp. abyssinicum and the development of genotypic definition for this subspecies. Pisum Genet 33:21–25
Weeden, NF, Brauner, S, Przyborowski, JA (2002) Genetic analysis of pod dehiscence in pea (*Pisum sativum* L.). Cell Mol Biol Lett 7:657–663
Weigelt K, Kuster H, Rutten T et al (2009) ADP-glucose pyrophosphorylase-deficient pea embryos reveal specific transcriptional and metabolic changes of carbon-nitrogen metabolism and stress responses. Plant Physiol 149:395–411
Welham T, Domoney C (2000) Temporal and spatial activity of a promoter from a pea enzyme inhibitor gene and its exploitation for seed quality improvement. Plant Sci 159:289–299
Weller JL, Hecht V, Liew LC et al (2009) Update on the genetic control of flowering in garden pea. J Expt Biol 60:2493–2499
Weller JL Liew LC, Hecht VFG et al (2012) A conserved molecular basis for photoperiod adaptation in two temperate legumes. Proc Natl Acad Sci U S A 109:21158
Wheeler T, von Braun J (2013) Climate change impacts on global food security. Science 341: 508–513
White Ö (1917) Studies of inheritance in *Pisum* II. The present state of knowledge of heredity and variation in peas. Proc Am Philos Soc 56:487–588

Xu X, Liu X, Ge S et al (2011) Resequencing 50 accessions of cultivated and wild rice yields markers for identifying agronomically important genes. Nat Biotechnol 30:105–111

Yarnell SH (1962) Cytogenetics of the vegetable crops. III. Legumes. Garden peas (*Pisum sativum* L.) Bot Rev 28:465–537

Young JPW, Matthews P (1982) A distinct class of peas (*Pisum sativum* L.) from Afghanistan that show strain specificity for symbiotic *Rhizobium*. Heredity 48:203–210

Zaytseva OO, Bogdanova VS, Kosterin OE (2012) Phylogenetic reconstruction at the species and intraspecies levels in the genus *Pisum* (L.) (peas) using a histone H1 gene. Gene 504:192–202

Zhang C, Tar'an B, Warkentin T et al (2006a) Selection for lodging resistance in early generations of field pea by molecular markers. Crop Sci 46:321–329

Zhang R, Hwang SF, Chang KF et al (2006b) Genetic resistance to *Mycosphaerella pinodes* in 558 field pea accessions. Crop Sci 46:2409–2414

Zhuang X, McPhee KE, Coram TE et al (2012) Rapid transcriptome characterization and parsing of sequences in a non-model host-pathogen interaction; pea-*Sclerotinia sclerotiorum*. BMC Genomics 13:668

Zohary D, Hopf M (2000) Domestication of plants in the old world. Oxford University Press, Oxford

Zong X, Guan JP, Wang SM et al (2008) Genetic diversity and core collection of alien *Pisum sativum* L. germplasm. Acta Agron Sin 34:1518–1528

Zong X, Redden R, Liu Q et al (2009) Analysis of a diverse *Pisum* sp. collection and development of a Chinese *P. sativum* core collection based on microsatellite markers. Theor Appl Genet 118:193–204

Chapter 3
Chickpea

Teresa Millán, Eva Madrid, José I. Cubero, Moez Amri, Patricia Castro and Josefa Rubio

1 Introduction

Chickpea (*Cicer arietinum* L.) is a unique cultivated species in the genus *Cicer*. It is an annual, self-pollinated crop adapted to post-rainy season either in spring sowing or summer-dominant rainfall regions (Berger and Turner 2007). It has been a food crop since ancient times in the Mediterranean basin from where it was dispersed to the Indian subcontinent becoming a basic constituent of Asian diets.

Nowadays, it is grown all over the five continents in around 50 countries, with 90 % of its cultivated area (around 13×10^6 ha) in developing countries. India ranks first in the world in respect of cultivated area (68.5 %) followed by Pakistan (8.7 %)

T. Millán (✉)
Department of Genetics, Campus de Rabanales, Edif C5 2ª planta, Córdoba, Spain
e-mail: teresa.millan@uco.es

E. Madrid
Institute for Sustainable Agriculture, CSIC, Avda. Menendez Pidal sn, Apdo 4084, Córdoba, Spain
e-mail: b62mahee@uco.es

J. I. Cubero
Departmento de Genetica, Universidad de Córdoba, Campus Rabanales, Córdoba, Spain
e-mail: ge1cusaj@uco.es

M. Amri
Regional Field Crop Research Center of Beja (CRRGC), PB 350, Beja 9000, Tunisia
e-mail: amrmoez@gmail.com

P. Castro
Genetic Improvement of Fruits and Vegetables Lab, Rm. 214, Bldg. 010A, BARC West, 10300 Baltimore Avevue, Beltsville, MD 20705-2350, USA
e-mail: prcastrolopez@gmail.com

J. Rubio
Area de Mejora y Biotecnologia, IFAPA Centro "Alameda del Obispo", Avda. Menendez Pidal s/n Apdo: 3092, Córdoba, Spain
e-mail: josefam.rubio@juntadeandalucia.es

A. M. De Ron (ed.), *Grain Legumes*, Handbook of Plant Breeding 10,
DOI 10.1007/978-1-4939-2797-5_3

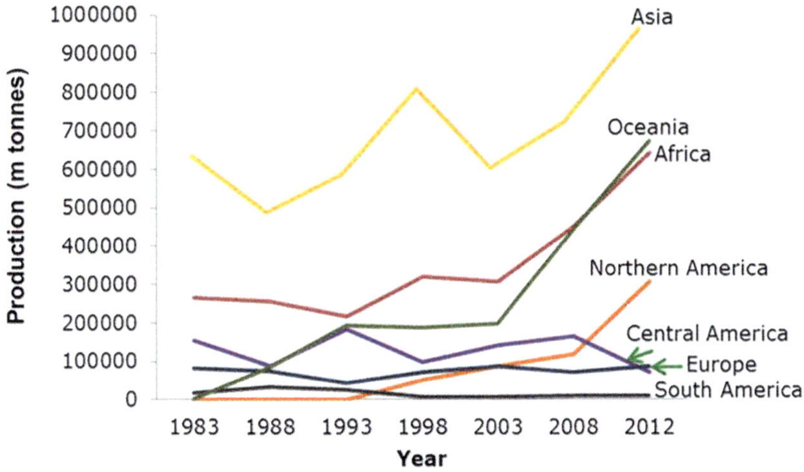

Fig. 3.1 Evolution of world chickpea production

as well as in total production (68 %) and followed by Australia (5.9 %), that is one of the major exporting countries. This crop is mainly grown under rainfed conditions with a world average yield of 931 kg ha^{-1} in 2012 but reaching a maximum of 6044 kg ha^{-1} in Israel where this crop is grown under irrigated conditions. Chickpea production has increased over the past 30 years from 6.1 to around 13×10^6 t. It has been particularly stunning that the increase of this crop in Africa and Oceania, where the total production was at minimum, doubled (Fig. 3.1). This increase can be explained by (i) the development of new high-yielding varieties tolerant/resistant to main diseases, pests and abiotic stresses and (ii) successful integrated crop management practices. The number of importing countries has increased from 65 in 1991 to 164 in 2011, suggesting the highest global demand for this crop in the past 20 years (Faostat 2014).

Chickpea cropping systems have been recently reviewed by Berrada et al. (2007). In the Mediterranean basin, springtime (February–mid-April) is the traditional sowing date, but with the new winter chickpea varieties, sowing can be advanced to January, whereas in the Indian subcontinent it is sown from mid-September to November after the rainy season. Chickpea has been mainly cultivated in rotation after multiple base crops such as wheat, teff, oat, rice, pearl millet, sorghum, maize or cotton, although intercropping has also been put into practice in subsistence farming areas. Production stability and yields have been limited by various constraints such as susceptibility to major worldwide-distributed diseases, Ascochyta blight and Fusarium wilt or terminal drought.

This crop, as a legume, improves soil fertility by fixing atmospheric nitrogen, meeting up to 80 % of its nitrogen requirement from symbiotic nitrogen fixation (Gaur et al. 2012). Chickpea-specific mesorhizobia are present in soils where chickpea has been traditionally grown but seed inoculation is required in the new chickpea-cultivated lands or in marginal soils. Better N_2 fixation can be achieved by selecting rhizobial strains of superior N_2-fixing capacity but also depends on chickpea cultivars (Kantar et al. 2007).

Chickpea is a high-quality and cheap source of protein for people in developing countries (particularly in South Asia), who are largely vegetarian. Additionally, it is rich in nutritionally important unsaturated fatty acids and vitamins such as riboflavin, niacin, thiamin, folate and the vitamin A precursor β-carotene. Different studies have provided some evidence to support the potential beneficial effects of chickpea components in lowering the incidence of various cancers, high-density lipoprotein (HDL) cholesterol, type-2 diabetes and heart diseases (Roy et al. 2010; Jukanti et al. 2012).

C. arietinum has a relatively low DNA content estimated to be 738.09 Mb (Varshney et al. 2013) distributed in eight pairs of chromosomes. Genomic tools in chickpea have been progressing very rapidly in the past years. Nowadays, reference genetic maps are available including cross-genome markers from the model species *Medicago truncatula,* which made it possible to assign chickpea linkage groups (LG) to *Medicago* chromosomes. (Millán et al. 2010; Nayak et al. 2010; Gujaria et al. 2011; Thudi et al. 2011). Furthermore, five out of the eight chickpea chromosomes could be isolated by flow-cytometry and specific sequence-tagged microsatellite site (STMS) markers amplified on sorted chromosomes and allowed to assign four linkage groups to particular chromosomes (Vláčilová et al. 2002; Zatloukalová et al. 2011). All this previous information, together with the recent publication of the first draft of the whole-chickpea genome sequence by two different groups (Jain Rajesh et al. 2013; Varshney et al. 2013), provides powerful tools to be used for searching genes involved in agronomic traits. The relatively simple genome of this species together with its importance to world agriculture and available genetic and genomic tools make this crop as a possible "model" for cool-season food-legume genomics.

2 Systematic and Origin

The genus *Cicer* belongs to the family Leguminosae, subfamily Papilionoideae and the monogeneric tribe *Cicereae* Alef., including 35 perennials, 8 annual and 1 unspecified wild species. It is a component of the Galegoid group (or cool season legumes) together with tribes *Viceae, Trifoliae* and *Loteae* (Zhu et al. 2005). The tribe *Cicerae* has been subclassified in sections *Cicer* (=*Monocicer*), *Chamaecicer*, *Polycicer* and *Acanthocicer* based on morphological traits and geographical distribution (van der Maesen et al. 2007). The single cultivated species *C. arietinum* belongs to the section *Cicer*.

Based on their crossability with chickpea, the wild *Cicer* annual species have been classified into gene pools reflecting their distance from the cultivated species as proposed by Harlan and de Wet (1971). According to this definition, the primary gene pool consists of *C. arietinum,* the wild annual progenitor, *Cicer reticulatum* Ladz. and the closely related *Cicer echinospermum* P. H. Davis. The secondary gene pool is composed by *Cicer bijugum* K. H. Rech, *Cicer pinnatifidum* Jaup and Spp. and *Cicer judaicum* Boiss. Sterility is associated with the first-generation hybrids for those species. Interspecific or wide hybridization has been identified as a potential means of increasing the genetic variation and introduction of resistance genes in cultivated species from wild species (Singh et al. 2008).

Chickpea was one of the first domesticated grain legumes together with other crops such as wheat, barley, rye, pea, lentil, flax and vetch in a reduced area of southeast of Turkey (Ladizinsky and Adler 1976; Abbo et al. 2003). The domestication seems to have occurred from the wild ancestor *C. reticulatum* (sin. *C. arietinum* subsp. *reticulatum*) with a monophyletic origin as revealed by the low genetic variation of the cultigen (*C. arietinum* subsp. *arietinum*) (Moreno and Cubero 1978). Only in the early Bronze period (fifth millennium BC) can chickpea be considered as an established crop in the Near East. A few seeds dated ca. 6000 BC have been found in Bulgaria, and two millennia later, in Greece. Thus, it seems that chickpea belonged to the first agricultural complex reaching Europe through the Black Sea (Zohary et al. 2012). Afterward, chickpea was likely taken to India by the Aryan tribes in the second millennium BC who probably brought the crop from Iranian tribes, as suggested by De Candolle on linguistic evidence. It was also spread over the Mediterranean basin and through the Nile River and was expanded to the east of Africa. Chickpea was taken by Spanish colonizers to the New World in 1492. It crossed the Atlantic Ocean in the first Columbus travel of discovery. Chickpea was always a humble crop that almost never appeared on the royal tables, being on the contrary a useful food for both humans and animals and a good companion of man around the globe.

3 Varietal Groups

C. arietinum is divided into two main cultivar groups for breeding purposes 'desi' and 'kabuli' types. This distinction is made mainly on the basis of a small number of morphological characters, principally the seed shape and colour. White flower, thin seed coat, large seed (200–680 mg) and more of them with a "ram's head" shape and cream-coloured seeds, smooth seed surface, lack of anthocyanin pigmentation and semi-spreading growth habit are present in the kabuli chickpea. On the other hand, pink flower, thick seed coat, small seed (100–200 mg) angular, dark seeds, anthocyanin pigmentation of stem, rough seed surface and either semi erect or semi-spreading growth habit are characteristics of desi ones (Pundir et al. 1985). This classification overlaps, to a certain extent, with the *macrosperma* and *microsperma* races proposed by Moreno and Cubero (1978) studying quantitative as well as qualitative traits. Additionally, a third type designated as pea-shaped characterized by medium to small seed size and cream-coloured seeds has been proposed (Upadhyaya et al. 2008a).

Desi types, mostly grown in India, Pakistan and East Africa, cover around 85% of chickpea cultivated area. In these countries, seeds are usually dehulled and split before cooking. Kabuli chickpeas are grown mainly in the Mediterranean basin, the Near East, Central Asia and America where whole seeds are used for human consumption after soaking and boiling. It is believed that the kabuli chickpea was introduced into India through Kabul, Afghanistan (therefore named kabuli), in the mid- to late-seventeenth century (Singh 1987). Kabuli probably evolved from the desi type in the Mediterranean basin and oligogenic traits like flower colour, seed coat thickness and seed size seem to have played an important role in its evolution (Moreno and Cubero 1978; Gil and Cubero 1993).

Significant differences in agronomic traits have been observed between these two groups. Ascochyta blight resistance, cold tolerance and erect growth habit are more frequently found in kabuli types, whereas Fusarium wilt resistance, heat and drought tolerance and early flowering are prevalent in desi types (Singh 1987). They also differ in quality components such as seed coat thickness, crude fibre content, mineral and trace element composition, polyphenolic content and in vitro digestibility (Jambunathan and Singh 1981; Gil et al. 1996). Kabuli types were reported to be nutritionally superior to desi types in terms of cooking time, biological value and sensory properties (Singh et al. 1991) and receive higher market price than desi types. The price premium in kabuli types generally increases as the seed size increases.

The distribution of genetic diversity in kabuli seems to be much narrower than in the desi predominant chickpea type. Both types are also nearly uniform in cytoplasm, indicating no evolution of hybridization barriers (Moreno and Cubero 1978). The desi and kabuli groups tend to have maintained distinct morphological types and may have different gene blocks for important yield components appearing as two separate groups when they are clustered based on molecular marker analysis (Iruela et al. 2002).

These differences have been employed in chickpea-breeding programmes using desi × kabuli (and vice versa) crosses to obtain new disease-resistant cultivars with higher yields, large seed size and vigour of desi types (Gaur et al. 2007).

4 Genetic Resources and Utilization

The genetic resources provide basic material for selection and improvement through breeding, leading to ensure food security needs of the world's rapidly rising population. They comprise diversity of genetic material contained in traditional varieties, modern cultivars, crop wild relatives and other wild species (Farshadfar and Farshadfar 2008; Upadhyaya et al. 2008b). The collections represent both insurance against genetic erosion and as sources of resistance/tolerance to diseases and pests, climatic and other environmental stresses.

The two major chickpea germplasm collections are maintained at the Consultative Group on International Agricultural Research (CGIAR) centres, the International Crops Research Institute for the Semi-Arid Tropics, http://www.icrisat.org (ICRISAT) and the International Center for Agricultural Research in Dryland Areas, http://www.icarda.org (ICARDA) with more than 20,000 and 13,000 accessions, respectively. ICRISAT mainly focusses on desi types while ICARDA maintains mostly kabuli types. Both centres, ICRISAT and ICARDA, maintain wild *Cicer* sp. Other important chickpea collections are conserved by the National Bureau of Plant Genetic Resources, India (NBPGR) with around 14,000 accessions, Centre for Legumes in Mediterranean Agriculture Australia (CLIMA) with approximately 8000 accessions and the United States Department of Agriculture, http://www.ars-grin.gov (USDA) with about 6000 accessions (Rubio et al. 2009; Upadhyaya et al. 2011). Other collections have been described by Gaur et al. (2012).

In spite of the large number of germplasm accessions available at the gene banks, there has been a very limited use of these accessions in chickpea breeding; thus, in India, only ten lines contributed to the 35 % of genetic base in chickpea (Upadhyaya et al. 2008b). The main reason for the modest utilization of germplasm is the lack of information on a large number of accessions. Thus, core and mini core collections (about 10 and 1 % of the total accessions, respectively) have been suggested as an opportunity for the utilization of genetic diversity in crop improvement (Upadhyaya and Ortiz 2001). In chickpea, core and mini core subsets, representative of the entire chickpea collections, have been obtained. Upadhyaya et al. (2001) developed in ICRISAT a chickpea core collection consisting of 1956 accessions, which represented the global chickpea germplasm collection. ICRISAT developed a reference set consisting of 300 accessions representing diversity from the entire spectrum of a composite collection made between ICRISAT and ICARDA (Upadhyaya and Ortiz 2001). Mini core collections have been useful to identify accessions with good agronomic traits and resistance to different diseases (Pande et al. 2006a) as well as in applied breeding for the development of broad-based elite breeding lines/cultivars with superior yield (Upadhyaya et al. 2008b).

The Generation Challenge Program (GCP; www.generationcp.org) is contributing to intensify the molecular characterization of core and mini core collection to identify genetically diverse parents for mapping and utilization in breeding programmes (Gaur et al. 2012).

5 Major Breeding Achievements and Specific Goals in Current Breeding

Breeding efforts have contributed substantially to improve chickpea yield potential but the lack of stable production still continues to be a major concern for the adoption of this crop by farmers. The major constraints limiting chickpea production include various abiotic and biotic stresses, particularly important are fungal diseases (Ascochyta blight and Fusarium wilt), pests (pod borer) and drought or cold stress. Parasitic plants could also be a big problem in such particular environments such as Mediterranean conditions (Gaur et al. 2007; Chen et al. 2011) (Fig. 3.2).

Fig. 3.2 Chickpea screening for resistance/tolerance to different diseases and parasitic plants in specific clinic fields in Tunisia (**a**) Ascochyta blight (**b**) Fusarium wilt (**c**) Broomrape (*Orobanche foetida*) (**d**) Dodder (*Cuscuta* spp.)

5.1 Biotic Stresses

Ascochyta blight, caused by the Ascomycota fungus *Ascochyta rabiei* (Pass.) Labr. (teleomorph *Didymella rabiei* (Kovatsch) Arx.), is one of the most serious diseases of chickpea worldwide, which affects all aerial parts of the plant. It reduces chickpea seed yield significantly reaching 100% of losses when favourable conditions such as cool (5–15 °C) and wet weather occurred (Pande et al. 2005). The use of varieties with improved levels of resistance is considered the most economical solution for long-term disease management. So that, efforts to develop and commercialized blight chickpea resistant cultivars have been made (Gaur et al. 2007; Taran et al. 2013). Currently, the development of blight-resistant lines has made possible the introduction of winter sowing in the Mediterranean region with the prospect of increasing chickpea production that could be doubled (Singh and Reddy 1996). However, reaching high levels of resistance to blight is complex because many genes with minor to moderate effects control this trait and only partial resistance is available (Pande et al. 2005; Taran et al. 2007; Bhardwaj et al. 2010).

The pathogen survives between crop seasons either on infected plant debris or in contaminated seeds. A wide pathogenic variation has been described and those are grouped in two or three pathotype systems (Udupa et al. 1998; Chen et al. 2004,

2011). Breeders try to search for different sources of resistance and pyramid different resistance genes into the same cultivar to improve both the resistance level and the durability of this resistance. Different sources of resistance/tolerance have been employed to achieve this goal, and as a result, a significant number of cultivars with improved Ascochyta blight resistance have been released. This information was summarized in Pande et al. (2010) and Millan et al. (2013). Additional sources of resistance have also been identified among wild *Cicer* species, including *C. judiacum, C. pinnatifidum, C. echinospermum* and *C. reticulatum* (Pande et al. 2006b)

Fusarium wilt, caused by the vascular pathogen *Fusarium oxysporum* Schlechtend: Fr. f. sp. *ciceris,* is, along with Ascochyta blight, the most important fungal disease in chickpea. Fusarium wilt epidemics can be devastating to individual crops and cause up to 100% loss under favourable conditions. Persistence of the pathogen in soil for many years even in the absence of the host renders its control difficult. Consequently, the use of resistant cultivars is the most economic, effective and ecofriendly method of controlling such pathogen. However, the effectiveness of resistant cultivars is limited by the existence of different races of pathogens. To date, eight physiological races (0, 1A, 1B/C, 2, 3, 4, 5 and 6) have been reported from India, Spain, Tunisia and the USA (Haware and Nene 1982; Jimenez-Gasco et al. 2004). Resistance to wilt in chickpea has been reported to be race-specific and controlled by major resistance genes (Gaur et al. 2007; Sharma and Muehlbauer 2007; Castro et al. 2012a).

The screening of international germplasm has led to the identification of sources of resistance to wilt in both cultivated (desi and kabuli types) and wild chickpea germplasm (Millan et al. 2013). However, the resistance sources in kabuli types are limited compared to the desi types. Haware et al. (1992) evaluated a world collection with 13,500 germplasm accessions for race 1, identifying 160 resistant accessions but only ten of them were kabuli type. The desi type accession WR315 is extensively recognized as possessing resistance to all known wilt races. It has been widely used in inheritance and mapping studies and is considered to be a very interesting source of wilt resistance for chickpea-breeding programmes (Haware 1998). Two kabuli accessions (ICC 14194 and ICC 17109) with complete wilt resistance and extra large seeds, a key quality determinant in the market, were detected by Gaur et al. (2006). Significant progress has been made in Fusarium research and cultivars including resistance to multiple races are now available (Malhotra et al. 2007; Singh et al. 2009).

Botrytis grey mould (BGM) caused by *Botrytis cinerea* Pers. ex. Fr. is the second most important foliar disease of chickpea after Ascochyta. BGM causes complete crop loss in several South Asian countries (Pande et al. 2006c). The limited reports available on genetics of BGM resistance in chickpea suggest that a few major genes control resistance in the host to BGM (Anuradha et al. 2011). There is no adequate level of genetic resistance to BGM in the cultivated genotypes. However, high levels of resistance have been found in the wild *Cicer* species, including *C. judaicum, C. bijugum, C. echinospermum* and *C. pinnnatifidum* (Pande et al. 2006c). Thus, several wide and intraspecific hybridizations have been carried out to transfer the identified disease resistance genes in wild types and land races to commonly adopted and widely grown chickpea cultivars.

Other fungal diseases considered of local importance could also affect chickpea productions. Rust (*Uromyces ciceris arietini*) has been reported to be a problem in central Mexico and Italy (Ragazzi 1982; Díaz-Franco and Pérez-García 1995). The resistance is controlled by a single gene (*Uca1/uca1*) (Madrid et al. 2008) and moderate levels of incomplete and partial resistance are available (Rubiales et al. 2001). Phytophthora root rot (caused by *Phytophthora medicaginis*) affects cool season plantings (Chen et al. 2011). Until now, five genotypes (FLIP 97-132C, FLIP 97-85C, FLIP 98-53C, ILC-5263 and NCS 9905) evaluated under controlled conditions in Pakistan exhibited highly resistant response to the disease (Akram et al. 2008). *C. echinospermum* appears to be the most promising source of resistance in wild species after field and controlled condition evaluations in Australia (Knights et al. 2008).

Parasitic plants as broomrape (*Orobanche crenata* and *Orobanche foetida*) may cause serious losses in chickpea productions in winter sowing under Mediterranean conditions (Rubiales et al. 1999; Roman et al. 2007). Sources of resistance to *O. crenata* in Spain (Rubiales et al. 1999) and to *O. foetida* in Tunisia (Amri personal communication) have been identified. Despite the low broomrape infestation levels observed in chickpea compared to other grain legume species, rapeseed or wild species recently, more aggressive and virulent new *Orobanche* populations are arising (Amri et al. 2009). Field dodder (*Cuscuta* spp.) is another parasite that was reported damaging chickpea production in many regions in the world (Goldwasser et al. 2012a; Chen et al. 2014). In highly infested fields, this parasite can cause up to 100 % loss in grain production (Singh et al. 2007). Sources of resistance for *Cuscuta campestris* (field dodder) (ICCV 95333 and Hazera 4) exhibiting high resistance were identified in Israel (Goldwasser et al. 2012b).

In addition, efforts have been made to identify sources of resistance to both pests pod borer (*Helicoverpa armigera* Hübner) and leaf miner (*Liriomyza cicerina* Rondani) in the cultivated and wild species at ICRISAT and ICARDA (Gaur et al. 2007).

5.2 *Abiotic Stresses*

Terminal drought is globally the most serious abiotic stress to chickpea productivity and the most important factor for instability of yield in major production countries as Asia and Africa, where chickpea is mainly grown as a rainfed crop on residual moisture. Cultivars may escape (early maturity) or tolerate terminal drought increasing the efficiency of water use. Promising accessions (ICC 4958, ICC 1882, ACC 316 and ACC 317) and varieties with a vigorous and deeper root system to improve drought tolerance have been developed (Saxena et al. 1993; Gaur et al. 2008; Cancy and Toker 2009). In addition, transgenic plants have been developed at ICRISAT having either a dehydration responsive element or a gene that increases proline accumulation in the plant (Gaur et al. 2007).

Salt stress imposes a significant limitation of productivity related to the adverse effects on the dry weights of both shoots and roots and also on nodulation and nitrogen fixation (Manchanda and Garg 2008). Limited efforts to identify salinity tolerance within chickpea indicated low genotypic variation and few varieties with tolerance to moderate levels of salinity (ECe ranging from 4 to 6 dS/m) have been developed. Karnal Chana 1 (CSG 8962) and Genesis 836 (ICCV 96836) were developed in India and Australia, respectively (Maliro et al. 2004). Recently, ICRISAT identified several lines that gave a higher yield than the salinity tolerant cultivar Karnal Chana 1 (Krishnamurthy et al. 2011; Gaur et al. 2012).

Finally, both freezing (<−1.5 °C) and chilling (between −1.5 and 15 °C) are known to affect chickpea at various development stages from germination to maturity (Croser et al. 2003) and should be considered in chickpea breeding for winter sowing in Mediterranean environments. Two chilling tolerant cultivars ('Sonali' and 'Rupali') have been released in Australia (Clarke et al. 2005) and should be included in winter sowing breeding programmes to avoid problems in pod filling in fresh spring. Also, ICARDA and ICRISAT breeding programmes developed cold-tolerant cultivars adapted to winter sowing (Gaur et al. 2007).

5.3 Phenological Characters

Flowering time is influenced by photoperiod and temperature and is a major task to improve crop adaptation. Early flowering associated with early maturity is preferred to escape from terminal drought, high temperature or frost at the end of the season (Gaur et al. 2007). A moderate and positive genetic correlation between days to flowering and seed weight was reported by Hovav et al. (2003), suggesting that it is difficult to breed early-flowering cultivars without compromising seed weight. However, ICRISAT developed two extra large and early-maturing kabuli types from Mexican origin, which suggests that it is possible to breed early varieties with extra large seeds (Gaur et al. 2006). The first landmark variety was ICCV 2, which matures in about 85 days, and it is perhaps the world's earliest maturing variety of kabuli chickpea. Two super-early desi chickpea lines, ICCV 96029 and ICCV 96030, which mature in 75–80 days in southern India were developed by Kumar and Rao (1996). Further advancements have been made in breeding for super-earliness and several short-duration high-yielding varieties of chickpea, both in desi and kabuli types (Gaur et al. 2012).

6 Breeding Methods

Productivity, yield stability in different environment conditions and resistance/tolerance to main damaging diseases are the major goals in chickpea-breeding programmes. These constraints become more pronounced especially with the climate change affecting both the development of the crop and its enemies (development of new pathogens/pests and changing in the aggressiveness and virulence of others). The development of chickpea crop in a sustainable agricultural system is facing several challenges: (i) being more efficient in the development of new varieties resistant/tolerant to main biotic and abiotic stresses and (ii) strict adoption of these new developed varieties by farmers that could result in an improved productivity, reducing yield fluctuations.

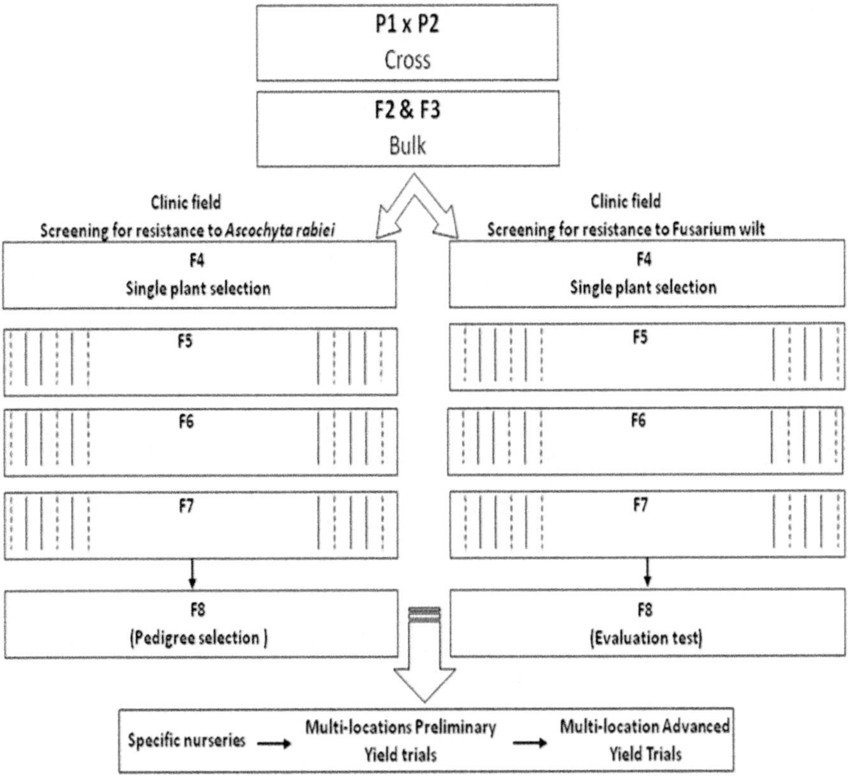

Fig. 3.3 Combined bulk and pedigree method in chickpea-breeding for tolerance/resistance to both Ascochyta blight and Fusarium wilt diseases (chickpea-breeding programme—Tunisia). Seeds coming from the same selected plant are sown in both blight and wilt clinic fields during the same cropping season. Crosses are mainly performed at ICARDA

Any breeding programme can be achieved through three main steps or components: (i) the genetic variation which is the base of the breeding programme, (ii) strict and rigorous selection within that variation and (iii) evaluation and confirmation of the selected lines (Salimath et al. 2007). Chickpea can be considered a strict self-pollinated crop. Hence, inbreeding to fix genes and develop pure-lines cultivars are the main breeding objectives. Mass or pure-line selection from landraces was the simplest method initially employed. Later, crossing programmes and subsequent different modification of pedigree methods and backcrossing programmes for qualitative traits have been developed (Gaur et al. 2012). Most chickpea-breeding programmes have been confined to intraspecific hybridization that includes desi × desi, kabuli × kabuli or desi × kabuli crosses (Gaur et al. 2007). As mentioned in Sect. 3 of this chapter, desi and kabuli types have different genetic backgrounds differing in disease resistance, tolerance to abiotic stresses and quality components. Depending on the objective of the breeding programme, single, three-way or multiple crosses are used. Single crosses have been the most widely adopted. Figure 3.3 shows an example of the methodology followed in the national chickpea-breeding programme in Tunisia at the National Institute of Agronomy Researches/Regional Field Crop Research Center of Beja (INRAT/CRRGC) while the crosses are mainly performed at ICARDA. Hybridization also allowed the obtention of recombinant inbred lines (RILs) by single-seed descent (SSD) methods (Johnson and Bernard 1962) where plant materials combining interesting traits have been selected (Rubio et al. 2006). Three-way or multiple cross approaches provide additional opportunities to study gene interactions with the aim of combining different traits from different parents in a new cultivar. Hybridization allows the exploitation of transgressive inheritance of some characters such as yield or seed size.

Singh (1987) summarized the effectiveness and specificity of various breeding methods in respect to the desired specific traits and reported that: (i) pedigree method could be used for resistance to biotic stresses; (ii) bulk-pedigree method for drought tolerance and winter hardiness; (iii) modified bulk method for abiotic stresses, seed size, earliness and plant type; (iv) backcross method for inter specific hybridization and (v) limited backcross for desi × kabuli introgression and resistance breeding. Pedigree methods are not frequently used due to the cumbersome data collection and the fact that this approach limits the breeding programme to only a few crosses. The combination of bulk and pedigree methods is widely used among many chickpea-breeding programmes (Gaur et al. 2012). The backcrossing is in general applicable to incorporate one or few oligogenic characters into a well-adapted elite variety. Thus, tolerance to Ascochyta blight and double podding were successfully introgressed into CDC Xena, CDC Leader and FLIP98-135C using markers linked to the quantitative trait loci (QTLs) for both traits (Taran et al. 2013).

Introgression from the wild related species into the cultigens has been also employed. Beneficial traits such as cold tolerance and a high degree of resistance to wilt, root rot, and *Botrytis* grey mould have been introgressed from *C. reticulatum* and *C. echinospermum* into cultivated chickpea (Singh et al. 2005; Ramgopal et al. 2012). Similarly, the *C. reticulatum* accession ILWC 119 was included in a crossing programme giving two cyst nematode resistant chickpea germplasm lines ILC 10765 and ILC 10766 (Malhotra et al. 2002).

7 Integration of New Biotechnologies in Breeding Programmes

The integration of genomic technologies in chickpea breeding will greatly improve the efficiency and time required of breeding programmes in the development of better cultivars (Gaur et al. 2012; Castro et al. 2013).

7.1 Genetic Maps

Molecular markers are considered valuable tools for crop improvement due to their usefulness in characterizing and manipulating genetic loci responsible for monogenic and polygenic traits. Markers have made it possible for the development of genetic maps, the necessary framework for any marker-assisted selection (MAS) programme. Availability of well-saturated genetic linkage maps is a prerequisite for tagging traits with molecular markers, thus enabling their use in MAS and positional cloning of genes of interest.

In chickpea, the first published maps were developed analyzing isozymes in F_2 populations derived from interspecific crosses (Gaur and Slinkard 1990; Kazan et al. 1993). Since then, tremendous advances in DNA marker technology and map development have been achieved, allowing for the identification of markers close to genes or genomic regions with agronomical importance. Even though maps are still incomplete, the chances of finding new polymorphic markers have been considerably increased, essentially due to the development of STMS markers (Hüttel et al. 1999; Winter et al. 1999; Sethy et al. 2003, 2006; Lichtenzveig et al. 2005; Choudhary et al. 2006; Nayak et al. 2010). The use of STMS markers in different populations gave the possibility of exchanging information between maps. These markers have also made it possible to build a consensus chickpea map (Millán et al. 2010) following the nomenclature previously proposed by Winter et al. (2000), which has been considered as the map of reference. Nayak et al. (2010) completed the map published by Winter et al. (2000) adding 175 new markers and provided anchor points for comparing chickpea linkage groups with linkage groups in the model species *M. truncatula*. More recently, the development of next-generation sequencing technologies allowed the obtaining of the first chickpea transcriptome (Hiremath et al. 2011). Large-scale molecular markers were obtained using the transcriptome information and comprehensive genetic maps were developed (Gujaria et al. 2011; Thudi et al. 2011; Hiremath et al. 2012; Jhanwar et al. 2012). High-density genetic maps of gene-based markers represent a powerful resource to enhance genome analysis, thus providing an important opportunity to directly tag genes related to agronomical traits. Due to the low levels of genetic diversity present within the gene pool of cultivated chickpea, the high-density genetic maps have been developed in the interspecific cross ICC 4958 × PI 489777 (Table 3.1), considered as

Table 3.1 Second generation genetic maps developed in chickpea recombinant inbred lines (RIL) populations

Reference[a]	Newly developed markers	No. of loci	Coverage (cM)	Average inter-markers distance (cM)
Gujaria et al. (2011)[a]	SSR, GMMs, CISR	300	766.56	2.55
Thudi et al. (2011)[a]	BES-SSR, DarT	1291	845.56	0.65
Choudhary et al. (2012)[a]	EST–SSR, ITPs, SNPs	406	1497.7	3.68
Gaur et al. (2012)[a]	SNPs, SSR	368	1808.7	1.7
Hiremath et al. 2012)[a]	CKAMs, TOGs-SNPs	1328	788.6	0.59
Stephens et al. (2014)[b]	SSRs, SNPs	401/417	658.7/752	1.74/2.16

[a] RIL derived from *Cicer arietinum* (ICC 4958) × *C. reticulatum* (PI 489777)

[b] RIL derived from Lasseter × ICC3996/ S95362 × Howzat

GMMs genic molecular markers, *SSR* simple sequence repeat, *EST* expressed sequence tag, *SNP* single-nucleotide polymorphism, *CKAMs* chickpea KASPar assay markers, *TOGs* tentative orthologous genes, *ITPs* intron targeted primers, *BES* bacterial artificial chromosome (BAC)-end sequences the reference mapping population. As far as we know, this population was only phenotyped for wilt (races 4 and 5, Winter et al. 2000). So, the high-resolution map developed in this cross could not be used for tagging the genomic regions associated to other agronomic traits, such as Ascochyta blight, which is the most important constraint in chickpea. Recently, the construction of two intraspecific genetic maps has been reported (Stephens et al. 2014) using the RIL populations Lasseter × ICC 3996 and S95362 × Howzat, both segregating for blight resistance. In addition to the reference interspecific map, it may be very useful performing a high-resolution mapping in new crosses segregating for more characters valuables for breeding applications.

Moreover, the availability of the draft genome in desi and kabuli chickpea has opened the possibility of anchoring genetic maps and positioning QTL on the physical one (Jain Rajesh et al. 2013; Varshney et al. 2013; Madrid et al. 2014). The identification of markers with complete association with QTLs will boost the development of "perfect" markers in pulses (Kumar et al. 2011). Such markers are extremely useful for guiding the introgression of multiple genes, because they increase selection efficiency and avoid recombination events between markers and QTLs (Hospital 2009).

Major efforts of breeding programmes are concentrated in the development of resistant lines to the main fungi affecting crops (Ascochyta blight and Fusarium wilt). The majority of authors consider the resistance to blight as a quantitative trait and several QTLs have been identified in the chickpea genetic map (Millan et al. 2013). QTLs for resistance to blight have been located and validated on linkage groups LG4 (QTL_{AR1} and QTL_{AR2}), LG2 (QTL_{AR3}), LG3 (QTL_{AR4}) and LG8 (QTL_{AR5}) of the chickpea map employing different mapping populations (Millan et al. 2013). Another QTL was also detected in LG6 using the cross ICCV 96029 × CDC Frontier (Anbessa et al. 2009).

Resistance to Fusarium wilt in chickpea has been described to be race specific and controlled by major resistance genes, the majority of which are recessive in nature. Resistance genes to races 0, 1, 2, 3, 4 and 5 (*foc-0*$_2$*, foc-1, foc-2, foc-3, foc-4 and foc-5)* have been found to form a cluster located on LG2 of the chickpea map (Sharma and Muehlbauer 2007; Cobos et al. 2009; Gowda et al. 2009; Halila et al. 2010). However, one of the two resistance genes for race 0 (*foc-0*$_1$) was found in LG5 (Cobos et al. 2005) confirming that the resistance is controlled by two independent genes (*foc-0*$_1$ and *foc-0*$_2$), as Rubio et al. (2003) reported before by classical genetic studies.

Markers associated with other diseases as rust and BGM have been localized. A gene that controls resistance to chickpea rust *(Uca1/uca1)* has been located in LG7 tightly flanked by two STMS markers (Madrid et al. 2008). On the other hand, three genomic areas controlling resistance to BGM have been identified by Anuradha et al. (2011). QTL1, sited in LG6, explained 12.8 % of the total phenotypic variation while QTL2 and QTL3 explained 9.5 and 48 %, respectively, both of them located in LG8 (names of the groups are referred to chickpea consensus map).

In addition, several important characteristics such as quality components and agronomic traits have also been mapped and flanking markers were identified.

Pigmentation of the flower (pink/white = *P/p* = *B/b*), brown/yellow seed testa *(T3/t3)*, purple/green epicotyl *(Gst/gst)* and seed coat thickness *(Tt/tt)* are some of the phenotypic traits with simple inheritance mapped in the chickpea genetic map and located in LG4 (Tekeoglu et al. 2002; Cobos et al. 2005). Erect/prostrate plant growth habit *(Hg/hg)* is another trait with simple inheritance which has been reported to be located in LG3 (Winter et al. 2000; Cobos et al. 2009). Double pod is a mutation controlled by a single recessive gene designated as *s* or *sfl* (Muehlbauer and Singh 1987) linked to TA80 (Rajesh et al. 2002) in LG6 (Cho et al. 2002; Cobos et al. 2005). Regarding yield, different QTLs have been identified in LGs 2, 4, 5 and 7 by different authors (Cobos et al. 2007; Gowda et al. 2011). Early flowering is another trait related to yield improvement. It seems to have a positive effect on yield under the Mediterranean environment (Siddique et al. 2003; Rubio et al. 2004). QTLs controlling this trait have been detected in LGs 1, 2, 3, 4 and 6. The QTLs with high limit of detection (LOD) values in LG3 have special interest for breeding applications as they were validated in different environments and in both intra- and interspecific populations (Cho et al. 2002; Cobos et al. 2009). A recent study published studying drought tolerance identified 45 robust main-effect QTLs (M-QTLs) explaining up to 58.20% phenotypic variation and 973 epistatic QTLs (E-QTLs) explaining up to 92.19% phenotypic variation for several target traits. Nevertheless, there is a *QTL-hotspot* in LG4 explaining about 58.20% phenotypic variation containing seven markers (ICCM0249, NCPGR127, TAA170, NCPGR21, TR11, GA24 and STMS11) that could be further used in MAS (Varshney et al. 2014b)

7.2 Marker-Assisted Breeding

Molecular markers closely linked to a particular agronomic trait facilitate the detection of favourable alleles in breeding programmes. MAS is particularly useful in the case of breeding for disease resistance in order to avoid complex and time-consuming evaluations as well as for pyramiding different resistance genes in the same genotype. However, the efficacy of MAS relies on the saturation of genomic areas of interest with robust, highly polymorphic, easy to interpret and cost-effective markers (Collard and Mackill 2008).

Despite the effort carried out during the past years to saturate genetic linkage maps and identified markers tightly linked to traits of interest in chickpea, the adoption of MAS in chickpea breeding has not been widely employed. Most cultivars of chickpea are the results of conventional plant-breeding programmes, where trait evaluation and phenotypic selection under field or greenhouse conditions are the routine procedure. With the advent of molecular markers and genetic maps, there has been an increased interest in the use of marker technology to facilitate chickpea crop improvement. As previously mentioned, molecular markers have been used for identification and mapping of genes and QTLs for agriculturally important traits in chickpea. However, the extent to which markers have to be employed in chickpea-breeding programmes has not been clearly determined. To date, only few studies reported the employment of MAS in conventional chickpea-breeding programmes (Taran et al. 2013; Varshney et al. 2014a). In addition to the employment of molecular markers in breeding programmes, they can be used to develop new genomic resources to be used in genomics studies or to introduce new genetic combinations in the programmes. For example, Castro et al. (2010) employed the most associated marker with the *Foc5* gene located in LG2 (TA59) to assist the selection of resistant or susceptible genotypes in order to develop near isogenic lines (NILs).

Regarding Ascochyta blight, the sequence characterized amplified region (SCAR) markers $SCY17_{590}$ and $SCAE19_{336,}$ tightly linked to $QTL_{AR2,}$ were successfully employed to tag a source of Ascochyta blight resistance in a collection of chickpea genotypes (Imtiaz et al. 2008). Recently, a new codominant molecular marker (CaETR) was developed by Madrid et al. (2013) based on allelic sequence length polymorphism in an ethylene receptor-like gene located in the QTL_{AR1} (Madrid et al. 2012). It was probed for the usefulness of the markers $SCY17_{590}$ and CaETR to discriminate between resistant and susceptible chickpea genotypes. Those markers have been used in other studies in order to check its efficiency in MAS of blight-resistant genotypes in different breeding programmes (Bouhadida et al. 2013; Castro et al. 2013). These two markers contribute efficiently in the selection of new chickpea varieties with better combinations of alleles to ensure resistance to Ascochyta blight.

Recently, the use of molecular markers in a backcrossing breeding programme (MABC) was reported. Markers were applied to introgress blight resistance (LG4b and LG8) and double podding (LG6) into adapted chickpea cultivars (Taran et al. 2013). In addition, MABC has been used to introgress resistance to Fusarium wilt race 1 and Ascochyta blight in C214, an elite cultivar of chickpea (Varshney et al. 2014a). This approach permits the selection of plants with more than one set of QTL for resistance to blight and double podding without phenotyping.

In spite of these efforts of using molecular markers in breeding programmes, it is imperative to characterize molecular markers useful for MAS targeting the most important agronomic traits, such as Fusarium, Ascochyta, yield, growth habit, etc.

On the other hand, the possibilities for genetic improvement and selection approaches are limited in chickpea when stress tolerance is present in sexually incompatible gene pools. To solve this problem, transgenic chickpeas have been developed. Transgenic chickpeas expressing either the *cry1Ac/b* or the *cry2Aa* gene and the bean-amylase inhibitor gene are resistant to *Helicoverpa* and bruchids, respectively, but these chickpeas have yet to be commercialized. Unfortunately, attempts to generate transgenic chickpeas with increased tolerance to drought and salinity or with increased methionine content have been less successful (Acharjee and Sarmah 2013).

7.3 Functional Genomics

The aim of functional genomics is to discover the biological function of genes and to determine how sets of genes and their expressed products interact in a particular phenotype. Understanding the functional characteristics and expression of a particular trait may also involve techniques such as expressed sequence tag (EST) library construction, mutant identification, RNA interference (RNAi) experiments and overexpression studies.

Preliminary investigations have been carried out in chickpea to determine important functional genes involved in traits such as abiotic tolerances (Mantri et al. 2007), seed quality (Gremigni et al. 2004) and biotic disease resistances (Jaiswal et al. 2004; Coram and Pang 2005a, b). Until now, transformation studies on chickpea are preliminary (Indurker et al. 2010; Acharjee and Sarmah 2013; Kanakala et al. 2013), and no RNAi essays have been performed. The most detailed functional studies have been made using an enriched library of EST sequences, microarray experiments or Supersage studies (Coram and Pang 2005a, b, 2006; Molina et al. 2008, 2011). Nevertheless, these studies did not establish correlation between the genes identified and the genomics regions genetically mapped related with agronomic traits. With the advent of the complete genome sequence, the identification of genes directly located on these regions will be within reach in an easy way.

One of the most common techniques to perform functional genomic studies is the real-time quantitative polymerase chain reaction (qPCR) technology. It has emerged as the most accurate and sensitive method for gene expression analyses (Derveaux et al. 2010). To perform these studies it is necessary to validate internal control genes first. In chickpea, different combinations of reference genes were given depending on the stress under study (Castro et al. 2012b) enabling an accurate and reliable normalization of qPCR results.

8 Seed Production

In the past years, new successful chickpea varieties have been originated over the world mainly by international or national research institutions or growers associations. The profit margin from chickpea seeds is low and, generally, does not attract private sector investment because chickpea is highly self-pollinated and many farmers use their own seeds stored on farm (Van Gastel et al. 2007). This is the common situation for small farmers in developing countries, where food legumes are very important in family nutrition, but, generally, they do not have access to seeds from improved food legume varieties. In contrast, developed countries such as the USA, Canada or Australia, mainly exporters, require high-quality seeds to be able to provide homogeneous raw material to be processed by the industry. Typically, seed quality parameters in chickpea were focused on seed size, shape and seed coat colour, but nowadays the demand of new varieties suitable for pre cooked or processed chickpea seeds is increasing. Chickpea seed production has been widely reviewed by Van Gastel et al. (2007), who described seed classes following the Organization for Economic Co-operation and Development (OECD) nomenclature. In general, new varieties obtained in chickpeas are pure lines, but still a small amount of heterozygosity could be present in the breeder or foundation seed. Around 500 selected plants from each variety should be harvested and threshed separately to initiate variety seed production. The next step should be sowing seeds from each plant in a single progeny row in order to discard rows with off-type plants.

Careful crop management practices such as sowing in uniform fields should be applied. In addition, requirements for previous cropping in the seed field should specify the crops that should not be grown for a limited time preceding the production of the seed crop. In chickpea, the land selected to produce seeds should be free of any other chickpea variety for at least 2 years for pre-basic and basic seeds. For certified seeds, only 1 year between two crops of different varieties is required (Van Gastel et al. 2007). A minimum isolation distance of 1–2 m between two fields is considered to be enough. However, slightly longer isolation distances are recommended for pre-basic seed and 3 m for basic and certified seeds. It is also suggested touse a relatively high plant population density to improve the competitive ability of chickpea plants to weed (Van Gastel et al. 2007).

Seed storage conditions are other important factors to take into account. Reduced moisture and low temperature increase the longevity of the seed. Storing seeds at less than 13 % moisture, however, has adverse effects on viability (Siddique and Krishnamurthy 2014). Seed standards (physical purity, percentage of germination, pest and diseases) have not exactly the same parameters in each country. Harmonizing seed certification procedures to develop a flexible or internationally acceptable seed certification scheme should be desirable for the benefit of the national seed industries (Van Gastel et al. 2007).

Possibly, chickpea seed producers associations will play in the future a major role in enhancing adoption of improved chickpea cultivars in developing countries as occurred in Ethiopia, the largest producer, consumer and exporter of chickpea in Africa. In this country, 90 % of the seed demand is being met by the farmers organized as seed growers (Fikre 2014).

9 Conclusions and Future Prospects

Chickpea crop has a promising future. It is already a basic food in many Asian countries and it is recognized as a source of biologically active compounds (Roy et al. 2010). It is also a crop with low inputs adapted to low water requirements. However, chickpea belongs to the category of "low value seed crop" because it is highly self-pollinated; so in most cultivated areas, farmers continue to grow old varieties and landraces using their own sowing seeds (Gaur et al. 2010). Research on chickpea crop has been done with successful results in international and national institutes but it is necessary to solve the transference of knowledge to private sector and to solve commercialization of the new varieties.

In spite of the progress made in the last years in the developing of genetic maps and the identification/location of different genes/QTLs related with the main agronomic traits affecting chickpea, the molecular basis of these traits remains unknown. Isolation and validation of genes underlying the genes/QTL for the traits of interest is an essential step to determine gene function. Development of a genome-wide physical map or local physical map around the gene/QTL region and then sequencing those are the next steps in this direction (Gaur et al. 2012). The availability of the reference genome for desi and kabuli types is facilitating this approach (Madrid et al. 2014) and will allow the development of diagnostic markers enhancing the adoption of molecular breeding for increasing chickpea productivity.

Acknowledgments Authors would like to acknowledge research funding support from the Spanish Ministry of Science and Innovation (MICINN, project RTA2010-00059), co financed with European Regional Development Fund (FEDER). E. Madrid is a researcher funded by the 'Juan de la Cierva' programme of the Ministry of Science and Innovation.

References

Abbo S, Shtienberg D, Lichtenzveig J et al (2003) The chickpea, summer cropping, and a new model for pulse domestication in the ancient near east. Quart Rev Biol 78:435–448

Acharjee S, Sarmah BK (2013) Biotechnologically generating 'super chickpea' for food and nutritional security. Plant Sci 207:108–116

Akram A, Iqbal SM, Rauf CA et al (2008) Detection of resistant sources for collar rot disease in chickpea germplasm. Pakistan J Bot 40:2211–2215

Amri M, Mlayeh O, Kharrat M (2009) Pathogenicity of different broomrape populations on five host plant species. In: Proc of the 10th world congress of parasitic plants, international parasitic plant society kusadasi, Turkey, 8–12 June, p 63

Anbessa Y, Taran B, Warkentin TD et al (2009) Genetic analyses and conservation of QTL for Ascochyta blight resistance in chickpea (*Cicer arietinum* L.). Theor Appl Genet 119:757–765

Anuradha C, Gaur P, Pande S et al (2011) Mapping QTL for resistance to botrytis grey mould in chickpea. Euphytica 182:1–9

Berger JD, Turner NC (2007) The ecology of chickpea. In: Yadav SS, Redden R, Chen W, Sharma B (eds) Chickpea breeding and management. CABI, Wallingford, pp 47–71

Berrada AF, Shivakumar BG, Yaduraju NT (2007) Chickpea in cropping systems. In: Yadav SS, Redden R, Chen W et al (eds) Chickpea breeding and management. CABI, Wallingford, pp 193–212

Bhardwaj R, Sandhu JS, Kaur L et al (2010) Genetics of Ascochyta blight resistance in chickpea. Euphytica 171:337–343

Bouhadida M, Benjannet R, Madrid E et al (2013) Efficiency of marker-assisted selection in detection of Ascochyta blight resistance in Tunisian chickpea breeding lines. Phytopathol Mediterr 52:202–211

Cancy H, Toker C (2009) Evaluation of yield criteria for drought and heat resistance in chickpea (*Cicer arietinum* L.). J Agron Crop Sci 195:47–54

Castro P, Piston F, Madrid E et al (2010) Development of chickpea near-isogenic lines for Fusarium wilt. Theor Appl Genet 121:1519–1526

Castro P, Rubio J, Millán T et al (2012a) Fusarium wilt in chickpea: general aspect and molecular breeding. In: Ferrari Rios T, Reyes E (eds) Fusarium epidemiology, environmental sources and prevention. Nova Sci, Hauppauge, pp 101–122

Castro P, Román B, Rubio J et al (2012b) Selection of reference genes for expression studies in *Cicer arietinum* L.: analysis of cyp81E3 gene expression against *Ascochyta rabiei*. Mol Breed 29:261–274

Castro P, Rubio J, Madrid E et al (2013) Efficiency of marker-assisted selection for Ascochyta blight in chickpea. J Agr Sci. doi:10.1017/S0021859613000865

Chen W, Coyne CJ, Peever TL et al (2004) Characterization of chickpea differentials for pathogenicity assay of Ascochyta blight and identification of chickpea accessions resistant to *Didymella rabiei*. Plant Pathol 53:759–769

Chen W, Sharma HC, Muehlbauer F (2011) Compendium of chickpea and lentil diseases and pests. The American Phytopatologist Society (APS), St. Paul

Chen W, Dugan FM, McGee R (2014) First report of dodder (*Cuscuta pentagona*) on chickpea (*Cicer arietinum*) in the United States. Plant Dis 98:165

Cho S, Kumar J, Shultz J et al (2002) Mapping genes for double podding and other morphological traits in chickpea. Euphytica 128:285–292

Choudhary S, Sethy NK, Shokeen B et al (2006) Development of sequence-tagged microsatellite site markers for chickpea (*Cicer arietinum* L.). Mol Ecol Notes 6:93–95

Choudhary S, Gaur R, Gupta S et al (2012) EST-derived genic molecular markers: development and utilization for generating an advanced transcript map of chickpea. Theor Appl Genet 124:1449–62

Clarke HJ, Siddique KHM, Khan TN (2005) Chickpea improvement in Southern Australia: breeding for tolerance to chilling at flowering. Indian J Pulses Res 18:1–8

Cobos MJ, Fernández MJ, Rubio J et al (2005) A linkage map of chickpea (*Cicer arietinum* L.) based on populations from Kabuli × Desi crosses: location of genes for resistance to Fusarium wilt race 0. Theor Appl Genet 110:1347–1353

Cobos MJ, Rubio J, Fernández-Romero MD et al (2007) Genetic analysis of seed size, yield and days to flowering in a chickpea recombinant inbred line population derived from a Kabuli × Desi cross. Ann Appl Biol 151:33–42

Cobos MJ, Winter P, Kharrat M et al (2009) Genetic analysis of agronomic traits in a wide cross of chickpea. Field Crops Res 111:130–136

Collard BCY, Mackill DJ (2008) Marker-assisted selection: an approach for precision plant breeding in the twenty-first century. Phil Trans R Soc B 363:557–572

Coram TE, Pang ECK (2005a) Isolation and analysis of candidate Ascochyta blight defence genes in chickpea. Part I. Generation and analysis of an expressed sequence tag (EST) library. Physiol Mol Plant Pathol 66:192–200

Coram TE, Pang ECK (2005b) Isolation and analysis of candidate Ascochyta blight defence genes in chickpea. Part II. Microarray expression analysis of putative defence-related ESTs. Physiol Mol Plant Pathol 66:201–210

Coram TE, Pang ECK (2006) Expression profiling of chickpea genes differentially regulated during a resistance response to *Ascochyta rabiei*. Plant Biotechnol J 4:647–666

Croser JS, Clarke HJ, Siddique KHM et al (2003) Low temperature stress: implications for chickpea (*Cicer arietinum* L.) improvement. Crit Rev Plant Sci 22:185–219

Derveaux S, Vandesompele J, Hellemans J (2010) How to do successful gene expression analysis using real-time PCR. Methods 50:227–230

Díaz-Franco A, Pérez-García P (1995) Control químico de la roya y la rabia del garbanzo y su influencia en el rendimiento de grano. Rev Mex Fitopatol 13:123–125

Faostat (2014) http://faostat.fao.org/. Accessed Oct 2014

Farshadfar M, Farshadfar E (2008) Genetic variability and path analysis of chickpea (*Cicer arietinum* L.) Landraces and Lines. J Appld Sci 8:3951–3956

Fikre A (2014) An overview of chickpea improvement research program in Ethiopia. Legume Perspect 3:47–49

Gaur PM, Slinkard AE (1990) Genetic control and linkage relations of additional isozyme markers in chick-pea. Theor Appl Genet 80:648–656

Gaur PM, Pande S, Upadhyaya HD et al (2006) Extra-large kabuli chickpea with high resistance to Fusarium wilt. Chickpea Pigeonpea Newsl 13:5–7

Gaur PM, Gowda CLL, Knights EJ et al (2007) Breeding achievements. In: Yadav SS, Redden R, Chen W, Sharma B (eds) Chickpea breeding and management. CABI, Wallingford, pp 391–416

Gaur PM, Krishnamurthy L, Kashiwagi J (2008) Improving drought-avoidance root traits inchickpea (*Cicer arietinum* L.): current status of research at ICRISAT. Plant Prod Sci 11:3–11

Gaur PM, Tripathi S, Gowda CLL et al (2010) Chickpea seed production manual. ICRISAT, Patancheru

Gaur PM, Jukanti AK, Varshney RK (2012) Impact of genomic technologies on chickpea breeding strategies. Agron 2:199–221

Gil J, Cubero JI (1993) Inheritance of seed coat thickness in chickpea (*Cicer arietinum* L.) and its evolutionary implications. Plant Breeding 111:257–260

Gil J, Nadal S, Luna D et al (1996) Variability for some physico-chemical characters in desi and kabuli chickpea types. J Sci Food Agr 71:179–184

Goldwasser Y, Miranda Sazo MR, Lanini WT (2012a) Control of field dodder (*Cuscuta campestris*) parasitizing tomato with ALS-inhibiting herbicides. Weed Technol 26:740–746

Goldwasser Y, Miryamchik H, Sibony M et al (2012b) Detection of resistant chickpea (*Cicer arietinum*) genotypes to *Cuscuta campestris* (field dodder). Weed Res 52:122–130

Gowda SJM, Radhika P, Kadoo NY et al (2009) Molecular mapping of wilt resistance genes in chickpea. Mol Breeding 24:177–183

Gowda SJM, Radhika P, Mhase LB et al (2011) Mapping of QTLs governing agronomic and yield traits in chickpea. J Appl Genet 52:9–21

Gremigni P, Furbank RT, Turner NC (2004) Genetic manipulation of seed size in chickpea. In: Legumes for the benefit, agriculture, nutrition and the environment: their genomics, their products, and their improvemnet. Conference handbook of the joint 5th European conference on grain legumes and 2nd international conference on legume genomics and genetics. Dijon, France, 7–11 June, p 141

Gujaria N, Kumar A, Dauthal P et al (2011) Development and use of genic molecular markers (GMMs) for construction of a transcript map of chickpea (*Cicer arietinum* L.). Theor Appl Genet 122:1577–1589

Halila I, Rubio J, Millán T et al (2010) Resistance in chickpea (*Cicer arietinum*) to Fusarium wilt race '0'. Plant Breeding 129:563–566

Harlan JR, de Wet JMJ (1971) Towards a rational classification of cultivated plants. Taxon 20:509–517

Haware MP (1998) Diseases of chickpea. In: Allen DJ, Lenne JM (eds) The plant pathology of food and pasture legumes. CABI, Wallinford, pp 473–516

Haware MP, Nene YL (1982) Races of *Fusarium oxysporum* f. sp. *ciceris*. Plant Dis 66:809–810

Haware MP, Nene YL, Pundir RPS et al (1992) Screening of world chickpea germplasm for resistance to Fusarium wilt. Field Crop Res 30:147–154

Hiremath PJ, Farmer A, Cannon SB et al (2011) Large-scale transcriptome analysis in chickpea (*Cicer arietinum* L.), an orphan legume crop of the semi-arid tropics of Asia and Africa. Plant Biotechnol J 9:922–931

Hiremath PJ, Kumar A, Penmetsa RV et al (2012) Large-scale development of cost-effective SNP marker assays for diversity assessment and genetic mapping in chickpea and comparative mapping in legumes. Plant Biotechnol J 10:716–732

Hospital F (2009) Challenges for effective marker-assisted selection in plants. Genetica 136:303–310

Hovav R, Upadhyaya KC, Beharav A et al (2003) Major flowering time gene and polygene effects on chickpea seed weight. Plant Breeding 122:539–541

Hüttel B, Winter P, Weising K et al (1999) Sequence-tagged microsatellite site markers for chickpea (*Cicer arietinum* L.). Genome 42:210–217

Imtiaz M, Materne M, Hobson K et al (2008) Molecular genetic diversity and linked resistance to Ascochyta blight in Australian chickpea breeding materials and their wild relatives. Austral J Agr Res 59:554–560

Indurker S, Misra H, Eapen S (2010) *Agrobacterium*-mediated transformation in chickpea (*Cicer arietinum* L.) with an insecticidal protein gene: optimisation of different factors. Physiol Mol Biol Plants 16:273–284

Iruela M, Rubio J, Cubero JI et al (2002) Phylogenetic analysis in the genus *Cicer* and cultivated chickpea using RAPD and ISSR markers. Theor Appl Genet 104:643–651

Jain Rajesh M, Misra G, Patel RK et al (2013) A draft genome sequence of the pulse crop chickpea (*Cicer arietinum* L.). Plant J 74:715–729

Jaiswal P, Singh A, Kumar K et al (2004) Functional genomics of chickpea in response to ascochyta infection. 12th Plant and Animal Genome Conference. San Diego, USA, 10–15 January, 2014

Jambunathan R, Singh U (1981) Studies on desi and kabuli chickpea (*Cicer arietinum*) cultivars-3. Mineral and trace elements composition. J Agric Food Chem 29:1091–1093

Jhanwar S, Priya P, Garg R et al (2012) Transcriptome sequencing of wild chickpea as a rich resource for marker development. Plant Biotechnol J 10:690–702

Jimenez-Gasco MM, Navas-Cortes JA, Jimenez-Diaz RM (2004) The *Fusarium oxysporum* f. sp. *ciceri/Cicer arietinum* pathosystem: a case study of the evolutionof plant-pathogenic fungi into races and pathotypes. Int Microbiol 7:95–104

Johnson HW, Bernard RL (1962) Soybean genetic and breeding. Adv Agron 14:149–221

Jukanti AK, Gaur PM, Gowda CLL et al (2012) Nutritional quality and health benefits of chickpea (*Cicer arietinum* L.): a review. Brit J Nutr 108:S11–S26

Kanakala S, Verma HN, Vijay P et al (2013) Response of chickpea genotypes to agrobacterium-mediated delivery of chickpea chlorotic dwarf virus (CpCDV) genome and identification of resistance source. Appl Microbiol Biotechnol 97:9491–9501

Kantar F, Hafeez FY, Shivakumar BG et al (2007) Chickpea: rhizobium management and nitrogen fixation In: Yadav SS, Redden R, Chen W et al (eds) Chickpea breeding and management. CABI, Wallingford, pp 179–192

Kazan K, Muehlbauer FJ, Weeden NE et al (1993) Inheritance and linkage relationships of morphological and isozyme loci in chickpea (*Cicer arietinum* L.). Theor Appl Genet 86:417–426

Knights EJ, Southwell RJ, Schwinghamer MW et al (2008) Resistance to *Phytophthora medicaginis* Hansen and Maxwell in wild *Cicer* species and its use in breeding root rot resistant chickpea (*Cicer arietinum* L.). Austral J Agr Res 59:383–387

Krishnamurthy L, Gaur PM, Basu PS et al (2011) Large genetic variation for heat tolerance in the reference collection of chickpea (*Cicer arietinum* L.) germplasm. Plant Genet Resources 9:59–69

Kumar J, Rao BV (1996) Super early chickpea developed at ICRISAT Asia Center Int. Chickpea Pigeonpea Nwsl 3:17–18

Kumar J, Choudhary AK, Solanki RK et al (2011) Towards marker-assisted selection in pulses: a review. Plant Breeding 130:297–313

Ladizinsky G, Adler A (1976) The origin of chickpea *Cicer arietinum*. Euphytica 25:211–217

Lichtenzveig J, Scheuring C, Dodge J et al (2005) Construction of BAC and BIBAC libraries and their applications for generation of SSR markers for genome analysis of chickpea, *Cicer arietinum* L. Theor Appl Genet 110:492–510

Madrid E, Rubiales D, Moral A (2008) Mechanism and molecular markers associated with rust resistance in a chickpea interspecific cross (*Cicer arietinum* × *Cicer reticulatum*). Eur J Plant Pathol 121:43–53

Madrid E, Rajesh PN, Rubio J et al (2012) Characterization and genetic analysis of an EIN4-like sequence (*CaETR-1*) located in QTL_{AR1} implicated in Ascochyta blight resistance in chickpea. Plant Cell Rept 31:1033–1042

Madrid E, Chen W, Rajesh PN et al (2013) Allele-specific amplification for the detection of Ascochyta blight resistance in chickpea. Euphytica 189:183–190

Madrid E, Seoane P, Claros MG et al (2014) Genetic and physical mapping of the QTLAR3 controlling blight resistance in chickpea (*Cicer arietinum* L). Euphytica. doi:10.1007/s10681–014-1084–6

Malhotra RS, Singh KB, Vito M et al (2002) Registration of ILC 10765 and ILC 10766 chickpea germplasm lines resistant to cyst nematode. Crop Sci 42(5):1756

Malhotra RS, Bejiga G, Anbessa Y et al (2007) Registration of 'Ejere': a Kabuli chickpea cultivar. Crop Sci 1:112

Maliro MFA, McNeil D, Kollmorgen J et al (2004) Screening chickpea (*Cicer arietinum* L.) and wild relatives germplasm from diverse country sources for salt tolerance. In: Proc of the international crop science congress, Brisbane, Australia, 26–1 October

Manchanda G, Garg N (2008) Salinity and its effects on the functional biology of legumes. Acta Physiol Plant 30:595–618

Mantri NL, Ford R, Coram TE et al (2007) Transcriptional profiling of chickpea genes differentially regulated in response to high-salinity, cold and drought. BMC Genomics 8:303

Millan T, Madrid E, Imtiaz M et al (2013) Disease resistance in chickpea in translational genomics for crop breeding Volume 1-Biotic Stress. Wiley-Blackwell, Oxford, pp 201–209

Millán T, Winter P, Jungling R et al (2010) A consensus genetic map of chickpea (*Cicer arietinum* L.) based on 10 mapping populations. Euphytica 175:175–189

Molina C, Rotter B, Horres R et al (2008) SuperSAGE: the drought stress-responsive transcriptome of chickpea roots. BMC Genomics 9:553

Molina C, Zaman-Allah M, Khan F et al (2011) The salt-responsive transcriptome of chickpea roots and nodules via deepSuperSAGE. BMC Plant Biol 11:31

Moreno MT, Cubero JI (1978) Variation in *Cicer arietinum* L. Euphytica 27:465–485

Muehlbauer FJ, Singh KB (1987) Genetic of chickpea. In: Saxena MC, Singh KB (eds) The chickpea. CABI, Wallingford, pp 99–125

Nayak S, Zhu H, Varghese N et al (2010) Integration of novel SSR and gene-based SNP marker loci in the chickpea genetic map and establishment of new anchor points with *Medicago truncatula* genome. Theor Appl Genet 120:1415–1441

Pande S, Siddique KHM, Kishore GK et al (2005) Ascochyta blight of chickpea biology, pathogenicity and disease management. Aust J Agr Res 56:317–332

Pande S, Kishore GK, Upadhyaya HD et al (2006a) Identification of sources of multiple disease resistance in mini-core collection of chickpea. Plant Dis 90:1214–1218

Pande S, Ramgopal D, Kishore GK et al (2006b) Evaluation of wild *Cicer* species for resistance to Ascochyta blight and botrytis gray mold in controlled environment at ICRISAT, Patancheru, India. ICRISAT 2(1)

Pande S, Galloway J, Gaur PM et al (2006c) Botrytis grey mould of chickpea: a review of biology, epidemiology, and disease management. Austral J Agr Res 57:1137–1150

Pande S, Sharma M, Gaur PM et al (2010) Host plant resistance to Ascochyta blight of chickpea. Information Bulletin N°82. ICRISAT, Patancheru, p 40

Pundir RPS, Rao NK, Van Der Maesen LJG (1985) Distribution of qualitative traits in the world germplasm of chickpea (*Cicer arietinum* L.). Euphytica 34:697–3

Ragazzi A (1982) Un grave attacco di ruggine su foglie discese. (A serious attack of rust on *Cicer arietinum* leaves). Informatore Fitopatológico 2:41–43

Rajesh PN, Tekeoglu M, Gupta VS et al (2002) Molecular mapping and characterization of an RGA locus RGAPtokin1–2171 in chickpea. Euphytica 128:427–433

Ramgopal D, Srivastava RK, Pande S et al (2012). Introgression of Botrytis gray mold resistance gene from *Cicer reticulatum* (*bgmr1cr*) and *C. echinospermum* (*bgmr1ce*) to chickpea. Plant Genet Resources 11:212–216

Roman B, Satovicb Z, Alfaro C et al (2007) Host differentiation in *Orobanche foetida* Poir. Flora 202:201–208

Roy F, Boye JI, Simpson BK (2010) Bioactive proteins and peptides in pulse crops: pea, chickpea and lentil. Food Res Intl 43:432–442

Rubiales D, Sillero JC, Moreno MT (1999) Resistance to *Orobanche crenata* in chickpea. In: Cubero JI, Moreno MT, Rubiales D, Sillero JC (eds) Resistance to *orobanche*: the state of the art. Junta de Andalucia, Sevilla, pp 55–62

Rubiales D, Moreno I, Moreno MT et al (2001) Identification of partial resistance to chickpea rust (*Uromyces ciceris-arietini*). Proc 4th European Conference on Grain Legumes. Cracow, Poland, 8–12 July, 2001, pp 194–195

Rubio J, Hajj-Moussa E, Kharrat M et al (2003) Two genes and linked RAPD markers involved in resistance to *Fusarium oxysporum* f. sp. *ciceris* race 0 in chickpea. Plant Breed 122:188–191

Rubio J, Flores F, Moreno MT et al (2004) Effects of the erect/bushy habit, single/double pod and late/early flowering genes on yield and seed size and their stability in chickpea. Field Crops Res 90:255–262

Rubio J, Moreno MT, Moral A (2006) Registration of RIL58-ILC72/Cr5, a chickpea germplasm line with rust and Ascochyta blight resistance. Crop Sci 46:2331–2332

Rubio J, Gil J, Cobos MJ et al (2009) Chickpea In: Perez de la Vega M, Torres AM, Cubero JI, Kole Ch (eds) Genetics, genomics and breeding of cool season grain legumes. Science, Enfield, pp 205–236

Salimath PM, Toker C, Sandhu JS et al (2007) Conventional breeding methods In: Yadav SS, Redden R, Chen W, Sharma B (eds) Chickpea breeding and management. CABI, Wallingford, pp 369–390

Saxena NP, Krishnamurthy L, Johansen C (1993) Registration of drought-resistant chickpea germplasm. Crop Sci 33:1424

Sethy NK, Shokeen B, Bhatia S (2003) Isolation and characterization of sequence-tagged microsatellite sites markers in chickpea (*Cicer arietinum* L.). Mol Ecol Notes 3:428–430

Sethy N, Shokeen B, Edwards K et al (2006) Development of microsatellite markers and analysis of intraspecific genetic variability in chickpea (*Cicer arietinum* L.). Theor Appl Genet 112:1416–1428

Sharma KD, Muehlbauer FJ (2007) Fusarium wilt of chickpea: physiological specialization, genetics of resistance and resistance gene tagging. Euphytica 157:1–14

Siddique KHM, Krishnamurthy L (2014) Chickpea production technology. Legume Perspect 3:29–32

Siddique KHM, Loss SP, Thomson BD (2003) Cool season grain legumes in dryland Mediterranean environments of Western Australia: significance of early flowering. In: Saxena NP (ed) Management of agricultural drought. Science, Enfield, pp 151–161

Singh G, Chen W, Rubiales D et al (2007) Diseases and their management. In: Yadav SS, Redden R, Chen W, Sharma B (eds) Chickpea breeding and management. CABI, Wallingford, pp 497–519

Singh KB (1987) Chickpea breeding. In: Saxena MC, Singh KB (eds) The chickpea. CABI, Wallingford, pp 127–162

Singh KB, Reddy MV (1996) Improving chickpea yield by incorporating resistance to Ascochyta blight. Theor Appl Genet 92:509–515

Singh NP, Shiv S, Iquebal MA et al (2009) Improved varieties of chickpea in India. Technical Bulletin, IIPR, Kanpur

Singh R, Sharma P, Varshney RK et al (2008) Chickpea improvement: role of wild species and genetic markers. Biotechnol Genet Eng 25:267–314

Singh S, Gumber RK, Joshi N et al (2005) Introgression from wild *Cicer reticulatum* to cultivated chickpea for productivity and disease resistance. Plant Breeding 124:477–480

Singh U, Subrahmanyam N, Kumar J (1991) Cooking quality and nutritional attributes of some newly developed cultivars of chickpea. J Food Sci Agr 55:37–46

Stephens A, Lombardi M, Cogan NI et al (2014) Genetic marker discovery, intraspecific linkage map construction and quantitative trait locus analysis of Ascochyta blight resistance in chickpea (*Cicer arietinum* L.). Mol Breeding 33:297–313

Taran B, Warkentin T, Tullu A et al (2007) Genetic mapping of Ascochyta blight resistance in chickpea (*Cicer arietinum* L.) using an SSR linkage map. Genome 50:26–34

Taran B, Warkentin TD, Vandenberg A (2013) Fast track genetic improvement of Ascochyta blight resistance and double podding in chickpea by marker-assisted backcrossing. Theor Appl Genet 126:1639–1647

Tekeoglu M, Rajesh PN, Muehlbauer FJ (2002) Integration of sequence tagged microsatellite sites to the chickpea genetic map. Theor Appl Genet 105:847–854

Thudi M, Bohra A, Nayak SN et al (2011) Novel SSR markers from BAC-end sequences, DArT arrays and a comprehensive genetic map with 1,291 marker loci for chickpea (*Cicer arietinum* L.). PLoS ONE 6(11):e27275. doi:10.1371/journalpone.0027275

Udupa S, Weigand MF, Saxena MC et al (1998) Genotyping with RAPD and microsatellite markers resolves pathotype diversity in the Ascochyta blight pathogen in chickpea. Theor Appl Genet 97:299–307

Upadhyaya HD, Ortiz R (2001) A minicore subset for capturing diversity and promoting utilization of chickpea genetic resources in crop improvement. Theor Appl Genet 102:1292–1298

Upadhyaya HD, Bramel PJ, Singh S (2001) Development of a chickpea core subset using geographic distribution and quantitative traits. Crop Sci 41:206–210

Upadhyaya HD, Dwivedi SL, Baum M et al (2008a) Genetic structure, diversity and allelic richness in composite collection and reference set in chickpea (*Cicer arietinum* L.). BMC Plant Biol 8:106–118

Upadhyaya HD, Gowda CLL, Sastry DVSSR (2008b) Management of germplasm collection and enhancing their use by minicore and molecular approaches. APEC-ATCWG Workshop. Capacity building for risk management systems on genetic resources, pp 35–70

Upadhyaya HD, Thudi M, Dronavalli N et al (2011) Genomic tools and germplasm diversity for chickpea improvement. Plant Gen Resour Char Util 9:45–48

Van der Maesen LJG, Maxted N, Javadi F et al (2007) Taxonomy of the genus *Cicer* revisited. In: Yadav SS, Redden R, Chen W, Sharma B (eds) Chickpea breeding and management. CABI, Wallingford, pp 14–46

Van Gastel AJG, Bishaw Z, Niane AA, Gregg BR, Gan Y (2007) Chickpea seed production. In: Yadav SS, Redden R, Chen W, Sharma B (eds) Chickpea breeding and management. CABI, Wallingford, pp 417–444

Varshney RK, Song C, Saxena RK et al (2013) Draft genome sequence of chickpea (*Cicer arietinum*) provides a resource for trait improvement. Nat Biotechnol 31:240–248

Varshney RK, Mohan SM, Gaur PM et al (2014a) Marker-assisted backcrossing to introgress resistance to Fusarium wilt race 1 and Ascochyta blight in C214, an elite cultivar of chickpea. Plant Gen 7:1–11

Varshney RK, Thudi M, Nayak S et al (2014b) Genetic dissection of drought tolerance in chickpea (*Cicer arietinum* L.). Theor Appl Genet 127:445–462

Vláčilová K, Ohri D, Vrána J et al (2002) Development of flow cytogenetics and physical genome mapping in chickpea (*Cicer arietinum* L.). Chrom Res 10:695–706

Winter P, Pfaff T, Udupa SM et al (1999) Characterization and mapping of sequence-tagged microsatellite sites in the chickpea (*Cicer arietinum* L.) genome. Mol Gen Genet 262:90–101

Winter P, Benko-Iseppon AM, Hüttel B et al (2000) A linkage map of the chickpea (*Cicer arietinum* L.) genome based on recombinant inbred lines from a *C. arietinum* × *C. reticulatum* cross: localization of resistance genes for Fusarium wilt races 4 and 5. Theor Appl Genet 101:1155–1163

Zatloukalová P, Hřibová E, Kubaláková M et al (2011) Integration of genetic and physical maps of the chickpea (*Cicer arietinum* L.) genome using flow-sorted chromosomes. Chromosome Res 19:729–739

Zhu H, Choi HK, Cook DR et al (2005) Bridging model and crop legumes through comparative genomics. Plant Physiol 137:1189–1196

Zohary D, Hopf M, Weiss E (2012) Domestication of plants in the old world, 4th edn. Oxford University Press, Oxford

Chapter 4
Lentil

Thomas R. Stefaniak and Kevin E. McPhee

1 Introduction

Lentil (*Lens culinaris* Medik.) is one of the first, if not the very first, domesticated grain crop. If the beginning of agriculture define humanities transition from wandering hunter-gatherers to members of a civilization, then lentil shares credit with only a few other species for making this transition possible. Archeologists found carbonized remains of lentil seed along with einkorn, emmer, and barley; suggesting that rotation with cereals was a feature of agricultural systems from the very beginning of early agriculture. The domestication of lentil and the development of agriculture made it possible for humans to become sedentary and turn from hunt and gathering to agricultural production.

Lentil plants are indeterminate semiprostrate members of the Leguminosae or Fabaceae family. Legume etymology is probably from the Latin word *legere,* which means to gather. Lentil leaflets are pinnately compound with one to eight pairs of leaflets. The older leaves terminate with a prehensile tendril. Lentil plants can have one to many primary branches depending on genotype and population density. Lentil flowers can be white, pink, and purple to pale blue in color. The fruit of the lentil plant is a pod, which is a defining feature of all legumes. The lentil pod generally contains two round convex lens-shaped seeds. Indeed, the English word *lens,* comes from the Latin word for this plant: *lens, lentis,* and *lentil.* Lentil roots form a slender taproot system that can range from shallow with many branching to deep with little branching depending on genotype and growing conditions. As

T. R. Stefaniak (✉)
North Central Research Extension Center, North Dakota State University,
5400 Highway 83 S, Minot, ND 58701, USA
e-mail: Thomas.stefaniak@ndsu.edu

K. E. McPhee
Department of Plant Sciences, North Dakota State University,
370G Loftsgard Hall, P.O. Box 6050, Fargo, ND 58108, USA
e-mail: kevin.mcphee@ndsu.edu

A. M. De Ron (ed.), *Grain Legumes,* Handbook of Plant Breeding 10,
DOI 10.1007/978-1-4939-2797-5_4

is the case with all legumes, the roots form a symbiosis with rhizobacteria that fix atmospheric nitrogen into the soil. This trait allows the plant to grow on nitrogen-poor soils.

Lentil seed is an excellent source of energy, dietary protein, fiber, and micronutrients. Lentil is similar to the other pulses in total energy content and lower than soybean (Urbano et al. 2007). However, this difference with soybean is due to the substantially higher fat content of soybean (Gebhardt and Thomas 2002). Protein content in lentil is the highest among food legumes consumed without industrial processing (Sharma 2009). Protein in lentil can range from 21.4 to 25.5 % (Rasheed et al. 2010). The necessity of combining lentil or other legumes with cereals in a vegetarian diet to promote good human health was recognized thousands of years ago. Cereals lack adequate amounts of the essential amino acid lysine. The amino acid profile of lentil has favorable leucine/isoleucine and leucine/lysine ratios ranking lentil highly as a vegetable protein source (Fernandez et al. 1996; Nestares et al. 1996). The protein profile as well as the high vitamin and mineral content of lentil complements that of cereals in a vegetarian diet (Urbano et al. 2007).

Lentil grain is harvested after the crop has reached physiological maturity resulting in hard unpalatable seed. Eating uncooked or undercooked mature lentil seeds in sufficient quantities will cause illness in monogastric animals including humans, as is the case with all the other major legume crop species. This is because lentil contains anti-nutritional constituents including phytic acid, trypsin inhibitors, and various tannins among others. These constituents are inactivated with cooking or by germinating the seeds. One of the traits highly favored in lentil is the rapidity with which it can be cooked relative to other whole grain legumes.

Due to a combination of socioeconomic status, cultural/religious mores, and climate, the region of the globe that consumes the most lentils is south central to southwest Asia. Diets in this region are largely vegetarian. Whole lentil seed is usually prepared in soups, stews, or porridge-like dishes. Its popularity is also due to the fact that lentil softens with cooking even without presoaking. Cooked lentil is also puréed and served in porridge like stew such as dahl, which is a typical dish of India. Many other Asian and North African countries have their own version of dahl such as Turkey's Kirmizi Mercimek Corbasi (Yadav et al. 2007). In Ethiopia lentil and wheat flour are moistened and baked into a type of bread called Sambusa (Yadav et al. 2007). In northern India, lentil flour is sometimes baked with wheat in the traditional bread naan. The Egyptians boil lentils in water, drain, and serve alone or tossed with a variety of sauces. In North Africa, lentils are combined in soups with chickpeas and white dry bean. More recently, lentils have become popular and served as sprouts in salads especially in Western vegetarian diets.

Lentil production in the developing world is grown mostly for subsistence and not for export. The organization Consultative Group on International Agricultural Research (CGIAR) reported that about 70 % of lentils produced are consumed in the country in which they were grown (CGIAR 2014). Many of the varieties are landraces that are potentially quite ancient and do not respond well to external inputs. India traditionally produces the greatest amount of lentils; however, in 2005 Canada surpassed India and accounted for 12–30 % of global production (NDSU extension 2006). India, Turkey, Ethiopia, China, Syria, Iran, the USA, Canada, and Bangladesh are the top lentil-producing countries (CGIAR 2014).

Many of the most populated regions of the developing world are dependent on lentil production and consumption primarily for two reasons. First, lentil is well suited to production systems in these areas because the crop has evolved with relatively little human influence to be productive on marginal land. Like all legumes, lentil fixes atmospheric nitrogen into a form that makes it available to the crop. This makes the application of costly nitrogen fertilizer unnecessary. Second, just as lentil well compliments cereals in a human diet, they also do so in a crop rotation by reducing populations of insects and pathogens, and by leaving nitrogen in the soil.

The largest lentil-importing nations between 2000 and 2005 were in the developing world and include India, Bangladesh, Egypt, Colombia, Algeria, and Sri Lanka (Skyrpetz 2006). India is both a major importer and producer of lentil. Dependence on imported lentil in the developing world is due to a combination of a high human population to arable land ratio, and the low yield potential of the landraces of lentil grown in these areas.

By far, the largest lentil exporter in the world is Canada. The United Nations Food and Agriculture Organization (FAO) reported that 31.7 % of annual production is exported and that the top five exporters were Canada, India, Turkey, Australia, and the USA between 2001 and 2006 (FAO 2008). Indeed, collectively these top five exporters account for over 80 % of global export. North America and Australia are able to lead lentil exports due to a combination of a low human population to arable land ratio, and the higher yield potential of the cultivars of lentil which have improved response to inputs that are grown in these areas. Additionally, lentil consumption is much lower in North America and Australia when measured as availability for consumption.

While it is true that producers in developing regions do not have access to the level of input resources that their counterparts in wealthier nations do, much effort has and is being expended on improving varieties and cultural practices. The need for these improvements is critical as can be seen in the fact that yields in countries such as India, yield has remained stagnant for much of the past four decades while increases in cereal yields have been steady (McNeil et al. 2007). The International Center for Agricultural Research in Dry Areas (ICARDA), which is headquartered in Lebanon, is an organization that is engaged in extensive efforts to improve lentil production. Efforts in the area of breeding and genomics include projects to assemble and genotype a reference collection for association mapping to improve stress tolerances, as well as biotechnological approaches (ICARDA 2013).

In the past two decades, lentil production has increased substantially in developed countries, especially in Canada and the USA. Concurrently, research efforts have been ramped up to increase productivity. The need for this is further indicated in the fact that the increase in lentil harvested has been largely due to increased acreage, and not increasing yields over the past 20 years (McNeil et al. 2007). In Canada, considerable resources are devoted to lentil improvement. The primary player here is the Crop Development Center (CDC) at the University of Saskatchewan. In addition to developing superior lines for increased productivity, the CDC is heavily involved in improving nutritional traits in their breeding programs.

In the USA, the vast majority of lentil production is in the northern Midwest (Montana and North Dakota) over to the Palouse region of the Pacific Northwest. Consequently, the primary universities involved in breeding and production research are North Dakota State University (NDSU), the Montana State University (MSU), the University of Idaho, and Washington State University (WSU). Research in Canada and the USA has different priorities that address the use of greater inputs available to producers in these regions. Because lentil breeding in North America is relatively young compared to crops such as wheat and maize, considerable opportunities exist to better adapt lentil to North America. Specifically, positive responses to improved fertility, better water management, and chemical control of pests are all goals of Canadian and American breeding programs.

2 Origin and Systematics

2.1 Prehistorical Data

The domestication of lentil (*L. culinaris* Medik.) is as old as agriculture and indeed human civilization itself. This domestication event was possible because the two ingredients necessary for humanities transition from hunter-gatherer, to farmer came together in time and space. Those two ingredients were a favorable growing season/region, and plant species amenable to cultivation. In the Fertile Crescent region, reliable sources of moisture, good soil fertility, and species with potential for domestication were all available to the Neolithic tribes of the region.

The oldest evidence of the consumption of lentil traces back to 10,000 BC in Greece where carbonized remains have been found (Sandhu and Singh 2007). It is uncertain whether these seeds were gathered or harvested from cultivated plots as this predates when archeologists believe humans began farming around 8000 BC (Brown et al. 2009). Additionally, these lentil seeds were small and indistinguishable from seed from wild *Lens* species (Sandhu and Sing 2007).

In Israel lentil seed dating to 6800 BC were discovered that appeared to have been stored in large quantities, indicating that seed stock was being preserved for the purpose of future cultivation (Sandhu and Sing 2007). Further archeological evidence that lentil was being cultivated and traded around this time comes from carbonized lentil seeds from the relatively distant locations of Tell Ramad in Syria (6250–5950 BC), Beidha in Jordan, Hacilar in Turkey (5800–5000 BC), and Sabz in Iran (5500–5000 BC) (Van Zeist and Bottema 1971; Helbeck 1959, 1963, 1970) These discoveries represent the oldest evidence that humans had begun farming.

2.2 *Historical Data and Literature*

As is the case in modern times, legumes were regarded differently than cereals in the ancient world. Then as now, legumes are described mostly in terms of their nutritional value and not their delectability. Pliny describes the growing of lentils from seed and its varieties. He mentions its medicinal properties and a variety of ways of boiling or otherwise cooking lentils for various remedies (Wright 2001). They are seen as a poor man's source of protein when he cannot afford meat. This is seen in the fact that mention of legumes, including lentil, are absent from many of the accounts by ancient historians (Flint-Hamilton 1999). Indeed, it was not recommended that farmers produce a surplus of legumes as they are of little value for trade. However, though legumes generally receive little attention in ancient literature, the one that is probably mentioned to the greatest degree is lentil.

An early mention of lentil from the fifth century BC comes from the comic playwright Aristophanes (Flint-Hamilton 1999). He wrote in *Plutus* that Chremulos has no further need of lentils because he has found wealth (Flint-Hamilton 1999). Later in the second century, Athenaeus, in *Deipnosophistae,* uses the description of a meal of lentil and a foul-smelling perch to indicate the unsophistication of the hosts.

In the Judean/Christian tradition lentil is also mentioned. In Genesis (34), where Esau sells his birthright for a bowl of lentils. The passage states, "Then Jacob gave Esau bread and pottage of lentils; and he did eat and drink, and rose up, and went his way: thus Esau despised his birthright." In Ezekial 4:9, it is written, "And you, take wheat and barley, beans and lentils, millet and emmer, and put them into a single vessel and make your bread from them." 2 Samuel 17:28–29 says, "Brought beds, basins, and earthen vessels, wheat, barley, flour, parched grain, beans and lentils, honey and curds and sheep and cheese from the herd, for David and the people with him to eat, for they said, the people are hungry and weary and thirsty in the wilderness."

3 Origin and Domestication

Lentil is believed to have been domesticated in the Fertile Crescent region of the Middle East in what is present day Iraq. Indeed, archeological evidence confirms the presence of lentil as far back as 8500–6000 BC in the Turkey/Syria/Iraq region (Yadav et al. 2007). The first agricultural revolution transformed human societies in this region from hunter-gatherers to agriculturalists. As is true in modern times, people generally have preferred to eat animals over plants. Animal husbandry was probably the impetus to the raising of crops. This is because the raising of livestock in this hard-scrabble environment was relatively inefficient. Even in modern times, food from animals requires more energy input than production of food from plants. Lentil was among the first crops because not only could it be used as animal fodder but it also could be a good source of protein when meat was scarce.

As previously discussed, the nutritive aspects of lentil were important in influencing Neolithic man to begin its cultivation. Lentils were among the grain legumes consumed as an alternate protein source to animals that were not in great supply. Lentil was also identified as a crop that can improve the health of the soil. Though they did not know why, Greek and Roman farmers noticed that it could be grown in fields depleted by the more valued cereal grains. Additionally, ancient agronomists observed that the seed from a desirable plant tended to produce a plant that was very similar to the plant from whence the seed was collected. The ability of lentil to fix atmospheric nitrogen and its self-pollinating habit made it an attractive wild species for domestication by ancient agronomists. For all these reasons, from the very beginning of the agricultural revolution, lentil was an important part of cropping systems, along with wheat, barley, pea, flax, emmer wheat, and einkorn.

Because lentil is self-pollinated, domestication first took the form of selecting desirable plants and retaining their seed for the next growing season. In its most basic terms, a species is domesticated when its seed is easily established, and its harvested tissues can be collected efficiently. As is the case with most if not all crop species, seed dormancy was the primary phenotype to be improved. Along with dormancy, the removal of pod dehiscence was required. In lentil domestication, seed size was also important as it relates to ease of harvest and yield. Indeed, a major phenotypic difference between the cultigen *L. culinaris* and its wild relatives is seed size. The other traits that most obviously separate cultivated lentil from its wild relatives include greater leaf area, leaf number, plant height, number of flowers, larger pods, longer rachis, and shorter peduncle.

3.1 The Spread of Lentil

Lentil cultivation spread with the spread of agriculture starting in Southwest Asia and fanning out from there to Greece, Central and Western Europe via the Danube, south through Africa along the Nile, and eastward to India (Harlan 1992). This spread was rapid as can be evidenced from the fact that archeologists have found the remains of lentil as far west as the eastern Iberian Peninsula by about 5450 BC (Cubero et al. 2009).

Due to an environment not conducive to the preservation of botanical remains, it is unclear when lentil reached the Nile delta, though it was no doubt early in its domestication, because of this regions close proximity to the center of domestication for lentil. However, further up the Nile in the tomb of the twelfth dynasty (2400–2200 BC) remains of lentils have been recovered (Erskine et al. 2009).

Lentil reached the Indian subcontinent around 2000 BC (Cubero et al. 2009). Tradesmen likely following what would later become the silk road brought lentil to what now is Uzbekistan, Kyrgyzstan, Pakistan, Afghanistan, and into India itself as part of an Indo-European invasion. Evidence suggests that this introduction represented very little genetic diversity as seen in the lack of variability in the local landraces still in use based on molecular evidence. This is surprising when it is considered that India quickly became, and remains the largest lentil-growing region in the world.

4 Taxonomy

4.1 The Genus Lens

Cultivated lentil *(L. culinaris)* belongs to the genus *Lens* and early researchers used comparisons of morphology and "crossability" to connect or separate putatively distinct individuals. Molecular genetic techniques made possible the clustering of individuals based on actual DNA sequences and many early classifications were modified. A detailed description of the changes in *Lens* classification over the years is beyond the scope of this chapter. In summary, species included in the genus went from four in 1979; *L. culinaris, L. orientalis, L. nigrican,* and *L. ervoides* (Ladizinsky 1979), to two in 1984; *L. culinaris* (subspecies *culinaris, orientalis,* and *odemensis*), and *L. nigricans* (subspecies *nigricans* and *ervoides*), back to four in 2000; *L. culinaris* (subspecies *culinaris, orientalis, odemensis,* and *tomentosis*), *L. nigricans, L. ervoides,* and *L. lamottei.* The most recent classification scheme also established in 2000 contains six species and is as follows: *L. culinaris* (subspecies *culinaris* and *orientalis*), *L. odemensis, L. tomentosis, L. nigricans, L. ervoides,* and *L. lamottei* (Cubero et al. 2009).

4.2 Biological Species

When new specimens are identified, breeders are interested in whether they can be crossed and give rise to viable progeny and molecular characteristics are secondary. Experiments attempting to establish reproductive barriers between putative species and subspecies can further complicate classification. However, for breeding progress, because the cultigen of lentil is *L. culinaris* ssp. *culinaris;* what can be crossed to it represents the most readily available reservoir of genetic variability for breeding improvement.

Cultivated lentil was initially subdivided into two subspecies macrosperma (seed diameter 6–9 mm) and microsperma (seed diameter 2–6 mm) based on seed size by Barulina in 1930 (Sandhu and Singh 2007). Indeed, most if not all breeding progress has been made by utilizing alleles from accessions that fit into these classifications. The macrospermas have yellow cotyledons and little or no pigment in their flowers. The microspermas have red, orange, or yellow cotyledons and pigmented flowers.

More recently, Ladizinsky (1979) reported that the cultigen *L. culinaris* ssp. *culinaris* and *L. orientalis* share a common karyotype and cross freely with one another. This is why *L. orientalis* is now placed firmly as a subspecies of *L. culinaris*. *L. nigricans* has a slightly different karyotype and different accessions within that species cross with varying degree of success with *L. culinaris* ssp. *culinaris* (Balyan et al. 2002). As a result, those *L. nigricans* accessions that do cross with *L. culinaris* were reclassified as *L. culinaris* ssp. *odemensis* (Ladizinsky et al. 1984). Hybridization experiments tended to verify or reassign accessions from subspecies of a species to separate species and vice versa.

In conclusion, hybridization experiments have supported the most recent classification scheme of six species within the *Lens* genus (Cubero et al. 2009). In terms of gene pool levels, *L. orientalis* clearly belongs in the primary gene pool of *L. culinaris* and is one of its subspecies. *L. odemensis,* because viable progeny can be obtained through embryo rescue, is in the secondary gene pool of *L. culinaris* and remains a separate species. *L. tomentosis, L. nigricans, L. ervoides,* and *L. lamottei* belong to secondary or tertiary gene pools due to the requirement of embryo rescue to obtain hybrids.

5 Varietal Groups

Lentil is composed of several varietal groups or market classes based on testa color and cotyledon color. The color classes red and green each have within them three relative sizes. The seed sizes of red lentils place them in the smaller seeded microsperma botanical category and are graded in the USA as extra small, small, and medium. The sizes of green lentils originating from macrosperma types are graded as small, medium, and large. The cultivar CDC Lemay is an example of the specialty small-seeded type French Green. Large-seeded types range in millimeters of diameter from 6 to 9, medium from 5 to 6, and small from 3 to 5, with extra small being up to 3. Much of the production of large green lentils is in North America, and principal importers of large green types include Colombia and Western Europe (Saskatchewan Ministry of Agriculture 2013). Pardina or Spanish brown types have a brown speckled testa with yellow cotyledons (Fig. 4.1). Red lentils have traditionally been produced and consumed in India and the Middle East. More recently, the USA, Canada, and Australia have become major producers of red lentils (Muehlbauer et al. 2009).

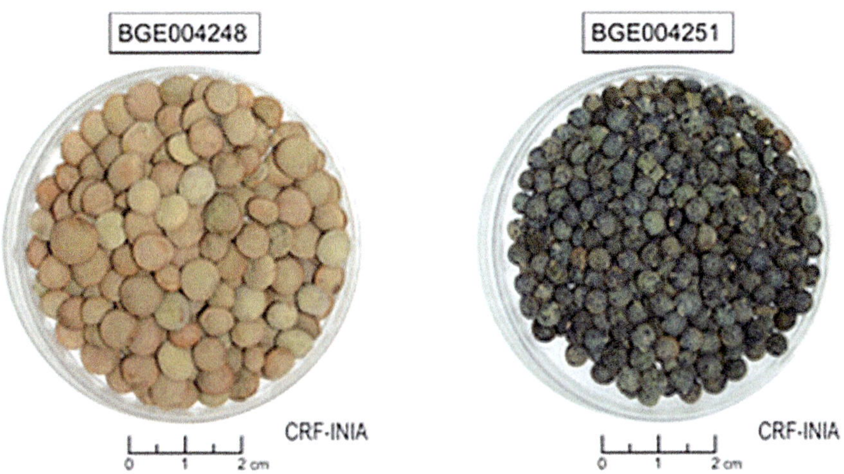

Fig. 4.1 'Armuñesa' *(left)* and 'verdina' *(small green, right)* lentil landraces. (Courtesy of the gene bank CRF-INIA, Alcalá de Henares, Spain)

6 Genetic Resources

Preserving and exploiting genetic diversity is of paramount concern to lentil breeders since genetic variation is required for breeding progress. Lentil is a relatively small-seeded grain species so ex situ storage of seed does not require unreasonably large storage facilities. The ICARDA, headquartered in Beirut, Lebanon, is well positioned geographically and culturally to be the primary institute for research on lentil improvement and; therefore, maintains the world collection of over 10,000 accessions of *Lens*. Included in these are almost 9000 accessions of cultivated *Lens* from 70 different countries representing four major geographic regions, 1373 ICARDA breeding lines, and 583 wild *Lens* taxa from 24 different countries (Furman et al. 2009). Most of these were obtained on collection missions conducted by ICARDA personnel. A large proportion of these are from Southwest Asia and North Africa, because as is the case with most species the greatest genetic variation can be found in or near the center of origin and domestication. Interestingly, heterogeneities within families descended from individuals from some landraces have resulted in significant phenotypic variation (Erskine and Choudhary 1986). These results suggest that useful genetic variation can be obtained from within particular landraces.

Four other institutes house substantial collections. They are the Australian Temperate Field Crops Collection (ATFCC) in Victoria, Australia, with 5250 accessions (Redden et al. 2007), the United States Department of Agriculture (USDA) in Pullman, USA, with 3011 accessions (USDA 2013); the N.I. Vavilov All-Russian Research Institute of Plant Husbandry (VIR) in St. Petesbourg, Russia, with 2687 accessions (VIR 2013), and the National Bureau of Plant Genetic Resources in New Delhi, India with 2212 accessions (Dwivedi et al. 2006). The genetic resources available to lentil researchers are substantial and would be unwieldy to utilize for breeding or genomic research in their entirety.

In order to make efficient use of the worldwide collections for improvement core collections have been assembled. These core collections are a subsample of an entire collection and represent the overall genetic, as well as agroclimatological diversity of lentil. Assembly of these collections have relied on efforts to cross-reference accession identifications with phenotypic and passport data. Maintaining passport data is simply an exercise in faithfully recording locations where accessions were obtained. However, those data are of little use without accurate characterization of phenotypes.

Comparison among accessions is most easily accomplished by assembling a panel including individuals from differing regions, growing them in a uniform environment, and recording their responses. Erskine et al. (1985) measured nine quantitative traits in 615 accessions from 13 lentil-growing regions. They found that seed size, pod height, and time to flowering were useful in grouping accessions into regional groups.

Studies of the magnitude of variability within regional collections have been conducted and are indicative of patterns of diversity. These studies can be regarded as representative of regions included in the larger collection held by ICARDA.

Sindhu and Mirsha (1982) evaluated 30 lines from centers in the All India Coordinated Pulse Improvement Program. They predicted substantial genetic advance for eight agronomic traits including grain yield. Ramgiry et al. (1989) evaluated the same agronomic traits in 21 Indian accessions and reported similarly high genetic variances. Lakhani et al. (1986) reported substantial heritabilities for germination traits using 100 Indian genotypes. They concluded that sufficient genetic variation existed in their sample to improve germination rates. Contradictory results were reported in an evaluation of 78 genotypes collected from around Bangladesh; where very little phenotypic variability was measured for five of six agronomic traits (Sarwar et al. 1982). Baidya et al. (1988) reported highly significant differences between 96 genotypes collected in Bangladesh, for days to flowering, plant height, seed weight per plot, and plant dry weight. It should be noted that in all the above experiments the phenotypic variances were substantially greater than the genotypic variances suggesting that environmental conditions could greatly hamper selection for the respective traits in the field.

In order to characterize their usefulness, genotypes in the ICARDA collection have been screened for particular tolerances such as drought (Sarker et al. 2005; Erskine et al. 1985), chilling (Erskine et al. 1981), and responses to regional pathogen populations (Erskine and Witcombe 1984). A seminal report published by ICARDA called the Lentil Germplasm Catalog is a summary of phenotypic data. Much of the overall genetic diversity is accounted for in the landraces and wild relatives collected near and within lentil's region origin. Additionally, breeding lines and germplasm that were derived in disparate climatic regions were included in order to ensure alleles for tolerance to a wide array of stresses were sampled. Genetic and statistical analyses that included hierarchical cluster analysis of agronomic traits and two step cluster analyses of agroclimatological data linked to geographical locations of collection were used to ensure diversity (Furman 2006).

7 Major Breeding Achievements

7.1 *Yield Potential*

The first concept that should be made clear in the beginning of any discussion of plant breeding is that the most important trait is yield for agronomic crops. Many treatments on plant breeding discuss tolerance to disease, extreme weather, insect predation, etc. Improvement of these traits is for the purpose of maintaining or increasing yield. This is true even in the case of quality traits, that is, breeders are trying to improve iron content in lentil. If lines are developed that have higher iron content with no grain yield penalty, then harvesting more iron per unit of crop area has been achieved. Lentil production has increased dramatically over the past 40 years (Fig. 4.2a, 4.2b, 4.2c). This increase is partly due to increased acreage and partly due to increased yield (Fig. 4.2a, 4.2b). Conventional wisdom holds that

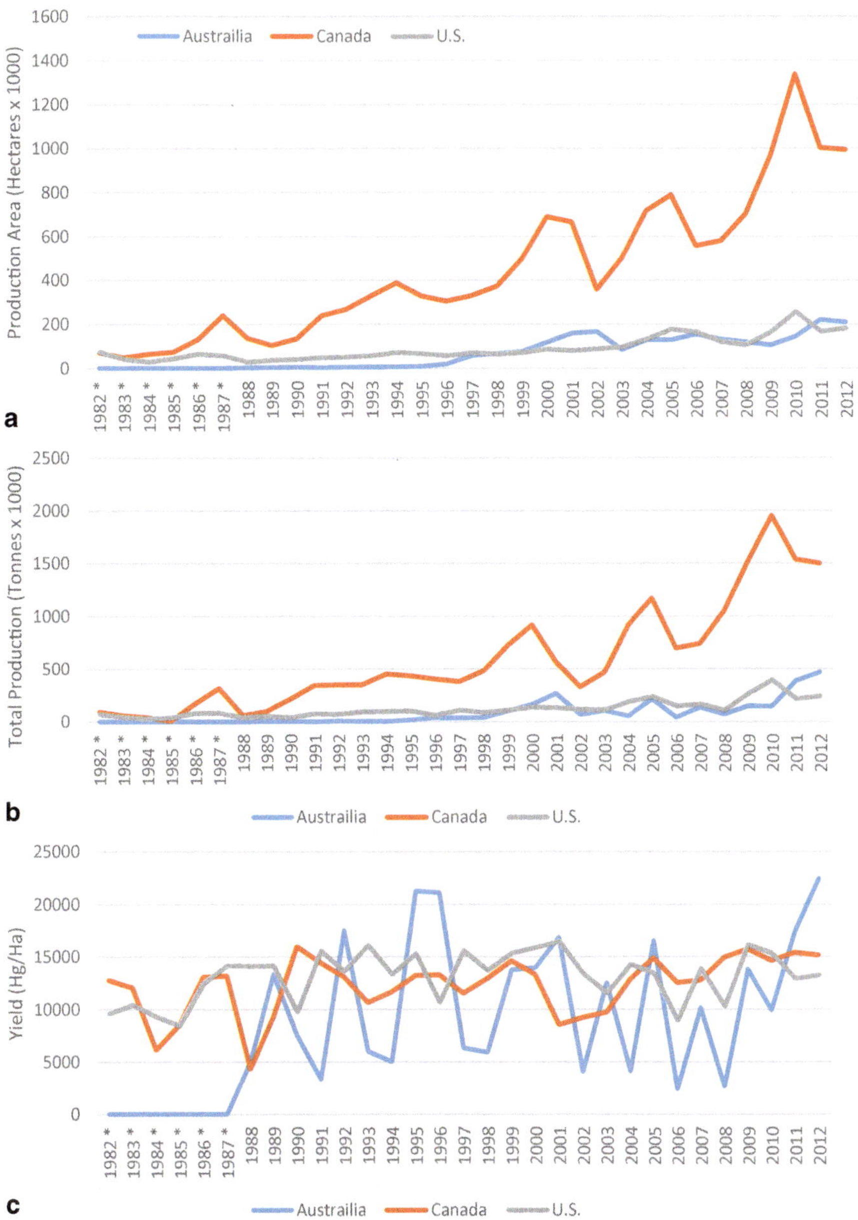

Fig. 4.2 Lentil production area (**a**), yield (**b**), and (**c**) total lentil production between 1982 and 2012 in three of the top producing developed countries. *Data for Australia are not available and assumed to be below 1000. (Adapted from http://faostat.fao.org/site/567/DesktopDefault.aspx?PageID=567#ancor)

about 50% of yield increases are due to breeding efforts and 50% are due to improved management practices.

Production has increased in some countries of the developing world and remained flat or even decreased in others over the past 40 years (Fig. 4.3). Yields have been flat or very erratic in developing world where domestic consumption is heavily reliant on local production. In this time period, India has at times been the top exporter and increased production has been solely a function of increased production area. Clearly, much opportunity exists for breeding improvement of Indian cultivars.

7.2 Genetic Variability Bottleneck

The history of lentil breeding surprisingly short given its ancient domestication. Before modern breeding techniques were developed, lentil production in South Asia was severely constrained by the unavailability of genotypes that possessed appropriate flowering and maturity traits. At this time, cultivars with little variability were available to farmers and breeders, which limited potential genetic improvement for a whole host of stress traits. Cultivars from this region are designated *pilosae* and have superficial characteristics that differentiate them from nearly all other lentils in other regions (Barulina 1930). When lentil was first brought to India by the Sanskrit-speaking race, it was virtually unknown to the local inhabitants (Cubero 1981). Attempts to grow this crop in a new and different environment placed substantial selection pressure, primarily for flowering time and maturity on the introduced landraces creating a genetic "bottleneck." This bottleneck was the impetus for major breeding efforts for the India subcontinent.

In the 1980s, ICARDA researchers began addressing the bottleneck by introducing several cultivars from Western Asia into India and Pakistan. However, it was observed that when these Western cultivars were only flowering, the indigenous genotypes were nearing maturity (Ceccarelli et al. 1994). Therefore, attempts at breaking the bottleneck by the simplest means, that is, introduction, has only been possible with Western lines having similar phenology. Linkage between maturity alleles and other important agronomic traits have hampered efforts to broaden genetic diversity in lentil. It is also important to note that tolerance to most if not all stresses is inherited quantitatively. Maturity is also a polygenic trait. This makes introgression of appropriate maturity alleles into germplasm with an array of stress tolerances challenging.

Effort by researchers at ICARDA beginning in 1981 to minimize the bottleneck effect focused on hybridizing later maturing West Asian lines with indigenous lines in order to introgress favorable maturity alleles into exotic germplasm (Erskine et al. 1998). These efforts represent major achievements for lentil breeders of the subcontinent because they allow for selections to be made in local nurseries without the need to manipulate photoperiod. High-yielding early-maturing cultivars that have been released include Barimasur-2 (Sarker et al. 1999a), and Barimasur-4 in Bangladesh (Sarker et al. 1999b).

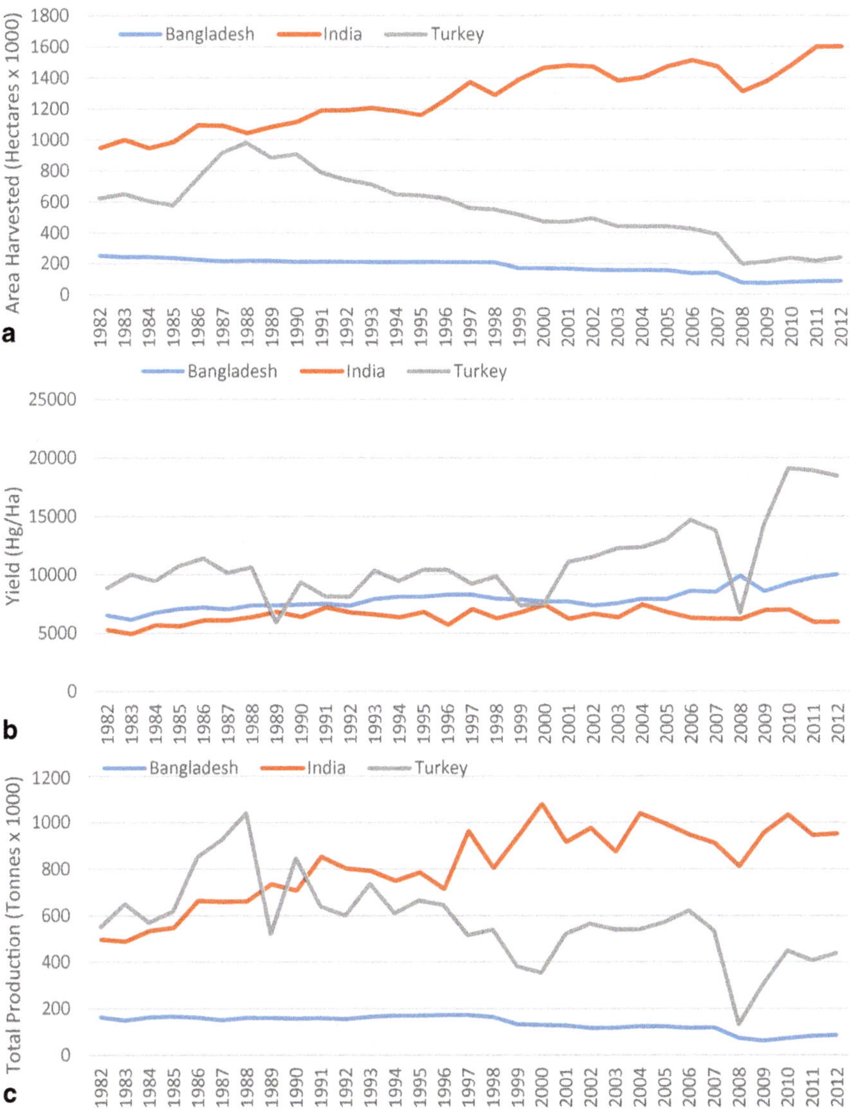

Fig. 4.3 Lentil production area (**a**), yield (**b**), and total lentil production (**c**) between 1982 and 2012 in three of the top producing developing countries, Bangladesh, India, and Turkey. (Adapted from http://faostat.fao.org/site/567/DesktopDefault.aspx?PageID=567#ancor)

Development of Barimasur-4 provides an illustrative example of the widening of the bottleneck in South Asian lentil. The cross was made at ICARDA in Aleppo, Syria using a locally adapted landrace ILL 5888 for the female parent and an exotic disease-resistant line FLIP 84–112 L (ILL 5782) for the male parent. A segregating F_3 population was sent to Bangladesh where single plant selections were made. Lines from the single plants were advanced to the F_6 using a family selection strategy, and advanced testing commenced. Barimasur-4 was eventually released and described as being an early-maturing high-yielding cultivar with good resistance to rust (*Uromyces viciae-fabae* (Pers.) de Bary, Pucciniaceae, Uredinales) and stemphylium blight (caused by *Stemphylium botryosum* Wallr.; Dematiaceae, Hyphales). Thusly, exotic disease resistance was assembled into a package with an adapted landrace. Substantial achievements have been made in breeding for tolerance to other diseases as well.

7.3 Biotic Stresses

Before resistance was available, Ascochyta blight caused by *Ascochyta lentis* and *Ascochyta fabae* were among the most serious disease threats to lentil production globally. Fortunately, breeding for tolerance has been one of the greatest successes for lentil breeders. Germplasm has been released that have been shown to be Ascochyta blight resistant including ILL 5588 (Erskine et al. 1996), "CDC Vantage" (Vandenberg et al. 2002a), and "CDC Plato" (Vandenberg et al. 2005) among many others. New species and pathotypes continue to emerge making breeding progress challenging. Ascochyta blight can affect not only the growing plant but also severely reduce seed quality. Ascochyta-infected seed can transmit the disease to the next growing season. The germplasm ILL 5588, previously mentioned, also has resistance to lentil vascular wilt (caused by *Fusarium oxysporum* Schlechtend.:Fr. f. sp. *lentis* (Vasudeva and Srinivasan) W. L. Gordon; Erskine et al. 1996). This germplasm was used to develop the Ascochyta-resistant cultivar "Northfield" (Ali 1995). Numerous other cultivars and germplasm have been released that have resistance to Ascochyta (Materne and McNeil 2007).

In South and West Asia as well as parts of North Africa, rust is the most important foliar disease of lentil (Erskine et al. 1994). In some epidemics complete crop losses have been observed. Genetic variation has been identified in several sources (Nene et al. 1975; Reddy and Khare 1984; Mishra et al. 2005; Singh and Sandhu 1988). Inheritance of resistance has been determined to be from a single dominant gene (Sinha and Yadav 1989). Released varieties resistant to rust include the above-mentioned 'Barimasur-2' (Sarker et al. 1999a), 'Barimasur-4' in Bangladesh (Sarker et al. 1999b), and 'Chakkouf' (Idrissi et al. 2012) among others.

Vascular wilt caused by *F. oxysporum* f. sp. *lentis* Vasudeva and Srinivasan is a serious constraint to lentil productivity. Tolerance to vascular wilt has been identified in wild species of *Lens* and introgressed *into L. culinaris*. Bayaa et al. (1995) identified resistant alleles in *L. montbretii* (Fisch et Mey) Davis et Plitm. This source of resistance was used to develop the germplasm ILL 5588 which was released and is available to the public from ICARDA (Erskine et al. 1996).

Stem rot caused by *Botrytis cinerea* infects lentil crops in many countries. Resistance to botrytis was discovered in the former Soviet Union (Khare 1981). As with Ascochyta this pathogen can also affect the seed and be transmitted to the next crop via infected seed. The first cultivar with resistance was released in Pakistan and called 'Masoor-93' (Tufail et al. 1995). The *Botrytis* species *B. fabae* also causes a stem rot, complicating resistance breeding (Materne and McNeil 2007). In Australia, genotypes were identified that had good resistance to both *B. fabae* and *B. cinerea.*

7.4 Abiotic Stresses

Increasing tolerances to extreme temperatures is a strategy that can improve productivity by allowing a crop to avoid stress during critical stages of development. In India, the lentil crop is planted after the rains of the monsoon in October to December at the beginning of the dry season. Therefore, it must have cold tolerance sufficient to establish a good stand in the winter and then appropriate phenology to complete its life cycle utilizing the moisture stored in the soil. Screening for cold tolerance is a challenge primarily because environmental conditions for field evaluations are not reliable. Ideally, the stressful temperatures would occur at a consistent phonological stage and at a constant rate of decline. Also, the lack of snow cover should be consistent. Screening for cold tolerance must be done in multiple locations and years due to the inconsistent environmental conditions (Erskine et al. 1981). Genetic variation for cold tolerance has been identified in 238 accessions through screening of a world collection of 3592 lines (Erskine et al. 1981).

Drought during the critical seed-fill stage of development can be avoided by sowing lentil in the winter. Substantial yield increases have been realized in research conducted in these countries by increasing cold tolerance so that the crop matures before the hot/dry months of the year (Muehlbauer and McPhee 2007). To accomplish this yield gain, phenology of the crop must also be altered so that pods are set before excessive heat causes pod abortion, and grain fill is completed before drought conditions occur. This strategy shows much promise, although it has yet to be implemented on a large scale. Data from yield trials corroborate this potential and cultivars as well as germplasm have been released. WA8449085, WA8449090, and WA8449041 are germplasm that are adapted to this strategy (Spaeth and Muehlbauer 1991). Two of these WA8649090 and WA8649041 (Spaeth and Muehlbauer 1991) were used as parents to develop the winter-hardy cultivar, Morton (Muehlbauer and McPhee 2007). In yield trials, Morton yielded 73 % more than the highest spring-sown variety (Muehlbauer and McPhee 2007). Additionally, Morton yielded 63 % better than the winter-hardy check.

Improvements in tolerance to inhospitable soils have been substantial in lentil. In Australia, lines have been identified that are extremely tolerant to boron toxicity, and it is projected that if superior alleles from these lines can be introgressed into adapted cultivars, yield could increase by 91 % (Hobson et al. 2006). Saline soils also constrain lentil production. Fortunately, significant genotypic difference to salt stress have been identified (Katerji et al. 2001; Ashraf and Waheed 1990).

8 Specific Goals in Current Breeding

8.1 Cooperations

One of the most innovative goals of plant breeding in lentil-producing regions of the developing world is to establish participatory plant breeding (PPB) and increasing popularity of participatory variety selection (PVS) networks. In PVS, local growers are recruited to test cultivars and provide feedback regarding their own wants and needs, and in PPB actually participate in conducting trials and making selections. These participations are particularly useful in the developing world because local producers tend to be isolated geographically from researchers. In India, where hundreds of languages and dialects are spoken, producers can be isolated linguistically. Consequently, lentil producers in these countries are slow to adopt newer improved cultivars. Additional reasons for this are insufficient capital, and cultural biases.

The four phases of PVS are: (1) identification of farmer preferences, (2) search for material to fill those needs, (3) testing of that material in local producer's fields or nearby research station, and (4) dissemination of information and cultivars to individuals in wider areas of the region (Solanki et al. 2007). Phase 1 is accomplished by making available an array of genotypes from local landraces to prereleased experimental lines to directly to the growers and they are queried for their impressions and results. Sometimes these demonstrations actually take place at a research station and the growers are brought to them for discussion. Phase 2 involves a search for available germplasm that more closely meets the grower's preferences. Phase 3 allows farmers to grow and compare the selected germplasm and compare their observations to those from phase 1. Finally, in phase 4, producers in the wider region who would likely benefit from the efforts of the first three phases are identified and seed is made available to them.

A good example of PVS was in 2012, when seed from high zinc and iron genotypes were distributed to over 3000 farmers in Bangladesh, India, and Nepal (ICARDA 2012). Information from these farmers will allow selection for adapted genotypes with these high-mineral traits. Concurrently, participating farmers are being made aware of the health benefits of eating these cultivars and that growing them also represents an opportunity for growing lentil destined for the export pipeline.

Compared to PVS, PPB is a more costly and time consuming activity. Farmer participation includes growing and selecting from segregating populations. It can be viewed as preliminary work that will result in material that would be entered into a PVS network. Currently, PPB in lentil remains a goal yet to be achieved or at least, documented (Solanki et al. 2007). It is anticipated that when progress from PVS nears exhaustion, PPB efforts will increase.

8.2 Abiotic Stress Tolerance

Because plants are unable to move away from stresses, a goal of plant breeding even more than animal breeding is adaptation, or ability to have high yield in a particular region. However, in the case of lentil breeding adaptation also includes making improvements of which the local producer can take advantage. Also important is that lentil growers from different regions can have vastly disparate levels of management resources available to them. This demonstrates why breeding objectives for lentil in the developed world focus much on adapting cultivars that have a maximum response to management inputs such as soil fertility and pesticide use.

Lentil is largely produced as a dry land crop; therefore, drought tolerance will continue to be an important trait worldwide. In lentil, much of the effort at mitigating aridity is focused around altering crop phenology avoid drought. Drought avoidance can be accomplished in both time and space. Plants have many mechanisms for avoiding drought including deep rooting, seed dormancy, and completion of the life cycle prior to drought, to name a few. The search for new sources of genetic variability continues to be a specific goal for lentil researchers at ICARDA, USDA, the University of Saskatchewan CDC, and Commonwealth Scientific and Industrial Research Organisation (CSIRO).

Improving drought avoidance has been most successful by altering crop phenology so that key developmental stages are completed in the absence of drought. In many parts of the lentil-growing world, this means that the crop must also be able to tolerate cold or freezing temperatures so that a good stand is established before the optimal temperatures for vegetative growth is perceived. Tailoring genotypes with combined cold tolerance and adapted phenology remains a major specific goal of lentil researchers.

Lentil plants must perceive a particular vernalization, temperature, and photoperiod regime to initiate reproductive growth (Summerfield et al. 1984; Erskine et al. 1989; McKenzie and Hill 1989). In the case of photoperiod, this response is known to be under genetic control (Summerfield et al. 1984). However, temperature and vernalization are known to influence the magnitude of the photoperiod response, and this magnitude varies regionally and presumably genetically (Saxena and Wassimi 1984; Erskine 1997). Clearly, the control of this aspect of lentil phenology is complicated.

Local lentil varieties grown in the Mediterranean climates of North Africa are sown in the spring and reproductive growth begins just before the solstice in early summer at day lengths at or near 12 h (Shrestha et al. 2009). These lentils ideally need to complete seed fill before the hot dry late summer months. Conversely, lentil grown in northern Argentina, South Asia, and Australia are planted in the fall, grow vegetatively during days of decreasing photoperiod and flower during much shorter days of about 11 h (Erskine 1983; Shrestha et al. 2009). In this situation, the crop will yield best if it completes seed fill just before the end of the spring rainy season. When lentil lines from the Mediterranean/West Asian regions are grown in South Asia, seed filling is diminished because of hot dry conditions resulting in lower yields (Erskine and Hawtin 1983; Shrestha et al. 2009). Erskine et al. (1994)

reported that 49% of the variation in yield is due to variation in flowering time, demonstrating the efficacy of phenotypes that were able to flower and fill pods prior to terminal drought stress. Therefore, the adaptation of local landraces has taught us that the breeding of new high-yielding cultivars must be done within the constraint of appropriate phenology for the region. Considerable effort has and is being expended to improve lentil performance using this strategy of drought avoidance. In Morocco, the cultivar Chakkouf was released and has been observed to yield 40% greater than the comparable check variety (Idrissi et al. 2012). At ICARDA, 'Idlib-3' was developed for use in Syria where it yielded 13.1% more than check varieties in advanced yield trials (El-Ashkar et al. 2004a).

Cold tolerance has been described as a quantitative trait in winter barley (*Hordeum vulgare* L.; Rohde and Pulham 1960), wheat (*Triticum aestivum* L.; Sutka 1994), and pea (*Pisum sativum* (L.) Mill.; Liesenfeld et al. 1986) among others species. Results from studies such as these suggest that the multitude of genes that affect this stress, do so by affecting not only physiological but also morphological traits. Genetic studies regarding the inheritance of cold tolerance are currently a major goal to augment breeding efforts in lentil and all the pulses.

Identification of genetic variability for cold tolerance has made possible the development of segregating populations for quantitative trait loci (QTL) analysis. Kahraman et al. (2004a) developed ten recombinant inbred line (RIL) populations from cold tolerant by sensitive parents. The population was evaluated for winter survival at one location in Washington State, and two locations in Turkey. They estimated heritabilites ranging from 15.9 to an extremely high 90.7. In one of their populations, they were able to map four QTL in three locations with one of them being identified in all the environments tested. Three QTL were detected in each of the two Turkey locations and one in Washington. Using the same population (Kahraman et al. 2010), a QTL for smaller leaf area that explained 20.5% of the phenotypic variation was subsequently identified in the same region indicating these two QTL could be for the same gene. The association between the two QTL was 0.75. These are encouraging results because they corroborate the efficacy of selecting for small leaf area to improve cold tolerance.

Some physiological mechanisms for drought tolerance have the potential for improvement in lentil. An increase in osmotic adjustment (OA) can enhance water use in lentil. OA means the plants cells increase their concentration of solutes which increases the cells' affinity for water. The result is less water leaving the cell and traveling through the apoplast and out of the plant through stomata. Substantial genotypic variation for this trait has been identified (Ashraf and Waheed 1990; Clements et al. 1997; Shreshta 2009). OA is a physiological trait that can also improve tolerance to salinity and freezing.

8.3 Biotic Stress Tolerance

Efforts to better understand the inheritance of disease tolerance are currently high priorities for many lentil breeders. Resistance to stemphylium blight caused by *S. botryosum* Wallr. has been shown to be inherited quantitatively (Saha et al. 2010). Disease resistance is frequently quantitative because the disease is often caused by multiple races or pathotypes, and numerous plant structures can potentially prevent infection. Saha et al. (2010) used a 206-family RIL population to study the inheritance of resistance to stemphylium blight in two successive growing seasons in Bangladesh. One QTL was detected in the 2006–2007 season, and three were detected in the 2007–2008. The one QTL from 2006–2007 was common to both years and explained 25.2 % of the phenotypic variation in that year and 46 % in the second. They concluded that this marker could be of substantial use to breeders upon verification in additional environments.

Many other foliar diseases infect lentil such as anthracnose caused by *Colletotrichum trifolii,* powdery mildew caused *Erysiphe polygoni,* and downy mildew caused by *Peronospora lentis,* among others (Khare 1981; Materne and McNeil 2007). Some root diseases that infect lentil are dampening off caused by *Pseudomonas, Fusarium,* and *Pythium* species and *Macrophomina phaseolina among* others (Khare 1981). *Aspergillus, Fusarium,* and *Helminthosporium* species are among the pathogens that cause seed spoilage diseases in lentil (Khare 1981). The above list includes diseases that researchers are currently working on to mitigate though resistance is by no means complete and further discussion of lentil pathology is beyond the scope of this chapter.

8.4 Plant Architecture

Lentil producers in West Asia and North Africa are interested in mechanizing their harvests. In fact, hand harvesting can represent up to 47 % of production costs for these farmers (Sarker et al. 2009). To achieve this goal, cultivars must be developed that combine the appropriate maturity of local landraces, with superior plant architecture of unadapted germplasm. In this case, superior plant architecture consists of good standing ability and pods that are higher off the ground. ICARDA has led the way in this effort and has reported reduction in harvest expenditures of between 17 and 20 %. Idlib-4 is a high-yielding lodging-resistant cultivar that has been released by ICARDA for use in Syria that yielded 20 % greater than the check variety in 14 on-farm yield trials (El-Ashkar et al. 2004b). Producers in Syria in particular are interested in using combine harvesters to reduce production costs. The quantitative inheritance of phenology and architecture has made progress laborious and costly. A specific goal for lentil breeders is to conduct genetic studies intended to map loci for these traits. Tullu et al. (2008) used a RIL population with 94 lines to investigate earliness and plant height in lentil. Phenotypic data were collected in two locations in western Canada. QTL for both traits were mapped to six linkage groups (LG), and they explained 37–46 % of the phenotypic variation for earliness and 31–40 % for height. More encouraging results from this study were that two QTL for earliness and two QTL for plant height were consistent across locations. These results indicate that these QTL would be good candidates for marker-assisted selection (MAS). Molecular mapping technologies are becoming less expensive daily and this will only augment current efforts to breed for improved phenology and architecture.

Lentil seed are rich in micronutrients. Breeding for the enhancement of some of these minerals would represent a cultivar with added value. ICARDA has been involved in this effort and has, or is in the process of fast-tracking the release of high iron and zinc lines to producers in Ethiopia, Bangladesh, Nepal, Syria, Turkey, Portugal, and Morocco (Sarker et al. 2009). Fortunately, significant genetic variation for these traits has been identified and are being exploited (Idrissi et al. 2012). The introduction of these value-added cultivars not only increase the worth of the crop but also open up new markets for producers in these countries. Additionally, these cultivars tend to perform better in some nutrient-deficient soils.

8.5 Inheritance of Important Traits

As an ancient self-pollinated crop species, the development of adapted or novel cultivars has two distinct phases: selection only followed by hybridization and selection. The first selections which defined domestication are listed above (plant height, seed size, flower number, etc.) and were made simply to increase harvestability and yield. Following domestication, selection took the form of selecting out of land races genotypes that were adapted to the regions where lentil was being introduced. This took the basic form of selecting lines that would flower and produce yield in a given environment. It is not until the early twentieth century that formal breeding of lentil began.

8.6 Qualitative Traits

Reports of formal breeding of lentil began with the introgression of particular cotyledon colors out of their "native" germplasm. In 1928, Tschermak observed a 3:1 ratio of red/orange to yellow cotyledons in reciprocal crosses between these types and reported that red/orange was dominant over yellow (Sharma 2007). In this report, it was noted that orange seed from heterozygous F_1 plants was indistinguishable from orange seed from homozygous plants. Therefore, he deduced that red/orange was not only completely dominant but also there was no cytoplasmic maternal effect on the trait. This was confirmed in later work using reciprocal crosses (Slinkard 1978) and the symbol *Yc* was proposed for the orange/red color class (Singh 1978). In the same crosses, Slinkard (1978) noted that yellow was completely dominant over green.

The genetic control of color of the seed coat or testa is not as clear as that for cotyledon color. The consensus presently is that there are four testa color classes: black, brown, grey, and green (Sharma 2007). Black testa color is controlled by one locus with either codominance (Vandenberg and Slinkard 1990) or a dosage effect (Emami and Sharma 2000; Sharma et al. 2004) that causes heterozygotes to be difficult to assigned to a discrete class. In either case, plants homozygous for black testa *(Blt)* have completely black seed coats that are easily scored (Sharma et al. 2004). *Bltblt* plants have seed that appear in a range between very dark brown (dense black spotting) to grey or green. To the breeder, the exact dominance state of seed color is not of practical importance because it is a monogenic trait that can be fixed with minimal inbreeding. Some other monogenic traits in lentil are seed hardness (Ladizinsky et al. 1984; Vaillancourt and Slinkard 1992), seedcoat spotting (Ladizinsky 1979; Vaillancourt et al. 1986), resistance to pea seed-borne mosaic virus (PSbMV; Haddad et al. 1978), number of flowers per inflorescence (Khosravi et al. 2010), growth habit (Ladizinsky 1979), pod shattering (Ladizinsky 1979), and zero seed tannin content (Vaillancourt et al. 1986).

8.7 Quantitative Traits

Many economically important traits in lentil are quantitatively inherited. These include seed size (Abbo et al. 1991), protein content (Chauhan and Singh 1995), winter hardiness (Kahraman et al. 2004a, b), primary branch number, seed number per pod, and seed yield (Goyal et al. 1976). An interesting result from the crosses studied in Goyal et al. (1976) is that as a whole the 7×7 diallel had highly significant general combining ability (GCA) and specific combining ability indicating that in some of the individual crosses, dominance and epistatic effects were important for yield and all its components, while in others the genetic effects were mostly additive. It is useful to have reliable estimates of genetic parameters in order to design efficient breeding methods for these traits.

9 Breeding—Improvement Methods

9.1 Pure-Line Selection

Many of the accessions in international lentil collections that represent available global genetic diversity are landraces, which themselves harbor substantial genetic variation. These landraces are surprisingly heterogeneous and many include individuals that meet many breeding objectives without the need of hybridization. Many cultivars and germplasm have been released simply by making selections from these land race populations using pure-line selection. Pure-line selection simply involves selecting a desirable individual plant, inbreeding until homozygosity is reached, increasing its seed, and then replicated testing in multiple locations and years. Selection in this strategy is very subjective and typically consists of just choosing the plant that looks best agronomically and is based on the observed phenotype. Results from the replicated tests confirm plant performance as a crop community. These results can be used by breeders to select desirable parents (germplasm) for breeding populations. Conversely, they can be used to describe the cultivars' predicted performance to producers directly. Cultivars released using this method include 'Crimson' (Muehlbauer 1991), Idlib-2 (El-Ashkar et al. 2003), 'Northfield' (Ali 1995) and 'Bichette' (Sakr et al. 2004). Germplasm released using pure-line selection include ILL 5582 (Erskine et al. 1996), ILL 5588 (Erskine et al. 1996), WH8449085, WH8449090, and WH8449041 (Spaeth and Muehlbauer 1991). In all the above selections, primary attention was paid to yield and growth habit appropriate to mechanized harvesting. Some examples of secondary selection parameters are Bichette and ILL 5588 selected for resistance to Ascochyta blight, Idlib-2 and ILL 5588 were selected for resistance to vascular wilt, and Northfield and Crimson were selected for maturity and seed type.

9.2 *Bulk Population Breeding*

Bulk population breeding (BPB) begins with a cross between parents whose traits the breeder would like to try to combine. After the initial cross, the progeny are harvested in bulk, randomly sampled, and then advanced to the next cycle. The goal is to inbreed the progeny to a certain level of homozygosity, usually the F_5 or F_6, and then make single plant selections at that time for increase. This method of breeding is simple and requires little record keeping. By the time the population has been advanced five or six generations, the ratio of superior to inferior yielding genotypes has increased through natural selection. BPB is popular among lentil breeders who are working with traits that are amenable to "natural" selection that occurs within research plot environment. The breeder selects individual superior plants for their desired traits such as seed size or quality once these traits are considered fixed. Replicated testing then commences and breeder seed is developed. A disadvantage to this method is that it cannot be done in greenhouses or off-season nurseries so progress may be slow. The cultivars 'Mason' (Muehlbauer 2002), 'Pennell' (Muehlbauer and McPhee 2004), and 'Merrit' (Muehlbauer and McPhee 2004) were developed using the bulk selection method. In all three of these examples, the breeders selected for particular seed size and color.

9.3 *Bulk Pedigree Selection*

Bulk pedigree selection (BPS) is a breeding method that is a modification of BPB and differs in the generation that selection is performed (Fig. 4.4). Most cultivars released thus far were developed using BPS. Typically, $F_{2:3}$ families are screened for

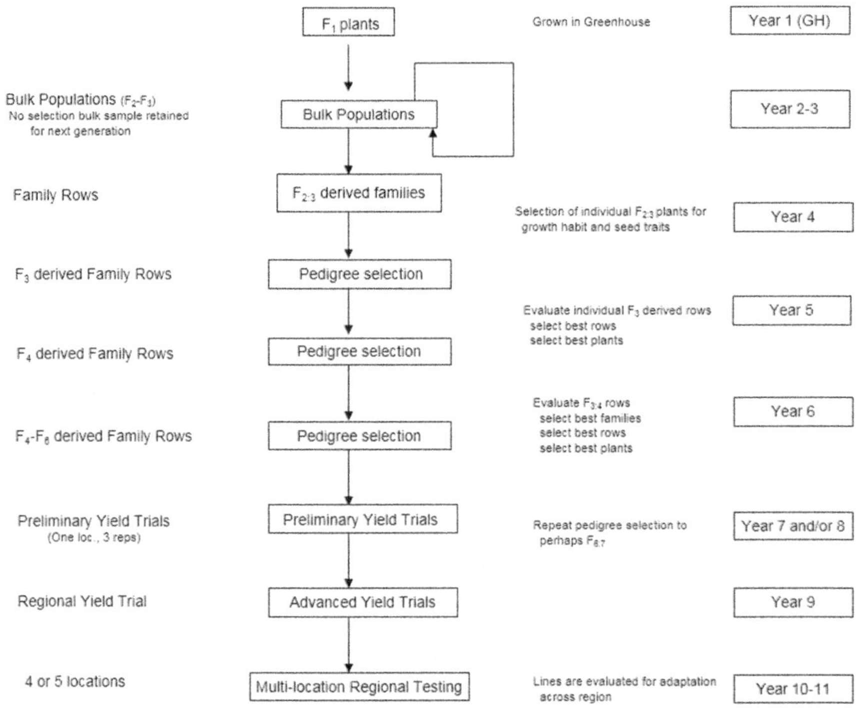

Fig. 4.4 Schematic representation of modified bulk population/pedigree selection breeding method

phenotypic response and single plants are selected. Family rows are evaluated for four to five generations, and the best rows and plants within rows are selected and advanced in bulk to a desired level of homozygosity, usually the F_6 or F_7. Preliminary replicated testing begins followed by advanced yield trials and finally multilocation regional testing.

The various public institutions engaged in lentil cultivar development modify BPS based on when familial generation pedigree selection begins and how many generations of pedigree selection are conducted before preliminary yield trials are conducted. Advantages of BPS are that selection of a plant in early generations can insure that a desired allele for a required phenotype will be present in that family and will be advanced through the program. This means that less land will be required because only those F_2- or F_3-derived families will need to be advanced. Disadvantages to this method is that the breeder must be able to observe the phenotype for selection in the F_2-derived lines using relatively few (~50) families. Additionally, bulking F_2-derived families would need to be done in the target environment limiting the number of generation per year that can be grown.

Cultivars released using this method include 'Emerald' and 'Brewer' (Muehlbauer 1987), Idlib-3 and Idlib-4 (El-Ashkar et al. 2004a, b), 'CDC Robin' (Vandenberg et al. 2002b), 'CDC Vantage' (Vandenberg et al. 2002a), and 'CDC Plato'

(Vandenberg et al. 2005). The $F_{2:3}$ family selected for the CDC cultivars were all chosen based on Ascochyta blight and anthracnose reaction as well as seed characters. The Idlib cultivars were selected for harvestability, disease reaction, and market traits reinforcing the more recent desire in Syria and Lebanon for mechanized harvest.

10 Integration of New Biotechnologies in Breeding Programs

Genetic engineering in lentil is not accepted in international markets, and this is not likely to change; therefore, no varieties have been released with artificially introduced traits. However, new molecular biology platforms and technology offer greater potential to develop saturated genetic maps and to study agronomically important traits in greater detail. This potential arises from two key limitations researchers currently encounter in mapping studies of lentil; optimal mapping populations have been difficult to assemble, and the number of informative molecular markers is small relative to other crop species.

The most recent and saturated genetic map for lentil was published in 2012 (Gupta et al. 2012). The map was a distance of 3843.4 cM which is considerably longer than the length of the pea genome (1100–1800 cM), which is predicted to be similar to lentil based on the two species' close phylogeny (Laucou et al. 1998). Additionally, this most recent map has 11 LGs and lentil is known to have a haploid chromosome number of $n=7$. Limitations of this study were that they were only able to screen the population with 1319 markers, and 523 of these were random amplified polymorphic DNA (RAPD) which are anonymous dominant markers. Of the 1319 markers used, 118 were found to be informative of which 79 were RAPDs and the rest were simple sequence repeat (SSR) or inter-simple sequence repeat (ISSR) markers. Also, the mapped population was at the F_2 generation which is particularly constraining because dominant markers cannot distinguish heterozygotes from homozygous dominant genotypes. Development of RIL populations from parents divergent for a host of agronomic traits will in the future be used to improve mapping studies. Additionally, the use of biotechnological techniques to develop more widely distributed markers, particularly SSRs and single nucleotide polymorphisms (SNPs), will also improve future genetic maps of lentil.

Though no cultivars developed using genetic transformation have been released, lentil has been successfully transformed with alien genes despite its recalcitrance to transformation. Initial attempts at lentil transformation have used variations of particle bombardment or *Agrobacterium tumefaciens* vectors with virtually no success. Akcay et al. (2009) described an *A. tumefaciens* protocol that has a transformation efficiency of 2.3 % using improved strains of *A. tumefaciens* and one Turkish lentil genotype. Chopra and Aparna (2012) described a procedure that was much less efficient (0.66 %), but that was successful using four Indian cultivars. Neither of the above reports utilized an agronomically beneficial gene in their inserted construct.

Khatib et al. (2011) developed a construct that included the DREB1 A gene. This gene has been reported to improve tolerance to drought, salinity, and freezing stress (Liu et al. 1998; Kasuga et al. 1999; Gilmour et al. 2000) in Arabidopsis presumably by means of OA. Transformants were induced to express the DREB1 A gene and demonstrated tolerance compared to the controls when watered with a saline solution of 50 mM NaCl (Khatib et al. 2011). Whether this work will result in field production of transgenic plants is uncertain. However, these techniques will be useful in generating genotypes that can serve as good parents for the development of mapping populations for QTL studies on stress tolerances.

11 Conclusions

Much opportunity exists for the breeding improvement of lentil for many reasons. Lentil has received very little if any attention from the major agribusiness conglomerates that dominate cultivar development in crops such as maize, soybean, cotton, sorghum, wheat, or barley. Substantial collections are being curated that harbor copious diversity for genetic improvement. Lentil is grown in both developed and developing countries making breeding objectives numerous. Finally, as a high-quality inexpensive source of protein, greater-yielding cultivars can diminish the need for animal protein and improve the diets of people worldwide.

References

Abbo S, Ladizinsky G, Weeden NF (1991) Genetic-analysis and linkage study of seed weight in lentil. Euphytica 58:259–266

Akcay UC, Mahmoudian M, Kamci H et al (2009) *Agrobacterium tumefaciens*-mediated genetic transformation of a recalcitrant grain legume, lentil (*Lens culinaris* Medik). Plant Cell Rpt 28:407–417

Ali SM (1995) Register of Australian grain legume cultivars *Lens culinaris* (lentil) cv. Northfield. Aust J Expt Agr 35:1181–1182

Ashraf M, Waheed A (1990) Screening of local/exotic accessions of lentil (*Lens culinaris* Medic.) for salt tolerance at two growth stages. Plant Soil 128:167–176

Baidya BN, Enus AN, Sen S (1988) Estimations of variability and correlation in yield and yield contributing characters in lentil (*Lens culinaris*). Environ Ecol 6:694–697

Barulina H (1930) Lentils of the USSR and other countries. Bull Appl Bot Genet Plant Breed 40:265–304

Bayaa B, Erskine W, Hamdi A (1995) Evaluation of a wild lentil collection for resistance to vascular wilt. Genet Resour Crop Evol 42:231–235

Balyan HS, Houben A, Ahne R (2002) Karyotype analysis and physical mapping of 18s-5.8s-25s and 5s ribosomal rna loci in species of genus lens miller (Fabaceae). Caryologia 55(2):121–128

Brown TA, Jones MK, Powell W et al (2009) The complex origins of domesticated crops in the fertile crescent. Trends Ecol Evol 24:103

Chopra R, Aparna SR (2012) Use of sonication and vacuum infiltration for agrobacterium—mediated transformation of an Indian lentil (*Lens culinaris* Medik.) cultivar. Sci Hortic 143:127–134

Chauhan MP, Singh IS (1995) Inheritance of protein content in lentil (*Lens culinaris* Medik). Legume Res 18:5–8
Ceccarelli S, Erskine W, Hamblin J et al (1994) Genotype by environment interaction and international breeding programs. Exp Agr 30:177–188
Clements JC, Haqqani AM, Siddique KHM, French RJ (1997) Drought tolerance in lentil (*Lens culinaris*). In: Abstracts of the International Food Legume Research Conferences III, 22–26 Sept 1997 Adelaide, Australia
Consultative Group on International Agriculture Research (CGIAR) (2014) Lentil. http://www.cgiar.org/our-research/crop-factsheets/lentil/. Accessed 7 Oct 2014
Cubero JI (1981) Origin, taxonomy and domestication. In: Webb C, Hawtin G (eds) Lentils. CAB, Slough, pp 15–21
Cubero JI, Perez de la Vega M, Fratini R (2009) Origin, phylogeny, domestication and spread. In: Erskine W, Muehlbauer FJ, Sarker A et al (eds) Lentil botany, production and uses. CABI International, Cambridge, pp 13–33
Dwivedi SL, Upadhyaya HD, Balaji J, Buhariwalla HK, Blair MW, Ortiz R, Crouch JH, Serraj R (2006) Using genomics to exploit grain legume biodiversity in crop improvement. Plant Breed Revi 26:171–357
El-Ashkar F, Sarker A, Haddad N et al (2003) Registration of 'Idlib-2' lentil. Crop Sci 43:728–729
El-Ashkar F, Sarker A, Erskine W et al (2004a) Registration of 'Idlib-3' lentil. Crop Sci 44:2261
El-Ashkar F, Sarker A, Erskine W et al (2004b) Registration of 'Idlib-4' lentil. Crop Sci 44:2261–2262
Emami MK, Sharma B (2000) Inheritance of black testa colour in lentil (*Lens culinaris* Medik.). Euphytica 115:43–47
Erskine W (1983) Perspectives in lentil breeding. In: Faba beans, kabuli chickpeas, and lentils in the 1980s, An international proceedings workshop International Center for Agricultural Research in the Dry Areas (ICARDA), Aleppo, Syria, pp 91–100
Erskine W (1997) Lessons for breeders from land races of lentil. Euphytica 93:107–112
Erskine W, Choudhary MA (1986) Variation between and within lentil landraces from Yemen-Arab-Republic. Euphytica 35:695–700
Erskine W, Hawtin GC (1983) Pre-breeding in faba beans and lentils. Genetika 15:287–294
Erskine W, Witcombe JR (1984) Lentil germplasm catalog. International Center for Agricultural research in the Dry Areas, Aleppo, Syria
Erskine W, Meyveci K, Izgin N (1981) Screening of a world lentil collection for cold tolerance. Lens 8:5–9
Erskine W, Williams PC, Nakhoul H (1985) Genetic and environmental variation in seed size, protein, yield, and cooking quality of lentils. Field Crops Res 12:153–161
Erskine W, Muehlbauer FJ, Short RW (1989) Food legume improvement program: annual report for 1989. International Center for Agricultural Research in the Dry Areas (ICARDA), Aleppo, Syria
Erskine W, Tufail M, Russell A et al (1994) Current and future strategies in breeding lentil for resistance to biotic and abiotic stresses. Euphytica 73:127–135
Erskine W, Bayaa B, Saxena MC (1996) Registration of ILL 5588 lentil germplasm resistant to vascular wilt and Ascochyta blight. Crop Sci 36:1080–1080
Erksine W, Muehlbauer F J, Sarker A, Sharma B (2009) Introduction. In: Erskine W, Muehlbauer FJ, Sarker A et al (eds) Lentil botany, production and uses. CABI International, Cambridge, pp 76–101
Erskine W, Chandra S, Chaudhry M et al (1998) A bottleneck in lentil: widening its genetic base in south Asia. Euphytica 101:207–211
Flint-Hamilton K (1999) Legumes in ancient Greece and Rome: food, medicine, or poison? Hesperia 68:371–385
Fernandez M, Lopez-Jurado M, Aranda P, Urbano G (1996) Nutritional assessment of raw and processed faba bean (*Vicia faba* L.) major in growing rats

Food and Agriculture Organization (FAO) (2008) FAOSTAT Statistical database of the United Nations food and agriculture organization Rome http://faostat.fao.org/site/342/default.aspx. Accessed 7 Oct 2014
Furman BJ (2006) Methodology to establish a core collection: case study in lentil. Plant Genet Res 4:2–12
Furman BJ, Coyne C, Redden B et al (2009) Genetic resources: collection, characterization, conservation and documentation. In: Erskine W, Muehlbauer FJ, Sarker A et al (eds) Lentil botany, production and uses. CABI International, Cambridge, pp 64–75
Gebhardt SE, Thomas RG (2002) Nutritive value of foods. In USDA Home and Garden Bulletin number 72. www.ars.usda.gov/SP2UserFiles/Place/12354500/Data/hg72/hg72_2002.pdf. Accessed 7 Oct 2014
Gilmour SJ, Sebolt AM, Salazar MP et al (2000) Overexpression of the Arabidopsis CBF3 transcriptional activator mimics multiple biochemical changes associated with cold acclimation. Plant Physiol 124:1854–1865
Goyal SM, Jaimini SN, Tikka SB (1976) Combining ability analysis for seed yield and some biometric characters in lentil. Lens Nwsl 4:10–13
Gupta M, Verma B, Kumar N et al (2012) Construction of intersubspecific molecular genetic map of lentil based on ISSR, RAPD and SSR markers. J Genet 91:279–287
Haddad NI, Muehlbauer FJ, Hampton RO (1978) Inheritance of resistance to pea seed-borne mosaic-virus in lentils. Crop Sci 18:613–615
Harlan JR (1992) Crops and man. ASA-CSSA, Madison
Helbeck H (1959) Domestication of food plants in old world. Science 130:365–372
Helbeck H (1963) Late Cypriote vegetable diet in Apliki. Act NSTIT Athen Reg Sueciae Ser 4:171–186
Helbeck H (1970) The plant industry in Hacilar. In: Mellert J (ed) Excavations at Hacilar Vol 1 Occasional publications No 9 of the British Institute of Archaeology, Ankara. Edinburgh University Press, pp 189–244
Hobson K, Armstrong R, Nicolas M, Connor D, Materne M (2006) Response of lentil (lens culinaris) germplasm to high concentrations of soil boron. Euphytica 151(3):371–382
International Center for Agricultural Research in the Dry Areas (ICARDA) (2012) Annual report South Asia and China regional program: annual report for 2012. Beirut, Lebanon, pp 47
International Center for Agricultural Research in Dry Areas (ICARDA) (2013) http://www.icarda.org/research-sub/crop-improvement. Accessed 7 Oct 2014
Idrissi O, Sakr B, Dahan R et al (2012) Registration of 'Chakkouf' lentil in Morocco. J Plant Regist 6:268–272
Khare MN (1981) Diseases of Lentils. In: Webb C, Hawtin G (eds) Lentils. Farnham Royal, UK, pp 163–172
Kahraman A, Kusmenoglu I, Aydin N et al (2004a) Genetics of winter hardiness in 10 lentil recombinant inbred line populations. Crop Sci 44:5–12
Kahraman A, Kusmenoglu I, Aydin N et al (2004b) QTL mapping of winter hardiness genes in lentil. Crop Sci 44:13–22
Kahraman A, Demirel U, Ozden M et al (2010) Mapping of QTLs for leaf area and the association with winter hardiness in fall-sown lentil. Afr J Biotechnol 9:8515–8519
Kasuga M, Liu Q, Miura S et al (1999) Improving plant drought, salt, and freezing tolerance by gene transfer of a single stress-inducible transcription factor. Nat Biotech 17:287–291
Katerji N, van Hoorn JW, Hamdy A, Mastrorilli M, Oweis T, Erskine W (2001) Response of two varieties of lentil to soil salanity. Agric Water Manage 47(3):179–190
Khatib F, Makris A, Yamaguchi-Shinozaki K et al (2011) Expression of the DREB1 A gene in lentil (*Lens culinaris* Medik. subsp *culinaris*) transformed with the *Agrobacterium* system. Crop Pasture Sci 62:488–495
Khosravi E, Khodambashi M, Mohammadi S (2010) A linkage group of four genes related to morphological traits in lentil (Lens culinaris Medik.). Int J Plant Prod 4:303–308
Ladizinsky G (1979) Genetics of several morphological traits in the lentil. J Hered 70:135–137
Ladizinsky G, Braun D, Goshen D et al (1984) The biological species of the genus *Lens*-L. Bot Gaz 145:253–261

Lakhani JP, Holker S, Mirsha R (1986) Genetics of seedling vigor and hard seed in lentil. Lens Nwsl 13:10–12

Laucou V, Haurogne K, Ellis N et al (1998) Genetic mapping in pea. 1. RAPD-based genetic linkage map of *Pisum sativum*. Theor Appl Genet 97:905–915

Liesenfeld DR, Auld DL, Murray GA et al (1986) Transmittance of winterhardiness in segregated populations of peas. Crop Sci 26:49–54

Liu Q, Kasuga M, Sakuma Y et al (1998) Two transcription factors, DREB1 and DREB2, with an EREBP/AP2 DNA binding domain separate two cellular signal transduction pathways in drought- and low-temperature-responsive gene expression, respectively, in *Arabidopsis*. Plant Cell 10:1391–1406

Materne M, McNeil D L (2007) Breeding methods and achievements. In: Yadav SS, McNeil DL, Stevenson PC (eds) Lentil: an ancient crop for modern times. Springer, Berlin, pp 173–188

McKenzie BA, Hill GD (1989) Environmental-control of lentil (lens-culinaris) crop development. J Agric Sci 113:67–72

McNeil DL, Hill GD, Materne M et al (2007) Global production and world trade. In: Yadav SS, McNeil DL, Stevenson PC (eds) Lentil: An Ancient Crop for Modern Times. Springer, The Netherlands, pp 95–105

Mishra SK, Singh BB, Basandrai AK, Sarker A, Dayachand (2005) Screening of ICARDA's lentil material for rust resistance in India. Indian J Plant Genet Resour 18(1):49–50

Muehlbauer FJ (1987) Registration of brewer and emerald lentil. Crop Sci 27:1088–1089

Muehlbauer FJ (1991) Registration of 'Crimson' lentil. Crop Sci 31:1094–1095

Muehlbauer FJ (2002) Registration of 'Mason' lentil. Crop Sci 42:301–302

Muehlbauer FJ, McPhee KE (2004) Registration of 'Pennell' lentil. Crop Sci 44:1488–1488

Muehlbauer FJ, McPhee KE (2007) Registration of 'Morton' winter-hardy lentil. Crop Sci 47:438–439

Muehlbauer FJ, Mihov M, Vandenberg A et al (2009) Improvement in developed countries. In: Yadav SS, McNeil DL, Stevenson PC (eds) Lentil: an ancient crop for modern times. Springer, Berlin, pp 137–154

Nene Yl, Kannaiyan J, Saxena GC (1975) Note on the performance of lentil variaties and germplasm cultures against *Uromyces fabae* (Pers.) de Bary. Indian J Agr Sci 45:177–178

N.I. Vavilov All-Russian Research Institute of Plant Husbandry (VIR) (2013) http://vir.nw.ru/data/dbf.htm. Accessed 7 Oct 2014

Nestares T, Lopez-Frias M, Barrioneuvo M, Urbano G (1996) Nutritional assessment of raw and processed chickpea (*Cicer arietinum*) protein in growing rats. Journal Agricultural and Food Chemistry 44:2760–2765

Ramgiry SR, Paliwal KK, Tomar SK (1989) Variability and correlations of grain yield and other quantitative characters in lentil. Lens Nwsl 16:19–21

Rasheed M, Jilani G, Shah I A, Najeeb U, Iqbal T (2010) Improved lentil production by utilizing genetic variability in response to phosphorus fertilization. Acta Agric Scand Sect B-Soil Plant Sci 60(6):485–493

Redden B, Maxted N, Furman B et al (2007) Lens biodiversity. In: Yadav SS, McNeil DL, Stevenson PC (eds) Lentil: an ancient crop for modern times. Springer, Berlin, pp 11–22

Reddy RR, Khare MN (1984) Further studies on factors influencing the mechanisms of resistance in lentil (*Lens culinaris* M.) to rust (*Uromyces fabae* (Pers.) de Bary. Lens Nwsl 11:29–32

Rohde CR, Pulham CF (1960) Heritability estimates of winter hardiness in winter barley determined by the standard unit method of regression analysis. Agron J 52:584–586

Saha GC, Sarker A, Chen WD, Vandemark GJ, Muehlbauer FJ (2010) Inheritance and linkage map positions of genes conferring resistance to stemphylium blight in lentil. Crop Sci 50(5):1831–1839

Sakr B, Sarker A, El Hassan H et al (2004) Registration of 'Bichette' lentil. Crop Sci 44:686–687

Sandhu JS, Singh S (2007) History and origins. In: Yadav SS, McNeil DL, Stevenson PC (eds) Lentil: an ancient crop for modern times. Springer, Berlin, pp 1–9

Sarker A, Erskine W, Hassan MS et al (1999a) Registration of 'Barimasur-4' lentil. Crop Sci 39:876–876

Sarker A, Erskine W, Hassan MS et al (1999b) Registration of 'Barimasur-2' lentil. Crop Sci 39:875–875
Sarker A, Erskine W, Singh M (2005) Variation in root and shoot traits and their relationship to drought tolerance in lentil. Genet Res Crop Evol 52:89–97
Sarker A, Aydogan A, Chandra S et al (2009) Genetic enhancement of yield and yield stability. In: Erskine W, Muehlbauer FJ, Sarker A et al (eds) Lentil botany, production and uses. CABI International, Cambridge, pp 102–120
Sarwar DM, Kaul AK, Quader M (1982) Correlation studies in lentil. Lens Nwsl 9:22–23
Saskatchewan Ministry of Agriculture (2013) www.agriculture.gov.sk.ca/ Default.aspx?DN=c5993bcc-009f-4031-b936-c52c992b9e7d. Accessed 7 Oct 2014
Saxena MC, Wassimi N (1984) Photoperiodic response of some diverse genotypes of lentil (*Lens culinaris* Med.). Lens Nwsl 11:25–29
Sharma B (2007) Genetics of economic traits. In: Erskine W, Muehlbauer FJ, Sarker A et al (eds) Lentil botany, production and uses. CABI International, Cambridge, pp 76–101
Sharma B (2009) Genetics of economic traits. In: Erskine W, Muehlbauer FJ, Sarker A et al (eds) Lentil botany, production and uses. CABI International, Cambridge, pp 76–101
Sharma B, Tyagi MC, Mishra SK (2004) Confirmation of inheritance pattern of black testa colour in lentil. J Lentil Res 1:19–20
Shrestha RK, Siddique HM, Turner DW et al (2009) Breeding and management to minimize the effects of drought and improve water use efficiency. In: Erskine W, Muehlbauer FJ, Sarker A et al (eds) Lentil botany, production and uses. CABI International, Cambridge, pp 172–193
Sindhu JS, Mirsha HO (1982) Genetic variability in Indian microsperma types of lentil. Lens Nwsl 9:10–11
Singh TP (1978) Inheritance of cotyledon color in lentil. Indian J Genet Plant Breed 38:11–12
Singh K, Sandhu TS (1988) Screening cultivars of lentil for resistance to rust. Lens Nwsl 15:28–29
Sinha RP, Yadav BP (1989) Inheritance of resistance to rust in lentil. Lens Nwsl 16:41
Skyrpetz S (2006) Lentils situation and outlook. Bi-weekly bulletin agriculture and agri-food Canada. May 12, 19(7), p 4
Slinkard AE (1978) Inheritance of cotyledon color in lentils. J Hered 69(2):139–140
Solanki IS, Yadav SS, Bahl PN (2007) Varietal adaptation, participatory breeding and plant type. In: Yadav SS, McNeil DL, Stevenson PC (eds) Lentil: an ancient crop for modern times. Springer, Berlin, pp 255–274
Spaeth SC, Muehlbauer FJ (1991) Registration of three germplasms of winter hardy lentil. Crop Sci 31:1395
Summerfield RJ, Muehlbauer FJ, Roberts EH (1984) Controlled environments as an adjunct to field-research on lentils (*Lens-culinaris*).3. Photoperiodic lighting and consequences for flowering. Exp Agr 20:1–18
Sutka J (1994) Genetic-control of frost tolerance in wheat (*Triticum aestivum* L). Euphytica 77:277–282
Tufail M, Ahmad M, Ali A (1995) Masoor-93: an ideal combination of characters for Punjab province, Pakistan. Lens Nwsl 22:50–52
Tullu A, Tar'an B, Warkentin T et al (2008) Construction of an Intraspecific linkage map and QTL analysis for earliness and plant height in lentil. Crop Sci 48:2254–2264
United States Department of Agriculture (USDA) Germplasm Resource Information Network (GRIN) (2013) http://www.ars-grin.gov/cgi-bin/npgs/swish/accboth?query=lens&si=0&start=3000. Accessed 7 Oct 2014
Urbano G, Porres JM, Frias J, Vidal-Valverde C (2007) Nutritive value. In: Yadav SS, McNeil DL, Stevenson PC (eds) Lentil: an ancient crop for modern times. Springer, Berlin, pp 47–93
Vaillancourt RE, Slinkard AE (1992) Inheritance of new genetic-markers in lentil (*Lens* miller). Euphytica 64:227–236
Vaillancourt R, Slinkard AE, Reichert RD (1986) The inheritance of condensed tannin concentration in lentil. Can J Plant Sci 66:241–246
Van Zeist W, Bottema S (1971) Plant husbandry in early Neolithic Nea Nikomedeia, Greece. Acta Bot Neerl 20:524–538

Vandenberg A, Slinkard A E (1990) Genetics of seed coat colour and pattern in lentil. Journal of Heredity 81:484–488
Vandenberg A, Kiehn FA, Vera C et al (2002a) CDC Vantage lentil. Can J Plant Sci 82:109–110
Vandenberg A, Kiehn FA, Vera C et al (2002b) CDC Robin lentil. Can J Plant Sci 82:111–112
Vandenberg A, Vera C, Buchwaldt L et al (2005) CDC Plato lentil. Can J Plant Sci 85:161–162
Wright C (2001) Mediterranean vegetables. The Harvard Common, Boston
Yadav SS, Stevenson PC, Rizvi AH et al (2007) Uses and consumption. In: Yadav SS, McNeil DL, Stevenson PC (eds) Lentil: an ancient crop for modern times. Springer, Berlin, pp 33–46

Chapter 5
Faba Bean

Gérard Duc, Jelena M. Aleksić, Pascal Marget, Aleksandar Mikić, Jeffrey Paull, Robert J. Redden, Olaf Sass, Frederick L. Stoddard, Albert Vandenberg, Margarita Vishnyakova and Ana M. Torres

1 Introduction

The faba bean (*Vicia faba* L.) is a rich protein grain legume belonging to the Fabaceae family with a long tradition of cultivation in the temperate zone of the northern hemisphere. Sometimes also referred to as horse bean or broad bean, it is mostly harvested as dry seeds for food or feeds, but its fresh seeds or pods can also be used as vegetables. Faba bean provides valuable ecological and environmental services in sustainable agriculture, diversity in cropping systems and host numerous associated organisms including pollinating insects. The capacity of this species to establish symbiosis with specific rhizobia bacteria results in biological nitrogen fixation which reduces the input of fertilizers in arable lands. The species has an intrinsic ability to adapt to diverse climates, but its low and unstable yields hamper its competitiveness as a crop.

The area grown with faba bean was only 1.2% of the 200 million ha of annual grain legume cultivation in the world in 2011 (Table 5.1). Among the grain legumes,

G. Duc (✉)
UMR1347 Agroecologie, INRA, 17 rue de Sully, BP 86510, 21000 Dijon, France
e-mail: gerard.duc@dijon.inra.fr

J. M. Aleksić
Laboratory for Plant Molecular Biology, Institute of Molecular Genetics and Genetic Engineering (IMGGE), University of Belgrade, Vojvode Stepe 444a, PO Box 23, Belgrade 11010, Serbia
e-mail: aleksic_jelena@yahoo.com.au; jelena.aleksic@imgge.bg.ac.rs

P. Marget
UMR1347 Agroecologie, INRA, 17 rue de Sully, BP 86510, 21000 Dijon, France
e-mail: pascal.marget@dijon.inra.fr

A. Mikić
Forage Crops Department, Institute of Field and Vegetable Crops,
Maksima Gorkog 30, Novi Sad 21000, Serbia
e-mail: aleksandar.mikic@nsseme.com

A. M. De Ron (ed.), *Grain Legumes*, Handbook of Plant Breeding 10,
DOI 10.1007/978-1-4939-2797-5_5

faba bean ranks at seventh position behind a top group including soya bean (*Glycine max* L.), groundnut (*Arachis hypogaea* L.) and common bean (*Phaseolus vulgaris* L.). In the EU27, faba bean covers 21.3 % of the 1.6 million ha of grain legume cultivation, ranking second behind pea and just ahead of soya bean. The major faba bean-producing countries at world level are presented in Table 5.2 (FAOstat 2013; Eurostat 2014). China, the leading producer, followed by Ethiopia, the UK, Australia and France were the five main producing countries in 2011, harvesting 72 % of the world production. No accurate statistics are available on fresh seed production as a vegetable, as a consequence of the predominantly small-scale gardening nature.

Research activity on faba bean can be measured by publication effort (Table 5.3). In a CAB abstracts query over all documents 2004–2013 (23 January 2014), a 10-year production on all grain legumes detected 63,970 papers, of which 4.8 % referred to *V. faba* when soya bean ranked first with 33 % of the publications. The distribution of disciplines allocated by journals slightly differed between faba bean and the assembly of grain legume species, by a lower proportion for genetics and breeding and for nonfood and non-feed uses, compensated by higher investments

J. Paull
School of Agriculture, Food and Wine, The University of Adelaide, PMB 1,
Glen Osmond, SA 5064, Australia
e-mail: jeffrey.paull@adelaide.edu.au

R. J. Redden
Agriculture Productivity. Department of Economic Development, Jobs, Transport & Resources,
Australian Temperate Field Crops Collection, Private Bag 260, Horsham VIC 3401, Australia
e-mail: bob.redden@ecodev.vic.gov.au

O. Sass
Norddeutsche Pflanzenzucht Hans-Georg Lembke KG,
Hohenlieth-Hof 1, 24363 Holtsee, Germany
e-mail: o.sass@npz.de

F. L. Stoddard
Department of Agricultural Sciences, University of Helsinki,
PO Box 27, (Latokartanonkaari 5-7), 00014 Helsinki, Finland
e-mail: Frederick.stoddard@helsinki.fi

A. Vandenberg
Department of Plant Sciences/Crop Development Centre, University of Saskatchewan,
51 Campus Drive, Room 2C24, Saskatoon, SK S7N 5A8, Canada
e-mail: albert.vandenberg@usask.ca

M. Vishnyakova
Genetic Resources of Grain Legumes, Vavilov Institute of Plant Industry,
Bolshaya Morskaya, 42, Saint-Petersburg, Russian Federation 190000
e-mail: m.vishnyakova@vir.nw.ru

A. M. Torres
Area de Mejora y Biotecnología, IFAPA Centro Alameda del Obispo,
Avenida Menéndez Pidal s/n, 14080 Córdoba, Spain
e-mail: anam.torres.romero@juntadeandalucia.es

Table 5.1 Area and production for major annual grain legumes at the world and European Union levels in 2011. (FAOstat 2013 data adjusted for Eurostat differences in faba bean and lupins)

	World		EU27	
	Production (1000 t)	Area (1000 ha)	Production (1000 t)	Area (1000 ha)
Oilseed				
Soya bean	262,038	103,605	1094	386
Groundnut	40,017	24,637	9	11
Pulses (dry seeds)				
Common bean	23,062	30,411	142	82
Chickpea	11,610	13,181	45	46
Cowpea	4952	10,640	0	0
Pea	9730	6141	1614	684
Faba bean	4686	2587	1168	411
Pigeon pea	4444	5863	0	0
Lentil	4404	4172	51	59
Lupin	1213	1115	237	251
Total	*366,156*	*202,351*	*4360*	*1926*
Annual grain legume area (% of arable land)		*0.145*		*0.016*

Table 5.2 Area of cultivation (ha) and production (Mg) of faba bean in major faba bean-producing countries at world level. (Eurostat data for Europe, FAOstat 2013 data for the rest of the world)

	Main countries	Area (ha)	Dry seeds (Mg)
	(>40,000 Mg)		
Asia		*919,000*	*1,655,000*
	China	872,000	1,550,000
Africa		*883,000*	*1,312,000*
	Ethiopia	460,000	698,000
	Egypt	55,000	175,000
	Morocco	200,000	171,000
	Sudan	72,000	155,000
	Tunisia	57,000	73,000
Europe		*410,000*	*1,168,000*
	France	91,000	345,000
	The UK	125,000	419,000
	Italy	50,000	96,000
	Germany	17,000	61,000
	Spain	35,000	55,000
Australia-New Zealand		*175,000*	*350,000*
	Australia	175,000	350,000
South America		*138,000*	*137,000*
	Peru	52,000	65,000
World		*2,587,000*	*4,690,000*

Table 5.3 CAB abstracts (www.cabi.org) query over all documents 2004–2013 (23 January 2014). Distribution of disciplines (%) among 63,970 papers on all grain legumes and 3053 papers on *Vicia faba* L.

Discipline	All grain legumes	Faba bean
Breeding and genetics	30.9	23.6
Physiology and biochemistry	20.6	24.5
Viral, bacterial and fungal diseases	12.9	13.4
Pests	10.6	12.6
Cropping systems	10.7	11.4
Soil biology	9.5	8.4
Forage and fodder crops	8.8	10.0
Feed composition and quality	1.8	3.8
Food composition and quality	7.7	7.2
Nonfood, non-feed uses	6.6	4.3

in plant physiology and biochemistry. Numerous papers refer to diverse ecosystem services provided by faba bean such as the potential for diversification of cropping systems, the establishment of the root symbioses with Rhizobiaceae bacteria *(Rhizobium leguminosarum* bv. *viciae)* and with vesicular-arbuscular mycorhiza (VAM), the impact on soil biology and fertility or with the capacity to host and feed pollinating insects. Apart from these issues, genetics and breeding progress is the main topic to improve yield reliability, bring competitiveness and ecosystem services, improve nutritional value and, as a result, increase the cultivated area of this species. This chapter reviews the advances in both faba bean genetics and breeding, underlining the significant progress currently made in molecular genetics which will help speed up the breeding efforts to improve targeted traits.

2 Origin and Systematics

Faba bean is among the most ancient crops of the Old World. Together with chickpea (*Cicer arietinum* L.), grass pea (*Lathyrus sativus* L.), lentil (*Lens culinaris* Medik.), pea (*Pisum sativum* L.) and bitter vetch (*Vicia ervilia* (L.) Willd.), it was one of the factors of the beginnings of early crop domestication. Faba bean played an important role in spreading agriculture during the Neolithic, Bronze and Iron Ages throughout Eurasia and North Africa, with numerous archaeological deposits. Very little is known of the exact origin and the single steps during the domestication of faba bean (Maxted et al. 1991). Its presumed domestication is in the Fertile Crescent with the most ancient fossils dated to 11,200 BP in Iraq (Ladizinsky 1998). Cubero (1973) suggested the Near East as its centre of origin, with four different directions from this centre: (1) to Europe, (2) along North African coast to the Iberian Peninsula, (3) along the Nile to the Ethiopian highlands and (4) from Mesopotamia to the Indian subcontinent.

Various trait analyses distinguished two main faba bean groups: the small-seeded forms from southwestern Asia and the large-seeded forms developed in the West. The Eastern group is very ancient and may be traced back to Neolithic culture. It has the greatest number of endemic forms and the greatest diversity of specific characteristics lacking in other groups, such as few to many pairs of leaflets, their glaucous and grey-green colour, shattering or non-shattering pods, the wide range of variation of the maturity period, seed size, colour and shape, leaflet size, stem height and branching of stem. The Mediterranean group, with a dense geographic concentration of the large-seeded forms, is regarded as a more recent secondary centre of the faba bean diversity. The remains of faba bean in the archaeological excavations in the Mediterranean and Central Europe are dated between the third and second millennia BC (Bond 1976; Cubero 1973). However, the broad bean (also known as *major* types, with more than 1g per seed) is thought to have originated only after 500 BC (Hanelt 1972).

Faba bean spread from the primary centre and from Mediterranean Europe, forming an important third centre of diversity in Ethiopia. The migration of faba bean to South America, especially to the Andes, probably occurred during the fifteenth century along with the Spanish and Portuguese conquests. Another relatively recent development is that of winter faba beans in the nineteenth century in Europe, bred from Russian and French small-seeded, winter-hardy populations (Bond and Crofton 1999). Finally, China seems to be another secondary centre of faba bean genetic diversity and reproductively isolated from the European and West Asian gene pools, especially the Chinese winter gene pool (Zong et al. 2009).

Since wild faba bean has not been found so far, and because faba bean does not cross with other *Vicia* species, its wild ancestor remains unknown (Muratova 1931; Hanelt 1972; Cubero 1982). As a consequence, it is not known how much of the diversity in the wild progenitor of faba bean has been lost. All botanically close faba bean relatives are diploid species with 14 chromosomes, while faba bean is a diploid with 12 chromosomes. A study of the nuclear DNA amounts and the chromosome number of 56 *Vicia* species (Raina and Rees 1983) showed that faba bean has a high content of DNA and a large metacentric pair of chromosomes, twice the size of the remaining five pairs of acrocentrics and probably evolved from an ancestral fusion. The size of the faba bean genome (1C = 13.3 pg) is very different from that of *Vicia sativa* (1C = 2.3 pg), *Vicia narbonensis* (1C = 8 pg; Raina and Rees 1983) and the model species in legume genomics, *Medicago truncatula* (1C = 0.48 pg).

Along with numerous archaeobotanical findings, there is rich linguistic evidence on the antiquity of the faba bean crop (Mikić 2012). It was known by the ancestors of nearly all present European nations, as suggested by the attested roots in the protolanguages of the existing language families. The words related to faba bean in the modern Indo-European languages and their dialects demonstrate a high degree of similarity, especially among and within the most abundant branches, such as Germanic, Italic, and Slavic. The migrations of the Indo-European tribes produced numerous derivations. The lexicological evidence is still preliminary and incomplete, but it may provide a valuable testimony to the role faba bean and other traditional grain legumes played in the everyday diets of the old and modern Europe peoples.

3 Varietal Groups

Various levels of tolerance to winter conditions helped to distinguish between spring-sown and autumn-sown cultivars. They differ in their level of frost hardiness, vernalization requirement, day-length response and flowering time and fungal disease resistance, but the boundary between the two cultivar types is not always clear cut and some spring cultivars of Northern Europe may be grown in winter of some Mediterranean zones.

It has long been recognized that faba bean falls into several distinct genetic pools. Moreno and Martinez (1980) recognized that Mediterranean faba beans differed from the rest. The distinction between autumn-sown and spring-sown faba beans was also clear (Link et al. 2010). Detailed genotype-by-environment analysis has shown that in both the autumn-sown and spring-sown classes, there are distinctions in adaptation to oceanic and continental climates (Flores et al. 2012, 2013). Thus, we may recognize five broad classes or gene pools of faba bean, at least in Europe: Mediterranean, spring-sown for the oceanic zone, spring-sown for the continental zone, autumn-sown for the oceanic zone and autumn-sown for the continental zone. The five pools differ with regard to optimum flowering time, duration of crop growth and tolerance to cool and warm weather. Furthermore, they are exposed to different types of changes in photoperiod, with both Mediterranean and autumn-sown cultivars establishing during shortening days, while spring-sown cultivars establish during lengthening days, and Mediterranean types ripen during lengthening days, while continental and oceanic cultivars ripen during shortening days. Flowering initiates during lengthening days, but, in the Mediterranean region, days are very short at the start of flowering shortly after midwinter (~ 10-h day length), and very long in the boreal end of the continental region, where it starts about midsummer (~ 18 h day length). Faba bean is generally considered day-neutral, but some accessions require at least 12-h photoperiod in order to flower (Stoddard 1993). Inter-pool crosses are valuable for expanding genetic diversity and bringing in some positive alleles (Link et al. 1996), but can bring in undesirable traits through linkage drag unless there is an appropriate amount of backcrossing and selection.

The genetic variability of the species is quite large and often described on the basis of differences in seed weight, shape and size. Muratova (1931) cited four botanical varieties: the large-seeded faba bean (*V. faba* sp. *faba* var. *major*), with a seed weight of more than 1g and developed in the South Mediterranean regions and China; *V. faba* sp. *faba* var. *equina,* with an intermediate seed size (0.45–1.1 g per seed), developed in the Middle East and North Africa; *V. faba* sp. *faba* var. *minor* (0.2–0.5 g per seed) found in the Ethiopian highlands, Sudan and in Northern Europe; and *V. faba* sp. *paucijuga* grown in Central Asia.

Seed size is important for meeting market and farmer needs, although it is a continuously variable property. Large-seeded cultivars are widely favoured for food use, whether as a fresh green vegetable or, in many cultures, dry. Small-seeded cultivars are preferred for the specialist market for feeding pigeons (for which 'Maris Bead' is still grown in the UK) and easy drying after harvest in high latitudes such as the Nordic region and the Canadian province of Saskatchewan. The most widely grown cultivars, mostly used for animal feed in European cultures, and for use as dry pulses in West Asia and North Africa, have medium-sized seeds. Breeding progress for yield tends to be fastest in this intermediate size class, perhaps because of the balance between high seedling vigour from relatively large seeds and high multiplication rate from relatively small seeds.

The absence of tannins in seeds, improving the protein digestibility in monogastric animals resulted in a "zero-tannin" group of cultivars while low contents of vicine and convicine in seeds, improving feed value in poultry and reducing favism risks in humans, define a "low vicine–convicine" group. The combination of both traits has received the generic name "fevita" (Crépon et al. 2010).

4 Genetic Resources and Utilization

Because the wild progenitor is unknown, all genetic diversity available is contained in ex situ collections and local populations maintained by farmers in traditional agricultures. Faba bean germplasm of over 38,000 accessions is conserved worldwide in at least 43 national gene banks as well as at the International Center for Agricultural Research in Dry Areas (ICARDA). ICARDA safeguards the largest collection (10,045 accessions) in the world with materials from 71 countries with a high percentage of unique accessions. In the global collection, 8628 accessions comprise the international collection held in trust for the global community. Worldwide, ca. 95 % of known accessions are maintained in 17 major national or international ex situ collections (Table 5.4).

Since the germplasm collections are large, new methods are needed for choosing which accessions to screen for any given trait. Faba bean was the successful test case for applying the focused identification of germplasm strategy (FIGS) to abiotic stress resistance (Khazaei et al. 2013b). In this method, all available data on the provenance of the accession, particularly climate data, are used to predict the probability of finding appropriate germplasm.

The adequate ex situ conservation of faba bean collections is limited by the outcrossing floral biology of the species and its low multiplication rate (Suso et al. 2011). Genetic contamination, allele losses and inbreeding should be minimized. Large open-field populations separated or bagged to prevent intercrossing between each other would offer a good protection of the genetic integrity of each accession, but it is practically and economically unachievable with thousands of genotype entries. Only a few collections maintain small plots under protection from pollinating insects by net cages or by inter-plot plant barriers (Suso et al. 2006).

Large investments in the discovery of genetic variability and also in breeding activity for traits of agronomic interest have been made for faba bean at the end of the twentieth century in European countries and also at ICARDA, Syria, in particular for tolerance to several biotic and abiotic stresses (Bond and Poulsen 1983; Duc 1997). A large genetic variability has been identified in terms of floral biology. Cytoplasmic and nuclear determinisms of male sterility, various flower architectures or colours and various levels of autofertility and attractivity for pollinating insects,

Table 5.4 Major *Vicia faba* ex situ collections worldwide in 2014

Country	Institute/city	Number of accessions
Syria	ICARDA/Aleppo	10,045
China	CAAS/Beijing	5900
Australia	Austalian Grain Gene Bank/ Victoria	2445
Germany	Gene Bank IPK/Gatersleben	1920
France	INRA/Dijon	1900
Russia	VIR/St Petersburg	1881
Italy	Genebank/Bari	1876
Morocco	INRA/Rabat	1715
Spain	CNR/Madrid	1622
Poland	IOPG-PA/Poznan	1258
Ethiopia	PGRC/Addis Ababa	1118
Spain	IFAPA/Cordoba	1091
Poland	PBAI/Radzikow	856
Portugal	INRB—IP/ Oeiras	788
USA	USDA/Pullman	750
The Netherlands	DLO/Wageningen	726
Bulgaria	IIPGR/Sadovo	692
Total world	More than 43 known collections	38,360 accessions

ICARDA International Institute for Agricultural Research in Dry Areas, *CAAS* Chinese Academy of Agricultural Sciences, *IPK* Institute of Plant Genetics and Crop Plant Research, *INRA Institut national de la recherche agronomique, CNR* Consiglio Nazionale delle Ricerche, *IOPG* Institute of Packaging, Ghana, *PGRC* Plant Genetic Resources Centre, *IFAPA Instituto de Investigación y Formación Agraria y Pesquera. Anterior, PBAI* Plant Breeding and Acclimatization Institute, *INRB* National Institute of Biological Resources, *USDA* US Department of Agriculture, *DLO* Dienst Landbouwkundig Onderzoek, *IIPGR* Institute for Introduction and Plant Genetic Resources

were described; this could help to manipulate the levels of outcrossing (Link et al. 1994; Duc 1997).

As a result, current national breeding programmes are delivering higher-yielding new cultivars with improved combinations of disease resistances and a stronger emphasis on market quality. The success of plant breeding highlights the importance of ex -situ gene banks in collecting and preserving local landraces with their associated ranges of adaptations to respective crop environments.

4.1 Plant Height and Branching

A set of ca. 5000 faba bean accessions from the National Genebank of China was randomly sampled and analysed for plant height, which averaged 78 cm (Zong et al. 2006). The shortest accession was only 10.3 cm in height, and the tallest was 201.5 cm. The number of effective branches (with seeded pods) ranged from 1.1 to 11.4. In the *Institut national de la recherche agronomique* (INRA)-F reference collection of 250 genotypes, plant height ranged 40–210 cm, and 20% of accessions had a single main stem, whereas 70 % carried 1–3 basal branches (Duc and Magnin-Robert pers. com.). Winter genotypes are more likely to develop basal branching.

4.2 Yield-Related Traits

The accessions from the National Genebank of China were sampled as well for characters of yield component analysis (Zong et al. 2006), the number of effective pods ranged from 1.1 to 93.7. The number of mature seeds per pod is from 0.8 to 6.1. The 100-seed weight of dry grain ranged from 6 to 240 g and the dry grain yield per plant from 1.2 to 127.0 g. The mean dry pod length was 6.5 cm (range 1.2–18.8 cm) and the width 1.6 cm (range 0.7–3.5 cm; Zong et al. 2006).

4.3 Flower and Seed Colour

Flower colour (Fig. 5.1) is subject to oligogenic determination of major traits. A black dot is often present on the wing petals, and the flowers can be pure white or with diffuse pigments on all petals (purple or dark brown; Picard 1976). Recessive alleles at either of two genes (*zt-1* and *zt-2*) can determine the pure white flower trait and have a pleiotropic effect on the seed coat composition determining the absence of tannins.

As shown in Fig. 5.2, faba bean seeds display large genetic variation in seed coat colour and pattern (spotted, marbled), hilum colour and cotyledon colour (yellow or

Fig. 5.1 Coloured flowers

Fig. 5.2 Variation in seed coat colour and pattern

green). Major oligogenic determinations have been described with a strong relationship with tannin content (Picard 1976; Duc et al. 1999).

4.4 Abiotic and Biotic Stress Resistance

Genetic resources are continuously under evaluation for escape or protection mechanisms against these stresses (see reviews by Khan et al. 2010; Duc et al. 2011). Major biotic stresses differ in diversity and damage according to geographic zones and date of sowing. Ascochyta blight, chocolate spot, rust, downy mildew, *Fusarium* spp., broomrapes, nematodes, aphids, sitona weevil and bruchids are the major parasites or pests so far reported for which sources of genetic resistance are needed. The history of wide dispersion of faba bean has resulted in regional adaptations to abiotic stresses and variation in disease resistances, and genetic resources collections are being evaluated to detect resistance sources. Resistance to chocolate spot has been found in Ecuador, tolerance to heat stress in Bangladesh, frost tolerance in winter types from Europe, drought tolerance in the Mediterranean region, plus genetic resistances to Ascochyta blight, nematodes, bean yellow mosaic virus (BYMV) and bean leaf roll virus (BLRV) and broomrapes. Genetic resistances of faba bean to its major fungal diseases in temperate regions: chocolate spot, Ascochyta blight, rust, downy mildew (reviewed by Sillero et al. 2010) and to its important parasitic plant *Orobanche* (Fernandez-Aparicio et al. 2012; Rubiales and Fernández-Aparicio 2012) have been identified. The use of these resistance sources in breeding programmes is described in Part 6 of this chapter.

4.5 Nutritional Composition

The nutritional composition of 1828 faba bean accessions from the National Genebank of China was determined (Zong et al. 2006). On a total seed dry matter basis, the crude protein content ranged 17.6–34.5 %. The total starch content ranged 33.2–53.4 %, while amylose content of starch ranged 6.0–27.9 %. The lipid content ranged 0.52–2.80 %. A similar range of variations was described for these main seed constituents in a European seed collection (Duc et al. 1999), where mean seed coat proportion ranged 11.0–14.8 %, (the zero tannin cultivars being 2 % lower than tannin-containing ones), mean neutral detergent fibre ranged 13.4–21.7 %. In a set of 72 accessions, amylose content ranged 17–29 % of starch, with a significant tendency for lower content in larger-seeded accessions.

By contrast with soya, trypsin inhibitor activity is generally low in faba bean seeds with a range of 0.3–5.3 units/mg measured on a European collection (Duc et al. 1999). In this collection, white-flowered genotypes were tannin-free, whereas coloured-flower genotypes contained 6–10 g/kg of condensed tannins, concentrated in the seed coat. An allele (*vc-*) of the *VC* gene has been discovered (Duc et al. 1989) which reduces by 10–20 times the content of vicine and convicine in the seed. Mean vicine–convicine content in conventional genotypes is 6–12 g/kg of seed dry matter (DM) when it is close to 0.5 g/kg in *vc*-homozygotes. These two products are concentrated in cotyledons and are responsible for favism risk in humans and lower production performances in chickens or laying hens (Arese and De Flora 1990; Crépon et al. 2010).

5 Major Breeding Achievements

The effort allocated to faba bean breeding is relatively small, and it is strongly associated with its relatively small area of cultivation. Major breeding programmes are located in several European countries, the ICARDA network, Egypt, Morocco, China, Canada and Australia. In the 2014 edition of the list A of the European catalogue, only 130 faba bean cultivars are listed in contrast to 397 field pea, 363 soya bean and 2109 bread wheat cultivars. This reflects the small size of the breeding programmes on faba bean in comparison with other major agronomic species. These programmes are mostly maintained by public institutions. In 2014, in Europe, fewer than ten private breeding programmes for faba bean were identified. Beyond the genetic improvement achieved by farmers through empiric mass selection over previous centuries, a burst of research and breeding activity occurred in Europe in the 1980s after a severe plant protein supply deficit occurred in 1973, which represents a relatively short history of breeding.

During the past 50 years, remarkable steps of improvement can be identified with representative genotypes which at a time have developed reduced stem growth preventing lodging and associated with an improved harvest index, earliness, yield stability and harvestability (cv. Aguadulce in the broad bean types, cvs Alfred and Espresso in the minor types); winter types with increased frost tolerance and various levels of earliness adapted to Northern Europe (cvs. Maris Beagle, Hiverna, Diva, Castel, Irena, and Wizard); winter types adapted to the Mediterranean zone (cvs. Alameda and Baraca); low vicine–convicine content in the seed (cv. Melodie and its successors); high tolerance to Ascochyta blight (cv. Irena); tolerance to *Orobanche* (Egyptian cv F402); outstanding spring types with short stems, low tannin and high earliness (cvs. Optica and Kompacta) or with small seeds (cv. Maris Bead; Duc 1997; Flores et al. 2012).

6 Specific Goals in Current Breeding

The traditional goal for a plant breeder is yield and yield stability. There are many ways of achieving that, considering phenology, adaptation and stress resistance. Moreover, the product has to be marketable, so key quality attributes have to be met. Lodging resistance, or standing power, is one of the top objectives and is affected by stem strength and stem length. The available dwarfing gene, used in the garden broad bean cultivar Sutton's Dwarf, reduces internode length too much for use in field-grown, machine-harvested cultivars. The Japanese autumn-sown cultivar 'Rinrei' has relatively short internodes and strong stems, but appears fairly unique in this regard. Reducing stem length by use of the terminal-inflorescence gene *ti* has been associated with too great a loss in yield in most environments, although it is used in Spanish cultivar 'Retaca' grown for vegetable use.

The indeterminate habit allows the crop to produce more reproductive nodes but is relatively sensitive to poor growing conditions, in comparison to other crops. This sensitivity equates broadly to stress susceptibility, so the various stresses are discussed next.

A major goal in sustainable crop production is intensification of the use of grain legumes in crop rotation for environmental contributions of nitrogen fixation and reduction of inputs. At the same time, the global discussion is expanding on the issues of food and nutritional security. Faba bean can play a significant role in both these areas, and it can also make a contribution to the expansion of the vegetable-based protein supply in temperate production regions. Breeding programmes will need to keep in mind the requirements for increased yield relative to other protein crops, while maintaining a focus on reduced input cost, on suitability for mechanised harvest and on improvement of protein quality and reducing vicine–convicine in seeds.

6.1 Biotic Stresses

Faba bean is subject to a range of diseases, parasites and pests, and resistance breeding is an important issue with priorities differing from region to region. As part of the levers for integrated disease management (reviewed in Stoddard et al. 2010), the resistances to pathogenic fungi reduce the need for application of fungicides but provide incomplete resistance. Therefore, they should be combined with practices such as crop rotation (duration and choice of non-host species), with sanitary control of certified seeds in the case of aschochyta blight, with reduction of relative humidity in the canopy (by lower sowing rate, good soil drainage, choice of plant architectures and prevention of lodging), with targeted period of sowing and plant cycle and with prevention of nutrient deficiencies, frost damage and weed infestations (Stoddard et al. 2010).

Of the pathogenic fungi, chocolate spot disease, caused by *Botrytis fabae* Sard., is the one on which there is the most literature (Stoddard et al. 2010). It is present in all regions where faba bean is grown and, because its expression depends particularly strongly on environmental conditions, it is one of the most difficult diseases to handle in a breeding programme. When conditions are just right for the fungus for about 48 h, with damp leaves, high atmospheric humidity and temperatures close to 20 °C, it grows rapidly from its small, brown spots scattered across leaves until the whole plant is dead. In any location, such conditions may occur only once every 10 years but when they do, the farmer has only a brief opportunity to use a fungicide before the crop is past saving. Resistance is quantitative, and the sources used by ICARDA breeding programmes are from South America (Tivoli et al. 2006), where faba bean has been cultivated for only five centuries.

Ascochyta blight, caused by *Ascochyta fabae* Speg. (teleomorph *Didymella fabae* Jellis and Punithalingam), is prevalent on autumn-sown faba bean, including Mediterranean climates. Breeding for resistance to this disease has been more successful than against chocolate spot, with major and minor genes identified (Kohpina et al. 2000) along with quantitative trait loci (QTLs; Román et al. 2003; Avila et al. 2004). Some accessions have shown stable resistance across environments, whereas others were more resistant in either the Mediterranean or continental environment (Rubiales et al. 2012).

Rust disease, caused by *Uromyces viciae-fabae* (Pers.) J. Schröt., is prevalent in warmer conditions than either of the preceding two. Rust resistance is generally quantitative or incomplete, but a hypersensitive response that slows rather than stops the spread of the disease has also been reported (Sillero et al. 2000). Combined disease resistance is obviously valuable, and it has been possible to select resistances to chocolate spot disease and rust disease together (Villegas-Fernández et al. 2011).

Resistance to downy mildew, caused by *Peronospora viciae* f. sp. *fabae* (Berk.) Caspary, has been identified in the UK germplasm (Thomas and Kenyon 2004) and resistance to Cercospora leaf spot, caused by *Cercospora zonata* G. Winter, is an objective in Australia (Kimber and Paull 2011).

Faba bean is susceptible to several generalist pathogens including *Fusarium oxysporum* Schl.; *Rhizoctonia solani,* J.G. Kühn; *Alternaria alternata* (Fr.) Keissl.; and *Sclerotinia sclerotiorum* (Lib.) de Bary; but there have been few advances in breeding for resistance to these. It is also attacked by the stem nematode (*Ditylenchus dipsaci* (Kühn) Filipjev species complex), root-knot nematodes (*Meloidogyne incognita* (Kofoid and White) Chitwood and *Meloidogyne javanica* (Treub) Chitwood) and root-lesion nematodes (several species of *Pratylenchus*). These are generally managed by rotation rather than resistance breeding, although it has been noted that some cultivars of faba bean are non-hosts of *Pratylenchus neglectus*

(Rensch) Filipjev and Schuurmans Stekhoven (Yunusa and Rashid 2007), and this may be a valuable avenue for further investigation.

Breeding for parasitic insects resistance has not been so far developed due to the lack of solid resistance sources. In the Mediterranean basin, broomrapes are the major limitation to growing faba bean. The main species of these root parasites are crenate broomrape (*Orobanche crenata* Forsk.), fetid broomrape (*Orobanche foetida* Poir.) and Egyptian broomrape (*Phelipanche aegyptiaca* Pers.) Pomel (syn. *Orobanche aegyptiaca* Pers.). All can attack other host species as well, but crenate broomrape is currently the most important and widespread. Field testing for resistance is especially difficult for a root parasite, as its seeds are unevenly distributed in the soil and individual host plants commonly escape infection. The resistance to *O. crenata* is generally quantitative (Díaz-Ruiz et al. 2010) and has been used in breeding programmes since the release of the Egyptian accession 'F402' in 1982.

6.2 Abiotic Stresses

Faba bean, like all crops, has optimum temperatures, water and mineral requirements for growth, and conditions outside these ranges cause stress. Climate change is likely to increase the frequency of heat and water deficit stresses, but may also affect the exposure to other stresses. Recovery from a transient stress is as important as survival, but it is less often examined in experiments.

Drought, or water deficit stress, is considered by many workers to be the major abiotic stress for most crops. In almost all conditions, transient water deficit occurs at some stage during the growing season, and, particularly in Mediterranean climates, terminal water deficit is normal. Drought responses are usually characterised as escape, avoidance and tolerance. Early maturity in a Mediterranean climate is an example of escape, improved water metabolism through better root systems and stomatal closure represents avoidance and protection of cell metabolism through an osmoprotective substance such as glycine betaine or proline provides tolerance. Escape is not a useful mechanism against transient, mid-season water deficit, and the other approaches are necessary. There is direct evidence for wide variation in stomatal traits, and there is indirect evidence from the same experiments for variation in the ability of roots to obtain water (Khazaei et al. 2013a). Canopy temperature depression is an effective, rapid and economical way to determine the water status of a faba bean plant (Stoddard et al. 2006; Khan et al. 2007, 2010; Khazaei et al. 2013a). Osmotic adjustment through the synthesis of protective substances has not been identified in faba bean.

Faba bean is well adapted to cool temperatures and seldom shows damage from temperatures down to −6 °C, although flowers are not tolerant of frost (Link et al. 2010). Autumn-sown beans in oceanic and especially continental regions have to withstand much harder frosts than this, often in repeated freeze–thaw cycles, and in some regions prolonged snow cover as well. The assemblage of traits required for winter tolerance is complex (Link et al. 2010) and has seen some attention to individual components such as frost tolerance (Arbaoui and Link 2008; Arbaoui et al. 2008) and tolerance to snow cover (Fukuta and Yukawa 1998).

There is little literature on heat stress as separate from water deficit stress in faba bean, but it has been suggested that selection for one would also select for the other. Calibration of the agricultural production systems simulator (APSIM) crop growth model suggested that the optimum temperature for growth in this species is 23 °C, whereas 27 °C was used for the CROPGRO model (Boote et al. 2002). The temperature at which processes such as pod setting fails ranges from 34 to 45 °C (Boote et al. 2002), and this implies that selection for heat tolerance may be required at many latitudes as climate changes.

Transient waterlogging may occur after a heavy downpour, during prolonged rainy periods, after snowmelt or as part of the difficulties associated with compacted soil. Faba bean was the most tolerant of seven grain legume species in a pot-based test at 35–42 days after sowing, and there was intraspecific variation in tolerance as well when tested 21–30 days after sowing (Solaiman et al. 2008).

Oxidative stress is a part of all abiotic stresses. General stress resistance (or avoidance) may therefore be conditioned by better methods of responding to oxidative stress. Superoxide dismutase (SOD) is one of the enzymes involved, and it is inducible by stress in this species, but no report showing significant genotypic variation in SOD activity has come to light (Khan et al. 2010). There are many articles on the induction of SOD activity by various metal ions (*Cd, Co, Al, La and* Pb, among others), and this method may be a simpler and more consistent way of inducing stress than extremes of water supply or temperature.

Early maturity is a desired attribute in other zones besides Mediterranean climates. Faba bean often matures after the small-grained cereals, whether spring sown in boreal climates or autumn sown in oceanic climates. There is a fundamental conflict between the aims of earliness and high yield, because the more radiation that a crop can intercept through a longer growing period, the greater its potential biomass production. Thus, the correlation between growing degree days and yield is often positive and significant (Fig. 5.3). Breeding progress for yield is easily made along the trend line, but is more worthwhile when it moves the line upward. Some of this increase presumably comes from improvement in radiation use efficiency, but there is remarkably little literature on this subject.

6.3 Targets Against Abiotic Stresses Associated with Climate Change

A reliable symbiotic activity in situations of abiotic stresses (drought, waterlogging and salt in particular) is crucial to assure stable yields. Rhizobial and VAM symbioses, each use 4–16 % of the assimilate produced by a faba bean plant, but the crop generally compensates for this by increasing its photosynthetic rate, so there is no net yield penalty to either of these symbioses (Kaschuk et al. 2009).

This indicates that faba bean yield is more sink limited than source limited, which has repercussions for other breeding objectives such as competitive ability and seed yield in a situation of stress, but it does not exclude that particular plant–symbiont combinations may be selected and monitored to improve the stress tolerance.

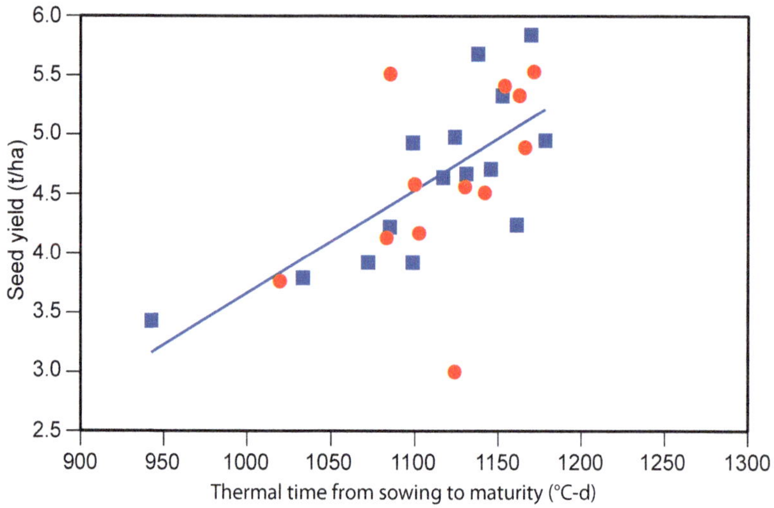

Fig. 5.3 Correlation between growing degree days and yield

Nutrient use efficiency is another important trait for the future, as we try to make crops thrive on reduced inputs. Faba bean releases citrate, malate and protons that solubilize many nutrients but gradually acidify the soil. Phosphorus uptake ability ranged threefold in a set of 50 faba bean accessions, attributed to variation in root exudates (Rose et al. 2010). A dedicated search should uncover variation in phosphorus utilization efficiency as well as uptake efficiency to be used in breeding programmes.

Winter faba bean offers the pertinent agronomic advantages of winter crops, namely higher probability of good sowing conditions, longer growth cycle, good root establishment, earlier flowering and seed-filling period, escape from the main risk of summer drought stress and early maturity. With the advent of climate change and some improvement of freezing resistance and general winter hardiness (Arbaoui and Link 2008), the zone of adaptation of winter faba bean is expected to expand to more continental regions in Europe.

European winter-hardy faba bean cultivars are *minor* types, some examples include Diva, Diver, Gladice, Hiverna, Husky and Wizard. Their level of resistance to freezing depends very much on (1) the basic resistance without hardening and on (2) the response to hardening. Hardening is also well known as cold acclimation and results from exposure to low, nonfreezing temperatures. For faba bean, about 5 °C and short day length (~8–10 h) were effective for hardening (Herzog 1987; Arbaoui and Link 2008). A slow dehardening will also be a key breeding trait required to tolerate more irregular winter temperatures.

Tolerance to drought, heat stress and salinity has been extensively investigated at ICARDA and elsewhere and were recently reviewed by Duc et al. (2011). Drought may be escaped by a deep root system together with earliness that concludes the reproductive phase before a terminal drought. Water loss is also minimized through stomatal control. Inbred line ILB938/2 and cultivar Melodie were the two best accessions at avoiding dehydration in this way (Khan et al. 2007). These two accessions differ at two loci affecting stomatal activity (Khazaei et al. 2014). The broad adaptation of the crop including its cultivation in tropical and subtropical regions, such as Bangladesh, suggests that many accessions of the species are tolerant of relatively high temperatures as long as water is in sufficient supply. Notable occurrences of salinity tolerance was found in accessions from China, Greece, Syria, Morocco and Australia (Leonforte at al. 2013). The mechanisms and genetic basis of these tolerances still remain to be analysed for their practical use in breeding.

Interestingly, grain legumes have the potential to decrease the greenhouse gas release in agriculture by reducing the production and use of chemical nitrogen fertilisers. Some rhizobia associated with legumes produce enough nitrous oxide reductase that greenhouse gases release cannot be detected (Sameshima-Saito et al. 2006), showing another way in which this environmental hazard can be reduced.

7 Breeding Methods and Specific Techniques

7.1 Evaluation

A high level of genotype × environment interaction has been demonstrated for faba bean, particularly for comparisons between germplasm pools and across mega-environments. There is also variation in the degree of adaptation across environments among cultivars of a germplasm pool, for example, autumn-sown cultivars for medium-duration winters and spring-sown cultivars where winters are long and severe, as in China. Much of this variation can be attributed to variation in the optimal phenology for different geoclimatic regions, but variation in stresses, such as the dominant diseases, between regions are also important. Trials conducted in Germany and ICARDA demonstrated very poor yields of Central European cultivars in Syria, whereas the Mediterranean cultivars yielded well, and were early flowering, in both environments (von Kittlitz et al. 1993). Similarly, the three European germplasm pools, minor, major and Mediterranean were adapted to different regions in trials in German and Mediterranean environments (Schill et al. 1998), while Mediterranean, spring and winter germplasm showed adaptation to specific regions in trials conducted in Mediterranean and continental European environments (Annicchiarico and Iannucci 2008). European spring faba bean production regions could be assigned to three mega-environments: continental, oceanic and Mediterranean, based on the genotype plus genotype × environment yield response of European spring faba bean cultivars across a range of test environments (Flores et al. 2013). These results demonstrate the importance of selection of trial locations that represent the target environments of the breeding programme when evaluating and selecting breeding lines. Genome-wide association studies combined with high throughput phenotyping studies for traits of interest represent new tools to evaluate genetic resources and to detect genes of interest for breeding.

7.2 New Variation

As previously described, faba bean is reproductively isolated from all other *Vicia* species. This limits the genetic variation available for crop improvement to existing germplasm collections, to new collection missions or to de novo approaches to generate genetic diversity. The two major methods to introduce novel variation in faba bean are mutation and transformation. Experiments undertaken by Sjodin (1971) with a range of mutagens, including X-rays, generated numerous mutant phenotypes, most of which were determined by recessive alleles against wild allele. Oldach (2011) listed 19 mutant faba bean cultivars released in the period 1959–2009. The mutant traits included early maturity, plant architecture, yield, dwarfism, protein content, disease resistance and lodging resistance. The major trait developed through mutagenesis and used in breeding programmes was the determinate or terminal-inflorescence growth habit conditioned by the recessive allele of the *ti* gene.

8 Integration of New Biotechnologies in Breeding Programmes

8.1 The Difficulty of Adaptation of In Vitro Methods to Vicia faba

The application of biotechnology in plant breeding requires an efficient in vitro regeneration procedure. Despite the intensive investigations since the 1960s, a reliable regeneration system for faba bean has been difficult to establish. Gnanasambandam et al. (2012) reviewed the attempts to cultivate faba bean in vitro, mainly focused on the optimal growth of callus tissue or suspension cultures rather than the induction of shoot morphogenesis and plant regeneration. Regeneration via organogenesis using in vitro seedling as explants by micro-cutting were followed by axillary bud proliferation or meristem tips as explants. Despite the use of different media compositions and explant sources, and optimizing the in vitro culture conditions, the morphogenic potential of faba bean cells was relatively low, and all attempts to initiate shoot regeneration remained unsuccessful (Gnanasambandam et al. 2012).

The major constraint in the process was the deterioration of the explants and cultivated tissues due to the action of phenolic compounds. Therefore, the effect of various chemical and physical factors in axillary shoot cultures was examined, revealing that low temperatures limit the formation of phenolics. Furthermore, plantlet regeneration from explants lacking pre-existing shoot meristems was reported. These achievements were followed by the regeneration of faba bean plants from protoplasts isolated from leaves and shoots, and from cell suspensions. Subsequently, somatic embryogenesis in callus and suspension cultures derived from immature cotyledons of faba bean and an improved protocol combined with the *Agrobacterium tumefaciens*-mediated gene transfer were also achieved (Gnanasambandam et al. 2012 and references herein). An efficient in vitro regeneration protocol has been reported using single cotyledon explants with half of the embryo axis (Anwar et al. 2011).

One of the most limiting factors to obtain successful faba bean regeneration is the difficulty of root induction from regenerated shoots. Root growth does not always occur in the earlier stages in plant cell culture, and it is an essential requirement for successful plant growth after the micropropagation procedure. This challenge seems to be overcome, as the in vitro root induction of faba bean was achieved from regenerated shoots (Ismail et al. 2011). Grafting techniques can also overcome the rooting difficulties. The efficient regeneration protocol reported so far allows for successful micropropagation of faba bean, a key issue for future genetic improvement of plants via transformation protocols.

8.2 Mutants and Transformants

The lack of an efficient protocol for the regeneration of transgenic plants was the main obstacle to faba bean transformation. The most widely used method to transfer foreign genes into dicotyledonous plant is *Agrobacterium*. However, only a few reports on faba bean using *Agrobacterium*-mediated transformation for producing transgenic plants based on stem segments, mature embryo discs or cotyledonary nodes are available (Bottinger et al. 2001; Hanafy et al. 2005). A regeneration and microprojectile-mediated transformation system was established using the biolistic bombardment gene delivery system (Metry et al. 2006).

Although the major obstacles in generating transgenic faba bean plants seem to be circumvented, the number of transgenes introduced so far is still limited. Two successful studies used this approach to improve protein quality (reviewed in Gnanasambandam et al. (2012)) and Hanafy et al. (2013) reported the first evidence that *PR10a,* a gene-enhancing salt and/or drought tolerance in potato, transferred into faba bean by *A. tumefaciens*-mediated transformation system, caused the same effect in this crop.

9 Molecular Characterisation of Genetic Resources

A burst in development of diverse molecular tools for molecular characterisation of *V. faba* germplasm, especially over the past decade, was followed by an increasing number of empirical studies analysing larger numbers of genotypes and/or accessions representing geographically broader diversity of *V. faba* (e.g. Kwon et al. 2010; Wang et al. 2012), its botanical varieties (e.g. Link et al. 1995; Sanz et al. 2007; Ouji et al. 2012) and/or winter versus spring forms of this crop (e.g. Zong et al. 2009, 2010; Wang et al. 2012). However, overall insights into the geographic structure of nuclear genetic diversity in *V. faba,* heterozygosity levels, delineation of core collections and broad utilization of these tools in breeding efforts and marker-assisted selection are still to come, and future large-scale surveys will provide information on the cytoplasmic diversity of the *V. faba* gene pool as well. Chloroplast DNA (cpDNA) and mitochondrial DNA (mtDNA) genomes, which are predominantly maternally inherited, are not only useful for genetic diversity surveys but are also markers of choice for both phylogenetic and phylogeographic studies that can shed more light on the origin and domestication of *V. faba*.

9.1 Molecular Characterisation of Cytoplasmic Diversity

Work on the *V. faba* chloroplast genome during the 1980s was mainly oriented towards understanding the evolution and organization of the genome (e.g. Ko et al. 1987), gene organization (e.g. Mubumbila et al. 1984) and gene mapping (e.g. Ko et al. 1983). The most striking discovery was that the *V. faba* chloroplast genome, whose size has been estimated to c. 123 kb (Raina and Ogihara 1994), does not have the quadripartite structure present in most angiosperms, that is, a large and a small single-copy region separated by a pair of c. 25-kb long inverted repeats comprising ribosomal RNA genes. One set of ribosomal RNA genes in an inverted repeat segment closest to the *psb*A gene and a quarter of the ancestral small single-copy region have been lost, and genetic organization of *V. faba* cpDNA genome was established via three inversions after the deletion event (Ko et al. 1987). Findings of a similar organization of the cpDNA genome in temperate legumes (e.g. *P. sativum, M. truncatula* and *L. sativus*), but not in tropical *Glycine max* and *Lotus japonicus* provide evidence that two large clades in Fabaceae are characterised by contrasting patterns of cpDNA evolution. Interestingly, the structural differences between the cpDNA of these two clades correspond to two contrasting nodulation types (Wojciechowski et al. 2004) resulting from plant-rhizobia coevolution.

The focus of studies on the *Vicia* cpDNA genome has shifted over the past 20 years towards utilization of nucleotide cpDNA diversity in phylogenetic surveys. Raina and Ogihara (1994) were the first to employ restriction-site variation (restriction fragmentation length polymorphisms, RFLPs) for assessing interspecific relations in the genus *Vicia* and delineation of closest wild relatives of *V. faba*. Fennell et al. (1998) expanded research to sequences of the *trn*L intron, while Potokina et al. (1999) studied the Restriction Fragment Length Polymorphism Analysis of PCR-Amplified Fragments (PCR-RFLP) of five chloroplast genes. Haider et al. (2012) demonstrated that PCR-RFLPs on nine chloroplast regions and a set of 12 consensus chloroplast simple sequence repeats (ccSSRs) can be used for identification of diverse *Vicia* species. Schaefer et al. (2012) provided the most comprehensive phylogenetic inference of the tribe *Fabeae* based on both chloroplast and nuclear data. Unfortunately, instead of providing unambiguous answers regarding the progenitor and allies of *V. faba,* these studies as well as molecular surveys based on nuclear and/or mitochondrial genomes (e.g. Potokina et al. 1999; Sanz et al. 2007) were largely incongruent with traditional morphology-based taxonomies and systematics (more than 20 since Linnaeus, reviewed in Maxted 1993) because morphologically similar species, such as *V. faba* and species from the Narbonensis complex, apparently are not close at the DNA level.

Furthermore, available molecular phylogenies are not consistent in delineating the closest relatives of *V. faba*. Potential reasons include convergent or homoplastic morphology and/or molecular characters, hybridisation followed by introgression, backcrossing following secondary contact, incomplete lineage sorting in recently diverged taxa or obstacles related to the methods used and/or low resolution of data sets of various sizes. The unquestioned outcome of diverse studies is the distinctiveness of *V. faba,* supported by cytological, karyological, hybridisation, biochemical and other studies and the justification for treating *Vicia paucijuga* as a subspecies of *V. faba* (Potokina et al. 1999; Schaefer et al. 2012).

The features of the *Vicia* cpDNA genome advantageous for phylogenetic surveys have posed a limitation on studies on the intraspecific level (Shiran and Mashayekh 2004; Haider et al. 2012). Diverse studies led to the conclusion that *V. faba* originated from a single ancestral maternal lineage that either possessed little cpDNA diversity at the origin or lost much of its variation during domestication through a severe bottleneck (Shiran and Mashayekh 2004). However, further studies based on broader samples are required for more general conclusions on evolutionary unfolding of *Vicia* and phylogeography of *V. faba.*

MtDNA diversity in *V. faba,* on the other hand, was observed rather early. Unlike cpDNA, angiosperm mtDNA genomes are rather dynamic and very variable in size (200–2500 kb), structure and sequence content. MtDNA mutations/rearrangements that can cause cytoplasmic male sterility (CMS) were observed in *V. faba* during the 1980s (e.g. Scalla et al. 1981; Nikiforova and Negruk 1983 among many other examples), and it has been shown that the nuclear context participates in the control of mitochondrial plasmid sequence, appearance, and copy number (Flamand et al. 1992).

Reports on organization and variability of *V. faba* mtDNA genes (e.g. MacFarlane et al. 1990) were confirmed when the sequence of the complete mtDNA genome of this crop became available (Negruk 2013). The size of the mitochondrial genome of cultivar Broad Windsor was estimated at 588 kb with 45.04 % guanine, cytosine (GC) content, with c. 40 % similarity with mtDNA of the legumes *Lotus japonica* and *Millettia pinnata* (Kazakoff et al. 2012). Furthermore, 45 % of *V. faba* mtDNA sequences were found to be homologous to the *Medicago truncatula* nuclear genome.

The availability of the complete mtDNA sequence of *V. faba* and other legumes opens a new chapter in phylogenetic and phylogeographic studies that may provide answers on the origin and domestication of *V. faba*. Using total mtDNA as a probe against mtDNA from 52 accessions of 12 *Vicia* species, Van de Ven et al. (1993)

were able to distinguish species but not cultivars representative of the different botanical varieties of *V. faba*. Furthermore, Scallan and Harmey (1996) found considerable mtDNA diversity at the species level, as they were able to divide nine *V. faba* commercial cultivars of the *major* type, which were all Western cultivars from a narrow geographical range, into two cytoplasmic types and at least six groups. These findings are supported by work in progress (Aleksić et al. personal communication).

Two major independent and spontaneous sources of cytoplasmic male sterility were identified in faba bean (350 and 447 cytoplasms), both of them displaying instability features incompatible with a hybrid seed production scheme (Duc et al. 1992). Markers of the male sterility trait at the level of mtDNA and RNA could not be detected (Flamand et al. 1992). Independently of mitochondria, RNA-containing particles present in the cytoplasm appeared to be accurate markers of the 447 determinism of male sterility (Scalla et al. 1981). A high content of plant tissues in these cytoplasmic particles was also a good marker of the stability of the 447 male sterile phenotype.

9.2 Genomics of Nuclear Genome

Recently, various molecular markers have been successfully used to characterise genetic diversity in faba bean accessions. Link et al. (1995) examined three groups of faba bean inbred lines from European and Mediterranean gene pools by randomly amplified polymorphic DNA (RAPD) assays. Amplified fragment length polymorphism (AFLP) diversity analysis of 22 *V. faba* elite cultivars from Europe found most to share genetic similarity among the spring types, with one winter cultivar classified as separate, and the closest relationship being between two advanced lines from the same parents (Zeid et al. 2001). A diverse collection of 20 faba bean accessions from a gene bank in Spain that were assayed with transposon-based molecular markers sorted into groups mostly unrelated to geographic origin or morphology (Sanz et al. 2007). Twenty Greek landrace populations were analysed with inter-simple sequence repeat (ISSR) by Terzopoulos and Bebeli (2008), with separation of small-seeded *minor* types from medium- to large-seeded Mediterranean types which were divided into two subgroups. Winter ecotype (Zong et al. 2009) and spring ecotype (Zong et al. 2010) faba bean germplasm sets from China were separately compared with accessions from the rest of the world using AFLP, revealing significant separations between Chinese accessions and the accessions from outside China. Kwon et al. (2010) evaluated a world collection of 151 faba bean accessions with target region amplification polymorphism (TRAP) markers, and showed a strong association with geographic origin, with five major groups identified with a dendrogram. These studies have thus shown that when studies include cross-continental sources of accessions, associations of genetic diversity with geographic diversity tend to be found, especially with spring and winter habits of land races reflecting the agriculture practices in respective diverse environments and geographic origins.

The genetic diversity and relationships of 802 faba bean accessions from different geographical locations of China and rest of the world (Asia except China, Europe and Africa) were examined using 11 ISSR primers (Wang et al. 2012). Accessions from northern China showed highest genetic diversity, while those from central China showed low diversity. Chinese spring faba bean germplasm was clearly separated from Chinese winter faba bean, based on principal component analysis and unweighted pair group method with arithmetic mean (UPGMA) clustering analysis. Winter accessions from Zhejiang (eastern China), Jiangxi (eastern China), Sichuan (southwestern China) and Guizhou (southwestern China) were quite distinct from those from other provinces in China. Great differentiation between Chinese accessions and those from rest, of the world was shown with a UPGMA dendrogram. Analysis of molecular variance (AMOVA) analyses demonstrated large variation and differentiation within and among groups of accessions from China. As a continental geographic group, accessions from Europe were genetically closer to those from North Africa. Based on ISSR data, grouping results of accessions from Asia, Europe and Africa were obviously associated with their geographical origin. The overall results clearly indicated that the genetic relationship of faba bean germplasm was closely associated with their geographical origin and their ecological habit. China is likely to be another secondary centre of diversity for faba bean, especially the Chinese winter gene pool, which has been reproductively isolated from the European and West Asian gene pools (Zong et al. 2009; Wang et al. 2012).

10 Integration of New Biotechnologies in Breeding Programmes

Complete exploitation of genomic and transcriptomic approaches and marker discovery to faba bean breeding is still to be achieved, but results obtained so far are promising. One of the main reasons for slow progress is the extraordinarily large size of the faba bean genome. Nevertheless, comprehensive genomic and post-genomic tools are being developed. These include several types of molecular marker sets, saturated genetic maps bearing relevant genes or QTLs and both cDNA libraries and transcriptome datasets for reverse genetics. A combination of approaches, candidate gene and colinearity with model legumes, is being pursued for the identification of genes underlying agronomically important traits. Finally, a functional consensus map of faba bean has been constructed integrating all the markers, genes and QTLs previously published (Román et al. 2004; Satovic et al. 2013). This section provides a review of the genetics and genomics tools available so far in this crop. New genomic advances, together with conventional breeding tools, will push forward future genomic-based faba bean breeding programmes.

10.1 Molecular Marker Development

The development of linkage maps in faba bean has been parallel to the progress and availability of different types of molecular markers. The first faba bean linkage groups only included morphological and isozyme loci but, with the advent of PCR-derived markers, RAPDs, microsatellites and, more recently, sequences or gene-derived markers based on single-nucleotide polymorphisms (SNPs), are part of the genetic linkage maps reported in the crop (reviewed in Torres et al. (2012)).

The first set of microsatellites (or simple sequence repeats; SSRs) from chromosome I DNA libraries were developed by Pozarkova et al. (2002). This number was enlarged by Zeid et al. (2009) and Gong et al. (2010). The recent screening of expressed sequence tags (EST) sequences in public databases led to the development of new SSR loci that have been validated among several faba bean accessions (Ma et al. 2011).

Faba bean has now achieved notable advances in marker repertoire, thanks to the synteny among closely related species. In the last decade, significant progress has been made in developing genomic resources in model species (*M. truncatula*), major legumes (soya bean and common bean, peanut and pea), and a number of so-called orphan crops (such as chickpea, cowpea, lentil and pigeon pea). As a result, reference genome sequences are now available for some of these species together with saturated genetic and physical maps. The availability of ESTs from *Medicago* or related legume crops constitutes a novel source of markers physically associated with coding regions that are being extensively exploited in faba bean for gene discovery and the macro- and microscale genomic comparison. A number of intron-targeted amplified polymorphic (ITAP) markers mapped in *M. truncatula, Glycine max* and *Lupinus* were used to develop the first gene-based genetic map of faba bean using a recombinant inbred line (RIL) population derived from cross Vf6 × Vf27 (Ellwood et al. 2008). This genetic map was further used to identify and validate QTLs controlling flowering time and other yield-related traits (Cruz-Izquierdo et al. 2012) and to study macrosyntenic relationships between *V. faba* and other related legume species. Within the framework of the Grain Legumes Integrated Project (GLIP), a large set of additional expressed sequence tag (EST)-derived markers were developed and mapped in the progeny of two faba bean inbred populations (Vf6 × Vf136 and 29H × Vf136) segregating for both broomrape and *A. fabae* resistance (Díaz-Ruiz et al. 2009, 2010; Gutierrez et al. 2013).

At present, one of the main pillars of genomic breeding is the development of high-throughput DNA sequencing methods (known as next-generation sequencing (NGS)). These technologies provide the mass sequencing of genomes and transcriptomes and deliver a vast array of genomic information for discovering genes and their positions. The lower cost, compared to traditional sequencing methods, is accelerating the analysis of less studied species, such as faba bean. Thus, using the Roche 454 GS FLX Titanium Platform sequencing, a genome library with 125,559 putative simple sequence repeat (SSR) sequences was constructed and characterised for repeat type and length from a mixture of spring- and winter-sown faba bean genotypes (Yang et al. 2012). A set of 94 primer pairs produced amplicons polymorphic among 32 faba bean genotypes selected from diverse geographical locations. Using the same technology, cDNA samples obtained from various faba bean tissue types allowed annotation of more than 18,000 faba bean unigenes. Primer pairs for SSR-containing expressed sequence tags (ESTs) were designed and a subset of 96 EST–SSR markers was screened for validation across different cultivars (Kaur et al. 2012). The high quality of novel SSRs developed via next-generation sequencing technology is a valuable source for future map construction and QTL mapping.

A new allele-specific PCR SNP genotyping system developed at KBiosciences (KASPar) has been also assayed to identify putative SNPs between sequenced seedling transcriptomes from two contrasting inbred lines of faba bean. Assays were designed for more than 800 loci, 85 % of which (757) showed robust amplification and were validated on a test panel of 33 inbred lines (Cottage et al. 2012). The approach was used to build a medium-density genetic map that will facilitate trait dissection in future faba bean breeding programmes and has been used to identify QTLs associated with stomatal response in a population of recombinant inbred lines (Khazaei et al. 2014).

10.2 Gene Tagging of Qualitative Traits

Some faba bean seed quality traits along with the hypersensitive resistance to rust are regulated by single genes. In these cases, bulk segregant analysis (BSA) has proved to be an effective method for the identification of markers tightly linked to the corresponding responsible genes (Avila et al. 2003; Gutierrez et al. 2006, 2007, 2008).

The bioactive compounds mostly targeted in faba bean breeding programmes are condensed tannins and vicine–convicine, two classes of antinutritional factor present in seeds. Two recessive genes (*zt-1* and *zt-2*) control the absence of tannins, and vicine–convicine content is decreased due to the presence of the *vc-* allele linked to the colourless hilum gene (Duc et al. 1989). BSA has been used to develop molecular markers tightly linked to the desired alleles (Gutierrez et al. 2006, 2007, 2008). These markers might facilitate gene pyramiding when producing new faba bean cultivars with improved nutritional value.

Avila et al. (2003) used BSA as well to detect RAPD markers tightly linked to one gene causing the hypersensitive resistance to rust (race 1) in the inbred line 2N52.

The candidate gene approach has also been used in faba bean. This methodology allows the development of disagnostic markers based on differences between alleles of the gene responsible for the trait. These markers have the significant advantage of being completely linked with the selected trait, thus providing 100 % efficiency in the selection. The strategy was used for the selection of the determinate growth habit, a trait suitable for the mechanical harvesting of the crop. A single gene, recognized as *CEN/TFL1*-homologous or *CEN/TFL1*-like gene, is responsible for the character and several orthologues have been described. Using this information, Avila et al. (2006, 2007) developed several PCR-based markers suitable for the unambiguous selection of determinate growth habit genotypes. Marker research for qualitative traits developed so far in faba bean has still not been adopted in any breeding programme. Nevertheless, the identification of candidate genes underlying these target traits is underway (Khazaei et al. 2015; Webb et al. 2015; Gutiérrez et al. 2015).

10.3 Genetic Maps

Genetic linkage maps are important tools for a wide range of genetics and breeding applications. The availability of high-density maps enhances breeding progress through the application of association analysis, map-based cloning or marker-assisted selection (MAS) methods. The approach is based on the selection of markers linked to the desired genes in segregating populations to trace or pyramid-favourable alleles in a given genome. As the introgression of target genes using conventional breeding strategies is long and complicated, MAS provides simplicity and reliability in identifying target individuals at early seedling stages, saving both time and resources.

So far, 14 relevant genetic maps have been developed in this crop (reviewed in Torres et al. 2012). Less accurate maps with dominant markers in F_2 were followed by more precise maps in the corresponding RIL populations, using codominant gene-based markers. Recently, an F_2 population from a large-seeded Chinese landrace and small-seeded inbred line was used to develop a map using exclusively SSR markers (Ma et al. 2013).

The major drawback of these maps, and the limitation to their practical application, is the small number of common markers and relatively low map density. The relationships between the maps of various populations had not been well studied, so a consensus genetic map has been lately constructed integrating most of the markers, genes and QTL information (Satovic et al. 2013). The map contains 729 loci and covers 4602 cM, representing around a threefold increase in the number of mapped markers and coverage, as compared with previous maps. The approach facilitated including a more consistent locus order, assisted in the assignment of anonymous linkage groups to chromosomes and updated the position of stable faba bean QTLs controlling resistance and yield-related traits (Satovic et al. 2013). This new map with the largest collection of molecular markers currently used in faba bean genome analyses constitutes a valuable resource for comparative genomics, genome organization and evolution studies, and it will provide a significant step forward for faba bean breeding and genomics.

10.4 QTL Analyses: Biotic, Abiotic Stresses and Yield-Related Traits

Disease resistances are important targets for MAS in view of the difficulties of manipulating these traits through conventional approaches. Major biotic constraints under multigenic control such as the resistance to crenate broomrape (*O. crenata*) and ascochyta blight (*A. fabae*) have been the subject of particularly intensive QTL studies (reviewed in Torres et al. 2010, 2012). Breeding for resistance to broomrape is challenging owing to the complexity of the disease evaluation and the polygenic nature of the trait. This holoparasitic angiosperm is considered the most damaging and widespread disease in the Mediterranean basin, and the identification of genes controlling resistance against it is a key subject in faba bean breeding.

Two progenies segregating for both pathogens (crosses Vf6 × Vf136 and 29H × Vf136), have been used for the identification and validation of relevant QTLs in different environments and genetic backgrounds. First, a set of 196 F_2 individuals derived from the cross Vf6 × Vf136 allowed detection of three putative QTLs (*Oc1, Oc2 and Oc3*) on chromosomes I, VI and II, respectively (Román et al. 2002). In order to validate both the presence and location of the F_2 QTLs, 165 RILs from this cross were used to construct a new linkage map (Díaz-Ruiz et al. 2010). RILs used in multi-environment trials increase the power for testing differences between genotypic classes and the precision of trait measurement compared with other types of progenies. This fact is of special significance for broomrape resistance, a trait highly influenced by the environment, resulting from the infectious ability of the parasite in a specific location together with a range of escape factors and resistance mechanisms acting at different levels of the infection process.

Validation experiments were carried out assessing the resistance of the RILs in three different environments. The major QTL, *Oc1,* was not detected in the RILs probably due to its overdominance in the F_2. In contrast, both *Oc2* and *Oc3* were consistent in F_2 and F_6 generations (Díaz-Ruiz et al. 2010), as were detected in at least two of the three environments, thus pointing to their suitability as targets for MAS. Two additional QTLs (*Oc4* and *Oc5*) with small effects were detected in chromosome I in one of the environments. A new RIL population derived from the cross 29H × Vf136 allowed identification of seven QTLs for *O. crenata* (*Oc7* to *Oc13*; Gutierrez et al. 2013). QTL *Oc7,* located on chromosome VI, may correspond to the previously reported *Oc2*. The addition of common markers in the 29H × Vf136 map is being performed in order to establish a clear homology between both regions. The outcomes achieved so far point to *Oc7* as a promising candidate for future MAS for broomrape resistance in faba bean. Saturation of this region will refine the QTL position and identify candidate genes underlying this resistance.

Similarly to crenate broomrape, the F_2 populations from both crosses were used to detect QTLs affecting ascochyta blight response (being Vf136 susceptible, Vf6 partially resistant and Vf29H resistant). Two QTLs (*Af1* and *Af2*) were detected on chromosomes III and II in the Vf6 × Vf136 F_2 population (Román et al. 2003). Avila et al. (2004) studied the resistance on leaves and stems in the F_2 population of cross Vf29H × Vf136, using two pathogenically distinct *Ascochyta* isolates. The study revealed six QTLs (*Af3* to *Af8*), and *Af1* (Román et al. 2003) and *Af3* (Avila et al. 2004) were homologous. QTL validation was undertaken next (Díaz-Ruiz et al. 2009) by studying 165 RILs from cross Vf6 × Vf136. The new linkage map confirmed two zones of putative QTL action, *Af1* and *Af2,* on chromosomes III and II, respectively. Fifteen common markers between maps facilitated the comparison of the homologous regions. The RIL population of cross Vf29H × Vf136 has recently been developed and the corresponding genetic map is being constructed to confirm the relevant QTLs across generations, environments and genetic backgrounds (Avila personal communication).

Recently, Kaur et al. (2012) developed a preliminary faba bean transcriptome and identified a large collection of EST-derived SNPs that were further used to develop a genetic map in a population varying for ascochyta blight resistance (Kaur et al. 2014). Four QTLs controlling resistance were identified in this map, one of which (QTL-3), appears to be identical with *Af2* identified in prior studies (Díaz-Ruiz et al. 2009).

Frost and drought tolerance have been as well the subject of QTL studies. An RIL population derived from the cross between two frost-tolerant lines, Côte d'Or 1 and BPL4628 was used to identify putative QTLs associated with frost tolerance and auxiliary traits (Arbaoui et al. 2008). Similarly, screening of an RIL population derived from two drought-tolerant lines, Melodie 2 × ILB938/2, identified two QTLs 100 cM apart on chromosome II that affected stomatal activity. In both cases candidate genes were identifiable from the synteny with the *M. truncatula* genome (Khazaei et al. 2014). Finally, an RIL population (Vf6 × Vf27) has been used to map orthologous legume cross-species markers. The study revealed the macrosyntenic relationships between *V. faba, M. truncatula, L. culinaris* and other related legume species and was used to identify and validate for the first time, QTLs controlling flowering time and other yield-related traits (Cruz-Izquierdo et al. 2012).

The prospects of MAS for quantitative traits in faba bean breeding programmes will depend on the extent to which QTL results can be extrapolated and validated among environments and genetic backgrounds. Mapping additional common anchor markers is a prerequisite to prove this conservation and determine a QTL homology or the existence of different resistance loci acting in each of the crosses. Moreover, marker density remains rather low for the identification of responsible genes and use of MAS approaches. Comparative genomics and transcriptomic analyses (see below) are providing new resources to saturate syntenic regions bearing relevant QTLs. Mapping new genes from *Medicago,* pea, lupin, lentil and soya bean will provide new anchor points for comparing the maps, refining the QTL positions and unravelling positional candidate genes responsible for these traits.

10.5 Transcriptomic Analysis

Recent advances in transcriptomics are emerging as promising genomic resources for breeding applications in faba bean. Using the 454 Roche GS FLX Titanium technology, Kaur et al. (2012) sequenced cDNA samples from various tissues and genotypes for a de novo assembly and gene annotation of a faba bean transcriptome. A total of 304,680 reads were assembled to generate a set of 60,440 unigenes. In comparison to *M. truncatula* and soya bean coding sequences, 10,179 and 19,497 unique hits, respectively, were obtained. A total of 18,052 faba bean unigenes was subsequently annotated from GenBank. Further comparison to the genome of soya bean resulted in 16,497 unique hits for faba bean, corresponding to ca. 30 % of the known gene space (Kaur et al. 2014).

Genome-wide transcription profiling by deepSuperSAGE has been used for quantifying the early transcriptional changes elicited by *A. fabae* in the resistant inbred line 29H and to identify key resistance factors steering plant responses to this stress (Madrid et al. 2013). A total of 1,313,009 tags were obtained, representing 51,485 unique sequences (UniTags) of which 2222 were expressed differentially between inoculated and control leaves. After gene ontology (GO) annotation, only 2143 of these matched database sequences. The most enriched GO terms corresponded to tags related with photosynthesis metabolism or structural components. Ten were associated with plant defence, due to their association with responses to the jasmonic acid pathway, pectin esterase activity, or gene silencing. Validation of the SuperSAGE data by qPCR of ten differentially expressed UniTags confirmed a rapid increase or decrease in messenger RNA (mRNA) 8–12 h after inoculation in most of the upregulated tags and, less consistently, in the downregulated ones (Madrid et al. 2013).

These studies provide a foundation for the identification of regulators associated with the host–pathogen interaction and also potential targets for molecular breeding in this crop.

11 Seed Production

Compared to other major crops, farmers need a relatively large amount of seed per ha for the establishment of the faba bean crop, so the seed costs are relatively high and farmers pay particular attention to quality when buying seed. The seeds of faba beans have a few specific features that make it a special challenge to produce vigorous seeds with a high level of germination. The high cost of seeds also explains why, in some countries, about 50 % of the seeds required for sowing are self-produced by farmers who accept the risks of lower sanitary control, lower germination quality and loss of genetic integrity.

Among the cultivated crops in most temperate farming systems, faba bean has the highest individual seed size. The thousand-seed weight (TSW) within breeding material can easily vary between 300 and 2000 g (China and Japan), and in released cultivars for Western Europe it ranges from 450 to 650 g. TSW is highly heritable, but the conditions during ripening determine whether the expression of the genetic seed size is at the higher end (optimal conditions) or the lower end (stress, drought). The variation within one cultivar can differ up to 150 g depending on the conditions of the production site. The large seed of faba bean offers a large surface area that is exposed to mechanical impact, and a large diameter that requires adequate time for moisture to move from the interior of the seed to the testa where it is released.

For companies involved in the seed production of faba bean seed, many specific requirements in the different steps of production need to be addressed. The partial allogamy of faba bean results in both self- and cross-fertilised seed on a single

Table 5.5 Regulations for seed production of faba beans in Germany. (Source: Sorten-und Saatgutrecht, 12th edition)

	Basic seed	Certified seed
Parameters for the seed		
Minimum germination	85 %	80 %
Maximum moisture content	15 %	15 %
Technical purity	98 %	98 %
Other species in 1000 g sample	0.30 %	0.50 %
Wild oats in a 1000 g sample	0 seeds	0 seeds
Rumex spec in a 1000 g sample	2 seeds	5 seed
Living adults of *Bruchus rufimanus*	0	0
Maximum number of *Ditylenchus dipsaci* in 300 seeds	5	5
Maximum percentage of *A. fabae* on germinating seeds	30 %	30 %
Parameters for the field		
Minimum distance to others *V. faba* fields	Field smaller 2 ha: 200 m	Field larger 2 ha: 100 m
Maximum number of offtypes on 150 m^2	5	15
Number of plants infected with *A. fabae* in 150 m^2	10	30
Number of plants infected with viruses in 150 m^2	10	30

plant and of plants of an individual breeding line when grown in open-pollinated conditions such as in a yield trial or in fields of the seed producer or the farmer. The importance of cross-fertilisation on the progeny is greater for traits that are under recessive genetic control (such as zero-tannin), and uncontrolled pollination will rapidly result in a loss of genetic integrity. Residual seed of the initial multiplication should be retained to enable further multiplications, and distances (from 300 m to 1 km) between multiplication fields must be assured.

11.1 Regulations and Practical Implications

The major regulations and thresholds which have to be observed in Germany are summarized in Table 5.5. In other countries, these figures may differ, but the table might serve as a template for the relevant parameters. The key factor for the handling of faba bean seeds is the moisture content. The threshold for seeds sold to the farmer is 15 %, but the optimal moisture content with respect to elasticity and tolerance to mechanical input is between 17 and 19 %. This means that harvest and transporting to the store should take place before the seed in the field is fully mature (15 % or higher). Especially under hot and dry conditions, the farmer must take care to start the harvest early enough and not wait until it is fully dry. If they miss this stage, it is often recommended to wait until the seed moisture content has been increased by either a little rain or early-morning dew. In western Canada, growers often apply harvest aids to faba bean crops to harvest, typically with diquat or other approved desiccants. This allows more rapid and uniform drying of stalks and seeds.

Once in the store, the commercial seed must be dried slowly down to the desired storing and packing moisture 14–15 % (which is higher than the targeted 7 % values frequently used by gene banks to allow long-term conservation). It is crucial that the drying process is slow and does not get beyond this crucial point of moisture content. It is strongly recommended to use air temperatures not higher than 35 °C. This usually allows a slow and continuous drying process. Especially if the seed arrives with moisture content above 20 %, it is recommended to allow for a break of at least 12 h during the drying process rather than aiming for the final moisture content in one step.

11.2 Machinery

Very often, equipment that is used for production of small grains such as cereals is not suitable for faba bean and needs adjustment or replacement. The setup specific for faba beans can be used for cereals but usually not vice versa. This results in specific investments for faba bean seed production. The sowing machinery must have sufficiently large coulters to allow uniform flow of large seeds, and the combine harvester should have its threshing concave opened wide to avoid broken, cracked and crushed seeds. As the pod shells open very easily, the drum speed should be set to the lowest possible speed. Driving during harvest should be relatively fast in order to get more biomass into the machine, which also has a buffer function for the seed inside the machine. The most gentle setup would be a combine with a tangential drum. Adjustments for the emptying auger should be set to run at lower revolutions per minute (RPM) in order to avoid seed damage. In addition, post-harvest handling systems must be adjusted to avoid drops from great height onto hard ground that would bruise the embryo or break the seed. Therefore, the belts for moving seeds should ideally be made of rubber, and special handling systems should be used in bins to avoid seed breakage during filling.

11.3 Sanitary Concerns

For seed production, the following diseases and pests are relevant:

- Black bean aphid (*Aphis fabae*, other species) transmit viruses in some environments, so control by suitable chemicals is recommended.

- Bean seed weevil *(Bruchus rufimanus)* larvae develop within the seeds and leave a hole in the seed. While this has relatively little impact on germination, adult weevils remain in the holes in the seeds, and removal will require chemical fumigation. Therefore, a strategy to control this pest is strongly recommended.
- The presence of stem nematode *(D. dipsaci)* precludes use of the seed. Close monitoring in the field together with appropriate crop rotation helps to control the nematodes.
- Ascochyta blight *(A. fabae)* disease starts on the leaves but can move to the pods and seeds. Infested seeds transmit the disease to the next field. Strict control by chemicals is highly recommended.
- Chocolate spot (*B. fabae*) fungus is not transmitted by seed. Severe infection can lead to early loss of the canopy resulting in small and shrivelled seeds. Timely fungicide application is recommended for control.

12 Conclusion

Faba bean breeding has a large number of objectives, and the priorities depend on regional stress factors and market forces. As a result of climate change, stronger demands exist for genetic resistance to abiotic stresses (early or late drought, heat, frost, waterlogging) as well as to pests and diseases (*O. crenata,* sitona weevils, leaf miners, bruchids and pathogenic fungi) in new geographic zones. Potential novel uses in foods also strengthen demands for low vicine–convicine, high-protein content and sometimes low tannin genotypes. Inevitably, when the breeder needs to add more objectives to an existing programme, either more resources have to be added to cover it or overall progress slows down. In addition to large, characterised and structured genetic resources collections, rapid breeding methods and translational omics will help to streamline the breeding progress, so more of the objectives can be met.

Progress in faba bean genomics is still behind that in other legume crops. Nevertheless, a wide range of genomic and post-genomic resources is being developed to boost genetic research and breeding applications. More than 14 mapping populations have been established to study the inheritance of important agronomic traits. Different molecular marker sets have been developed and used to construct more saturated linkage maps in order to identify genes or QTLs controlling major stress responses. Current efforts are focussed on the development of highly accurate selective breeding tools, using NGS methods. Transcriptomic analysis (SuperSAGE and cDNA libraries) is enriching the scarce faba bean EST libraries and provides additional resources for refining the maps with functional markers. Moreover, translational genomics, based on the collinearity with model and related species, opens the possibility to identify candidate genes underlying agronomically important traits. Finally, a faba bean functional consensus map has been constructed integrating all the genes and QTLs previously published. The combination of these and new tools together with a close link between academic research and commercially focused breeding programmes may help researchers to find genes of interest to speed the release of more competitive faba bean cultivars in the near future.

Acknowledgments This research was partly supported by the following projects: (1) FP7 LEGATO, European Union; Research Grant 173005 of the Ministry of Education, Science, and Technological Development of the Republic of Serbia, and the project of bilateral collaboration between the Republic of Serbia and the Republic of France "Pavle Savić", FABAGRALE 2012–2013, number 680-00-132/2012-09/15, (2) ECONET- PAVLE SAVIC projects, Ministry of Foreign Affairs—France and PARI-1 Agrale 6-2010-2013, Burgundy Council-France, (3) RTA2010-00059 and IPT-2011-1259-010000 Spanish projects, cofinanced by *Fonds Européen de Développement Économique et Régional* (FEDER) funds, (4) Grains Research and Development Corporation, Australia.

References

Annicchiarico P, Iannucci A (2008) Breeding strategy for faba bean in southern Europe based on cultivar responses across climatically contrasting environments. Crop Sci 48:983–991
Anwar F, Alghamdi SS, Ammar MH et al (2011) An efficient in vitro regeneration protocol for faba bean (*Vicia faba* L.). J Med Plants Res 5:6460–6467
Arbaoui M, Link W (2008) Effect of hardening on frost tolerance and fatty acid composition of leaves and stems of a set of faba bean (*Vicia faba* L.) genotypes. Euphytica 162:211–219
Arbaoui M, Link W, Satovic Z et al (2008) Quantitative trait loci of frost tolerance and physiologically related traits in faba bean (*Vicia faba* L.). Euphytica 164:93–104
Arese P, De Flora A (1990) Pathophysiology of hemolysis in glucose-6-phosphate dehydrogenasedeficiency. Semin Hematol 27:1–40
Avila CM, Sillero JC, Rubiales D et al (2003) Identification of RAPD markers linked to *Uvf-1* gene conferring hypersensitive resistance against rust (*Uromyces viciae-fabae*) in *Vicia faba* L. Theor Appl Genet 107:353–358
Avila CM, Satovic Z, Sillero JC et al (2004) Isolate and organ-specific QTLs for ascochyta blight resistance in faba bean. Theor Appl Genet 108:1071–1078
Avila CM, Nadal S, Moreno MT et al (2006) Development of a simple PCR-based marker for the determination of growth habit in *Vicia faba* L. using a candidate gene approach. Mol Breed 17:185–190
Avila CM, Atienza SG, Moreno MT et al (2007) Development of a new diagnostic marker for growth habit selection in faba bean (*Vicia faba* L.) breeding. Theor Appl Genet 115:1075–1082
Bond DA (1976) Field bean, *Vicia faba* (*Legumillosae papilionatae*). In: Simmonds NW (ed) Evolution of crop plants. Longman, London, pp 179–182
Bond DA, Crofton GRA (1999) History of winter beans in the UK. J R Agr Soc Engl 160:200–209
Bond DA, Poulsen MH (1983) Pollination. In: Hebblethwaite PD (ed) The faba bean (*Vicia faba* L.). Butterworths, London, pp 77–101
Boote KJ, Minguez MI, Sau F (2002) Adapting the CROPGRO-legume model to simulate growth of faba bean. Agron J 94:743–756
Bottinger P, Steinmetz A, Schieder O et al (2001) *Agrobacterium*-mediated transformation of *Vicia faba*. Mol Breed 8:243–254
Cottage A, Gostkiewicz K, Thomas JE et al (2012) Heterozygosity and diversity analysis using mapped SNPs in a faba bean inbreeding programme. Mol Breed 30:1799–1809
Crépon K, Marget P, Peyronnet C et al (2010) Nutritional value of faba bean (*Vicia faba* L.) seeds for feed and food. Field Crop Res 115:329–339
Cruz-Izquierdo S, Avila CM, Satovic Z et al (2012) Comparative genomics to bridge *Vicia faba* with model and closely-related legume species: stability of QTLs for flowering and yield-related traits. Theor Appl Genet 125:1767–1782

Cubero JI (1973) Evolutionary trends in *Vicia faba* L. Theor Appl Genet 43:59–65

Cubero JI (1982) Interspecific hybridization in *Vicia*. In: Hawtin G, Webb C (eds) Faba bean improvement. Martinus Nijhoff, The Hague, pp 91–108

Díaz-Ruiz R, Satovic Z, Avila CM et al (2009) Confirmation of QTLs controlling *Ascochyta fabae* resistance in different generations of faba bean (*Vicia faba* L.). Crop Pasture Sci 60:353–361

Díaz-Ruiz R, Torres AM, Satovic Z et al (2010) Validation of QTLs for *Orobanche crenata* resistance in faba bean (*Vicia faba* L.) across environments and generations. Theor Appl Genet 120:909–919

Duc G (1997) Faba bean (*Vicia faba* L.). Field Crops Res 53:99–109

Duc G, Sixdenier G, Lila M et al (1989) Search of genetic variability for vicine and convicine content in *Vicia faba* L. A first report of a gene which codes for nearly zero-vicine and zero-convicine contents. In: Huisman AJM, Van der Poel AFB, Liener IE (eds) Recent advances of research in antinutritional factors in legumes seeds. Pudoc, The Netherlands, pp 305–313

Duc G, Berthaut N, Le Guen J et al (1992) Hybrid breeding in faba bean (*Vicia faba* L.) In: AEP (ed) Proceeding 1st Conference on Grain Legumes, Angers, France, 1–3 June 1992, pp 45–46

Duc G, Moussy F, Zong X (1999) A single gene mutation for green cotyledons in faba bean (*Vicia faba* L.) marks embryonic phenotype. Plant Breed 118:577–578

Duc G, Link W, Marget P et al (2011) Genetic adjustment to changing climates: faba beanin crop adaptation to climate change. In: Rshyam JRP, Yadav S, Hatfield JL, Lotze-Campen H, Hall AE (eds) Crop Adaptation to Climate Change. Wiley-Blackwell, Oxford, pp 269–286

Ellwood SR, Phan HTT, Jordan M et al (2008) Construction of a comparative genetic map in faba bean (*Vicia faba* L.); conservation of genome structure with *Lens culinaris*. BMC Genomics 9:380

Eurostat (2014) ec.europa.eu/eurostat/fr

FAOSTAT (2013): http://faostat.fao.org/. Accessed Oct 2014

Fennell SR, Powell W, Wright F et al (1998) Phylogenetic relationships between *Vicia faba* (Fabaceae) and related species inferred from chloroplast trnL sequences. Plant Syst Evol 212:247–259

Fernandez-Aparicio M, Moral A, Kharrat M et al (2012) Resistance against broomrapes (*Orobanche and Phelipanche* spp.) in faba bean (*Vicia faba*) based in low induction of broomrape seed germination. Euphytica 186:897–905

Flamand MC, Goblet JP, Duc G et al (1992) Sequence and transcription analysis of mitochondrial plasmids isolated from cytoplasmic male-sterile lines of *Vicia faba*. Plant Mol Biol 19:913–923

Flores F, Nadal S, Solis I et al (2012) Faba bean adaptation to autumn sowing under European climates. Agron Sustain Dev 32:727–734

Flores F, Hybl M, Knudsen JC et al (2013) Adaptation of spring faba bean types across European climates. Field Crops Res 145:1–9

Fukuta N, Yukawa T (1998) Varietal differences in snow tolerance and growth characteristics of bread bean (*Vicia faba* L.). Jpn J Crop Sci 67:505–509

Gnanasambandam A, Paull J, Torres A et al (2012) Impact of molecular technologies on faba bean (*Vicia faba* L.) breeding strategies. Agron 2:132–166

Gong Y, Xu S, Mao W et al (2010) Generation and characterization of 11 novel EST derived microsatellites from *Vicia faba* (Fabaceae). Am J Bot 97:e69–e71

Gutierrez N, Avila CM, Duc G et al (2006) CAPs markers to assist selection for low vicine and convicine content in faba bean (*Vicia faba* L.). Theor Appl Genet 114:59–66

Gutierrez N, Avila CM, Rodriguez-Suarez C et al (2007) Development of SCAR markers linked to a gene controlling absence of tannins in faba bean. Mol Breed 19:305–314

Gutierrez N, Avila CM, Moreno MT et al (2008) Development of SCAR markers linked to *zt-2*, one of the genes controlling absence of tannins in faba bean. Aust J Agr Res 59:62–68

Gutierrez N, Palomino C, Satovic Z et al (2013) QTLs for *Orobanche* spp. resistance in faba bean: identification and validation across different environments. Mol Breed 32:909–922

Gutiérrez N, Avila CM, Torres AM (2015) A TTG1 transcription factos regulates the white flower and the absence of tannins in faba bean (zt1 gene). EUCARPIA International Symposium on Protein crops, 4–7 May 2015, Pontevedra, Spain

Haider N, Nabulsi I, Mirali N (2012) Identification of species of *Vicia* subgenus *Vicia* (Fabaceae) using chloroplast DNA data. Turk J Agr For 36:297–308

Hanafy M, Pickardt T, Kiesecker H et al (2005) *Agrobacterium*-mediated transformation of faba bean (*Vicia faba* L.) using embryo axes. Euphytica 142:227–236

Hanafy MS, Antar El-Banna A, Schumacher HM et al (2013) Enhanced tolerance to drought and salt stresses in transgenic faba bean (*Vicia faba* L.) plants by heterologous expression of the PR10a gene from potato. Plant Cell Rpt 32:663–674

Hanelt P (1972) Zur geschichte des anbaues von *Vicia fab*a L. und ihre verschiedenen formen. Kulturplanze 20:209–223

Herzog H (1987) A quantitative method to assess freezing resistance in faba beans. J Agron Crop Sci 158:195–204

Ismail RM, Elazab HEM, Hussein GMH et al (2011) In vitro root induction of faba bean (*Vicia faba* L.) GM Crops 2:176–181

Kaschuk G, Kuyper TW, Leffelaar PA et al (2009) Are the rates of photosynthesis stimulated by the carbon sink strength of rhizobial and arbuscular mycorrhizal symbioses? Soil Biol Biochem 41:1233–1244

Kaur S, Pembleton L, Cogan N et al (2012) Transcriptome sequencing of field pea and faba bean for discovery and validation of SSR genetic markers. BMC Genomics 13:104

Kaur S, Kimber RB, Cogan N et al (2014) SNP discovery and high-density genetic mapping in faba bean (*Vicia faba* L.) permits identification of QTLs for ascochyta blight resistance. Plant Sci 217–218:47–55

Kazakoff SH, Imelfort M, Edwards D et al (2012) Capturing the biofuel wellhead and powerhouse: the chloroplast and mitochondrial genomes of the leguminous feedstock tree *Pongamia pinnata*. PLoS One 7:e51687

Khan HR, Link W, Stoddard FL (2007) Evaluation of physiological traits for improving drought tolerance in faba bean (*Vicia faba* L.). Plant Soil 292:205–217

Khan HR, Paull JG, Siddique KHM et al (2010) Faba bean breeding for drought-affected environments: a physiological and agronomic perspective. Field Crops Res 115:279–286

Khazaei H, Street K, Santanen A et al (2013a) Do faba bean (*Vicia faba* L.) accessions from environments with contrasting seasonal moisture availabilities differ in stomatal characteristics and related traits? Genet Resour Crop Ev 60:2343–2357

Khazaei H, Street K, Bari A et al (2013b) The FIGS (Focused Identification of Germplasm Strategy) approach identifies traits related to drought adaptation in *Vicia faba* genetic resources. PLoS ONE 8:e63107. doi:10.1371/journal.pone.0063107

Khazaei H, O'Sullivan DM, Sillanpää MJ et al (2014) Use of synteny to identify candidate genes underlying QTL controlling stomatal traits in faba bean (*Vicia faba* L.). Theor Appl Genet. doi:10.1007/s00122-014-2383-y

Khazaei H, O'Sullivan D, Jones H (2015) Flanking SNP markers for vicine-convicine concentration in faba bean (*Vicia faba* L.). Mol Breed 35–38

Kimber RBE, Paull JG (2011) Identification and genetics of resistance to cercospora leaf spot (*Cercospora zonata*) in faba bean (*Vicia faba*). Euphytica 177:419–429

Ko K, Straus NA, Williams JP (1983) Mapping the chloroplast DNA of *Vicia faba*. Curr Genet 7:255–263

Ko K, Orfanides AG, Straus NA (1987) A model for the evolution of the *Vicia faba* chloroplast genome. Theor Appl Genet 74:125–139

Kohpina S, Knight R, Stoddard FL (2000) Genetics of resistance to ascochyta blight in two populations of faba bean. Euphytica 112:101–107

Kwon SJ, Hu J, Coyne CJ (2010) Genetic diversity and relationship among 151 faba bean (*Vicia faba* L.) germplasm entries as revealed by TRAP markers. Plant Genet Resour 8:204–213

Ladizinsky G (1998) Plant evolution under domestication. Kluwer, Dordrecht

Leonforte A, Forster J, Redden R, Nicolas M, Salisbury P (2013) Sources of high tolerance to salinity in pea (*Pisum sativum* L.). Euphytica 189:203–216

Link W, Ederer W, Metz P et al (1994) Genotypic and environmental variation for degree of cross-fertilization in faba bean. Crop Sci 34:960–964

Link W, Dixkens C, Singh M et al (1995) Genetic diversity in European and Mediterranean faba bean germ plasm revealed by RAPD markers. Theor Appl Genet 90:27–32
Link W, Schill B, Barbera AC et al (1996) Comparison on infra and inter-pool crosses in faba beans (*Vicia faba* L.). Hybrid performance and heterosis in Mediterranean and German environments. Plant Breed 115:352–360
Link W, Balko C, Stoddard FL (2010) Winter hardiness in faba bean: physiology and breeding. Field Crops Res 115:287–296
Ma Y, Yang T, Guan J et al (2011) Development and characterization of 21 EST-derived microsatellite markers in *Vicia faba* (faba bean). Am J Bot 98:e22–24
Ma Y, Bao S, Yang T et al (2013) Genetic linkage map of Chinese native variety faba bean (*Vicia faba* L.) based on simple sequence repeat markers. Plant Breed 132:397–400
MacFarlane JL, Wahleithner JA, Wolstenholme DR (1990) A gene for cytochrome c oxidase subunit III (COXIII) in broad bean mitochondrial DNA: structural features and sequence evolution. Curr Genet 17:33–40
Madrid E, Palomino C, Plötner A et al (2013) DeepSuperSage analysis of the *Vicia faba* transcriptome in response to *Ascochyta fabae* infection. Phytopathol Mediterr 52:166–182
Maxted N (1993) A phenetic investigation of *Vicia* L. subgenus *Vicia* (*Leguminosae Vicieae*). Bot J Linn Soc 111:155–182
Maxted N, Khattab MA, Bisby A (1991) Domesticated legumes and their wild relatives: newly discovered relatives of *Vicia faba* L. do little to resolve the enigma of its origin. Bot Chron 10:435–465
Metry EA, Ismail RM, Hussien GM et al (2006) Regeneration and microprojectile mediated transformation in *Vicia faba* L. Arab J Biotech 10:23–36
Mikić A (2012) Origin of the words denoting some of the most ancient Old World pulse crops and their diversity in modern European languages. PLoS ONE 7:e44512
Moreno MT, Martinez A (1980) The divided world on *V. faba*. Fabis 2:18–19
Mubumbila M., Crouse EJ, Weil JH (1984) Transfer RNAs and tRNA genes of *Vicia faba* chloroplasts. Curr Genet 8:379–385
Muratova VS (1931) Common beans (*Vicia faba* L.). Bull Appl Bot Genet Pl Breed 50:1–298
Negruk V (2013) Mitochondrial genome sequence of the legume *Vicia faba*. Front Plant Sci 4:128
Nikiforova ID, Negruk VI (1983) Comparative electrophoretical analysis of plasmid-like mitochondrial DNAs in *Vicia faba* and in some other legumes. Planta 157:81–84
Oldach K.H. (2011) Mutagenesis in food legumes. In: Pratap A, Kumar J (eds) Biology and breeding of food legumes. CABI, Oxfordshire, p 418
Ouji A, El Bok S, Syed NH et al (2012) Genetic diversity of faba bean (*Vicia faba* L.) populations revealed by sequence specific amplified polymorphism (SSAP) markers. Afr J Biotechnol 11:2162–2168
Picard J (1976) Aperçu sur l'hérédité du caractère absence de tannins dans les graines de féveroles (*Vicia faba* L.). Annales de l'amélioration des plantes 26:101–106
Potokina E, Tomooka N, Vaughan DA et al (1999) Phylogeny of *Vicia* subgenus *Vicia* (Fabaceae) based on analysis of RAPDs and RFLP of PCR-amplified chloroplast genes. Genet Resour Crop Ev 46:149–161
Pozarkova D, Koblizkova A, Román B et al (2002) Development and characterization of microsatellite markers from chromosome 1-specific DNA libraries of *Vicia faba*. Biol Plant 45:337–345
Raina SN, Ogihara Y (1994) Chloroplast DNA diversity in *Vicia faba* and its close wild relatives: implications for reassessment. Theor Appl Genet 88:261–266
Raina SN, Rees H (1983) DNA variation between and within chromosome complements in *Vicia* species. Hered 51:335–346
Román B, Torres AM, Rubiales D et al (2002) Mapping of quantitative trait loci controlling broomrape (*Orobanche crenata* Forsk.) resistance in faba bean (*Vicia faba* L.). Genome 45:1057–1063
Román B, Satovic Z, Avila CM et al (2003) Locating genes associated with *Ascochyta fabae* resistance in *Vicia faba* L. Aust J Agr Res 54:85–90

Román B, Satovic Z, Pozarkova D et al (2004) Development of a composite map in *Vicia faba* L. breeding applications and future prospects. Theor Appl Genet 108:1079–1088
Rose TJ, Damon P, Rengel Z (2010) Phosphorus-efficient faba bean (*Vicia faba* L.) genotypes enhance subsequent wheat crop growth in an acid and an alkaline soil. Crop Pasture Sci 61:1009–1016
Rubiales D, Fernández-Aparicio M (2012) Innovations in parasitic weeds management in legume crops. A review. Agron Sustain Dev 32:433–449
Rubiales D, Avila CM, Sillero JC et al (2012) Identification and multi-environment validation of resistance to *Ascochyta fabae* in faba bean (*Vicia faba*). Field Crops Res 126:165–170
Sameshima-Saito R, Chiba K, Hirayama J et al (2006) Symbiotic *Bradyrhizobium japonicum* reduces N_2O surrounding the soya bean root system via nitrous oxide reductase. Appl Environ Microbiol 72:2526–2532
Sanz AM, Gonzalez SG, Syed NH et al (2007) Genetic diversity analysis in *Vicia* species using retrotransposon-based SSAP markers. Mol Genet Genomics 278:433–441
Satovic Z, Avila CM, Palomino C et al (2013) A reference consensus genetic map for molecular markers and economically important traits in faba bean (*Vicia faba* L.). BMC Genomics 14:932
Scalla R, Duc G, Rigaud J et al (1981) RNA containing intracellular particles in cytoplasmic male sterile faba bean *Vicia faba* L. Plant Sci Lett 22:269–277
Scallan CD, Harmey MA (1996) The use of mitochondrial DNA polymorphisms to distinguish commercial cultivars of the broad bean, *Vicia faba* L. Euphytica 87:225–235
Schaefer H, Hechenleitner P, Santos-Guerra A et al (2012) Systematics, biogeography, and character evolution of the legume tribe Fabeae with special focus on the middle-Atlantic island lineages. BMC Evol Biol 12:250
Schill B, Melchinger AE, Gumber RK et al (1998) Comparison of intra- and inter-pool crosses in faba beans (*Vicia faba* L.) II. Genetic effects estimated from generation means in Mediterranean and German environments. Plant Breed 117:351–359
Shiran B, Mashayekh AM (2004) Evaluation of the chlroplast DNA among *Vicia faba* L. Iranian J Sci Tech Transac A 28:51–55
Sillero JC, Moreno MT, Rubiales D (2000) Characterization of new sources of resistance to *Uromyces viciae-fabae* in a germplasm collection of *Vicia faba*. Plant Pathol 49:389–395
Sillero JC, Villegas-Fernandez AM, Thomas J et al (2010) Faba bean breeding for disease resistance. Field Crops Res 115:297–307
Sjodin J (1971) Induced morphological variation in *Vicia faba* L. Hereditas 67:155–I80
Solaiman Z, Colmer TD, Loss SP et al (2008) Growth responses of cool-season grain legumes to transient waterlogging. Aust J Agr Res 58:406–412
Stoddard FL (1993) Limits to retention of fertilised flowers in faba beans (*Vicia faba* L.). J Agron Crop Sci 171:251–259
Stoddard FL, Balko C, Erskine W et al (2006) Screening techniques and sources of resistance to abiotic stresses in cool season food legumes. Euphytica 147:167–186
Stoddard FL, Nicholas AH, Rubiales D et al (2010) Integrated pest management in faba bean. Field Crops Res 115:308–318
Suso MJ, Gilsanz S, Duc G et al (2006) Germplasm management of faba bean (*Vicia faba* L.): monitoring intercrossing between accessions with inter-plot barriers. Genet Resour Crop Evol 9:1–13
Suso MJ, Vishniakova M, Ramos A et al (2011) State of the art of grain legume management in gene banks. J Agr Sci Technol 1:975–981
Terzopoulos PJ, Bebeli PJ (2008) Genetic diversity analysis of Mediterranean faba bean (*Vicia faba* L.) with ISSR markers. Field Crops Res 108:39–44
Thomas J, Kenyon D (2004) Evaluating resistance to downy mildew (Peronospora viciae) in field peas (*Pisum sativum* L.) and field beans (*Vicia faba* L.). In: AEP (ed) Proceedings of the 5th European Conference on Grain Legumes and 2nd International Conference on Legume Genomics and Genetics, 7–11 June 2004, Dijon, France pp 81–82
Tivoli B, Baranger A, Avila CM et al (2006) Screening techniques and sources of resistance to foliar diseases caused by major necrotrophic fungi in grain legumes. Euphytica 147:223–253

Torres AM, Avila CM, Gutierrez N et al (2010) Marker-assisted selection in faba bean (*Vicia faba* L.). Field Crops Res 115:243–252

Torres AM, Avila CM, Stoddard FL et al (2012) Faba bean. In: Pérez de la Vega M, Torres AM, Cubero JI, Kole C (eds) Genetics, genomics and breeding in crop plants: cool season food legumes. Science, Plymouth, pp 50–97

Van de Ven WTG, Duncan N, Ramsay G et al (1993) Taxonomic relationships between *V. faba* and its relatives based on nuclear and mitochondrial RFLPs and PCR analysis. Theor Appl Genet 86:71–80

Villegas-Fernández AM, Sillero JC, Emeran AA et al (2011) Multiple-disease resistance in *Vicia faba*: multi-environment field testing for identification of combined resistance to rust and chocolate spot. Field Crops Res 124:59–65

Von Kittlitz E, Ibrahim KIM, Ruckenbauer P et al (1993) Analysis and use of interpool-crosses of Mediterranean and Central European faba beans (*Vicia faba* L.). Performance of Mediterranean and central European faba beans in Syria and Germany. Plant Breed 110:307–314

Wang H, Zong X, Guan J et al (2012) Genetic diversity and relationship of global faba bean (*Vicia faba* L.) germplasm revealed by ISSR markers. Theor Appl Genet 124:789–797

Webb A, Cottage A, Wood T et al (2015) A SNP-based consensus genetic map for synteny-based trait targeting in faba bean (*Vicia faba* L.). Plant Biotechnol J. doi:10.111/pbi.12371

Wojciechowski MF, Lavin M, Sanderson MJ (2004) A phylogeny of legumes (Leguminosae) based on analysis of the plastid *mat*K gene resolves many well-supported subclades within the family. Am J Bot 91:1846–1862

Yang T, Bao S, Ford R et al (2012) High-throughput novel microsatellite marker of faba bean via next generation sequencing. BMC Genomics 13:602

Yunusa IAM, Rashid MA (2007) Productivity and rotational benefits of grass, medic pastures and faba beans in a rainfall limited environment. Soil Till Res 97:150–161

Zeid M, Schön CC, Link W (2001) Genetic diversity in recent elite faba bean lines. Czech J Genet Plant Breed 37:34–40

Zeid M, Mitchell S, Link W et al (2009) Simple sequence repeats (SSRs) in faba bean: new loci from *Orobanche* resistant cultivar 'Giza 402'. Plant Breeding 128:149–155

Zong X, Cheng X, Wang S (2006) Food legume crops. In: Yuchen D, Diansheng Z (eds) Crops and its relative species in China-Grain crops. China Agriculture, Beijing, pp 406–479

Zong X, Liu X, Guan J et al (2009) Molecular variation among Chinese and global winter faba bean germplasm. Theor Appl Genet 118:971–978

Zong X, Ren J, Guan J et al (2010) Molecular variation among Chinese and global germplasm in spring faba bean areas. Plant Breed 129:508–513

Chapter 6
Lupins

Wojciech Święcicki, Magdalena Kroc and Katarzyna Anna Kamel

1 Introduction

The genus *Lupinus* is characterized by particularly broad diversity in terms of growth and development, colors, environmental requirements, distribution regions, number of chromosomes, type of usage, etc. Lupins are not only annual herbs but also trees grown from the tropics to Alaska. Fifteen-thousand-year-old seeds of *Lupinus arcticus* still germinating have been found, as well as seeds germinating from under ash after the eruption of the Mt. St. Helens volcano in 1980 (Findley 1981; Porsild et al. 1967). Some species are beautiful ornamentals; seeds can be used for feeding man and animals (32–45 % of protein), and the green mass can be used as animal fodder or organic manure. Therefore, lupins can be of great practical importance and also an intriguing research material.

Lupins were used by man in ancient times—Andean lupin (*Lupinus mutabilis* Sweet) in South America and white lupin (*Lupinus albus* L.) in the Old World. Even ancient man was aware of lupin's advantages—for soil improvement and soaked, debittered seeds (Cowling et al. 1998b).

The use of lupins in agriculture was caused by some factors. First of all, 50 % of the area of Poland and Belarus, the former Prussia, is covered by sandy soils. Moreover, the consequences of the Thirty Years' War, the collapse in yields in 1772, and the increasing population resulted in the introduction of new, leafy crops (beets, clover, potatoes, rape) to improve soils. The Prussian King, Friedrich II (1740–1784),

W. Święcicki (✉) · M. Kroc · K. A. Kamel
Department of Genomics, Institute of Plant Genetics, Polish Academy of Sciences, Strzeszyńska 34, 60-479 Poznan, Poland
e-mail: wswi@igr.poznan.pl

M. Kroc
e-mail: mkro@igr.poznan.pl

K. A. Kamel
e-mail: kkam@igr.poznan.pl

A. M. De Ron (ed.), *Grain Legumes*, Handbook of Plant Breeding 10,
DOI 10.1007/978-1-4939-2797-5_6

and the land owner, Carl von Wulffen, started (1816) to grow white lupin to improve soils (seeds were imported from Italy). Unfortunately, due to its soil requirements and late ripening, both projects collapsed. The successful introduction of yellow lupin as a green manure was effected for the first time in 1840 by J. N. Borchardt. He multiplied his own seeds and produced a higher yield of rye as a successive crop after lupin. The popularity of yellow lupin was transferred from Prussia to Poland and Russia (Brummund and Święcicki 2011).

Lupin domestication was connected with the removal of the hard seed coat, dehiscing pods, and decreasing alkaloid content (Święcicki and Święcicki 1995; Święcicki et al. 2000a). Elaboration of a screening method by von Sengbusch enabled the selection of the first "sweet" plants of yellow lupin (1927), narrow-leafed lupin (1927–1928), white lupin (1930–1931), and *Lupinus polyphyllus* (1931). The first, low alkaloid yellow lupin cv. "von Sengbusch Munchberger Gelbe, Grünfutter Süsslupine" is considered the beginning of fodder lupins. The importance of this type of use led to a growth in its popularity as a result of increasing demands for protein in feeding pigs and poultry. It was impossible to ignore plants with 32–45 % protein in seeds, useful for cropping on light soils in moderate climatic conditions.

There were some important steps in the improvement of lupin cultivars, for example, breeding of cultivars resistant to *Fusarium,* fast growing, thermoneutral, self-completing (restricted branching), and recently with improved resistance to anthracnose. The employment of molecular biology and biotechnology also enabled achievements in contemporary breeding programs. The history of lupin in agriculture was summarized by Brummund and Święcicki (2011). In addition to Germany, large lupin breeding projects were implemented after the Second World War in Belarus, Chile, Poland, Russia, Ukraine, and Australia (Kubok 1988; Kurlovich 2002; Lukashevitch et al. 2011; Kuptsov 2000). The last example should be underlined particularly (Cowling et al. 1998a). The narrow-leafed lupin in Western Australia stopped a decrease in the yields of wheat grown in monoculture and the acreage increased up to 1.0–1.5 bln of hectares in the nineteens of the twentieth century. The world lupin area harvested in 2012 is presented in Table 6.1. A substantially important role in the propagation of lupin use and breeding progress is performed by the International Lupin Association. An excellent source of information on these crops are proceedings of 13th International Lupin Conferences organized in different continents and in different countries between 1980 and 2011.

What is the main factor distinguishing lupins? The process of transformation of most species from a wild plant to a crop lasts hundreds or thousands of years. In the case of lupin, it took only 1–2 generations of breeders. Outstanding people and their contributions were discussed by Brummund and Święcicki (2011).

2 Origin and Systematics

The genus *Lupinus* belongs to the Papilionoideae subfamily of legumes where it is classified among the tribe Genisteae within an early-branching phylogenetic clade Genistoid (Lavin et al. 2005; Cardoso et al. 2013). It is generally accepted that

Table 6.1 The lupin area harvested (FAOSTAT 2012, http://faostat.fao.org/)

Country	Area (ha)
Argentina	110
Australia	689,064
Austria	98
Belarus	20,735
Chile	21,467
Ecuador	4500
Egypt	764
France	2553
Germany	17,900
Hungary	54
Italy	5000
Lativa	100
Lebanon	50
Lithuania	5100
Peru	9656
Poland	49,221
Russian Federation	17,800
Slovakia	80
South Africa	12,000
Spain	6700
Switzerland	49
Syrian Arab Republic	13
Ukraine	24,000
Total	*887,014 (100 %)*
South America	35,733 (4 %)
Australia	689,064 (78 %)
South Africa	12,000 (1.4 %)
Other Africa	764 (0.09 %)
Europe	149,390 (17 %)

Lupinus evolved out of the tropical and subtropical representatives of the primitive papilionoid tribe Sophoreae living in tertiary (Gladstones 1998a; Lavin et al. 2005). *Lupinus* is a diverse and widespread genus comprising 275 annual and perennial species found in both lowland and mountain regions (Cardoso et al. 2013; Drummond et al. 2012). There are two geographically isolated groups within the genus: the Old World and New World lupins (Cowling 1999). The genus *Lupinus* originated in the Old World and subsequently dispersed to the New World (Aïnouche et al. 2004; Drummond et al. 2012).

The group of the Old World lupins is composed of 13–15 species (Table 6.2) found in the Mediterranean region and northern Africa (Cowling et al. 1998a; Święcicki et al. 1996; 1999; Pascual 2004; Gladstones 1974). These are all annual herbaceous, mostly autogamous species characterized by digitate leaves (Aïnouche et al. 2004). Twelve Old World lupin species were described by Gladstones (1974), and one of them, *Lupinus somaliensis*, is probably extinct. Two species—

Table 6.2 Old-world lupin species list with the division according to the seed coat texture

No	Species	Chromosome no
	Smooth-seeded group	
1	*L. albus* L.	48
2	*L. angustifolius* L.	40
3	*L. hispanicoluteus* Swiec. et Swiec.	52
4	*L. hispanicus* Boiss. & Reut	52
5	*L. luteus* L.	52
6	*L. micranthus* Guss	52
	Rough-seeded group	
7	*L. anatolicus* Swiec. et Swiec.	
8	*L. atlanticus* Glads	38
9	*L. consentini* Guss	32
10	*L. digitatus* Forsk	36
11	*L. palaestinus* Boiss	42
12	*L. pilosus* Murr	42
13	*L. princei* Harms	38
14	*L. somaliensis* Baker (extinct)	
	Intermediate species	
15	*L. mariae-josephi* H. Pascual	52

Lupinus anatolicus (Święcicki et al. 1996) and *Lupinus mariae-josephi* (Pascual 2004; Pascual et al. 2006)—have been fairly recently discovered, and one species—*Lupinus* × *hispanicoluteus* (Święcicki et al. 1999)—was created as an interspecific hybrid of two Old World lupins. The group of the Old World lupins has been further divided into two sections according to seed coat texture: the smooth-seeded (section *Malacospermae*) and the rough-seeded (section *Scabrispermae*) species (Table 6.2; Gladstones 1974). Representatives of the *Scabrispermae* could successfully be crossed, confirming their genetic relationship beyond similarities in terms of morphological, chemical, and cytological characteristics (Aïnouche et al. 2004; Naganowska et al. 2003; Przybylska and Zimniak-Przybylska 1995; Gupta et al. 1996; Wolko and Weeden 1990; Gladstones 1974). They also form a strongly supported monophyletic group in phylogenetic analyses despite their variable genome size and chromosome number (Aïnouche et al. 2004). The smooth-seeded lupins form a heterogeneous group of species separated by major genetic barriers (Cowling et al. 1998a; Plitmann and Pazy 1984). The three cultivated lupin species—*L. angustifolius, L. albus,* and *L. luteus*—are the smooth-seeded lupins. Two Old World lupins do not fall into the classic division scheme. One of them is *L. anatolicus*, morphologically similar to *Lupinus micranthus* and *Lupinus pilosus* and described as a smooth-seeded species on the basis of its macroscopic seed coat texture (Święcicki et al. 1996, 2001). The distinctness of the Anatolian lupin has been supported by chemotaxonomical analyses, including protein, oil, and alkaloid profiles as well as isozyme and seed globulin patterns (Przybylska and Zimniak-Przybylska 1995; Świecicki et al. 2001). Additional studies have shown that based on micromorphological seed coat structure, *L. anatolicus* is clearly related to the

Scabrispermae section, while internal transcribed spacers (ITS) sequence data have moreover revealed its explicit relationship to *L. pilosus* and *Lupinus palaestinus*. At the same time, the differences of *L. anatolicus* ITS sequence from the latter two species exhibiting the same ITS sequence are evident, indicating the divergence of the Anatolian lupin (Aïnouche and Bayer 2000). On the other hand, there have also been reports questioning the distinctness of this new species. Clements et al. (1996) considered it a smooth-seeded accession of *L. pilosus*, as such a mutation was found in the natural *L. pilosus* population. These authors report successful crosses between the Anatolian lupin and the related *L. pilosus*. The second new species, *L. mariae-josephi,* is intermediate between the smooth-seeded and rough-seeded lupins based on the micromorphological seed coat texture which also differs from any known New World species. Phylogenetic data show that it is a distinct line within the Old World lupins without clear affinity to either smooth-seeded or rough-seeded groups. Another unique feature of this species is its occurrence which is restricted to calcareous basic soils, exceptional within the genus (Mahé et al. 2011).

The majority of species within *Lupinus* occur in the New World. They are distributed in both North and South America, with the highest concentration of species being endemic to mountain habitats (Aïnouche et al. 2004; Drummond 2008). The New World species are both annuals and herbaceous and woody perennials with unifoliolate or compound leaves (Aïnouche et al. 2004; Drummond et al. 2012). Phylogenetic analyses based on nuclear and chloroplast DNA have shown that the New World lupins exhibit the highest known speciation rate among land plants (Hughes and Eastwood 2006; Drummond 2008). Further studies have also proved that the rapid diversification within the genus was promoted by evolutionary transition from annuality to perenniality coinciding with colonization of montane territories and dispersion from North America to South America and Mexico (Drummond 2008; Drummond et al. 2012).

3 Varietal Groups

For ordering of species variation, a division to groups of different types is quite often used, for example, systematic (for species with a rich variation, botanical varieties are separated on the basis of monogenic differences) or genetic divisions (for monogenic characteristics not considered in taxonomy for different reasons). In the genus *Lupinus,* an additional reason are the regions of species origin and some characteristics which appeared during domestication and cultivar improvement. These divisions have a great practical importance in lupin gene resource characterization and usage.

The first division was effected on the basis of origin and distribution regions (New World lupins/American continents and Old World lupins/the Mediterranean basin). The next division considers seed characteristics—size and seed coat structure. New World lupins have small seeds, with the Andean lupin (*L. mutabilis)* being an exception. All Old World lupins have large seeds. These species are divided into two groups, on the basis of the type of seed coat—smooth seeded (*L. albus, L. angustifolius, L. luteus, L. hispanicus, L. micranthus*) and rough seeded (*L. cosentinii, L. palaestinus, L. pilosus, L. atlanticus, L. princei, L. digitatus*; Cowling et al. 1998a).

For increasing food and fodder value during lupin domestication, decreasing alkaloid content in seeds was very important. Seeds of wild lupins can contain up to several percent in dry mass. There is no clear cut level of alkaloids from which seeds can be used in feeding. At the beginning, a level of 0.1 % in dry mass was accepted for fodder lupins. According to Römer and Jahn-Deesbach (1988), sweet lupins should contain less than 0.05 % of alkaloids, but Gladstones (1988) suggested 0.02 % in cultivar seeds as being reasonably safe for feeding. Today, different minimum levels are accepted in different countries, but the newest cultivars contain even less than 0.01 %. This criterion still exists in practice, and lupins are divided into two groups using popular names—high alkaloid (bitter) and low alkaloid (sweet) lupins.

There are some other important characteristics of variety ideotype which divide lupins into groups, for example, vegetation self-completing (restricted branching), thermoneutrality, and the type of usage (for seeds and green forage; Święcicki and Święcicki 1995; Święcicki et al. 2000a).

4 Genetic Resources and Utilization

The main aim of collections and gene banks is to maintain the genetic variation (to protect against genetic erosion) of crop plants and their wild relatives. In the case of the genus *Lupinus*, influencing factors are present needs.

The large number of taxa inside the genus *Lupinus*, for example, differentiated regions of origin and secondary distribution, and also the environmental requirements of four lupin crops have led to a broad interest in these genetic resources as a material for basic research and applied breeding. This has generated a particular interest in the enlargement and valorization of resources as a result of the increasing importance of lupins—a relatively new, modern crop (Brummund and Święcicki 2011). As a consequence, some large and active collection institutions work in close contact with breeders.

For the genus *Lupinus* (annual and perennial species, from acaulescent to tree-like shrubs), two centers of origin and secondary distribution are considered—American continents (the New World—200–300 small-seeded species and one lupin crop, large-seeded the Andean lupin, *L. mutabilis*) and the Mediterranean basin (13–15 large-seeded species including three crops—yellow lupin/*L. luteus*, narrow-leafed lupin/*L. angustifolus*, and white lupin/*L. albus*). The first collecting mission for the genus *Lupinus* was the Vavilov's expedition to the Mediterranean basin in 1926. His, and also further, collections are maintained at the Vavilov's institute at St. Petersburg, Russia (including original paper bags with Vavilov's handwriting), but the availability of information is restricted by the Russian alphabet.

Of substantial importance for *Lupinus* genetic resource, users are the International Lupin Association, the Biodiversity/International Plant Genetic Resources Institute, and the lupin project of the Department of Western Australia/University of Western Australia at Perth. These organizations have strongly supported not only the inventorization and elaboration of descriptors (IBPGR 1981) and databases but also the collection, characterization, and valorization of samples. At the Fifth International Lupin Conference in Poznań (Poland), in 1988, managers for collection databases of respective lupin crops were elected (Święcicki and Leraczyk 1994). Then, during the first meeting of the Legume Working Group of the European Cooperative Programme (ECP)/Genetic Resources (GR) at Copenhagen (Denmark) in 1995, the manager for the European *Lupinus* Collections Database was elected. The first version of this database was presented during the Ninth International Lupin Conference in Klink (Germany) in 1999 (Święcicki et al. 2000b) and the updated version at the meeting of the Legume Working Group of the ECP/GR in Novi Sad (Serbia) in 2013. Permanently useful for lupin collection curators and users are the references of Gladstones (1974) and Cowling et al. (1998a) on taxonomy, regions of origin, and distribution of the Old World lupins as well as gene resource characterization, collection gaps, and agronomy problems. The Australian lupin project is a model example of how direct practical benefits can be obtained thanks to investment in gene resources. One result was a dramatic increase in the narrow-leafed cropping area from 97,000 ha in 1981 to 1,151,000 ha in 1994. Wolko et al. (2011) summarized the world holdings of *Lupinus* germplasm accessions on the basis of the International Plant Genetic Resources Institute (IPGRI) Directory of Germplasm Collections. Data cover 36,854 lupin accessions gathered in 42 centers in 20 countries including the Americas and Australia. For the common, passport database of lupin collections, data from 13 European centers (57 % of accessions), Australia, and the USA (43 % of accessions) were obtained for 13,964 accessions (Tables 6.3 and 6.4). Table 6.4 shows the numbers of accessions gathered in respective countries and for respective *Lupinus* species. As many as 96 % of accessions are collected in six countries, 76 % are lupins of the Old World and 71 % belong to lupin crops (including wild populations as well as genotypes created by man). Wild populations of lupin crops protected in Spain and Portugal are very important as, in particular, is the Australian collection resulting from collecting expeditions together with a characterization of collection sites and their environmental conditions (Cowling et al. 1998a). Table 6.4 also suggests directions and regions for collecting expeditions and shows the restricted availability of the New World lupin gene resources and some wild species of the Old World. On the one hand, we have a resistance to *Colletotrichum lupini* and alkaloid content analyzed in hundreds of accessions of lupin crops within Wiatrowo's collection, and on the other hand, there is no accession for *Lupinus somaliensis* or just one accession of *L. anatolicus* and six accessions of *L. princei*.

Table 6.3 Donor institutions of the *Lupinus* collections passport data

Donor institutions	
AGRITEC, Ltd. Sumperk, Czech Republic	CZE
INRA—Dijon, France	FRA
Leibniz Institute of Plant Genetics and Crop Plant Research (IPK), Gatersleben, Germany	DEU
Centre for Genetic Resources (CGN), University and Research Centre, Wageningen, The Netherlands	NLD
Poznan Plant Breeders Ltd., Wiatrowo, Poland	POL
Instituto Superior de Agronomia, Departamento de Botânica e Engenharia Biológica (DBEB), Tapada da Ajuda, Lisboa, Portugal	PRT 1
Instituto Nacional de Recursos Biologicos (INRB) INIA—Oeiras, Quinta do Marquês, Portugal	PRT 2
Estacao Nacional de Melhoramento de Plantes, Elvas, Portugal	PRT 3
ISOPlexis—Banco de Germoplasma, Centro de Estudos da Macaronésia, Universidade da Madeira, Campus da Penteada, Funchal, Portugal	PRT 4
Plant Production Research Center, Piešťany, Slovak Republic	SVK
Instituto Nacional de Investigacion y Tecnologia Agraria y Alimentaria, Centro Nacional de Recursos Fitogeneticos, Alcala de Henares, Madrid, Spain	ESP
Western Regional Plant Introduction Station, Pullman, Washington, USA	USA
University of Western Australia, Centre for Legumes in Mediterranean Agriculture, Crawley, Western Australia	AUS

5 Major Breeding Achievements

In the familiar speech of a farmer, the term "lupin" in the singular or plural is often conversely used to name the crop. Meanwhile, there are at least four lupin crops each with clearly differentiated characteristics, including environmental and agronomic requirements. Each of them has a different number of chromosomes; they do not cross with each other; and two are self- and two cross-pollinating. For example, white and yellow lupin are as different as wheat and rye. As a consequence, the progress of breeding for respective species has been different in different countries. Obviously, these crops are closely related, resulting in the chance to select a Vavilov's homologous order of desirable, hitherto not existing characteristics (e.g., the mutation of the *rb* gene). Seeds of a white and Andean lupin were used in the distant past, but the real transition from a wild plant to a modern crop was achieved by a maximum two generations of breeders (Brummund and Święcicki 2011). Despite the genetic improvement of the above four lupins, the domestication of some others has also been conducted.

*5.1 Andean Lupin, Tarwi (*L. mutabilis *Sweet)*

Breeding achievements for this crop are clearly smaller than in Old World lupin crops. Species domestication is a more appropriate term than breeding improvement

Table 6.4 *Lupinus* species collected in gene banks

Species	NLD	SVK	FRA	ESP	USA	DEU	AUS	CZE	POL	PRT1	PRT2	PRT3	PRT4	Total
affinis					1			4						5
affinus							1							1
albicaulis					5	1								6
albifrons					5		1							6
albo-coccineus							1							1
albus	13	28	253	732	347	233	979	39	362	375	305	11		3677
anatolicus									1					1
andersonii					3									3
angustifolius		8		542	190	279	2165	17	361	41	291			3894
arboreus					4	3	1							8
arbustus					13									13
arcticus					1									1
argenteus					48									48
arizonicus					2									2
arizonicus					1		1							2
atlanticus					4	4	153							166
benthamii						1								1
bicolor					11	1					4			16
bracteolaris					1									1
campestris					1									1
citrinus					1									1
coccinnus					3									3
cosentinii				17	5	11	251		5		22			311
cryptantus									1					1
densiflorus						3	1							4
digitatus					3		4		1					8

Table 6.4 (continued)

Species	NLD	SVK	FRA	ESP	USA	DEU	AUS	CZE	POL	PRT1	PRT2	PRT3	PRT4	Total
douglasi									1					1
elegans					2				1					3
exaltus							2							2
excubitus					1									1
formosus					1									1
garfieldensis					2									2
gibertianus							1							1
graecus									10					10
gredensis				107										107
havardii					2									2
hirsutissimus					2	1	2							5
hirsutus									2					2
hispanicus				102	45	48	98		16	3	103			415
hispanicoluteus									7					7
hybrid					5									5
hybridus lem.							1							1
interspecific cross							8							8
latifolius					7		1							8
lepidus					20									20
leucophyllus					55	1								56
liniifolius									4					4
littoralis					6									6
longifolius							1							1
luteolus					4									4
luteus	56	18		303	86	132	463	26	354	78	283			1799

Table 6.4 (continued)

Species	NLD	SVK	FRA	ESP	USA	DEU	AUS	CZE	POL	PRT1	PRT2	PRT3	PRT4	Total
mariae-josephi H. Pascual				2										2
mexicanus					8	5	6							19
micranthus				12		20	51		1		10			94
microcarpus					10		3							13
multiflorus					1									1
mutabilis				20	79	30	221	2	17	150				519
nanus					5	3	1	1	3					13
nootkatensis					15	2								17
pachylobus					1									1
palaestinus						3	10		6					19
paniculatus									1					1
parviflorus					1									1
perennis				1	1	1								3
pilosus					61	14	189		6	1				271
polycarpus						1	1							2
polyphyllus				1	49	28	12	17	2					109
princei							6							6
pubescens						15	1		1					17
pusillus					3									3
rivularis					6	1								7
rothmaleri											58			58
rotundiflorus							1							1
sericeus					60		2							62
sparsiflorus					1		1							2

Table 6.4 (continued)

Species	NLD	SVK	FRA	ESP	USA	DEU	AUS	CZE	POL	PRT1	PRT2	PRT3	PRT4	Total
stiversii					3									3
subcarnosus					8	8	2							18
subvexus						1								1
succulentus					4	10	3	1	1					19
sulphureus					11									11
texensis					2		1							3
truncatus						1	6							7
variicolor					1	1								2
lupinus sp.			1	4	75	1905	13				1		7	2006
Total	69	54	254	1843	1293	2767	4665	107	1169	648	1077	11	7	13,964

(von Baer 2002). Moreover, achievements in a center of diversity (Andean countries) are different in comparison to other, particularly European countries (Römer and Jahn-Deesbach 1988). The main, exciting advantages of the species are the high protein (over 50 %) and oil (16–20 %) content in seeds. Disadvantages include the high alkaloid content in seeds and late maturity under long photoperiod and humid weather conditions. A definite difficulty in breeding is cross-pollination—above 10 % (Römer and Jahn-Deesbach 1988) or even 18 % (Gnatowska et al. 2000).

In Andean countries with extensive agriculture, the technology of seed debittering (decreasing the alkaloid content from 3 to 0.003 %) was more popular than the breeding of low alkaloid cultivars. In these conditions, the selection of local varieties was more effective, although one undoubted achievement was the breeding of the low alkaloid (0.0075 %) cultivar Inti in Chile, than was the case with participation in breeding projects in other countries (Römer and Jahn-Deesbach 1988).

In European countries interested in the advantages of the Andean lupin, a serious disadvantage is the indeterminate growth habit resulting in endless flowering under humid weather conditions during the ripening period. Early maturity was looked for in different ways—collections of genetic resources were analyzed, distant crosses were performed (using *L. albus. L. elegans,* and *L. polyphyllus*), and mutations were induced. The restricted branching mutant described by Römer (1994) seems to be promising.

A comparison of Römer and Jahn-Deesbach (1988) and von Baer (2011) presentations allows the following achievements in the improvement of Andean lupin to be mentioned:

- A high protein and oil content in seeds was combined with a low alkaloid content, but this is difficult to maintain because of the polygenic inheritance and high degree of cross-pollination,
- Hybrids *L. polyphyllus* × *L. mutabilis* (von Baer 2002),
- A valuable source of earlier maturity is the restricted branching mutant selected and described by Römer (1994).

Despite the above achievements, valuable cultivars useful for modern agriculture have not been bred so far. Natural species adaptation to environmental Andean conditions resulted in barriers which are difficult to overcome. Andean lupin cultivars will not quickly be able to compete with European and Australian narrow-leafed and yellow lupin cultivars.

5.2 *White Lupin (*L. albus *L.)*

White lupin is grown on a limited area, although it could be an important fodder crop. Seeds contain 32–36 % protein and 8–14 % oil. It requires better soil than yellow and narrow-leafed lupin, but seed yields are higher (3–4 t/ha). In east-central Europe, it is used as a spring crop, but in southern countries, it is a winter crop. The main disadvantages are the long vegetation period and susceptibility to anthracnose. Long vegetation results in late harvesting, and seeds are infected by fungi and mycotoxins. The small acreage is a result of limited cultivar value, and vice versa breeding is less advanced (a limited number of breeding centers) because of the limited usage of the crop. In spite of the above, hitherto breeding work has resulted in remarkable effects. For example, a breeding project in Poland started in 1946, and the first cultivars were released in 1965 (cv. Kali) and 1976 (cv. Kalina). In successive cultivars, that is, cv. Wat (1978) and cv. Hetman (1980), the harvest index was substantially improved (Kubok 1988).

In white lupin breeding, some achievements should be underlined: decrease in alkaloid content in seeds (although not as much as in narrow-leafed and yellow lupin), improvement of harvest index (vegetative growth decreased), shortening of vegetation period, selection of restricted branching, and partially thermoneutral mutants and finally introducing these characters to cultivars (Święcicki 1986; Boersma et al. 2007b). The newest Polish cultivars, Butan (normal growth, released in 2000) and Boros (restricted branching, released in 2003), are suitable for modern farming, but a radical increase of white lupin acreage depends on a substantial shortening of vegetation and improving resistance to anthracnose *(C. lupini)*. The difficulty lies in the selection of both characteristics from genetic resources and their transfer to cultivars (Cowling et al. 1998a; Adhikari et al. 2009). Work on white lupin suitability to winter cropping in south European countries was conducted by Huyghe and Papineau (1990). Cold resistance is the most important characteristic. Experiments on the role of different parts of plants in cold resistance show that a large root and root parenchyma are required. High vernalization needs are necessary but not sufficient condition. The ability to acclimatize to the cold, which partially depends on the above characteristics, is very important. The above research was extended by Annicchiarico et al. (2011).

*5.3 Yellow Lupin (***L. luteus** *L.)*

The species is one of the most important legumes for sandy soils. N fixation, an improvement of soil fertility and high protein content in seeds (40–46 %), caused a large interest in east-central and northern European countries and in Western Australia. Yellow lupin was sown for the first time in Germany in 1840 as green manure, and then it spread rapidly in Prussia and north-central Europe as a forecrop to fertilize sandy soils and increase the yields of successive crops (Brummund and Święcicki 2011). It appeared in modern farming after 1945 together with chemical weed control, defoliation, and the use of combine harvesters. Improved cultivars were adopted to two types of usage—for seeds and green mass (as green manure and forage).

The first achievements which were important for modern breeding were the removal of wild characteristics (hard seed coat and pod shattering), selection of low alkaloid plants/seeds by Sengbusch in 1927, and then sowing 2 ha of sweet lupin culture in 1931 (Brummund and Święcicki 2011). Great breeding progress achieved in a relatively short period was presented by Święcicki (1986), Brummund (1988), and Kubok (1988). Święcicki et al. (2000a) described the range of expression and the practical value of 43 alleles at 20 loci controlling alkaloid content, pod shattering, growth rhythm, color of plants, flowers and seed coat, plant branching patterns and resistance to diseases and abiotic stresses. From the above review, it is possible to select some characteristics; the introduction of which to the cultivar ideotype was an outstanding achievement in breeding.

In the 1950s/1960s, a serious menace for yellow lupin cropping was the *Fusarium*. Sometimes, an outbreak completely destroyed the crop in a field. Fortunately, genetic sources of resistance controlled by the gene Fus_1 were available. The first *Fusarium*-resistant cultivar, Refusanova bred in Germany in 1965, became a source of resistance in further breeding (Cowling et al. 1998b). In a relatively short time, new resistant cultivars were introduced to cropping. This achievement should be underlined. In the breeding of other crops, there are not so many examples of stable resistance to a disease (that still exists), caused by a pathogen with a few races and controlled by a single gene.

The next milestone was the introduction of thermoneutrality to cultivars. Yellow lupin plants need vernalization in an early growth stage (2–4 °C for about 2 weeks) to reach the generative stage and produce a high seed yield. Selected plants with lower thermal requirements have a shorter initial growth stage, start to flower earlier, and mature earlier and more uniformly (Święcicki et al. 2000a; Adhikari et al. 2012; Fig. 6.1). Thanks to a faster plant development, an escape resistance to bean yellow mosaic virus is observed (incompatibility between aphids and host plant development).

Fig. 6.1 Selection of thermoneutral yellow lupin genotypes (flowering plots) in late sowing conditions (without vernalization)

The cultivar ideotype used for seeds and green forage was the same for a long period of time. In the case of seed harvesting, a disadvantage is indeterminate plant growth. Successive branches (secondary or even tertiary laterals) cause nonuniform seed maturation and make harvesting more difficult. In the 1970s, a spontaneous mutation with only one main shoot and single flowers instead of branches was selected (Troll 1967). This appeared to be controlled by the recessive gene *rb* (restricted branching). The *rb* gene creates a new cultivar ideotype—early and uniformly maturing (the so-called self-completing cultivar) cropped for seeds (Święcicki and Święcicki 1994). The strongly reduced vegetative mass (improved harvest index) should give an increased seed yield, particularly in regions with a short vegetation season and higher rainfall. The first *rb* cultivar, that is, Manru (released in Poland 1990), yielded higher than controls in state trials.

*5.4 Narrow-Leafed Lupin (***L. angustifolius** *L.)*

The reviews by Gladstones (1988), Święcicki and Święcicki (1995), and Brummund and Święcicki (2011) discuss the most important achievements in the breeding of this crop, particularly when we take into account the uncommon career in Western Australia where its acreage increased from 97,000 to 1,151,000 ha between 1981 and 1994.

Similar to the yellow lupin, a crucial role was played by the selection by von Sengbusch in winter 1927/1928 of low alkaloid seeds/plants as sources of the trait for modern cultivars. Investigations have revealed that a low alkaloid content in seeds is controlled by three genes: *iuc, depr,* and *es*. Suggestions of breeders and the Australian National Health and Medical Research point to a "normal" cultivar level of about 0.020 % for reasonable safety (Gladstones 1988). However, use of gas chromatography in alkaloid content screening allowed cultivars to be bred with less than 0.01 % seed dry mass.

Local varieties and the first cultivars were characterized by pod shattering, particularly in dry and hot weather during maturation and harvest. This led to as much as an 80 % yield loss. In removing or at least clearly improving this character, two genes were used. The gene *le* (*lentus*) causes more elastic walls preventing their shrinking at drying, and the gene *ta* (*tardus*) is responsible for strong sutures.

The importance of restricted branching is differently estimated depending on growing conditions. In the Middle East and North Europe, it is similar to yellow lupin. In narrow-leafed lupin, there are broader possibilities because more cases of mutation, spontaneous and included, were selected with different levels of lateral branch reduction, controlled by alleles of the *rb* gene or by another locus (Gawłowska et al. 1999). Restricted branching cultivars were bred yielding traditional but earlier and more uniform ripening (Fig. 6.2). However, the winter growing season in south-west Australia shows that a fully determinant type is too restricted in its branching as ripening is premature, and the pod fills poorly.

In the review by Święcicki and Święcicki (1995), the importance of 35 genes controlling important plant characters was presented. For example, in cv. Svalöf

Fig. 6.2 Narrow-leafed lupin—improved harvest index and earliness of restricted branching *(right)* versus traditional cultivar *(left)*

Böre, a spontaneous mutant was found with very early flowering and lower vernalization requirements. Gladstones (1977) defined this trait as thermoneutrality controlled by the gene *Ku* which then was introduced to Australian and Polish cultivars. *Ku* cultivars have faster initial growth and show the so-called escape resistance to cucumber mosaic virus (CMV).

At the end of the last century, the CMV and leaf spot *(Stemphylium vesicarium)* were still considered the most important diseases. Anthracnose *(C. lupini)* was deemed less significant in Central Europe. Unfortunately, a highly virulent strain of *C. lupini* has spread around the world in lupin (Cowling et al. 1998a). This disease appeared to be the most important in breeding because it caused a substantial acreage decrease. Additionally, sources of resistance and pathogen biology were unknown. Intensive work by breeders and pathologists and then crop rotation and seed production monitoring resulted in resistant cultivars and hopefully elimination of the epidemic.

6 Specific Goals in Current Breeding

Despite a number of common characteristics, different goals can be determined for each lupin species. Thanks to biotechnological and molecular techniques, these goals should be achieved faster.

Andean lupin in regions of natural distribution (Peru, Bolivia, Equador) should be adopted to extensive agriculture. Therefore, the morphotype of the local variety should not be changed drastically. To decrease alkaloid content, the technological procedure of debittering was suggested by Römer and Jahn-Deesbach (1988). It is also possible to improve cultivars genetically using available sources of low alkaloid content, but a high degree of cross-pollination and unmonitored wild Andean lupin plants could cause difficulties in maintaining the stability of characteristics.

The next, important aim even for local varieties is resistance to anthracnose. There are two ways: breeding improvement similar to narrow-leafed lupin or using 2-year-old seeds for sowing, because storing seeds for 18 month results in practical elimination of infection (Cowling et al. 1998a).

An exciting target seems to be a European crop made from the Andean lupin. Over 50 % protein and about 20 % oil in seeds would result in an excellent component for feeding pigs and poultry. Unfortunately, the adaptation of this plant to European environmental conditions, drastic improvement of the harvest index, and resistance to anthracnose are extremely difficult aims and are unlikely to be realized in the near future.

White lupin (32–36 % protein, 8–14 % oil, and potentially the highest yield of European lupins) provides a more realistic possibility. The cultivar ideotype should contain high seed yield with low alkaloid content (below 0.02 %), possibly the highest content of protein and oil and, most importantly, resistance to anthracnose plus a vegetation period shortened by at least 10–14 days. Present cultivars give satisfactory seed yield with 0.02–0.03 % alkaloids and 8–11 % oil content. The biggest disadvantages are late maturing and susceptibility to anthracnose. An improvement of both characteristics is the main condition for an increase in commercial crop importance and acreage. Collections of genetic resources contain valuable initial material for most of the required characteristics (Cowling et al. 1998a; Adhikari et al. 2009; Kurlovich 2002). For similar results to those obtained in *L. angustifolius,* a large project of resistance breeding against *C. lupini* is indispensible. More difficult for another reason (no source of earliness) seems to be the shortening of vegetation for earlier and more uniform ripening. A precise review of available gene resources (Cowling et al. 1998b; Święcicki et al. 2000b) and mutation breeding is necessary. The gene *rb* (restricted branching) could be useful, as this strongly changes the harvest index value. First *rb* cultivars yield about 10 % lower than traditional ones. However, this gene influences earlier and more uniform maturity. It must be remembered that the introduction of strong, even monogenic changes into a cultivar ideotype quite often disturbs genotype balance, and its reconstruction is time-consuming. For example, the *afila* characteristic in pea was described in 1954. Then, a period of pea cultivar improvement in Poland using the *afila* gene from releasing the first, high-yielding cultivars Sum and Wasata to all *afila* cvs. in the national register lasted from 1979 to 2002.

Achievement of the above mentioned goals in white lupin breeding justifies every expense and is much more practicable than an introduction of Andean lupin to European agriculture.

Fig. 6.3 *Left* and *Right,* symptoms of anthracnose *(Colletotrichum lupini)* on lupin plants

For yellow and narrow-leafed lupin, specific goals are partially similar, but the scale of importance is somewhat different. In yellow lupin improvement, the most important is a resistance to anthracnose, which is clearly worse than in narrow-leafed lupin (Fig. 6.3). Moreover, it is necessary to decrease the alkaloid content in seeds and exclude the gramine which in some cultivars is present in substantial quantities.

In narrow-leafed lupin breeding, an important goal is resistance to *Fusarium* up to a level similar to that in yellow lupin. In both lupin crops, yield stability is very important. Here, of undoubted influence are resistance to draught and hitherto unknown factors causing flower abortion—the most important disadvantage for all legumes.

Lupin fields are quite often destroyed by wild animals (stags, deer and hares). Therefore, if possible, it would be desirable to breed cultivars with high alkaloid content in their green mass and low alkaloid seeds.

Among the specific goals for future breeding, attention must be drawn to two subjects. Increased yields have been hitherto achieved via improving harvest index. Without any doubt, unused reserves are available in activity/efficiency of plant physiological processes. Additionally, mobile equipment for mass screening in the field is available. However, in terms of seed quality, improvement of the content of protein and antinutritional compounds (alkaloids) is considered exclusively. Experts should establish if more substances are present in lupin seeds where the improvement could increase seed value or at least feeding efficiency. For example, the most recent literature data show the particular properties of a white lupin protein in curing diabetes and atherosclerosis.

Trials are undertaken also to domesticate new wild lupins, for example, *L. atlanticus*, *L. cosentinii*, *L. hispanicus*, *L. pilosus*, or *L. polyphyllus* (Cordero et al. 1988; Buirchell 2000a, b; Campos-Andrada et al. 2000; Kurlovich et al. 2008). The aim is to use their adaptability to untypical or even extreme environmental conditions. For example, *L. pilosus* is the most lime tolerant of the lupin species.

7 Breeding Methods and Specific Techniques

The domestication period of lupin crops is over. In lupin breeding, most important is the cross-pedigree method. In a breeding procedure, a larger size of family (an offspring of a single plant) and later beginning of selection in segregating progenies (F_3 the earliest) must be considered because of lupin polyploidy. Cross-pollination of *L. luteus* and *L. mutabilis* causes some technical difficulties—the isolation of single plants selected from a segregating progeny or the isolation of a whole breeding material under a net tent. In the final stage of maintaining breeding, the homogeneity/purity of a given strain/cultivar can be maintained in a spatial isolation.

The selection of a plant morphotype, growth and development, as well as yield creating factors is conducted in field-natural conditions. For certain characteristics, it is indispensable to test samples in parallel (representing progenies sown in a field), and these are then destroyed (e.g., estimation of alkaloid content or resistance to pathogens with artificial infection). A pedigree scheme allows us to estimate the value of a plant material sown in a field on the basis of the laboratory or greenhouse testing of a sister part of the progeny.

The initial material is of substantial importance for breeding effectiveness. Collections of gene resources as well as genotypes obtained in pre-breeding are considered to be a so-called breeding gene pool. Exchange between lupin companies is not easy and is restricted by the standard material transfer agreement (SMTA). Moreover, there are not so many lupin centers in the world. Therefore, characterization and valorization of gene resources are important, as are convergent crossing programs (unfortunately, these are multiyear) for combining new characteristics on a valuable genotypic background. For example, sources of desirable and available characteristics/genes in lupin collections are given by Święcicki and Święcicki (1995), Święcicki et al. (2000a), and Kurlovich (2002).

Sources of genetic variation are mutations and recombinations. The natural variation of lupins is not too broad from a breeding point of view. Therefore, in the event of a lack of desirable characteristics, a useful tool is mutation induction. For an efficient project of mutation breeding, the introductory work is important—mutagene and dose optimization and suitably large pedigree population for mutant selection (= a number of families/offsprings of treated seeds × number of plants per family). Because of lupin polyploidy, the number of plants per family must be bigger than for diploids, and the selection should start in M_3, at the earliest. In the case of looking for one particular characteristic, it is possible to sow a mixture in a suitable generation (instead of a pedigree population), but when a mutant is low yielding or lost, there will be no possibility of its reselection.

Different genotypes have been treated by different mutagenes in lupins, and different results have been obtained (Klamroth et al. 2011; Rudloff 2011; Stawinski and Rybinski 2000; Micke and Święcicki 1988; Święcicki and Olejniczak 1999). The most spectacular seems to be a selection of restricted branching mutants—induced in Andean, white, and narrow-leafed lupins (Micke and Święcicki 1988; Römer 1994; Gawłowska et al. 1999) and spontaneous in yellow and narrow-leafed lupin (Gawłowska et al. 1999; Troll 1967).

In lupin breeding, similar to other crops, mass-screening and selecting techniques are very important. These techniques should accept micro samples and not destroy them. Desirable is also mobile field equipment allowing selection in early stages of plant growth. For certain characteristics (e.g., resistance to diseases and unfavorable environmental conditions), selection is often conducted under conditions with a great intensity of a stress factor (10 years of continually cropping lupins or very late sowings).

Lupins have their own, particular environmental requirements and show weak growth in greenhouse conditions including in vitro regeneration. Therefore, effective methods to shorten multiyear breeding cycles are not available. Surma et al. (2013) tried to elaborate the single-seed descent (SSD) technique using in vitro cultures of lupin embryos (*L. luteus* and *L. angustifolius*). It appeared that lupin roots and shoots grow well in vitro, but further plantlet acclimatization ex vitro is rather difficult. Nevertheless, it is possible to obtain 2.5–3 generations per year.

In recent years, a lot of results have been published on the possibilities of using molecular markers in the selection of agricultural characteristics (see the next section).

8 Integration of New Biotechnologies in Breeding Programs

8.1 Biotechnology

The progress in lupin breeding is undeniable, although many problems affecting yield still apply and could be solved by combining genotypes with desired traits in specific sexual hybridization between species (Rybczynski and Podyma 1993a). The high level of species incompatibility within the genus *Lupinus* significantly limits the production of viable seeds through interspecific crosses using conventional hybridization methods (Święcicki et al. 2000a). Employment of biotechnology in contemporary breeding programs may allow the creation of new germplasm, significantly broadening the genetic variability. Plant tissue culture is also useful in vegetative propagation to increase the number of individuals with the same genotype. Although research towards tissue culture manipulation and transformation has long been conducted, in the case of lupins, the scope of achievements is still limited (for a review, see also Sator (1990), Atkins and Smith (1997), Święcicki et al. (2000a), and Wolko et al. (2011)).

Like many large-seeded legumes, lupins are considered difficult to manipulate in culture (Nadolska-Orczyk 1992). The significant variation in the morphogenic potential of lupin species has been emphasized by Rybczynski and Podyma (1993b) and Zgagacz and Rybczynski (1996). The key requirement for successful plant tissue culture is an optimal media composition and a protocol developed for the species/genotype of interest. Therefore, the attention of many scientists has been drawn to the adjustment of proper culture conditions to the culture aim. The research undertaken concerned mainly micropropagation, somatic embryogenesis, embryo rescue, protoplast culture, androgenesis and transformation. Various lupin explants derived from different stages of plant ontogenesis have been analyzed.

Experiments on lupin plant restoration (*L. albus*) were started by Ball (1946, 1960). Sroga (1983) successfully produced callus for *L. angustifolius* hypocotyl explants which was further used for the induction of suspension cultures. The research on regeneration from leaf and hypocotyl explants of *L. polyphyllus, L. hartwegii, L. angustifolius,* and *L. luteus* resulted in successful callus induction but very limited plant regeneration (Sator 1985b). In contrast, organogenesis into complete plants via callus culture in *L. angustifolius* and *L. polyphyllus* was reported by Sroga (1987), although these results could not be repeated by Nadolska-Orczyk (1992). Callus culture formation from seedling explants of *L. mutabilis* was also investigated by Phoplonker and Caligari (1993) who, by testing different culture conditions and protocols, concluded that stem and hypocotyl explants produced the greatest quantity of callus.

Regeneration and propagation of plants bypassing callus formation have also been explored in lupins. Micropropagation of *L. texensis* hypocotyl explants has led to shoot formation while the rooting is rather weak (Upadhyaya et al. 1992). On the other hand, complete, blooming and seed-setting plants of *L. luteus* have been obtained from hypocotyl segments (Daza and Chamber 1993). Rybczynski and Podyma (1993a) regenerated root systems of the shoot tip explants of *L. albus, L. luteus, L. angustifolius, L. hispanicus* and *L. polyphyllus*. Vegetative propagation based on apical meristem cultures was successful for *L. hispanicus* and using cotyledonary node explants for *L. hispanicus* and *L. albus*. The plantlets were further successfully adapted to in vivo conditions (Rybczynski and Podyma 1993a). The complete long-term micropropagation protocol for the four lupin crops (*L. angustifolius*, *L. albus*, *L. luteus*, and *L. mutabilis*) through axillary meristem multiplication was fairly recently reported (Pniewski et al. 2002). Regenerated shoots could be cultured for at least ten passages in good plant conditions. Moreover, further rooting experiments carried out on material from the second regeneration round demonstrated the good rooting ability of all the species. Since the potential of root formation decreased in succeeding regeneration rounds, grafting was performed as an alternative method of transferring plants to greenhouse conditions (Pniewski et al. 2002).

In vitro plant multiplication might also be carried out through somatic embryogenesis. Successful direct somatic embryogenesis from immature cotyledons of *L. angustifolius, L. albus* and *L. mutabilis* has been achieved, although the authors give no details on plant regeneration (Nadolska-Orczyk 1992). Similar studies have been carried out for *L. albus, L. luteus, L. angustifolius* and *L. hispanicus* presenting effective somatic embryogenesis for *L. albus* only; however, the obtained plantlets were further easily adapted to greenhouse conditions (Rybczynski and Podyma 1993b). The response of immature embryo-derived explants to different culture media was also tested by Zgagacz and Rybczynski (1996). The experiment performed on *L. albus, L.mutabilis* and *L. hispanicus* showed that the most responsive part of the immature embryo was cotyledons and depending on the species only somatic embryos or shoots could be observed in culture (Zgagacz and Rybczynski 1996).

Embryo rescue culture might be an alternative approach to overcome crossability barriers. Several attempts at embryo rescue and its modification have been undertaken in the case of lupins. An embryo rescue protocol and culture condition have been evaluated for *L. luteus*, *L. angustifolius*, *L. albus*, *L. mutabilis* and *L. polyphyllus* as a basis for future development of interspecific hybrids (Wilson et al. 2008; Kasten and Kunert 1991). In the trial of interspecific crosses between *L. albus, L. mutabilis* and *L. angustifolius*, hybrid embryo rescue was carried out and its further in vitro culture led to the development of plants (Przyborowski 1997). Embryo rescue has also been applied to successfully produce F_1 plants of *L. mutabilis* × *L. hartwegii* (Schaefer-Menuhr et al. 1988; Clements et al. 2008; Kasten et al. 1991), *L. angustifolius* × *L. luteus* (Kasten et al. 1991) as well as *L. mutabilis* × *L. mexicans* and *L. arizonicus* (Clements et al. 2008).

In vitro research in lupins also targets protoplast culture, focusing not only on plant regeneration attempts but also on protoplast fusion. Babaoglu (2000) investigated various cultivation systems and factors on the yield, viability and division of protoplasts from *L. mutabilis,* but the progress in determining optimum explant and culture conditions was limited. Sinha et al. (2003/2004) optimized the protocol for routine production of highly viable cotyledonary protoplasts of white lupin and further elaborated the plating environment stimulating protoplast elongation and division (Sinha and Caligari 2005). Their results established benchmarks for future regeneration and hybridization studies. The procedure for protoplast isolation derived from hypocotyls, cotyledons and young leaves, as well as the effects of various culture conditions on protoplasts development have also been elaborated for yellow lupin by Wiszniewska and Pindel (2009, 2013). These authors also reported protocolony formation in yellow lupin protoplast culture (Wiszniewska and Pindel 2009). Successful protoplast fusion of *L. angustifolius* and *L. subcarnosus* as well as subsequent shoot regeneration of hybrid callus was recently reported by Sonntag et al. (2009).

Androgenesis plays an important role in double haploid (DH) line development. Fast recovery of fully homozygous inbred lines could significantly accelerate modern breeding in new cultivar production (Ormerod and Caligari 1994). Up until now, several studies on androgenesis induction in lupins have been reported, but no DH lines have yet been produced (Skrzypek et al. 2008). Callus from another culture of *L. polyphyllus* was obtained by Sator et al. (1983). Unfortunately, the chromosome count of plantlets regenerated afterwards showed their diploid characteristic (Sator 1985a). Ormerod and Caligari (1994) presented the results of spontaneously released microspores in liquid medium of *L. albus* anthers and further regeneration of embryo-like structures from microspores. A protocol for potentially large-scale production of *L. albus* and *L. angustifolius* pro-embryos (multicellular structures) has also been reported (Bayliss et al. 2004). Androgenetic induction to anther-derived callus of *L. luteus*, *L. angustifolius,* and *L. albus* has been obtained, however, no plant regeneration has been achieved (Skrzypek et al. 2008). Further studies to determine the optimum conditions for another culture initiation in *L. angustifolius* have concerned not only the localization and size of buds, color, and size of anthers but also the influence of different factors on microspore embryogenesis induction (Kozak et al. 2012). Moreover, for all tested *L. angustifolius* genotypes, callus could be obtained that further continued its growth and produced roots (Kozak et al. 2012).

An alternative method to DH aiming at receiving homozygous lines is the "SSD technique," which might significantly shorten the breeding cycles. A combination of the SSD method and in vitro culture of embryos was recently proposed for lupins (Surma et al. 2013). As a first step towards that goal, culture conditions for *L. angustifolius* and *L. luteus* embryos dissected from fully developed, green seeds have been established. Regeneration of shoots and roots has been observed. Moreover, some of the regenerated plants have survived the transfer to greenhouse conditions where they flowered and set pods/seeds. These results constitute a very optimistic step towards the rapid attainment of succeeding generations via the SSD technique (Surma et al. 2013).

Transformation systems might serve as a method of direct introduction of agronomically useful genes into the genomes of interest. Several successful transformation protocols have been published for lupins. Mugnier (1988) obtained hairy root cultures of *L. albus* and *L. pollyphyllus* by inoculation with *Agrobacterium rhizogenes*. Routine transformation of *L. angustifolius* to herbicide resistance was performed by Pigeaire et al. (1997). The shoot apices transformation was mediated by *Agrobacterium tumefaciens* carrying the *bar* gene, and transformants were regenerated from axillary buds (Pigeaire et al. 1997). T1 and T2 plants were stable in a glasshouse, and molecular analyses confirmed foreign gene integration and its stable expression (Pigeaire et al. 1997). A trial at improving narrow-leafed lupin nutritive value by means of expressing a sulfur-rich, sunflower seed albumin (SSA) gene was undertaken by Molvig et al. (1997). The authors demonstrated a stably transformed line with an SSA gene expressed at high level in seeds. The SSA accounted for 5 % of the extractable seed protein, and the conducted feeding trial showed an increase in the nutritive values of transgenic versus wild-type seeds (Molvig et al. 1997). *A. tumefaciens*-based transformation has also been attempted to improve narrow-leafed lupin resistance to fungal pathogens (Wijayanto et al. 2009). Several transgenic lines expressing the anti-apoptotic baculovirus gene p35 have been characterized with reduced disease symptoms (Wijayanto et al. 2009). An attempt to increase narrow-leafed lupin yield by preventing flower abortion was also undertaken with the aid of transgenic approach (Atkins et al. 2011). Lines stably transformed with a gene involved in cytokine synthesis from *A. tumefaciens* were obtained, and some of the transgenic lines were characterized by a higher pod number (Atkins et al. 2011). The first transgenic plants of *L. mutabilis* were reported by Babaoglu et al. (2000). *A. tumefaciens*-mediated transformation of shoot apices led to the development of kanamycin-resistant plants expressing β-glucuronidase (Babaoglu et al. 2000). Transgenic *L. luteus* plants with herbicide resistance were obtained by infection of apical meristem explants with *A. tumefaciens* (Li et al. 2000). *Agrobacterium*-based transformation of yellow lupin was also optimized by Pniewski et al. (2006) with the aim of a callus/tumor tissue generation capable of hepatitis B virus (HBV) surface antigen production useful in oral immunization. Achievements in genetic transformation have also been attained in the case of white lupin. In the course of study on the formation and function of cluster roots formed as a consequence of adaptation to phosphorous deficiency, *A. rhizogenes*-mediated root transformation systems were established (Uhde-Stone et al. 2005; Cheng et al. 2011).

8.2 Molecular Breeding and Marker-Assisted Selection

Field-based selection for agricultural traits is often time-consuming, inconvenient, or even simply not possible. Marker-assisted selection (MAS) is a very promising tool to improve efficiency and accelerate selection in breeding programs, although its potential is still under-evaluated. The prospect of being able to select desirable lines based on genotype using simple molecular markers at the seedling stage in early generations and parallel screening of multiple traits is very attractive to plant breeders (Young 1999; Holland 2004).

Development of molecular markers linked to genes controlling desirable traits is the main challenge in MAS. In order to routinely use a molecular marker in breeding selection, the following requirements should be met: (1) the marker is closely linked or co-segregates with a gene of interest, (2) the screening techniques should be efficient in a large population and highly reproducible and (3) the application of the marker should be more economical and user friendly than conventional selection methods (Gupta et al. 1999). Nowadays, simple polymerase chain reaction (PCR)-based markers, sequence-tagged microsatellite site (STMS), sequence-characterized amplified region (SCAR), sequence-tagged site (STS), and allele-specific PCR (AS-PCR) markers best conform to the marker requirements for selection in plant breeding (Wolko et al. 2011).

In recent years, a large expansion of next-generation sequencing (NGS) technologies enabling polymorphism identification and marker development can be observed. NGS methods have also been successfully applied for lupins including: restriction-site associated DNA sequencing—RAD-seq (Yang et al. 2012, 2013a) and transcriptome sequencing—RNA-seq (Parra-Gonzalez et al. 2012). A great advantage of using NGS technology is the possibility of the easy conversion of candidate markers into simple and inexpensive PCR-based markers useful in molecular breeding (Yang et al. 2012).

8.2.1 Narrow-Leafed Lupin

Narrow-leafed lupin, as the most popular lupin crop, is characterized with rich molecular achievements including draft genome sequence (Yang et al. 2013b) and well-saturated genetic maps (Nelson et al. 2006, 2010; Kroc et al. 2014; Yang et al. 2013b). In that species, the microsatellite-anchored fragment length polymorphism (MFLP) technique plays a pivotal role in selection marker development (Yang et al. 2001). MFLP markers have been used in conversion into many simple PCR-based markers for MAS (Li et al. 2011, 2012b; Yang et al. 2008, 2010; Boersma et al. 2007b). Moreover, NGS technologies, especially the RAD-sequencing approach, have also recently been applied in the process of marker generation (Yang et al. 2012, 2013a).

8.2.1.1 Pod Characteristics

The trait of the pod-shattering in narrow-leafed lupin is regulated by two major genes: *lentus* and *tardus* (Święcicki and Święcicki 1995). Only plants possessing recessive alleles of both genes are fully non-shattering, since each of them controls only partially reduced pod shattering. The recessive allele of the *lentus* gene modifies the orientation of the sclerified endocarp of the pod, which manifests itself in a change in internal pod pigmentation. This modification results in a reduction of torsional forces upon drying and causes reduced pod shatter (Gladstones 1967). Three simple PCR-based molecular markers closely linked to *lentus* gene have been developed for narrow-leafed lupin (Boersma et al. 2007c; Li et al. 2012b). Two of them are dominant markers (LeM1 and LeM2), unable to discriminate between homozygous and heterozygous genotypes (Boersma et al. 2007c). Recently, Li et al. (2012b) reported the development of a new codominant marker (LeLi) flanking the *lentus* gene. This marker is more useful than the previous ones in the discrimination of heterozygous genotypes in the *lentus* locus. The overall matching rate between the LeLi marker genotype and plant phenotype was 60.67% on tested material, including Australian cultivars and accessions of the Australian Lupin Collection (Li et al. 2012b).

The second gene necessary to determine the fully non-shattering effect in narrow-leafed lupin is the *tardus* gene. The recessive allele of this gene affects the sclerenchyma strips of the dorsal and ventral pod seams, fusing the two pod halves in a way that their separation is impeded. Its expression is dependent on environmental factors, especially temperature and humidity (Gladstones 1967). In 2009, Boersma et al. (2009) developed three locus-specific markers (TaM1, TaM2 and TaM3) closely linked to the *tardus* gene. These markers showed only a 24–39%

correlation between plant phenotypes and marker genotypes on wild accessions of the Australian Lupin Collection, so further development of markers linked to the *tardus* gene was still needed. A new codominant PCR-based marker flanking *tardus* gene was designed by Li et al. (2010). The marker TaLi had a very high association (94 %) between marker genotype and phenotype on tested domesticated cultivars and the lupin core collection and may be widely used for selection in breeding programs (Li et al. 2010).

8.2.1.2 Structure of Seed Coat

A hard seed coat, impermeable to water, prevents germination and causes dormancy of seeds. This adaptation is a very important strategy, ensuring species survival, especially to long drought periods lasting for several months (Święcicki and Święcicki 1995). A single recessive gene, *mollis,* resulting in soft seed coat development has been identified so far (Mikołajczyk 1966). Up until now, two PCR-based markers flanking the *mollis* gene have been designed (Boersma et al. 2007a; Li et al. 2012a). The first one (MoA) was a codominant SNP-based marker, although the detection method based on single-strand conformation polymorphism (SSCP) was very time-consuming and the marker was not routinely used in MAS (Boersma et al. 2007a). A recently published molecular marker (MoLi) closely linked to the *mollis* gene had a very high correlation (91.3 %) between phenotypes and marker genotypes on tested Australian cultivars and accessions of the core collection. Moreover, the marker MoLi is a codominant, length-polymorphism marker, which makes its detection more effective and the marker in general more useful in narrow-leafed lupin large-scale selection (Li et al. 2012a).

8.2.1.3 Earliness of Flowering

Many of the wild type narrow-leafed lupins require vernalization in early growth to promote flowering (Gladstones 1970). This requirement is not desired in domesticated cultivars, especially in areas with short growing seasons. It is therefore necessary to introduce a single dominant *Ku* gene responsible for the early-flowering effect in worldwide modern lupin breeding (Brien et al. 2000; Gladstones and Hill 1969). A sequence-specific, size-based marker (KuHM1) closely linked to the *Ku* gene was reported by Boersma et al. (2007b). A correlation between the markers for genotype and phenotype was tested only on parental lines and 106 F8 recombinant inbred lines (RILs) of the narrow-leafed lupin mapping population (83A:476 × P27255) and it showed perfect compatibility (Boersma et al. 2007b). Unfortunately, a matching test has not been carried out on cultivars and accessions of the lupin core collection, so the application of the KuHm1 marker in breeding selection is limited.

8.2.1.4 Alkaloid Content

A characteristic trait of lupins is a production of quinolizidine alkaloids. Due to their toxicity and bitter taste, a decrease in alkaloid content is especially important for propagation of lupins in animal feed and human consumption (Gladstones 1998b). Three recessive genes—*iucundus, esculentus,* and *depressus*—controlling low alkaloid content are known in narrow-leafed lupin (Święcicki and Święcicki 1995; Kurlovich 2002). It is also recognized that alkaloid content is dependent on environmental factors such as weather or growing conditions (Barbacki 1952) and soil nutritional imbalances including phosphorus and potassium deficiency (Gremigni et al. 2003). Conventional measurement methods assessing general alkaloid content only (high/low) require the usage of the Dragendorff reagent (Harrison and Williams 1982). On the other hand, more sensitive gas chromatography enables both quantitative and qualitative alkaloid estimation, but due to the equipment requirements and quite long procedure it is not widely used by breeders (Ruiz 1978; Wink et al. 1995).

The major gene regulating seed alkaloid content is the *iucundus* gene (Gremigni et al. 2003). The first sequence-specific size-based marker (IucLi) closely linked to the *iucundus* gene was reported by Li et al. (2011). Application of this codominant marker enables the recognition of homozygous and heterozygous genotypes, which is not possible using conventional breeding methods. Reported compatibility between genotypes and phenotypes of tested cultivars and accessions of the Australian Lupin Collection was 88.67% (Li et al. 2011). The number of "false-positive" results encountered in the analyzed material indicates that the marker is not anchored but only linked to the *iucundus* gene, which may limit its routine application in breeding selection (Li et al. 2011).

8.2.1.5 Fungal Disease Resistance

Anthracnose caused by the fungal pathogen *C. lupini* is a worldwide problem in lupin cultivations, with yield losses up to 30% (Nirenberg et al. 2002). Therefore, a creation of anthracnose resistant cultivars is one of the most important aims in lupin breeding. One single dominant gene *Lanr1* determining resistance to anthracnose has so far been identified and extensively applied in Australian breeding programs (Yang et al. 2004).

To date, three MFLP-derived, sequence-specific markers (AntjM1, AntjM2, and AnManM1) linked to *Lanr1* gene have been developed (Yang et al. 2004, 2008; You et al. 2005). Unfortunately, a certain genetic distance between the markers and the gene of interest has often caused "false-positive" results, thus limiting their usability (Yang et al. 2008). Recently, Yang et al. (2012) reported five new sequence-specific PCR-based markers (AnSeq1, AnSeq2, AnSeq3, AnSeq4, AnSeq5) tightly linked to the *Lanr1* gene developed using the RAD-sequencing approach. Three of those new markers (AnSeq1, AnSeq3, and AnSeq4) were more closely linked to the target gene than all those previously published and replaced former markers in the Australian national lupin breeding programs (Yang et al. 2012).

Another important disease of lupins is phomopsis stem blight (PSB) caused by the fungal pathogen *Diaporthe toxica* (Shankar et al. 1996). Two PSB-resistance genes—*Phr1* (breeding line 75A:258) and *Phr2* (Merrit cultivar)—are known (Shankar et al. 2002). The *Phr1* gene has been found to be flanked by two MFLP-derived PCR-based markers (Ph258M1 and Ph258M2; Yang et al. 2002). Owing to the fact that this gene has not yet been integrated into commercial cultivars in Australia, further PBS-resistance sources and linked markers are still desirable (Yang et al. 2013a).

In 2013, Yang et al. developed seven new simple PCR-based markers (PhtjM1, PhtjM2, PhtjM3, PhtjM4, PhtjM5, PhtjM6 and PhtjM7) using RAD-sequencing technology. One of these markers (PhtjM3) was co-segregating with the putative PSB-resistance gene R (cultivar Tanjil) and the other six were closely linked to this gene. All the new markers were tested on historical and current Australia cultivars. The marker PhtjM3 showed high compatibility with resistant cultivars, but had several "false-positive" results. Three other markers (PhtjM4, PhtjM5, and PhtjM7) showed perfect correlations (100 %) between markers for genotype and plants phenotype, however, these require further verification on the wider gene pool, including accessions of the Australian Lupin Collection (Yang et al. 2013a).

8.2.2 White Lupin

The research towards marker development for practical MAS is much less advanced in white lupin. Up until now, only four MFLP-derived simple PCR-based markers closely linked to agricultural traits have been reported (Lin et al. 2009; Yang et al. 2010).

8.2.2.1 Alkaloid Content

In the white lupin, nine recessive genes—*exiguus, mitis, nutricius, pauper, primus, quintus, reductus, suavis,* and *tertius*—determining low alkaloid content are known (Kurlovich 2002); however, all the Australian cultivars possess the *pauper* gene for maintaining low alkaloid levels (Lin et al. 2009). One sequence specific codominant PCR-based marker (PauperM1) linked to the *pauper* gene conferring low alkaloid content was reported by Lin et al. (2009). This marker was able to distinguish the *pauper* gene from the other low alkaloid genes *exiguus* and *nutricius*, however, during the marker validation process, a few "false positive" were also observed (Lin et al. 2009). Furthermore, the *pauper* marker detection format demands the use of a radioisotope or fluorescence labeling primers, which precludes its routine application in MAS.

8.2.2.2 Anthracnose Resistance

Field disease resistance tests have suggested that anthracnose resistance in the white lupin is polygenically controlled. Moreover, the influence of environmental factors has also been observed (Yang et al. 2010). Two quantitative trait loci (QTLs) controlling anthracnose resistance have been identified on two linkage groups, LG4 and LG17, explaining 31 and 26% of the phenotypic variance (Phan et al. 2007). In 2010, Yang et al. (2010) developed three codominant sequence-specific PCR-based markers (WANR1, WANR2, and WANR3). The marker WANR1 was linked to a major anthracnose resistance QTL, while the second QTL was flanked by markers WANR2 and WANR3. Utilization of these three markers jointly explains 53% of phenotypic variation and has been applied in MAS since 2008 in Western Australia (Yang et al. 2010).

9 Seed Production

Seed production is a part of the seed industry in which several institutions and organizations are involved. There are breeding and seed companies, organizations (e.g., STV in Germany, SICASOV in France and the Seed Agency in Poland) which monitor royalties accrued by seed companies on behalf of breeders, and seed producers and institutions dealing with cultivar registration and protection as well as accepting seeds as sowing material. The breeder has an exclusive right to the cultivar, its multiplication, seed processing, selling, export/import and storage and can authorize a seed company to multiply, process, and sell cultivars. However, the aim of seed production is to transfer a biological progress from a cultivar (breeder's seeds) to seeds obtained by a farmer. Breeder's seeds are multiplied into pre-basic (PB), basic (B) and then qualified seeds (C1), and sometimes to C2. Cultivar and sowing material usage are monitored by the respective institutions and regulations at a national and international level. One of them, the Organisation for Economic Co-operation and Development (OECD), defines the rules for seed trade. The International Union for the Protection of New Varieties of Plants (UPOV) establishes rules for cultivar usage (the cultivar is the exclusive property of a breeder with two exceptions: (1) it can be used for crossings, (2) the farmer is allowed to use seeds produced by himself on his own field by paying half of the requisite royalties) and, among other activities, elaborates guidelines for the conduct of tests for the distinctness, uniformity and stability (DUS) of the cultivars of a given species. The Community Plant Variety Office (CPVO) in Europe adjudicates on the law to protect cultivars and manages its own catalogue, and the International Seed Testing Association (ISTA) elaborates methods for laboratory seed testing. However, regulations related to the production and quality of seeds (for field and laboratory use) obliging in seed turnover are determined in the EU by other appropriate guidelines.

The European Seed Certification Agencies Association (ESCAA) also participates in the regulation of the seed industry. The ESCAA states that the lupin seed production area in Europe in 2013 was about 6330 ha with Poland and Germany being the biggest producers. Unfortunately, no data are available on east European countries (Belarus, Russia, Ukraine) or Australia where large breeding projects exist and where the total harvested area of lupin in 2012 was 20,735, 17,800, 24,000 and 689,064 ha, respectively (the Food and Agriculture Organization Corporate Statistical Database, FAOSTAT, Table 6.1). There are further countries with larger areas of lupin harvest, for example, Chile (21,467 ha), South Africa (12,000 ha), Peru (9656 ha), Spain (6700 ha), and Lithuania (5100 ha) giving a global harvested area of lupin of about 900,000 ha. To sow this area, 10 % (about 90,000 ha) needed to be set aside for seed production.

UPOV guidelines to estimate the DUS of three European lupins (Andean lupin is not considered) cover 20 quantitative and qualitative characteristics describing plant alkaloid content, the color of leaves, stems, flowers and seeds, the size of leaves, pods and seeds, growth type and plant height in different growth stages, the height of the insertion of the first inflorescence and the earliness of some phases of plants growth and development. The technical questionnaire for cultivar registration as a rule covers fewer characteristics (alkaloid content, color of stem and flower wings, term of flowering, and growth type) and additional factors in which a new cultivar differs from similar ones. These characteristics are inherited in different ways, can be mono- or polygenic and different variants of a characteristic can be controlled by multiple alleles of one locus. The influence of the environment on character expression can be different, and the mode of inheritance of all lupin characteristics is not yet known (Święcicki and Święcicki 1995; Święcicki et al. 2000a). Investigations into the expression of a single gene in segregating progeny show how strong the influence of genotypic background and/or environment can be on a given character expression controlled by a single allele, even in the case of a marker gene/character (Święcicki 1989, 2001). However, for obvious reasons, cultivar DUS estimation carries substantial importance in field as well as laboratory classification. The above should be considered by a breeder before cultivar registration to avoid complications during the qualification of seed materials. For example, in lupins, difficulties can be caused by the distribution and density of ornamentation on seed coats. These are characteristics which are very difficult (sometimes impossible) to keep uniform, even when inherited from one gene. Sunshine can influence their expression (color intensity and density of ornamentation). Pod position in relation to sunshine causes a different ornamentation on both seed sides. Ornamentation uniformity can be more complicated when a cultivar genotype includes additional, pleiotropic genes. A similar phenomenon exists in pea (e.g., expression of genes *Obs, U, F, Fs;* Blixt 1972) and probably in other legumes with colored seed coats. Here, a question arises as to whether marker characteristics for a cultivar should be considered in the breeding process. For example, an untypical color for a seed coat can be a useful cultivar marker, but it would be very convenient to have all cultivars with white, non-ornamented seed coats.

In the estimation of growth type (determinate–indeterminate), the mode of inheritance of the *restricted branching* characteristic and its differing expression in a respective lupin crop (particularly in narrow-leafed lupin) must be considered (Gawłowska et al. 1999).

UPOV guidelines divide lupin cultivars into two groups based on the presence of bitter substances (alkaloids) in seeds—with or without—listing the example cultivars for a given lupin species. For the narrow-leafed lupin, an example cultivar with bitter substances is Azuro, and without such substances—Bordako. The division is based on the grain-cut method (Eggebrecht 1949). Current lupin breeding achieves the lowest possible alkaloid content based on gas chromatography. This method shows that the bitter cv. Azuro contains 1.381 % alkaloids in seeds and sweet cv. Bordako 0.016 %.

Precise requirements are elaborated for producing/multiplying lupin sowing material. They define the lowest seed grade (C/2), the number of field inspections (first—in full flowering, second—in pod setting), type of forecrop (minimum 3 years break in lupin cropping), spatial isolation (200 m for basic seeds of a cross-pollinating yellow lupin and 2 m minimum as a technological belt for self-pollinating narrow-leafed lupin), cultivar (1 plant/30 m^2 of basic material) and species purity, the presence of weeds (qualifying fields should be free from weeds, the seeds of which are difficult to remove during purification process), and the presence of diseases and pests. The monitoring of the spatial isolation in a cross-pollinating yellow lupin is very important for maintaining the DUS. However, there are no possibilities for crossings between fields with different lupin species (spatial isolation is not needed). The interspecific hybrid between yellow and narrow-leafed lupin would be a sensation and success, most probably worth the disqualification of the seed material. In the case of diseases and pests, there are general requirements (a disqualification can be justified when the presence of diseases and pests reduces lupin seed development or makes field inspection difficult) or more precise ones for dangerous fungi diseases (anthracnose, fusariose) defining the maximum number of infected plants or seeds. It is worth considering that to protect against *C. lupini* infection, it is sufficient to sow 2-year-old seeds. According to Cowling et al. (1998a), storing seeds for 18 months results in the practical elimination of infection.

A difficulty in lupin seed production is the maintenance of the high energy of seed germination. Before harvesting, it is necessary to avoid overdrying seeds, for example, when in branching cultivars one waits too long for seeds to ripen on secondary branches, seeds on the main stem will be much too dry (overdried) and sensitive to mechanical damage during threshing and cleaning.

The above section has provided some conclusions on the current state of lupin seed production. Global lupin seed production is rather limited; at least, no precise data are available. It is important to consider that there are three different lupin crops (narrow-leafed, white and yellow lupin, no data on Andean lupin seed production in South America) and they do not cross with each other and to some extent have different biology and genetics. This must be considered in cultivar description and DUS estimation. An excellent example of using genetics in enhancing the precise identification and classification of cultivars was presented for pea by Winfield and Green (1986).

References

Adhikari KN, Buirchell BJ, Thomas GJ et al (2009) Identification of anthracnose resistance in *Lupinus albus* L. and its transfer from landraces to modern cultivars. Crop Pasture Sci 60:472–479

Adhikari KN, Buirchell BJ, Sweetingham MW (2012) Length of vernalization period affects flowering time in three lupin species. Plant Breed 131:631–636

Aïnouche A, Bayer R (2000) Genetic evidence supports the new anatolian lupine accession, *Lupinus anatolicus*, as an old world "rough-seeded" lupine (section *Scabrispermae*) related to *L. pilosus*. Folia Geobot 35:83–95

Aïnouche A, Bayer RJ, Misset MT (2004) Molecular phylogeny, diversification and character evolution in *Lupinus* (Fabaceae) with special attention to Mediterranean and African lupines. Plant Syst Evol 246:211–222

Annicchiarico P, Manuza P, Proietti S (2011) White lupin tolerance to winter cold, terminal drought and soil lime: patterns of genetic variation and their exploitation in breeding for southern Europe. In: Naganowska B, Wolko B, Kachlicki P (eds) Lupin crops—an opportunity for today, a promise for the future. 13th International Lupin conference, Poznań, 6–10 June 2011. International Lupin Association, Canterbury, pp 99–103

Atkins CA, Smith PMC (1997) Genetic transformation and regeneration of legumes. In: Legocki A, Bothe H, Pühler A (eds) Biological fixation of nitrogen for ecology and sustainable agriculture. NATO ASI series. Springer, Heidelberg, pp 283–304

Atkins CA, Emery RJ, Smith PM (2011) Consequences of transforming narrow leafed lupin (*Lupinus angustifolius* [L.]) with an *ipt* gene under control of a flower-specific promoter. Transgenic Res 20:1321–1332

Babaoglu M (2000) Protoplast isolation in lupin (*Lupinus mutabilis* Sweet): determination of optimum explant sources and isolation conditions. Turk J Bot 24:177–185

Babaoglu M, Mccabe M, Power JB et al (2000) *Agrobacterium*-mediated transformation of *Lupinus mutabilis* L. using shoot apical explants. Acta Physiol Plant 22:111–119

Ball E (1946) Development in sterile culture of stem tips and subjacent regions of *Tropaeolum majus* L. and of *Lupinus albus* L. Am J Bot 33:301–318

Ball E (1960) Sterile culture of the shoot apex of *Lupinus albus*. Growth 24:91–110

Barbacki S (1952) Lupin. PWRiL, Warsaw (in Polish)

Bayliss KL, Wroth JM, Cowling WA (2004) Pro-embryos of *Lupinus* spp. produced from isolated microspore culture. Aust J Agr Res 55:589–593

Blixt S (1972) Mutation genetics in *Pisum*. Agr Hort Genet 30:1–293

Boersma JG, Buirchell BJ, Sivasithamparam K et al (2007a) Development of a PCR marker tightly linked to *mollis*, the gene that controls seed dormancy in *Lupinus angustifolius* L. Plant Breed 126:612–616

Boersma JG, Buirchell BJ, Sivasithamparam K et al (2007b) Development of a sequence-specific marker linked to the *Ku* gene which removes the vernalization requirement in narrow-leafed lupin. Plant Breed 126:306–309

Boersma JG, Buirchell BJ, Sivasithamparam K et al (2007c) Development of two sequence-specific PCR markers linked to the *le* gene that reduces pod shattering in narrow-leafed Lupin (*Lupinus angustifolius* L.). Genet Mol Biol 30:623–629

Boersma JG, Nelson M, Sivasithamparam K et al (2009) Development of sequence-specific PCR markers linked to the *Tardus* gene that reduces pod shattering in narrow-leafed lupin (*Lupinus angustifolius* L.). Mol Breed 23:259–267

Brien SJ, Cowling WA, Potter RH et al (2000) A molecular marker for early maturity (Ku) and marker-assisted breeding of *Lupinus angustifolius*. In: Van Santen E, Wink M, Weissmann S et al (eds) Lupin, an ancient crop for the new millennium. 9th International Lupin conference, Klink/Muritz, 20–24 June 2000. International Lupin Association, Canterbury, pp 204–205

Brummund M (1988) Progress in the breeding of yellow lupin. In: Twardowski T (ed) 5th International Lupin conference Poznań, 5–8 July 1988. PWRiL, pp 25–39

Brummund M, Święcicki WK (2011) The recent history of lupin in agriculture. In: Naganowska B, Wolko B, Kachlicki P (eds) Lupin crops—an opportunity for today, a promise for the future. 13th International Lupin conference, Poznań, 6–10 June 2011. International Lupin Association, Canterbury, pp 15–23

Buirchell BJ (2000a) Yield trials with full domesticated *L. atlanticus* (atlas lupin) in Western Australia. In: Van Santen E, Wink M, Weissmann S et al (eds) Lupin, an ancient crop for the new millennium. 9th International Lupin conference, Klink/Muritz, 20–24 June 2000a. International Lupin Association, Canterbury, pp 158–159

Buirchell BJ, Cowling W (2000b) Domestication of *Lupinus pilosus* Murray through mutation. In: Van Santen E, Wink M, Weissmann S et al (eds) Lupin, an ancient crop for the new millennium. 9th International Lupin conference, Klink/Muritz, 20–24 June 2000b. International Lupin Association, Canterbury, pp 156–157

Campos-Andrada MP, Baeta JP, Harris D et al (2000) Phenotypic variation in Portugal of three *L. cosentinii* Guss germplasm accessions of different genetic origin. In: Van Santen E, Wink M, Weissmann S et al (eds) Lupin, an ancient crop for the new millennium. 9th International Lupin conference, Klink/Muritz, 20–24 June 2000. International Lupin Association, Canterbury, pp 160–161

Cardoso D, Pennington RT, De Queiroz LP et al (2013) Reconstructing the deep-branching relationships of the papilionoid legumes. South African J Bot 89:58–75

Cheng L, Bucciarelli B, Liu J et al (2011) White lupin cluster root acclimation to phosphorus deficiency and root hair development involve unique glycerophosphodiester phosphodiesterases. Plant Physiol 156:1131–1148

Clements JC, Buirchell BJ, Cowling WA (1996) Relationship between morphological variation and geographical origin or selection history in *Lupinus pilosus*. Plant Breed 115:16–22

Clements JC, Prilyuk L, Quealy J et al (2008) Interspecific crossing among the new world lupin species for *Lupinus mutabilis* crop improvement. In: Palta JA, Berger JB (eds) Lupins for health and wealth. 12th International Lupin conference, Fremantle, Western Australia, 14–18 Sept 2008. International Lupin Association, Canterbury, pp 324–327

Cordero SA, Arrieta V, Muzquiz M (1988) *Lupinus hispanicus* breeding. In: Twardowski T (ed) 5th International Lupin conference, Poznań, 5–8 July 1988. PWRiL, pp 156–169

Cowling WA (1999) Pedigrees and characteristics of narrow-leafed lupin cultivars released in Australia from 1976 to 1998. Bulettin 4365. Agriculture Western Australia, Perth

Cowling WA, Buirchell BJ, Tapia ME (1998a) Lupin. *Lupinus* L. Promoting the conservation and use of underutilized and neglected crops. 23. International Board for Plant Genetic Resources (IBPGR), Rome

Cowling WA, Huyghe C, Święcicki W (1998b) Lupin breeding. In: Gladstones JS, Atkins CA, Hamblin J (eds) Lupins as crop plants: biology, production, and utilization. CAB International, Wallingford, pp 93–120

Daza A, Chamber M (1993) Plant regeneration from hypocotyl segments of *Lupinus luteus* cv. L. Aurea. Plant Cell Tiss Org 34:303–305

Drummond CS (2008) Diversification of *Lupinus* (Leguminosae) in the western new world: derived evolution of perennial life history and colonization of montane habitats. Mol Phylogenet Evol 48:408–21

Drummond CS, Eastwood RJ, Miotto ST et al (2012) Multiple continental radiations and correlates of diversification in *Lupinus* (Leguminosae): testing for key innovation with incomplete taxon sampling. Syst Biol 61:443–60

Eggebrecht H (1949) Methodenbuch B and V. Die Untersuchung von Saatgut. Neumann-Neudamm, Radebeul

Findley R (1981) The day the sky fell (Mount St Helens). Natl Geogr 159:50–65

Gawłowska M, Świecicki WK, Święcicki W et al (1999) Inheritance of restricted branching in the narrow-leafed lupin (*L. angustifolius* L.). In: Prusiński J (ed) Lupin in Polish and European agriculture. International Lupin conference, Przysiek, 2–3 Sept 1999. Przysiek Regional Advisory Centre for Agriculture and Rural Development, pp 159–162

Gladstones JS (1967) Selection for economic characters in *Lupinus angustifolius* and *L. digitatus*. 1. Non-shattering pods. Aust J Exp Agr Anim Husb 7:360–366

Gladstones JS (1970) Lupins as crop plants. Field Crops Abstr 23:123–148

Gladstones JS (1974) Lupins of the Mediterranean region and Africa. Department of Agriculture of Western Australia, South Perth

Gladstones JS (1977) The narrow-leafed lupin in Western Australia. Western Australian department of agriculture. Perth Bul 3990:1–39

Gladstones JS (1988) More important problems in *Lupinus angustifolius* breeding. In: Twardowski T (ed) 5th International Lupin conference Poznań, 5–8 July 1988. PWRiL, pp 15–24

Gladstones JS (1998a) Distribution, origin, taxonomy, history and importance. In: Gladstones JS, Atkins CA, Hamblin J (eds) Lupins as crop plants: biology, production, and utilization. CAB International, Oxon, pp 1–36

Gladstones JS (1998b) Lupin breeding. In: Gladstones JS, Atkins CA, Hamblin J (eds) Lupins as crop plants: biology, production, and utilization. CAB International, Oxon, pp 93–120

Gladstones JS, Hill GD (1969) Selection for economic characters in *Lupinus angustifolius* and *L. digitatus*. 2. Time of flowering. Aust J Expt Agr Anim Husb 9:213–220

Gnatowska M, Święcicki WK, Wolko B (2000) Preliminary data on the outcrossing rate in *L. mutabilis*. In: Van Santen E, Wink M, Weissmann S et al (eds) Lupin, an ancient crop for the new millennium. 9th International Lupin conference, Klink/Muritz, 20–24 June 2000. International Lupin Association, Canterbury, pp 167–168

Gremigni P, Hamblin J, Harris D et al (2003) The interaction of phosphorus and potassium with seed alkaloid concentrations, yield and mineral content in narrow-leafed lupin (*Lupinus angustifolius* L.). Plant Soil 253:413–427

Gupta S, Buirchell BJ, Cowling WA (1996) Interspecific reproductive barriers and genomic similarity among the rough-seeded *Lupinus* species. Plant Breed 115:123–127

Gupta PK, Varshney RK, Sharma PC et al (1999) Molecular markers and their applications in wheat breeding. Plant Breed 118:369–390

Harrison JM, Williams W (1982) Genetical control of alkaloids in *Lupinus albus*. Euphytica 31:357–364

Holland JB (2004) Implementation of molecular markers for quantitative traits in breeding programs—challenges and opportunities. In: Fischer T, Turner N, Angus J et al (eds) New directions for a diverse planet. 4th International Crop Science Congress, Brisbane, 26 Sept–1 Oct 2004, pp 1–13

Hughes C, Eastwood R (2006) Island radiation on a continental scale: exceptional rates of plant diversification after uplift of the Andes. Proc Natl Acad Sci U S A 103:10334–10339

Huyghe C, Papineau J (1990) Winter development of autumn sown white lupin: agronomic and breeding consequences. Agron 10:709–716

IBPGR (1981) Lupin descriptors. IBPGR, Rome

Kasten W, Kunert R (1991) A culture method for isolated immature embryos of different *Lupinus* species. Biol Zent Bl 110:290–300

Kasten W, Paradies T, Kunert R et al (1991) Progress in realization of interspecific hybrids in the genus *Lupinus* by means of an embryo rescue technique. Biol Zent Bl 110:301–309

Klamroth AK, Dieterich R, Rudloff E (2011) Mutagenic treatment leads to increased genetic diversity in narrow-leafed lupin (*L. angustifolius* L.) In: Naganowska B, Wolko B, Kachlicki P (eds) Lupin crops—an opportunity for today, a promise for the future. 13th International Lupin conference, Poznań, 6–10 June 2011. International Lupin Association, Canterbury, pp 40–42

Kozak K, Galek R, Waheed M et al (2012) Anther culture of *Lupinus angustifolius*: callus formation and the development of multicellular and embryo-like structures. Plant Growth Regul 66:145–153

Kroc M, Koczyk G, Święcicki W et al (2014) New evidence of ancestral polyploidy in the Genistoid legume *Lupinus angustifolius* L. *(narrow-leafed lupin)*. Theor Appl Genet 127:1237–1249

Kubok I (1988) The history of lupine breeding in Poland. Plant Breeding and Acclimatization Institute, Radzików

Kuptsov N (2000) Strategies, results and perspectives of breeding narrow-leafed lupin in Belarus. In: Van Santen E, Wink M, Weissmann S et al (eds) Lupin, an ancient crop for the new millennium: 9th International Lupin conference, Klink/Muritz, 20–24 June 2000. International Lupin Association, Canterbury, pp 145–146

Kurlovich BS (2002) Lupins (Geography, classification, genetic resources and breeding). OY International North Express, St. Petersburg

Kurlovich BS, Stoddard FL, Earnshaw P (2008) Potential and problems of *Lupinus pollyphyllus* Lindl domestication. In: Palta JA, Berger JB (eds) Lupins for health and wealth. 12th International Lupin conference, Fremantle, 14–18 Sept 2008. International Lupin Association, Canterbury, pp 304–307

Lavin M, Herendeen PS, Wojciechowski MF (2005) Evolutionary rates analysis of Leguminosae implicates a rapid diversification of lineages during the tertiary. Syst Biol 54:575–594

Li H, Wylie SJ, Jones MGK (2000) Transgenic yellow lupin (*Lupinus luteus*). Plant Cell Rep 19:634–637

Li X, Renshaw D, Yang H et al (2010) Development of a co-dominant DNA marker tightly linked to gene *tardus* conferring reduced pod shattering in narrow-leafed lupin (*Lupinus angustifolius* L.). Euphytica 176:49–58

Li X, Yang H, Buirchell B et al (2011) Development of a DNA marker tightly linked to low-alkaloid gene *iucundus* in narrow-leafed lupin (*Lupinus angustifolius* L.) for marker-assisted selection. Crop Pasture Sci 62:218–224

Li X, Buirchell B, Yan G et al (2012a) A molecular marker linked to the *mollis* gene conferring soft-seediness for marker-assisted selection applicable to a wide range of crosses in lupin (*Lupinus angustifolius* L.) breeding. Mol Breed 29:361–370

Li X, Yang H, Yan G (2012b) Development of a co-dominant DNA marker linked to the gene *lentus* conferring reduced pod shattering for marker-assisted selection in narrow-leafed lupin (*Lupinus angustifolius*) breeding. Plant Breed 131:540–544

Lin R, Renshaw D, Luckett D et al (2009) Development of a sequence-specific PCR marker linked to the gene "*pauper*" conferring low-alkaloids in white lupin (*Lupinus albus* L.) for marker assisted selection. Mol Breed 23:153–161

Lukashevitch MI, Ageeva PA, Sviridenko TV et al (2011) Results and perspectives of lupin breeding in Russia. In: Naganowska B, Wolko B, Kachlicki P (eds) Lupin crops—an opportunity for today, a promise for the future. 13th International Lupin conference, Poznań, 6–10 June 2011. International Lupin Association, Canterbury, pp 118–121

Mahé F, Pascual H, Coriton O et al (2011) New data and phylogenetic placement of the enigmatic Old World lupin: *Lupinus mariae-josephi* H. Pascual. Genet Res Crop Evol 58:101–114

Micke A, Święcicki, W. (1988) Induced mutations in lupins. In: Twardowski T (ed) 5th International Lupin conference Poznań, 5–8 July 1988. PWRiL, pp 110–127

Mikołajczyk J (1966) Genetic studies in *Lupinus angustifolius*. II. Inheritance of some morphological characters in blue lupine. Genet Pol 7:153–180

Molvig L, Tabe LM, Eggum BO et al (1997) Enhanced methionine levels and increased nutritive value of seeds of transgenic lupins (*Lupinus angustifolius* L.) expressing a sunflower seed albumin gene. Proc Natl Acad Sci U S A 94:8393–8

Mugnier J (1988) Establishment of new axenic hairy root lines by inoculation with *Agrobacterium rhizogenes*. Plant Cell Rpt 7:9–12

Nadolska-Orczyk A (1992) Somatic embryogenesis of agriculturally important lupin species (*Lupinus angustifolius*, *L. albus*, *L. mutabilis*). Plant Cell Tiss Org 28:19–25

Naganowska B, Wolko B, Sliwinska E et al (2003) Nuclear DNA content variation and species relationships in the genus *Lupinus* (Fabaceae). Ann Bot 92:349–55

Nelson MN, Phan HT, Ellwood SR et al (2006) The first gene-based map of *Lupinus angustifolius* L.-location of domestication genes and conserved synteny with *Medicago truncatula*. Theor Appl Genet 113:225–238

Nelson MN, Moolhuijzen PM, Boersma JG et al (2010) Aligning a new reference genetic map of *Lupinus angustifolius* with the genome sequence of the model legume, *Lotus japonicus*. DNA Res 17:73–83

Nirenberg HI, Feiler U, Hagedorn G (2002) Description of *Colletotrichum lupini* comb. nov. in modern terms. Mycologia 94:307–320

Ormerod AJ, Caligari PDS (1994) Anther and microspore culture of *Lupinus albus* in liquid culture medium. Plant Cell Tiss Org 36:227–236

Parra-Gonzalez L, Aravena-Abarzua G, Navarro-Navarro C et al (2012) Yellow lupin (*Lupinus luteus* L.) transcriptome sequencing: molecular marker development and comparative studies. BMC Genomics 13:425

Pascual H (2004) *Lupinus mariae-josephi* (Fabaceae), nueva y sopredente especie descubierta en Espana [*Lupinus mariae-josephi* (Fabaceae), new and surprising species discovered in Spain]. Anales Jardín Botánico Madrid 61:69–72

Pascual H, Muguerza NB, Laberge G (2006) *Lupinus mariae-josephi* H. Pascual—a new lupin species adapted to calcareous soils. Grain Legumes 46:8–9

Phan HT, Ellwood SR, Adhikari K et al (2007) The first genetic and comparative map of white lupin (*Lupinus albus* L.): identification of QTLs for anthracnose resistance and flowering time, and a locus for alkaloid content. DNA Res 14:59–70

Phoplonker MA, Caligari PDS (1993) Cultural manipulations affecting callus formation from seedling explants of the pearl lupin (*Lupinus mutabilis* Sweet). Ann Appl Biol 123:419–432

Pigeaire A, Abernethy D, Smith P et al (1997) Transformation of a grain legume (*Lupinus angustifolius* L.) via *Agrobacterium tumefaciens*-mediated gene transfer to shoot apices. Mol Breed 3:341–349

Plitmann U, Pazy B (1984) Cytogeographical distribution of old world *Lupinus*. Webbia 38:531–539

Pniewski T, Kapusta J, Legocki A (2002) In vitro micropropagation of four lupin species. Acta Physiol Plant 24:417–424

Pniewski T, Kapusta J, Plucienniczak A (2006) *Agrobacterium*-mediated transformation of yellow lupin to generate callus tissue producing HBV surface antigen in a long-term culture. J Appl Genet 47:309–18

Porsild AE, Harington CR, Mulligan GA (1967) *Lupinus arcticus* Wats. Grown from seeds of Pleistocene Age. Science 158:113–114

Przyborowski JA (1997) Embryo development after interspecific hybridization of *Lupinus albus* L., *L. mutabilis* Sweet. and *L. angustifolius* L. J Appl Genet 38:131–141

Przybylska J, Zimniak-Przybylska Z (1995) Electrophoretic patterns of seed globulins in the old-world *Lupinus* species. Genet Res Crop Evol 42:69–75

Römer P (1994) A determinated mutant of *L. mutabilis* as a possible source of early maturity. In: Neves-Martins J, Beirao Da Costa ML (eds) Advances in Lupin research. 7th International Lupin conference, Evora, 18–23 April 1994. ISA Press, Lisbon, pp 90–91

Römer P, Jahn-Deesbach W (1988) Developments in *Lupinus mutabilis* breeding. In: Twardowski T (ed) 5th International Lupin conference, Poznań, 5–8 July 1988. PWRiL, pp 40–50

Rudloff E (2011) EMS-induced mutants—a valuable genetic pool for the breeding of narrow-leafed sweet lupin (*L. angustifolius* L.). In: Naganowska B, Wolko B, Kachlicki P (eds) Lupin crops—an opportunity for today, a promise for the future. 13th International Lupin conference, Poznań, 6–10 June 2011. International Lupin Association, Canterbury, pp 92–98

Ruiz LP (1978) Alkaloid analysis of "sweet" lupin seed by GLC. New Zeal J Agr Res 21:241–242

Rybczynski JJ, Podyma E (1993a) Micropropagation of some *Lupinus* species from seedling explants. Genet Pol 34:237–247

Rybczynski JJ, Podyma E (1993b) Preliminary studies of plant regeneration via somatic embryogenesis induced on immature cotyledons of white lupin (*Lupinus albus* L.) Genet Pol 34:249–257

Sator C (1985a) Plant regeneration for anthers of lupins. Landbauforschung Volkenrode 35:5–7

Sator C (1985b) Studies on shoot regeneration of lupins (*Lupinus* spp.). Plant Cell Rept 4:126–128

Sator C (1990) Lupins (*Lupinus* spp.). In: Bajaj YPS (ed) Legumes and oilseed crops I. Springer, Berlin, pp. 288–311

Sator C, Mix G, Menge U (1983) Investigations of anther culture of *Lupinus polyphyllus*. Plant Res Dev 18:37–46

Schaefer-Menuhr A, Busmann A, Czerwinski T (1988) Embryo rescue of interspecific hybrids. In: Twardowski T (ed) 5th International Lupin Conference Poznań, 5–8 July 1988. PWRiL, pp 424–428

Shankar M, Cowling WA, Sweetingham M (1996) The expression of resistance to latent stem infection by *Diaporthe toxica* in narrow leafed-lupin. Phytopathol 86:692–697

Shankar M, Sweetingham MW, Cowling WA (2002) Identification of alleles at two loci controlling resistance to *Phomopsis* stem blight in narrow-leafed lupin (*Lupinus angustifolius* L.). Euphytica 125:35–44

Sinha A, Caligari PDS (2005) Enhanced protoplast division by encapsulation in droplets: an advance towards somatic hybridisation in recalcitrant white lupin. Ann Appl Biol 146:441–448

Sinha A, Wetten AC, Caligari PDS (2003/2004) Optimisation of protoplast production in white lupin. Biol Plant 47:21–25

Skrzypek E, Czyczyło-Mysza I, Marcińska I et al (2008) Prospects of androgenetic induction in *Lupinus* spp. Plant Cell Tiss Org 94:131–137

Sonntag K, Ruge-Wehling B, Wehling P (2009) Protoplast isolation and culture for somatic hybridization of *Lupinus angustifolius* and *L. subcarnosus*. Plant Cell Tiss Org 96:297–305

Sroga GE (1983) Callus and suspension culture of *Lupinus angustifolius* cv. turkus. Plant Sci Lett 32:183–192

Sroga GE (1987) Plant regeneration of two *Lupinus* spp. from callus cultures via organogenesis. Plant Sci 51:245–249

Stawinski S, Rybinski, W. (2000) The analysis of M1 progeny in lupin after chemomutagen treatment. In: Van Santen E, Wink M, Weissmann S et al (eds) Lupin, an ancient crop for the new millennium. 9th International Lupin conference, Klink/Muritz, 20–24 June 2000. International Lupin Association, Canterbury, pp 128–131

Surma M, Adamski T, Święcicki W et al (2013) Preliminary results of in vitro culture of pea and lupin embryos for the reduction of generation cycles in single seed descent technique. Acta Soc Bot Pol 82 231–236

Święcicki W (1986) Developments in breeding *L. luteus* and its relatives. 4th International Lupin conference, Geraldton, 15–22 Aug 1986, pp 20–24

Święcicki WK (1989) Linkages in *Pisum* V. New gene for orange cotyledons (*Orc*) and intragenic recombination between heteroalles. Genet Pol 30:155–163

Święcicki WK (2001) Supplemental data on fasciata genes in *Pisum* resources. Pisum Genet 33:19–20

Święcicki WK, Leraczyk K (1994) *Lupinus luteus* database. In: Neves Martins JM, Beirao Da Costa ML (eds) Advances in lupin research. 7th International Lupin conference, Evora, 18–23 April 1994. ISA Press, Lisbon, pp 70–72

Święcicki WK, Olejniczak J (1999) Variability of white lupin induced by sodum azide (SA). In: Hill GD (ed) Towards the 21st century. 8th International Lupin conference, Asilomar, 11–16 May 1999. International Lupin Association, Canterbury, pp 370–373

Święcicki W, Święcicki WK (1994) Self-completion and thermoneutrality for seed yield improvement. In: Dracup M, Palta J (eds) 1st Australian Lupin symposium, Perth, 17–21 Oct 1994, p 314

Święcicki W, Święcicki WK (1995) Domestication and breeding improvement of narrow-leafed lupin (*L. angustifolius* L.). J Appl Genet 36:155–167

Święcicki W, Święcicki WK, Wolko B (1996) *Lupinus anatolicus*—a new lupin species of the Old World. Genet Res Crop Ev 43:109–117

Święcicki W, Święcicki WK, Nijaki T (1999) *Lupinus* × *hispanicoluteus*—an interspecific hybrid of Old World lupins. Acta Soc Bot Pol 68:217–220

Święcicki W, Rybczyński J, Święcicki WK (2000a) Domestication and genetics of the yellow lupin (*Lupinus luteus* L.) and the biotechnological improvement of lupins. J Appl Genet 41:11–34

Święcicki WK, Cowling WA, Buirchell BJ (2000b) Databases and genetic resources evaluation in *Lupinus* collections. In: Van Santen E, Wink M, Weissmann S et al (eds) Lupin, an ancient crop for the new millennium. 9th International Lupin conference, Klink/Muritz, 20–24 June 2000b. International Lupin Association, Canterbury, pp 132–137

Święcicki WK, Wolko B, Naganowska B (2001) *Lupinus anatolicus* and *L.* × *hispanicoluteus*—morphological, biochemical and cytological characterization. In: Święcicki W, Naganowska B, Wolko B (ed) Broad variation and precise characterization—limitation for the future. XVI EUCARPIA genetic resources section workshop, Poznań, 16–20 May 2001. pp 281–284
Troll HJ (1967) Phylogenetically interesting mutants of *Lupinus luteus*. Erwin Baer Gedächtniss-Vorlesungen IV:131–138
Uhde-Stone C, Liu J, Zinn KE et al (2005) Transgenic proteoid roots of white lupin: a vehicle for characterizing and silencing root genes involved in adaptation to P stress. Plant J 44:840–53
Upadhyaya A, Davis TD, Sankhla D et al (1992) Micropropagation of *Lupinus texensis* from cotyledonary node explants. HortScience 27:1222–1223
Von Baer E (2002) The potential of crossing *L. polyphyllus* and *L. mutabilis*. In: Van Santen E, Hill GD (eds) Wild and cultivated lupins from the Tropics to the Poles. 10th International Lupin conference, Laugarvatan, 19–24 June 2002. International Lupin Association, Canterbury, pp 75–76
Von Baer E (2011) Domestication of Andean lupin (*L. mubabilis*). In: Naganowska B, Wolko B, Kachlicki P (eds) Lupin crops—an opportunity for today, a promise for the future. 13th International Lupin conference, Poznań, 6–10 June 2011. International Lupin Association, Canterbury, pp 129–132
Wijayanto T, Barker SJ, Wylie SJ, Gilchrist DG, Cowling WA (2009) Significant reduction of fungal disease symptoms in transgenic lupin (*Lupinus angustifolius*) expressing the anti-apoptotic baculovirus gene p35. Plant Biotechnol J 7:778–790
Wilson JG, Clements JC, Quealy J et al (2008) Development of an interspecific hybridisation protocol for *Lupinus*. In: Palta JA, Berger JB (eds) Lupins for health and wealth. 12th International Lupin conference, Fremantle, 14–18 Sept 2008. International Lupin Association, Canterbury, pp 147–151
Winfield PJ, Green FN (1986) The role of genetics in defining characters for the identyfication and classification of pea cultivars (*Pisum sativum* L.). In: Styles BT (ed) Interspecific classification of wild and cultivated plants. Clarendon, Oxford, pp 317–330
Wink M, Meißner C, Witte L (1995) Patterns of quinolizidine alkaloids in 56 species of the genus *Lupinus*. Phytochemistry 38:139–153
Wiszniewska A, Pindel A (2009) Improvement in *Lupinus luteus* (Fabaceae) protoplast culture—stimulatory effect of agarose embedding and chemical nursing on protoplast divisions. Austral J Bot 57:502–511
Wiszniewska A, Pindel A (2013) Explant-dependent receptivity to isolation and a cell-wall resynthesis in protoplast culture of recalcitrant yellow lupin. J Central European Agr 14:143–156
Wolko B, Weeden NF (1990) Relationships among lupin species as reflected by isozyme phenotype. Genet Pol 31:189–197
Wolko B, Clements JC, Naganowska B et al (2011) Lupinus. In: Kole C (ed) Wild crop relatives: genomic and breeding resources. Springer, Berlin, pp 153–206
Yang H, Sweetingham MW, Cowling WA et al (2001) DNA fingerprinting based on microsatellite-anchored fragment length polymorphisms, and isolation of sequence-specific PCR markers in lupin (*Lupinus angustifolius* L.). Mol Breed 7:203–209
Yang H, Shankar M, Buirchell J et al (2002) Development of molecular markers using MFLP linked to a gene conferring resistance to *Diaporthe toxica* in narrow-leafed lupin (*Lupinus angustifolius* L.). Theor Appl Genet 105:265–270
Yang H, Boersma JG, You M et al (2004) Development and implementation of a sequence-specific PCR marker linked to a gene conferring resistance to anthracnose disease in narrow-leafed lupin (*Lupinus angustifolius* L.). Mol Breed 14:145–151
Yang H, Renshaw D, Thomas G et al (2008) A strategy to develop molecular markers applicable to a wide range of crosses for marker assisted selection in plant breeding: a case study on anthracnose disease resistance in lupin (*Lupinus angustifolius* L.). Mol Breed 21:473–483
Yang H, Lin R, Renshaw D et al (2010) Development of sequence-specific PCR markers associated with a polygenic controlled trait for marker-assisted selection using a modified selective

genotyping strategy: a case study on anthracnose disease resistance in white lupin (*Lupinus albus* L.). Mol Breed 25:239–249

Yang H, Tao Y, Zheng Z et al (2012) Application of next-generation sequencing for rapid marker development in molecular plant breeding: a case study on anthracnose disease resistance in *Lupinus angustifolius* L. BMC Genom 13:318

Yang H, Tao Y, Zheng Z et al (2013a) Rapid development of molecular markers by next-generation sequencing linked to a gene conferring phomopsis stem blight disease resistance for marker-assisted selection in lupin (*Lupinus angustifolius* L.) breeding. Theor Appl Genet 126:511–522

Yang H, Tao Y, Zheng Z et al (2013b) Draft genome sequence, and a sequence-defined genetic linkage map of the legume crop species *Lupinus angustifolius* L. PLoS ONE 8

You M, Boersma JG, Buirchell BJ et al (2005) A PCR-based molecular marker applicable for marker-assisted selection for anthracnose disease resistance in lupin breeding. Cell Mol Biol Lett 10:123–34

Young ND (1999) A cautiously optimistic vision for marker-assisted breeding. Mol Breed 5:505–510

Zgagacz SE, Rybczynski JJ (1996) Different in vitro responses of three species of Lupine. Genet Pol 37A:133–135

Chapter 7
Cowpea

Ousmane Boukar, Christian A. Fatokun, Philip A. Roberts, Michael Abberton, Bao Lam Huynh, Timothy J. Close, Stephen Kyei-Boahen, Thomas J.V. Higgins and Jeffrey D. Ehlers

1 Introduction

Cowpea (*Vigna unguiculata* (L.) Walpers) is a commonly grown and consumed grain legume in sub-Saharan Africa (SSA). It is particularly well adapted to the dry savanna region of SSA where many other crops could fail or perform very poorly due to water stress caused by irregular and short duration rainfall as well as poor soil fertility. The grains, which are the main product of the crop, contain between 22 and 30% protein thus making it a good source of quality food especially among the rural dwellers and urban poor. Cowpea grains are consumed in different forms. They are eaten boiled, fried (as akara), or steamed (as moi moi). In addition to the high protein content, cowpea grains are high in complex carbohydrates. Cowpea haulms (dried leaves, stems, and pod walls) are also a good source of quality fodder for livestock especially ruminants. In some parts of East Africa, notably Kenya and Tanzania, young succulent leaves of cowpea, also characterized by high protein and mineral nutrient contents, are picked and eaten as pot herbs.

O. Boukar (✉)
International Institute of Tropical Agriculture, Cowpea Breeding Unit,
Sabo Bakin Zuwo Road, Kano, Kano 3112, Nigeria
e-mail: o.boukar@cgiar.org

C. A. Fatokun
International Institute ot Tropical Agriculture (IITA), PMB 5320, Ibadan, Oyo State, Nigeria
e-mail: c.fatokun@cgiar.org

P. A. Roberts
Department of Nematology, University of California—Riverside, 900 University Avenue,
Riverside, CA 92521, USA
e-mail: philip.roberts@ucr.edu

M. Abberton
International Institute of Tropical Agriculture, PMB 5320, Ibadan, Nigeria
e-mail: mabberton@cgiar.org

A. M. De Ron (ed.), *Grain Legumes*, Handbook of Plant Breeding 10,
DOI 10.1007/978-1-4939-2797-5_7

Cowpea is grown mainly as an intercrop along with sorghum and millet in the dry savannas but is also intercropped with maize in the moist savannas. Only very few farmers in SSA grow cowpea as a sole crop. It is grown in wide spacing when intercropped such that plant population density is usually low, perhaps around 1000 plants/ha or even less. However, when grown sole the population density is much higher and this is reflected in higher grain yield. Cowpea, like many other legumes, is able to contribute to the sustainability of the soil in SSA farmers' fields. Being a legume, cowpea is capable of fixing atmospheric nitrogen in its root nodules hence it has little or no need for nitrogen fertilizer application. It can fix up to 240 kg/ha and leave between 60 and 70 kg/ha in the soil after harvest (Rachie 1985). The following crop can therefore benefit from this left over nitrogen.

Cowpea is grown in no less than 45 countries across the globe on about 14.5×10^6 ha. A total of 6.2 million metric tons (MMT) of grains are produced annually implying an average yield of 454 kg/ha. Nigeria and Republic of Niger produce about 45 and 15 % of total world cowpea followed by Burkina Faso with about 6 %. The bulk of cowpea production as well as consumption are in West Africa. Another major producing country is Brazil, but the quantity they produce is not correctly reported in the Food and Agriculture Organization (FAO) statistics. The projected annual production rate of growth for cowpea in SSA is expected to be 3 %, which means 8×10^6 t by 2020 (Abate et al. 2012). Demand, however, will increase at the rate of 5 % per year in West Africa and this has implications for the people in West Africa especially Nigerians. Demand for cowpea grain is expected to decline in Kenya and South Africa during this same period (Abate et al. 2012). Cowpea does not feature in international trade but trade between neighboring countries such as Niger

B. L. Huynh
Department of Nematology, University of California—Riverside,
3401 Watkins Drive, Boyce Hall 1463, Riverside, CA 92521, USA
e-mail: baolam.huynh@ucr.edu

T. J. Close
Department of Botany and Plant Sciences, University of California,
900 University Avenue, Riverside, CA 92521-0124, USA
e-mail: timothy.close@ucr.edu

S. Kyei-Boahen
International Institute of Tropical Agriculture (IITA), Av. Eduardo Mondlane 326,
2nd Floor Room 210, Nampula 709, Mozambique
e-mail: s.boahen@cgiar.org

T. J.V. Higgins
CSIRO Agriculture Flagship, Clunies Ross Street, Black Mountain, Canberra,
ACT 2601, Australia
e-mail: tj.higgins@csiro.au

J. D. Ehlers
Department of Agricultural Development, Bill and Melinda Gates Foundation,
500 N. 5th Ave N, Seattle, WA 98102, USA
e-mail: jeff.ehlers@gatesfoundation.org

and Nigeria takes place. There is an annual deficit of over 0.5×10^6 t in Nigeria and supplies from Niger and Cameroon have made up for this shortfall. There is a need for expansion in the production of cowpea if the projected deficit is to be adequately forestalled. The bulk of the growth in cowpea recorded over years is attributable to increase in land area cultivated to the crop. Technologies that will lead to increased productivity per unit area of land now need to be developed and promoted if food security is to be ensured.

There is a need for the application of agrochemicals especially insecticides to the cowpea crop. Farmers who grow cowpea in intercrop usually do not give any protection to the crop against insect pests and apply no fertilizer. However, the few farmers who grow cowpea as a sole crop try to apply insecticides to provide protection against insect pests that otherwise cause significant grain yield losses. In many instances, the insecticides applied may not be effective against all of the insects that limit the crop's productivity. Different insects attack cowpea plants at various stages of the crop's life cycle. Aphids (*Aphis craccivora*) attack cowpea and cause the most damage when the plants are in the seedling stage while flower bud thrips (*Megalurothrips sjostedti*) cause flower buds to abort prematurely thereby preventing them from reaching anthesis. The legume pod borer (*Maruca vitrata*), the most cosmopolitan of cowpea insect pests, damages flowers and developing pods and seeds. A complex of pod-sucking bugs (e.g., *Clavigralla tomentosicollis, Anoplocnemis curvipes,* and *Riptortus dentipes*) feeds on both mature and immature pods and seeds leading to shrinking, deformity, and nonviability of the seeds. Such deformed seeds are not fit for consumption and therefore not marketable. Cowpea weevil (*Callosobruchus maculatus*) feeds on stored seeds, which is why most farmers sell off the seeds shortly after harvest at fairly low and noncompetitive prices to avoid storage losses caused by the weevil. From the foregoing, it is obvious that insects are capable of wreaking immense damage to productivity of cowpea if not adequately controlled. For now, the application of insecticides seems to be the only method for control of some of the cowpea pests.

Generally, the traditional farmers' cowpea varieties are late maturing (>90 days to flowering) and characterized by spreading growth habit. On the other hand, most of the improved varieties are erect to semierect in growth habit and could be early (60–65 days) or medium maturing (75–80 days). The early maturing erect cowpea lines are well suited to sole cropping and could be planted at high population density, while the spreading type seems to be more adapted to intercropping systems. Studies have shown that some spreading-type cowpea lines such as 'Dan Ila' are able to withstand shading better than non-spreading types (Terao et al. 1997). Farmers in the dry savanna areas still grow their traditional varieties because even when insects have caused grain yield losses these varieties still are able to get fodder which they harvest and sell for income or use as quality feed for their livestock.

The response of cowpea plants to photoperiod has been described as being typical of quantitatively short-day implying that photoperiod beyond a critical value can only delay but not prevent flowering (Njoku 1958; Lush et al. 1980). While most of the farmers' traditional varieties belong to this category, that is, day-length sensitive, there are some lines which are day neutral (i.e., length of days does not influence time to flower). Most of the improved cowpea varieties being grown presently are day neutral in addition to being erect or semierect in growth habit.

2 Origin and Systematics

Cowpea is an indigenous crop in SSA. It has been reported that the immediate progenitors of cultivated cowpea such as *V. unguiculata* ssp. *dekindtiana/spontanea* are widely distributed across Africa including Madagascar (Ng and Singh 1997). Ng and Maréchal (1985) suggested that cultivated cowpea moved from West to East Africa from where it was taken to Europe. It was recognized by the Romans as far back as 2300 before present (BP). It probably moved from Europe to India in 2200 BP and to the Americas by Spanish and Portuguese traders in the seventeenth century. The greatest amount of genetic diversity in cultivated cowpea has been found to exist in West Africa especially the dry savanna regions of Cameroon, Niger, Nigeria, Burkina Faso, Benin, and Togo. However, the origin of wild cowpea has been traced to southern Africa particularly the area covering from Namibia, Transvaal to Swaziland. It is in this subregion of Africa that the highest amount of genetic diversity for wild *V. unguiculata* and *V. rhomboidea* has been detected (Ng and Singh 1997). The wild cowpea lines found in southern Africa usually have small seeds when compared with those found in West Africa such as subspecies *dekindtiana/spontanea,* which have slightly larger seeds. This probably confirms the claim that *V. unguiculata* ssp. *dekindtiana/spontanea* is an immediate progenitor of cultivated cowpea that is characterized by a large-seed size. The different wild cowpea relatives were previously regarded as independent species but Maréchal et al. (1978) merged all of them into a single species (*V. unguiculata*). Further, taxonomic efforts have subdivided these into various subspecies and varieties. Examples are *V. unguiculata* ssp. *unguiculata* var. *spontane*a, ssp. *dekindtiana* var. *dekindtiana,* ssp. *pubescens,* and ssp. *protracta* var. *protracta* among others. The taxonomy of the genus *Vigna* and particularly cowpea and its relatives was recently reviewed by Pasquet and Padulosi (2012). There is yet to be a well-defined and universally agreed classification of the wild cowpea relatives. So far, there are no distinct classifications into primary, secondary, and tertiary gene pools for cowpea. Varying levels of difficulties are encountered when crossing some of the wild *V. unguiculata* subspecies with cowpea and even among themselves. For example, embryo rescue was necessary for a successful cross between a cowpea line and a line of *V. unguiculata* ssp. *pubescens* (Fatokun and Singh 1987). That the wild cowpea relatives have hardly been used in the genetic improvement of the crop may have contributed to the low interest in defining the crop's gene pools. Desirable genes conferring resistance to many insect pests that cause damage to cowpea yield are present in some wild *Vigna* species such as *V. vexillata,* which have resistance to aphids, Maruca pod borer, and some others. However, strong incompatibility barriers prevent successful crossing of *vexillata* and cowpea thus making it impossible to transfer such useful genes to cowpea through conventional breeding methods.

3 Varietal Groups

Cultivated cowpea and its cross-compatible wild relatives belong to the section *Catiang* of the genus *Vigna*. All of the cultivated cowpea lines are classified as *V. unguiculata* subspecies *unguiculata*. Cultivated cowpea is subdivided into four cultivar groups (cv.-gr.), namely, Biflora, Textilis, Sesquipedalis (yard-long- bean), and Unguiculata/Melanophthalmus (Westphal 1974; Marechal et al. 1978). Each of these cultivar groups is distinct from the others. For example, Textilis is characterized by long peduncles, which are a good source of fiber used in the textile industry, while Sesquipedalis, the yard-long bean, has long, fleshy, and pendulous pods. The yard-long bean whose pods can be as long as 90 cm or more is consumed as a vegetable especially in Asia. Yard-long bean with long pods may have evolved from regular cowpea due to selection pressure exerted in Asia, where its consumption as a vegetable is very popular. Despite the length of the pods, the number of seeds per pod is usually not more than is found in cowpea which belongs to cultivar group cv.-gr. Melanophthalmus. The cv.-gr. Unguiculata/Melanophthalmus comprises the cultivated cowpea with most number of accessions. The protein-rich grains are the most economically important part of the crop hence seeds are large and crowded in the pods. This probably explains why cowpea is also referred to as crowder bean in some communities. Dual purpose varieties are noted for their grain and fodder yield and should be attractive to people in East Africa who consume cowpea leaves as vegetables as well as livestock farmers in the dry savanna regions of SSA. Many cowpea farmers in the dry savanna areas of SSA get almost the same amount of income from sales of fodder as from grains.

4 Genetic Resources: Conservation and Utilization

The collection, conservation, characterization, documentation, and distribution of genetic resources (germplasm) are important, and the diversity of germplasm gathered in ex situ collections, or gene banks, is a key underpinning of current and future breeding programs.

The most extensive collection of cowpea germplasm (15,371 accessions) is held by the Genetic Resources Center (GRC) of the International Institute of Tropical Agriculture (IITA), Ibadan, Nigeria. The IITA collection contains germplasm from 90 countries but around half of the collection (7912 accessions) is derived from West Africa, the center of diversity for the crop. This collection, along with others in the Consultative Group for International Agricultural Research (CGIAR) gene banks, was designated by FAO as held “in trust” for the international community, a status reinforced under Article 15 of the International Treaty on Plant Genetic Resources for Food and Agriculture, which entered into force in 2004. Germplasm from this collection is distributed to all those requesting it for research and breeding for food, feed, and agriculture, free of charge under the standard material transfer agreement. Cowpea seed is conserved under medium-term storage at 5 °C for the active collection, from which seeds will be distributed, and in long-term storage at −20 °C. Other significant collections are held in the USA with the US Department of Agriculture (USDA), Griffin and with the University of California Riverside (UCR), (around 10,000 and 6000 accessions, respectively). It should be noted that IITA and these two USA collections have a combination of duplicate and unique cowpea accessions.

Collection of genetic resources is accompanied by "passport data" giving details of collecting site and other information recorded by collectors as well as accession identifiers. Germplasm is characterized primarily by a number of morphological "descriptors" of plant features which are relatively constant under different environments. A descriptor list developed by the International Board for Plant Genetic Resources (IBPGR 1983) and modified by IITA and Bioversity International (Dumet et al. 2010) is used for the characterization of cowpea germplasm; this includes vegetative and floral parts and seed. Progress in the development of improved cowpea varieties that would be well adapted to different agroecologies in the tropics depend largely on the genetic resources being conserved. Besides adaptation to different agroecologies, many new varieties with specialty traits and consumer preferences may also be developed using these resources (Table 7.1).

The core collection of a germplasm set is often developed to capture a high proportion of the diversity in a number of accessions that can be more manageably phenotyped, typically 5–10 % of the total collection. In many cases, both agro-morphological characters and molecular markers are used to develop this. IITA has developed a "core-collection" of 2062 accessions of cultivated cowpea using geographical, agronomic, and botanical descriptors (Mahalakshmi et al. 2007). In a further refinement of this core collection concept, under the auspices of the Generation Challenge Program (GCP) of the CGIAR (http://www.generationcp.org/research/research-projects), a "mini-core" collection of cowpea was developed with 374 accessions. The continuing development and application of genomic tools, including a draft cowpea genome sequence, will underpin new approaches to the characterization of cowpea genetic resources and enhance their utility for breeders.

Hearne et al. (2012) studied a subset of the core collection comprising 86 accessions plus 10 gene bank accessions, representing 84 countries of origin. Fourteen SSR markers were used in the study, based on high polymorphism rates for two alleles per marker and good technical resolution, to assess the levels of inbreeding and heterogeneity within this group. The study revealed that inbreeding was not as complete as previously assumed and that up to five plants per accession would provide more accurate measures of diversity (Hearne et al. 2012).

A recent study of cowpea diversity analyzed the genetic relatedness of 433 cowpea landraces collected from 56 countries and 46 accessions of wild cowpeas using a set of > 1200 genome-wide single-nucleotide polymorphism (SNP) markers (Huynh et al. 2013a). Among the landraces, 323 were from North, West, Central, East, southeastern, and southern Africa, and 99 were representative of the rest of the world. The wild cowpea accessions represented three countries in West Africa and five countries in eastern and southern Africa. The genotyping was conducted using the 1536-SNP cowpea Illumina GoldenGate assay (Muchero et al. 2009a).

Table 7.1 Cowpea germplasm accessions with desirable traits

Resistant/tolerant	Germplasm accessions	References
Diseases		
Fusarium wilt	TVu 109-2, TVu 347, TVu 984, TVu 1000	Singh et al. (1983)
Scab	TVu 853, TVu 1404, TVu 1433.	Singh et al. (1983)
Septoria	TVu 456, TVu 483-2, TVu 486, TVu 1433, TVu 11761, TVu 12349	Singh et al. (2002)
Bacterial blight	TVu 347, TVu 410, TVu 483-2, Danilla (Nigerian landrace)	Singh et al. (1983)
BICMV	TVu 2480; TVu 2657, TVu 3433	Taiwo et al. (1982); Bashir (1992)
CABMV	TVu 401, TVu 1582	Bashir (1992)
CPMV	TVu 227, TVu 345, TVu 612, TVu 2331	Patel (1982)
CPMoV	TVu 3901	Allen et al. (1982)
Striga and Alectra	B301; TVu 14676	Lane et al. (1997); Ouedraogo et al. (2012)
Insect pests		
Aphid	TVu 36, TVu 62, TVu 408, TVu 410, TVu 801, TVu 2896, TVu 3000	Singh et al. (1983)
Flower bud thrips	TVu 1509, Sanzi (Ghanaian land race)	
Leafhoppers	TVu 59, TVu 123, TVu 662,	Singh et al. (1983)
Bruchid	TVu 2027, TVu 11952, TVu 11953	Singh et al. (1983)
Drought	TVu 11979, TVu 14914, Danilla	Watanabe et al. (1997), Agbicodo (2009)

TVu tropical *Vigna unguiculata* are germplasm lines available at the genetic resources center of IITA

BICMV Blackeye cowpea mosaic virus, *CABMV* Cowpea aphid-borne mosaic virus, *CPMV* Cowpea mosaic virus, *CPMoV* Cowpea mottle virus

The diversity analysis using Bayesian inference identified two distinct cowpea gene pools in Africa, one centered in western Africa and a second gene pool centered in eastern Africa. Each gene pool was most closely related to the wild cowpeas collected from the same geographic region. These results indicate that a process of divergent domestication has occurred leading to the formation of the two gene pools. Genetic variation was found to be slightly higher among the group of accessions from non-African countries than among the African accessions, and accessions from Asia and Europe were more related to those from western Africa while accessions from the Americas were closer to the eastern Africa gene pool (Huynh et al. 2013a). The overlap in distribution and the chronological sequence of domestication events is difficult to interpret precisely due to the lack of historical records of human involvement in domestication and geographical movement of early cowpea forms. However, these diversity studies are valuable in guiding introgression decisions in breeding programs and for enhancing utilization of cowpea germplasm collections.

Availability of seed for distribution depends on number of seeds, their quality, and health status. At IITA, the viability of seed under storage is maintained at 85% or above and regeneration of accessions is carried out when viability falls below this level to ensure availability of sufficient seed of good quality for distribution.

An important aspect of the conservation and distribution of cowpea seed is health testing, particularly virus indexing. Guidelines for the regeneration of cowpea were developed by Dumet et al. (2008).

The cowpea breeding program at IITA makes extensive use of the cowpea core collection from its GRC. In the last few years, accessions from the core collection have been evaluated for a wide range of traits by the Cowpea Breeding Unit. These include resistance to biotic stresses: Striga, aphids, bacterial blight, fusarium, smut, brown blotch, and several viruses. Accessions were also evaluated for grain protein and mineral content (Boukar et al. 2011) and drought tolerance (Fatokun et al. 2012a). For a number of these traits, important sources of genes were identified, and following crossing, lines were advanced for international trials in many countries of SSA where lines with superior performance have been identified and released.

Cowpea has a number of related subspecies and species, which may be valuable sources of important agronomic traits. The IITA's GRC maintains about 2000 accessions of cowpea wild relatives. The genetic resources of these wild relatives are not well represented in ex situ collections. The cultivated cowpea and some of its wild relatives belong to the section *Catiang* of the genus *Vigna*. Previous activities aimed at introgressing desirable genes through hybridization from some wild *Vigna* species have shown that crosses are possible only among members of this section. However, varying levels of compatibility have been observed when crosses were made between cultivated cowpea and some of these wild relatives. Some wild *Vigna* species such as *V. vexillata* have been found to show high levels of resistance to the major insect pests (cowpea aphid, flower bud thrips, legume pod borer, pod-sucking bugs, and cowpea bruchid) that cause immense grain-yield reduction in cultivated cowpea (Singh et al. 1992). The efforts devoted to making interspecific crosses between cowpea and *vexillata* did not yield any hybrid (Fatokun 2002). In addition, experience has shown that seeds of wild cowpea relatives are very small, with hard testa and unattractive color and texture. Breeders have tended to shy away from utilizing the crop's wild relatives. However, recent developments in the new genomic tools for cowpea may change this attitude.

To avoid risk of loss of valuable germplasm it is good practice to "safety duplicate" in another gene bank preferably in another country. The "safety backup" for many important collections is the Svalbard global seed vault in Norway. The current status of this vault was reviewed by Westengen et al. (2013) and 14,099 of the IITA cowpea collection are currently held at Svalbard with the great majority of these also safety duplicated in another gene bank in addition to that of IITA.

In 2008, the Global Crop Diversity Trust commissioned IITA to lead a survey and expert consultation on the development of a strategy for cowpea conservation, the results of which are summarized in Dumet et al. (2012). This highlighted the need for capacity development of national systems, particularly in SSA.

The line B301 was collected from Botswana, and it has been a major source of genes for resistance to Striga and Alectra. It has therefore been used extensively as a donor for resistance to these two parasitic flowering plants. An improved cowpea breeding line TVx 3236 with tolerance to flower bud thrips was selected from a segregating population that resulted from a cross involving TVu1509 (Singh et al. 1992).

5 Major Breeding Achievements

The productivity of cowpea in SSA farmers' fields is very low with mean grain yield of less than 400 kg/ha, whereas in the USA yield is >5000 kg/ha. Several factors, notably an array of insect pests, diseases, and drought, militate against high grain yield in SSA. That cowpea is grown in SSA by peasant farmers who are resource poor and unable to procure the necessary pesticides to protect their crops in the field and grain in storage, contributes to the low productivity of the crop. Most of the breeding activities therefore focus on how to increase productivity by developing improved varieties with high yield potential and resistance to the various abiotic and biotic stresses. IITA has an active cowpea breeding program and very strong collaboration with cowpea breeders in the various National Agricultural Research Systems (NARS) of SSA. Besides, a number of advanced research institutions and universities collaborate with IITA scientists in all aspects of cowpea research. The cowpea germplasm lines available in the genetic resources center of IITA have continued to be the major source of genetic diversity upon which breeders depend for a continuous generation of improved breeding lines.

Thus far, only cultivated cowpea germplasm lines have been exploited in the development of improved varieties. Several improved cowpea breeding lines have been developed, many of which have been evaluated across various agroecologies in different countries and those with good performance and that are attractive to farmers and consumers have been released as varieties in various countries (Table 7.2). Through genetic improvement, the majority of cowpeas now adopted in SSA farmers' fields are varieties that are erect in growth habit and are day neutral. The grain yield of traditional cowpea varieties is inherently low. In addition, because they spread, farmers plant them at wide spacing thereby resulting in fewer plants per hectare as compared to the erect or semierect improved varieties. Table 7.2 lists cowpea varieties from IITA breeding nurseries that were released globally. Many of the varieties combine resistance to diseases, Striga, Alectra, and flower bud thrips.

Generally, the key achievements in breeding have focused on introgression of traits dealing with biotic and abiotic stresses to cowpea yield, combined with improved agronomic qualities of enhanced grain size, grain quality, including seedcoat color and texture, plant architecture, and time to maturity. Examples of genetic improvements in West African country breeding programs are provided by new cowpea variety releases in Burkina Faso and Senegal, with support from the US Agency for International Development (USAID) Bean/Cowpea and Dry Grain Pulses Collaborative Research Support Program (CRSP) programs. Institut Senegalais Recherches Agricoles (ISRA) of Senegal has released a series of varieties over the past 20 years, which combines targeted biotic stress resistances with enhanced yield and grain qualities preferred by consumers and shortened maturity times to hedge against drought years. These include Melakh, Mouride, Yacine, Pakau, and in 2013 three new lines with large white grain types. These varieties have increased yields over the national average by up to about 20% and typically combine resistance or tolerance to one or more cowpea insect pests such as flower thrips or

Table 7.2 List of cowpea varieties released in different countries. (Adapted and updated from Singh et al. 2002)

Country	Variety released	Country	Variety released
Angola	TVx 3236	Argentina	IT82D-716
Australia	IT82E-18 (as Big Buff)	Belize	VITA-3, IT82D-889, IT82E-18
Benin Republic	IT81D-1137	Bolivia	IT82D-889, IT83D-442
	IT84S-2246-4,IT95K-193-12		
Botswana	ER-7, TVx 3236	Brazil	VITA-3, VITA-7, TVx 1836-013J
Burkina Faso	TVx 3236, KN-1		
	IT99K-573-2-1, IT98K-205-8	Myanmar	VITA-4 (Yezin-1)
Cameroon	IT81D-985 (BR1)	Central African Republic	VITA-1, VITA-4, VITA-7, VITA-5
	IT81D-994 (BR2)		TVx 1948-01F, IT81D-1137, IT83S-818
	TVx 3236		IT82E-18, IT81D-994
	IT88D-363 (GLM-92)		
	IT90K-277-2 (GLM-93)	Colombia	IT83S-841
		Cote d'Ivoire	IT88D-361, IT88D-363
Cuba	IT84D-449 (Titan)	Democratic Republic of Congo.	IT89KD-349, IT89KD-389
	IT84D-666(Cubinata-666)		IT89KD-355
	IT86D-314 (Mulatina-314)		
	IT86D-368 (IITA-Precoz)		
	IT86D-782 (Tropico-782)		
	IT86D-792 (Yarey-792)		
	IT88S-574-3(OR 574-3)		
Equatorial Guinea	IT87D-885	Ethiopia	TVx 1977-01D, IT82E-16, IT82E-32
Egypt	Tvu 21, IT82D-716	Fiji	VITA-1, VITA-3
	IT82D-709, IT82D-812		
	IT82E-16	Gambia	IT84S-2049 (Sosokoyo), IT83S-728-13

Table 7.2 (continued)

Country	Variety released	Country	Variety released
Ghana	IT82E-16 (Asontem)		
	IT83S-728-13 (Ayiyi)	Guinea Conakry	IT81D-879, IT83D-340-5
	IT83S-818 (Bengpla)		IT82E-16, IT85F-867-5 (Poku Togboi),
	TVx 1843-1C (Boafa)		IT85F-2805, IT83S-990,
	TVx 2724-01F (Soronko)		IT87S-1463, IT84S-2246-4
	IT87D-611-3 (Nhyira), IT87D-2075 (Tona)		
Guinea Bissau	IT82E-889, IT82D-889	Haiti	VITA-4, IT87D-885
India	VITA-4, TVx 1502	Jamaica	VITA-3, ER-7, IT84S-2246-4
Lesotho	IT82E-889, IT87D-885		IT84E-124
	IT82E-16, IT82E-32		
		Liberia	IT82D-889, TVx 3236, VITA-5
Malawi	IT82D-889, IT82E-16		VITA-4, VITA-7
	IT82E-25, IT99K-494-6		
		Mali	TVx 3236, IT89KD-374 (Korobalen)
Mauritius	TVx 3236		IT89KD-245 (Sangaraka), IT97K-499-35 (Jiguiya), IT93K-876-30, IT82D-812, IT83S-818, IT85F-2020
		Mozambique	IT82E-16
			IT00K-1263, IT97K-1069-6
Nepal	IT82D-752 (Aakash)		
	IT82-889 (Prakash)		
		Nigeria	TVx 3236, IT81D-994, IT86D-719

Table 7.2 (continued)

Country	Variety released	Country	Variety released
Niger	IT89KD-374, IT90K-372-1-2		IT84S-2246-4, IT90K-76
	IT97K-499-35, IT97K-499-38		IT86D-721, IT88D-867-11
	IT98K-205-8, IT99K-573-1-1, IT96D-610		IT82E-60, IT89KD-374, IT90K-277-2
			IT90K-82-2, IT89KD-288,IT97K-499-35, IT89KD-391, IT99K-573-1-1
			IT99K-573-2-1
		Paraguay	IT86D-1010, IT87D-378-4
			IT87D-697-2, IT87D-2075
		Philippines	IT82D-889
Sierra Leone	TVx 1990-01E, IT86D-721	Senegal	TVx 3236
	IT86D-719, IT86D-1010		
	IT82E-32, TVx 3236, Tvu 1190	Somalia	TVx 1502, IT82D-889
	VITA-3		IT82E-32
		Sudan	IT84S-2163
South Africa	IT90K-59		(Daha ElGoz = Gold from Sand)
	IT82E-16 (Pannar 311)		
		Swaziland	IT82D-889 (Umtilane), IT82E-18
Sri Lanka	IT82D-789 (Wijaya)		IT82E-27, IT82E-71
	IT82D-889 (Waruni)		
	TVx 309-01EG, VITA-4	Thailand	VITA-3, IT82D-889
	TVx 930-01B (Lita)	Uganda	TVx 3236, IT82E-60
	IT86D-1010		
Suriname	IT82D-889, IT82D-789	USA	IT84S-2246-4, IT84S-2049 (for nematode resistance)

Table 7.2 (continued)

Country	Variety released	Country	Variety released
Tanzania	TKx 9-11D (Tumaini)		
	TVx 1948-01F (Fahari)		
	IT82D-889 (Vuli-1), IT85F-2020	Yemen	TVx 3236, IT82D-789, VITA-5
	IT99K-7-21-2–2-1, IT99K-573-1-1		
	(Vuli AR1), (Vuli AR2)	Venezuela	VITA-3, IT81D-795
			IT82D-504-4, TVx 1850-01E
Togo	VITA-5, TVx 3236		–
	IT81D-985 (VITOCO)	Zambia	TVx 456-01F, TVx 309-01G
			IT82E-16 (Bubebe)
		Zimbabwe	IT82D-889

pathogens such as virus or bacterial blight with higher innate yield potential. In Burkina Faso, the Institut de l'Environnement et de Recherches Agricoles (INERA) has released a series of cowpea varieties with larger grain size and resistance to the parasitic weed Striga plus resistance to aphids and viruses. Interestingly, the variety Melakh, which was bred and is now widely grown in Senegal, was found to be an excellent variety in areas of Burkina Faso with similar agroecologies, and this variety has also been released by INERA in Burkina Faso. This example of a line developed in Senegal being evaluated and released in Burkina Faso demonstrates the advantages of collaborative breeding programs and material exchange between scientists from different countries of the subregion.

In the USA, cowpeas are bred to meet markets for use as both a vegetable and as a dry bean. In California, the breeding program has focused on developing improved blackeye dry grain cowpea types. The focus for cowpea improvement at UCR has been to introgress resistance to *Fusarium* wilt and root-knot nematodes into high-yielding backgrounds with improved grain quality. A recent example is California Blackeye No. 50 (CB50) released in 2009, which has improved grain size and quality (brighter white seed-coat color) combined with resistance to *Fusarium* wilt races 3 and 4 plus strong resistance to the root-knot nematodes *Meloidogyne incognita* and *M. javanica* (Ehlers et al. 2009). An earlier variety of blackeye cowpea, California Blackeye No. 27 (CB27), which was released in 1999, was bred to combine *Fusarium* wilt and root-knot nematode resistance with heat tolerance (Ehlers et al. 2000). In the southern USA, a successful breeding focus has been to incorporate the persistent green seed-coat trait into high-yielding cowpea varieties for canning or as fresh-shelled peas for freezing. This breeding focus was described in Ehlers et al. (2002) and is based on the green cotyledon and green testa traits which result in a persistent green seed color. The green cotyledon trait is conditioned by a single recessive gene, with the symbol *gc* (Fery and Dukes 1994), and several successful varieties have been released including Bettergreen, Charleston Greenpack, Petite-n-Green, and Green Dixie (Ehlers et al. 2002).

6 Specific Goals in Current Breeding

The goals of cowpea genetic improvement change with time and usually depend on agreed priorities set by stakeholders. The stakeholders include farmers, extension agents, NGOs and donor representatives, seed companies, consumers, and researchers. Current cowpea breeding goals also vary with the target production areas but are based on enhancing yield and grain quality, largely through introgression of biotic and abiotic stress tolerance and resistance. The low productivity of cowpea in the subregion is of major concern and efforts are directed primarily at addressing the factors that appear responsible. In recent times, nutrition and health conscious individuals and organizations seek cowpea varieties with higher protein content than the levels in many of the available varieties which is around 25 %. Breeders aim to develop varieties that overcome some of the identified production constraints as follows.

6.1 Resistance to Insect Pests

From the seedling stage to time of harvest and even seed storage there are major insect pests that damage cowpea. Aphids *(Aphis craccivora)* attack cowpea plants in the field and if not controlled they can kill the plants especially the seedlings. They are particularly troublesome when there is a short spell of drought after seedling emergence. Earlier, a single dominant gene conferred resistance to aphids but those varieties have now succumbed to the insect. New races of the insect have evolved, so new sources of genes for resistance are being sought among cultivated and wild cross-compatible cowpea relatives. A few lines among wild cowpea have been found to be aphid-resistant. Molecular markers will be deployed to better understand the resistance and to facilitate marker-assisted selection (MAS).

In California, emphasis is also on cowpea aphid resistance and tolerance to Lygus bug using resistance and tolerance traits identified in African cowpea germplasm lines from IITA, including IT97K-556-6 for aphid resistance and IT93K-2046-1. The aphid resistance in IT97K-556-6 has been shown by quantitative trait loci (QTL) mapping to be inherited by one minor and one major QTL on cowpea linkage groups 1 and 7, respectively (Huynh et al. 2015). In California, Lygus bugs cause two types of yield loss in cowpea: First, feeding on young floral buds causes these buds to drop which drastically reduces pod set and grain yield, and second, feeding by Lygus bugs during pod and grain development leads to pitted and discolored grains. Conventional breeding of Lygus-tolerant blackeye pea is underway by pedigree selection from crosses between the African donor line and California Blackeye varieties, and field phenotyping for tolerance in insecticide protected and unprotected plot designs to assess grain yield and quality under natural Lygus bug infestation. The Lygus tolerance determinants have yet to be mapped and SNP-tagged within the cowpea genome, currently precluding MAS approaches for breeding.

A significant challenge for breeders is to better define traits for flower thrips resistance and resistance to pod-sucking bugs. Phenotypic screening efforts are underway in West African cowpea programs to provide genetic mapping data for QTL discovery for these traits. The critical yield losses caused by these insect groups make them a priority focus in cowpea breeding. In the case of flower thrips, Omo-Ikerodah et al. (2008) identified DNA markers associated with QTLs that have effects on resistance to flower bud thrips in a biparental mapping population derived from a cross that had Sanzi, the land race from Ghana, as one of the parents. Many other improved breeding lines with resistance to diseases, drought, Striga, and pests were derived from crosses that involved above listed (Table 7.1) and other germplasm lines.

6.2 Tolerance to Drought and Low Soil P

Cowpea is grown mainly in drought-prone areas of SSA. Compared to many other crops, cowpea is regarded as relatively drought tolerant. This notwithstanding, depending upon severity, drought can still cause yield reduction in cowpea and the variation observed among germplasm lines indicates the present level of drought tolerance in the crop can be enhanced (Fatokun et al. 2012a). Molecular markers have been identified that are associated with QTLs for drought tolerance (Agbicodo 2009). A series of QTLs have been identified following the analysis of different recombinant inbred lines (RIL) segregating for drought tolerance. Seedling-stage drought induced delayed senescence traits were identified in cowpea genotype IT93K-503-1 and others in both greenhouse and field phenotyping experiments, and reproducible QTLs for this trait were mapped in the cowpea genome (Muchero et al. 2008, 2009b). More recently, the staygreen phenomenon, a trait which enhances delayed senescence, biomass, and grain yield under drought stress, was characterized in cowpea through genetic mapping using SNP genotyping, field and greenhouse phenotyping, and linkage disequilibrium association mapping in conjunction with biparental QTL mapping (Muchero et al. 2013). Seven loci were identified; out of which five exhibited pleiotropy for delayed senescence, biomass, and grain yield. In particular, three of these putative staygreen QTLs were resolved at 3.2 cM or lower map distances and provide important targets for introgression through marker-assisted selection (MAS). In addition, co-location of these QTLs with those governing early vegetative delayed senescence provides a rapid screening approach by phenotyping plants at the seedling stage for drought response (Muchero et al. 2013). These markers will be useful in marker-assisted recurrent selection (MARS) for drought tolerance in cowpea.

Like most other legumes cowpea has a need for phosphorus to be able to form nodules adequately and fix nitrogen. The soils in SSA are generally low in phosphorus and farmers who grow cowpea usually do not apply fertilizer to their crops. Improved breeding and germplasm lines have been evaluated for tolerance to low soil P and differences were detected among them which are indications that lines with a need for low levels of soil P can be developed.

6.3 Heat

Heat stress in cowpea disrupts flowering and pod set. It is an abiotic stress for which genes for tolerance are available as targets for molecular breeding approaches. A set of five heat tolerance QTLs were identified through QTL mapping in a biparental RIL population developed from the heat-tolerant variety CB27 as one of the parents (Lucas et al. 2013a). These QTLs provide resources for incorporating heat tolerance into other elite heat-sensitive cowpea varieties using MAS.

6.4 Dual Purpose

Although grain is the most economically important product of cowpea, in some parts of SSA such as Kenya, Tanzania, and Mozambique, young green and succulent leaves are relished as pot herbs while the haulms are a source of quality fodder for livestock in the dry savannas of West Africa. The young leaves are known to contain high levels of protein while many farmers derive income from selling dried cowpea fodder. The development of dual purpose varieties for leafy vegetables as well as for grain could meet the needs of many more people across the African region. Some attention is being devoted to selecting lines with this dual attribute that will serve as a source of grains and leaves for human food and animal feed. Improved dual purpose breeding lines have been identified in the IITA cowpea breeding program and shared with collaborators for evaluation in their countries for acceptability to farmers and consumers.

6.5 High Protein Content in Grains

The protein content of the grain is a major reason why cowpea is popularly consumed at home in several SSA communities. It is also why cowpea is commonly referred to as poor man's meat. In SSA, the cost of meat is prohibitive and not affordable in the quantity that is needed for a balanced diet. Depending on the variety, cowpea grains contain between 18 and 29 % protein with a potential for 35 % (Duke 1981). Among 79 cowpea varieties studied by Evans and Boulter (1974), protein content ranged from 21 to 34 %. In another study involving 100 lines, Nielsen et al. (1993) found that protein content ranged from 22.9 to 32.9 % with a mean of 28.6 %. Boukar et al. (2011) identified the following germplasm lines as having high protein content—TVu 408, TVu 526, TVu 1820, TVu 2356, TVu 2508, TVu 2723, TVu 2880, TVu 3638, TVu 8810-1—which could be used in the development of improved breeding lines. The cowpea protein consists of 90 % salt-soluble globulins and 10 % water-soluble albumins (Duke 1981). The anti-nutritional factors found in cowpea grains such as hemagglutinins and trypsin inhibitors are heat labile and can be inactivated easily by heating, thus making cowpea protein readily digested and absorbed. This makes cowpea protein suitable for infants and formulations of baby foods containing cowpea should be encouraged and commercialized in SSA.

6.6 Resistance to Diseases

Many diseases afflict cowpea plants in the field. There are fungal, bacterial, and viral diseases that attack the plants. Since farmers do not apply chemicals to protect their cowpea crops the diseases are best controlled by planting varieties that are resistant. Among the most devastating of fungal diseases is ascochyta blight caused by *Ascochyta phaseolorum* Sacc., which is seed-borne (Emechebe and Shoyinka 1985). The disease causes severe defoliation and lesions on stem and pods and can lead to the death of susceptible plants. Line TVu 11761 was identified as a potential source of resistance to this disease (Singh et al. 2002). Brown blotch is another major fungal disease of cowpea in SSA. The causal organism is *Colletotrichum capsici* (Emechebe and Florini 1997). All plant parts above soil level show symptoms of the disease in susceptible lines. Such symptoms include failure of seeds to germinate, damping off of seedlings, girdling of stem and branches, and flower abortion, among others. Treating seeds with fungicides, such as benomyl or carbendazim, before sowing helps reduce incidence of the disease. However, the development of resistant varieties appears the most attractive option for most SSA farmers. Scab, smut, and Septoria caused by *Elsino phaseoli, Protomycopsis phaseoli,* and *Septoria vignicola*, respectively, are also important fungal diseases of cowpea and cause yield reductions.

The most important bacterial disease of cowpea is bacterial blight caused by *Xanthomonas campestris* pv. *vignicola*. It is a disease that has been reported on cowpea in different parts of the world. Disease symptoms include large irregular foliar lesions with yellow margins, stem cankers, and preemergence and postemergence seedling mortality (Emechebe and Florini 1997). Some germplasm lines have been found that are resistant to the disease and the genes conferring resistance have been transferred to several improved varieties. However, in view of the high rate of mutation in the bacterium, it is necessary to continue identifying additional sources of resistance.

Many viruses attack cowpeas and these can only be controlled by sowing resistant varieties. The cowpea aphid-borne mosaic virus (CABMV) and bean common mosaic virus (BCMV-BIC), both of which are seed-borne and occur worldwide, are two economically important cowpea pathogens (Huguenot et al. 1997). In addition to these two viruses, Hampton et al. (1997) reported that cucumber mosaic cucumovirus (CMV), cowpea mosaic (CPMV) and cowpea severe mosaic (CPSMV) comoviruses make up the most devastating viruses of cowpea. They are also seed-borne. Worrisome in cowpea production is the occurrence of mixed infections of these viruses. Mixed infections cause drastic disease symptoms and even death of plants in the field. Since there are no chemicals to control these pathogens the development of resistant varieties is the only option for their control. Sources of genes for resistance to several of the viruses have been identified and many have been incorporated in released varieties.

A recent review of the important biotic stress resistance traits with molecular marker-based associations is provided in Huynh et al. (2013b). Genetic map positions in the cowpea genome and linked, flanking SNP markers have been identified for resistance to root-knot nematodes, *Fusarium* wilt, *Macrophomina phaseolina* (ashy stem blight or charcoal rot), bacterial blight, several cowpea viruses (cowpea mosaic virus, cowpea severe mosaic virus, blackeye cowpea mosaic potyvirus), foliar thrips, cowpea aphid, and parasitic weed *Striga gesnerioides*. These resources have helped to better define the breeding targets in several US, African, and Asian cowpea-breeding programs. In California, the focus remains on *Fusarium* wilt and root-knot nematode resistance.

6.7 Large Seed Size

In recent times, consumers have shown a preference for cowpeas with large grain size. Breeding activities have been initiated towards developing varieties that meet this preference. Experience has, however, shown that small seed size in cowpea is dominant to large seed size hence there is need to embark on backcrossing to transfer the genes for large seed size to preferred varieties.

6.8 Adaptation to Intercropping

Most of the improved cowpea varieties that have been released have erect or semi-erect growth habit. They are, therefore, well adapted to sole cropping. Many farmers in SSA still prefer to intercrop cowpea with sorghum, millets, and other cereals. This habit is difficult to change because each farmer has access to a small land area where he plants all the crops that provide the family with food and some income. When intercropped, cowpea plants are readily shaded by the taller cereals. The traditional varieties which farmers grow under this cropping system spread on the ground and remain there until the cereals are harvested when they now receive more sunlight, flower, and set pods. New varieties can be developed that can adapt to intercrop conditions such that their flowering and podding are not adversely affected by shading.

6.9 Striga Resistance

In Africa, most programs are targeting Striga resistance, in combination with drought-tolerance traits as well as virus and insect resistance. For example, in Burkina Faso, any new variety released must contain Striga resistance (Drabo personal communication 2014).

7 Breeding Methods and Specific Techniques

Cowpea breeding methods are similar to those of other self-pollinated crops such as peanut, soybean, wheat, and barley. Cowpea breeders depend upon the germplasm available in the different collections described earlier. The IITA GRC is a source of genes of interest in both domesticated and wild cross-compatible cowpea relatives. Several decades of breeding effort by different institutions also provide the chance to build on the numerous improved breeding lines. The IITA cowpea breeding nursery distributes many breeding lines annually for testing, adoption as released varieties or as parents to be used in the importing countries' breeding programs. Most cowpea breeding activities are focused on development of improved varieties with farmers' and consumers' preferred traits. Breeding programs include the development of populations segregating for desirable traits and from which selections are made for both simple and complex characteristics (e.g., disease resistance, drought tolerance, grain yield). Additional activities include determination of inheritance, creation, and evaluation of new genetic variability and production of specific genotypes.

Cowpea variety development programs follow the conventional breeding steps for self-pollinated crops (pure-line breeding, pedigree breeding, single-seed descent, etc.). Parents characterized by important traits of interest are chosen and used to generate genetically variable populations through artificial hybridization followed by selection. Appropriate selection pressure that favors identification of desired traits such as imposition of disease pathogens, insect pests, drought, heat, low phosphorus, etc. is exerted on the segregating populations. In addition, the lines to be selected are assessed for agronomic and quality characteristics. As segregating populations advance, homozygosity also increases. The lines selected at this stage on the basis of good performance are homozygous and ready for replicated performance trials across multiple locations and cropping seasons. Seeds of the best one to five percent of lines with superior agronomic performance or quality characteristics are multiplied under controlled conditions for variety release and to maintain their genetic purity.

As in all breeding programs, the exact techniques used in cowpea cultivar development vary widely. In the case of simply inherited traits such as disease resistance, the backcross method is used to introgress the associated gene(s) into existing cultivars that are lacking the trait. When several traits are being moved from two or more parents, hybrids from single, double, three-way, or other complex crosses are advanced through any of several methods that support the acceleration of homozygosity. Commonly used procedures include bulk population, the pedigree method, single-seed descent and modifications of these methods as necessary. The ultimate product of all the methods is a group of homozygous lines, which only vary in the time frames during which selection pressures are applied. Cowpea breeders have relied primarily on pedigree breeding to combine favorable traits from two parents and by recurrent backcrossing to introgress a major trait from a donor line into an elite recurrent parent which is usually a preferred current variety. Considerable success has been achieved by both approaches to improve cowpea grain quality and grain and biomass yield by the combining of traits determining grain size, texture and color, drought and heat tolerance, and resistance to a range of pests and diseases.

Cowpea being primarily self-pollinated, hybridization between parents usually involves emasculation (removal of anthers) from flowers of one parent (female or seed parent) and artificial transfer of pollen from the alternate (male) parent. Cowpeas are easier to cross than many other grain legumes due to the large size of the flowers and to the fact that the keel is straight, beaked, and not twisted. Cowpea flowers have few floral nodes per raceme and tend to have a lower rate of abortion than many other species. Rapid and effective methods of hand emasculating and crossing cowpeas were described by Myers (1996).

Although emasculation and pollination can be carried out all day, hybridization of cowpea is less effective when the temperature is high. Night temperatures greater than 20 °C reduce microsporogenesis leading to formation of indehiscent anthers and pollen with low viability (Warrag and Hall 1983; Ahmed et al. 1992). Thus, moderate temperature and increased humidity appear to increase the percentage of pod set following hand-emasculated crosses. In general, the rate of such pod setting varies enormously with environmental conditions, genotype, and manipulative techniques.

In IITA, crossing activities are conducted in mesh houses or greenhouses to allow: (1) good control of insect pollen vectors, major pests, and diseases; (2) good plant development (water, fertilizer, etc.); and (3) easy manipulation of plants during crosses. When planted in the screenhouse, most cowpea lines tend to climb and stakes are, therefore, needed which also help position the flowers at a height that is comfortable for the person making the crosses. In the planning of crossing activities, photosensitivity and number of days to flowering of the parental genotypes are always considered to avoid asynchronous flowering. Use of different planting dates, removal of developing flowers and fruits, using black polythene to cover plants from afternoon to next morning for a number of days (to reduce length of days) and the use of cuttings are some of the techniques to ensure synchronous flowering by the parental lines.

A hybrid plant reproduces to form a segregating population (segregation and recombination of genes). Development of a new variety usually involves inbreeding of a segregating cowpea population for three to seven generations, during which selection is applied and individuals in the population become increasingly homozygous (true breeding). Narrow crosses between closely related parents normally require fewer generations of inbreeding than wide crosses (very different parents) to become true breeding.

In IITA, screenhouses are used for rapid advancement of cowpea breeding populations with the possibility of 3–4 generations per year. When photoperiod-sensitive parents are involved in the crosses, breeding programs can only have 2–3 generations per year. Generally, a high level of homozygosity is attained from F_5 to F_7 since variation within rows derived from single selected plants will be relatively small. Bulked F_6/F_7 populations are grouped according to maturity, plant type, seed quality, and resistance to major pests and diseases. Farmers' involvement through participatory variety selection is encouraged. Homozygous materials are evaluated in initial evaluation trials (IET) at 2–3 locations without replication. Selected lines from IET are tested in preliminary variety trials and advanced variety trials (AVT) in 3 replications across 4 locations representing different agroecological zones: (a) Ibadan (7 ° 25′ N, 3 ° 37′ E) derived savanna with bimodal rainfall (1500 mm); (b) Samaru (11 ° 10′ N, 7 ° 38′ E) in the northern Guinea savanna (1000 mm rainfall); (c) Minjibir (12 ° 08′ N, 8 ° 40′ E) in Sudan savanna with about 700 mm rainfall; and (d) Toumnia (13 ° 58′ N, 9 ° 01′ E) or Malamadori representing the Sahel with about 350 mm rainfall. High-performing lines from AVT are compiled into cowpea international trials which are sent to collaborators for testing in their local environments. The best lines are released as varieties or used in their breeding programs for further improvement.

In some cases, this traditional approach to cowpea breeding which is based on phenotypic selection of progenies carrying the desired traits is still continuing in most programs. This is particularly the case where molecular markers for genotyping are unavailable, or where the marker-trait locus associations have not been identified. The advent of new marker technologies for genotyping, explained in detail in the next few paragraphs, is expected to facilitate a more efficient breeding approach than the previous reliance on phenotypic selection which is time-consuming. Thus, marker-assisted pedigree breeding (MAPB) and marker-assisted backcrossing (MABC) have been incorporated into several cowpea breeding programs by taking advantage of the new marker-driven selection tools. For MAPB, the design of an "ideotype" as the breeding goal is a useful first step in targeting the genotype of the combined set of favorable alleles to be donated by the two parents. Ehlers et al. (2012) provided a detailed example of this approach based on experience with two African cowpea populations derived from "elite" × "elite" crosses produced by ISRA, Senegal (Mouride × IT84S-2246-4) and INERA, Burkina Faso (IT93K-503-1 × IT84S-2246-4). The ideotype design is the full combination of the target QTLs all in the homozygous condition for the favorable alleles. Molecular breeding software programs (see next section) can then be used to analyze the genotypic data to identify and rank families or individual plants in progenies with the appropriate molecular scores (ideotype = maximum score or 100 %).

In MABC, the marker genotyping application has two components, namely "foreground" selection for the presence of the target trait QTL using linked flanking markers at the trait QTL, and "background" selection using genome-wide markers to select for individual plants with the highest recurrent parent genotype profile. In the California Blackeye cowpea-breeding program, transferring aphid, nematode, and *Fusarium* wilt resistances into improved versions of the current varieties CB27, CB46, and CB50 is being achieved with a MABC approach. For example, the QTLs identified by Pottorff et al. (2012, 2014) for resistance to *Fusarium* wilt races 3 and 4 are being transferred through MABC into new lines. CB46 is race 3 resistant but lacks race 4 and aphid resistances, so markers for the resistance loci can be used to select the donated favorable alleles for resistances to aphid and *Fusarium* race 4 and also confirm the presence in the background of the favorable haplotypes for *Fusarium* race 3 and root-knot nematode resistances. In Africa, several breeders have started using MABC to add Striga and aphid resistances into elite local varieties through collaborations with advanced research institutions.

A third breeding approach is MARS, which is being tested in four African cowpea breeding programs at INERA, Eduardo Mondlane University, IITA, and ISRA in partnership with UCR. MARS has been used with success in some cereal breeding programs (Charmet et al. 2001). The goal of MARS is to combine multiple favorable traits from two parents in a complementary manner, in which early generation progenies with partial combinations of the full set of traits are intercrossed for as many as three cycles of recombination to maximize the pyramiding of QTLs for multiple traits into advanced breeding lines (Charmet et al. 2001). This approach overcomes the limitations which occur in pedigree breeding with manageable progeny sizes, especially when the favorable alleles at several QTLs need to be combined. The geometric increase of a target locus becoming homozygous for the unfavorable allele with each generation, renders individual homozygous positives for all target QTLs rare or absent (Ehlers et al. 2012).

A series of MARS populations were developed from crossing elite parents with complementary trait sets (Suvita 2 × IT97K-499-35, IT84S-2246-4 × IT98K-1111-1, CB27 × IT97K-499-35, IT93K-503-1 × Mouride) relevant to each target environment in the four SSA cowpea breeding programs, with a focus on drought tolerance, seed size, and color, yield gain and biotic stress resistance (Striga and root-knot nematode). About 300 F_2 seeds were derived from each biparental cross followed by selfing to produce about 300 $F_{2:3}$ or $F_{2:4}$ families. Each of the $F_{2:3}$ or $F_{2:4}$ populations was phenotyped in different field sites for yield and other agronomic traits. Leaf samples of F_2 individuals or $F_{2:3}$/$F_{2:4}$ bulks from the field were genotyped by KBioscience (www.lgcgenomics.com) using a customized list of polymorphic SNPs generated by SNP Selector (www.breedit.org) based on distance in cM between genome-wide markers, and between markers at known trait positions. QTLs were discovered at $F_{2:3}$ or $F_{2:4}$ generations. For recombination cycles, QTL indices were computed by OptiMAS (http://moulon.inra.fr/optimas/), and members of highest QTL-index families were then genotyped, selected, and intercrossed to recombine favorable alleles. The outcome of this MARS breeding plan is that advanced lines have been produced that are homozygous for the favorable alleles of the target QTLs for yield, seed size, heat tolerance, staygreen, and resistance to root-knot nematodes and Striga. Currently the 20–30 advanced lines with the QTLs confirmed by SNP-genotyping are in a 2-year performance testing phase in production field trials, from which new variety releases are expected, and which will provide elite lines for use as parents in further cowpea improvement.

8 Integration of New Biotechnologies in Breeding Programs

High levels of resistance to several insects and diseases exist in wild *Vigna* species, but cross incompatibility with cultivated lines is the biggest bottleneck limiting their exploitation for cowpea improvement through conventional breeding. Biotechnological approaches were suggested as ways to overcome these limitations. If useful genes can be isolated from wild *Vigna* species, a genetic transformation system is a prerequisite for their deployment in cultivated cowpea. Initial genetic transformation efforts using *Agrobacterium tumefaciens* as the gene vector were conducted by Garcia et al. (1986, 1987). This was followed by embryo imbibition with or without subsequent electroporation (Akella and Lurquin 1993; Penza et al. 1992).

In all these cases, the development of transgenic cowpea calli or chimeric plantlets from leaf discs, axilliary buds, or embryos were obtained but no mature transgenic plants could be generated. Microprojectile bombardment (biolistics) was also used by several researchers to achieve the introduction of foreign DNA into cowpea leaf tissues and embryos and to obtain high levels of transient expression of the ß-glucuronidase transgene, but regeneration of plantlets from the transformed cells was not possible (Kononowicz et al. 1997). The development of transformation systems using either microprojectile bombardment or *Agrobacterium* cocultivation gave some promising results with the coculturing of de-embryonated cotyledons with *A. tumefaciens* resulting in selection of four plants on hygromycin (Kononowicz et al. 1997). This last approach helped in the development of a system that was the first to be reproducible and that obeys Mendelian rules of inheritance (Popelka et al. 2006). Critical features of this system include suitable explants from cotyledonary nodes or embryonic axes and a tissue-culture regime without auxins, but which includes a cytokinin at low levels during shoot initiation. There are now several reports showing experimental evidence for reproducible gene transfer to cowpea including genes for resistance to pod borer (Higgins et al. 2012) and cowpea weevil (Solleti et al. 2008) as well as for weed control (Citadin et al. 2013) and a range of model genes to evaluate the technology (Citadin et al. 2011).

The development of cowpea with a *Bt* gene was carried out successfully in the Commonwealth Scientific and Industrial Research Organisation (CSIRO) in Australia. Field testing of these lines has been carried out in Nigeria, Burkina Faso and Ghana in a Pod Borer-Resistant (PBR) Cowpea Project led by the African Agricultural Technology Foundation (AATF) and supported by USAID. A selected *Bt* cowpea line with near complete resistance to Maruca pod borer is being used to introgress the *Bt* gene into farmer preferred varieties. Selection using molecular markers will expedite the rapid development of cowpea varieties with resistance to Maruca and incorporating other traits preferred by farmers. Encouraging results have been obtained by the relevant breeding programs in SSA, and the AATF is working towards commercializing and making the PBR cowpea available to farmers in SSA.

In addition to genetic transformation, molecular breeding for cowpea is also well advanced but requires a genotyping capability that is cost effective and efficient so that genotyping results can be generated and interpreted quickly enough to make breeding selection decisions for crossing or targeted phenotyping. The genomics revolution has had important positive impacts on modern cowpea breeding. SNP genotyping platforms were developed based on genic SNPs developed from expressed sequence tags. They have high rates of polymorphism among the primary cowpea germplasm sources in the target breeding programs. Muchero et al. (2009a) developed a 1536-SNP Illumina GoldenGate assay which included about 1100 genetically mapped SNPs. This platform was used for extensive QTL discovery and to develop a consensus genetic map for cowpea constructed from six RIL populations (Muchero et al. 2009a).

The cowpea consensus map has been improved several times (Diop et al. 2012; Lucas et al. 2011) and the current version constructed from 11 RIL populations and two breeding populations is available online via HarvEST:Cowpea (Close and Wanamaker 2001). The 1536-SNP platform and the genetic maps were used to identify the genomic positions of the many QTLs for important cowpea traits described earlier. The QTL discovery has been based on extensive phenotyping for agronomic as well as abiotic and biotic stress resistance traits in the genotyped biparental RIL populations and cowpea diversity panels (Huynh et al. 2013b; Muchero et al. 2013; Lucas et al. 2013b; Pottorff et al. 2014). The SNP calls associated with the favorable alleles at each QTL provide the marker haplotypes needed for positive trait selection for use in foreground selection. The genome-wide markers across the 680 cM genetic map, spaced on average, every 0.6 cM, provided the resource for background selection across the genome in the MABC breeding efforts.

Application of the SNP marker resource for cowpea breeders was further advanced by converting the mapped SNP markers to a flexible genotyping platform, using the Kompetitive Allele Specific PCR (KASP) technology of LGC Genomics (formerly KBiosciences). This platform enables choice in which and how many SNP markers the breeder would like to use in a given breeding project and flexibility in the number of DNA samples per genotyping run. This translates into a more cost-efficient genotyping capability than the fixed GoldenGate platform. A new 60,000 SNP genotyping platform has been developed using the Illumina Infinium iSelect technology (Close et al. 2015). The cowpea breeding programs with a strong focus on molecular breeding have taken advantage of outsourcing the genotyping work. In the programs at UCR, INERA, SARI, Eduardo Mondlane University, IITA, and ISRA, the molecular breeding is underpinned by the outsourcing operation (Ehlers et al. 2012). Leaf samples (leaf punches placed in 96-well plates and dried with silica gel for preservation) are collected in the greenhouse or field, then express-shipped to the genotyping facility (LGC Genomics in the UK or the USA for the KASP platform), where DNA is extracted and genotyped with a preselected subset of informative SNP markers. The data are made available usually within 4 weeks, which can then be interpreted to make breeding selection decisions for crossing or progeny selections.

Improvements in the workflow for molecular breeding have included development of some in-house software programs for SNP selection and data analysis. They are used in conjunction with the CGIAR IBP (Integrated Breeding Platform) Breeding Management System software programs for analyzing the genotype and phenotype data for QTL tracking and in the MARS, MAPB, and MABC breeding schemes.

9 Seed Production

Improved seeds constitute one of the most important farm inputs needed for increasing agricultural production. High quality seeds of improved varieties should, therefore, be available to farmers to ensure sustainable crop production. However, it has been observed that there is not much enthusiasm on the part of large seed companies to engage in grain legume seed enterprises because of low margin of profit, as farmers could recycle their own saved seed for up to 5 years (Abate et al. 2012). These authors reported that more than 70 % of farmers use their own saved seed across the thirteen countries where the Tropical Legumes II (TL II) project is being implemented. Kenya is the only exception where saved seed supplied just over 34 % of farmers' needs. In Mozambique, only 12 % of adopters of improved cowpea varieties bought improved seed from agro-dealers, with the rest using their own recycled seed (Fatokun et al. 2012b).

Strategies to ensure some level of production of good seeds include strengthening community-based and farmer-level seed production systems. Generally, the National Agricultural Research Institutes are responsible for the production of breeder and foundation seeds. Individual farmers and farmers' groups, agricultural universities, and small private seed companies also produce foundation seeds. The private sector and farmers' groups are generally responsible for certified seed production. In some countries, small-scale farmers and the public sector with the use of contract farmers also produce certified seeds. Other quality seeds are also produced by farmers' associations generally supported by NGOs. In Nigeria for example, breeder and foundation seeds of cowpea are produced mainly by research institutes such as IITA and the Institute for Agricultural Research while certified seeds are produced by seed companies (e.g., Maina Seeds, Alheri Seeds, Jikur Seed), Agricultural Development Projects, some NGOs, out-growers, and National Program for Food Security community seed growers.

Access to quality seed is a crucial factor in the adoption of improved technologies by farmers. Use of improved, modern varieties was generally low across some SSA countries following baseline studies conducted at the beginning of phase I of the TL II project (Abate et al. 2012). It was also reported by these authors that unavailability of improved seed and, in some cases, lack of access to credit were major bottlenecks for improved variety adoption. Fatokun et al. (2012b) noted that in Mozambique over 70 % of non-adopters of improved cowpea indicated lack of access to improved seeds as the major constraint. In Nigeria, 71 % of male-headed households complained about lack of cash availability to purchase seeds and other inputs. In their investigations of the cowpea seed subsector in Nigeria, they found that access to quality, and affordability, of improved seeds was of concern. Most cowpea producers (60 % in Kano and 86 % in Borno States of Nigeria) get information on availability of seeds of improved varieties through Ministries of Agriculture extension agents. Few producers get information from seed companies, research institutes' staff, fellow farmers, and NGOs in Kano State. These observations have implications for the adoption of new varieties.

The cowpea seed system receives very little attention from the formal seed industry consisting of public sector research institutions, seed companies, and organizations (the National Agricultural Seed Council) in almost all the countries of SSA. A larger proportion of the smallholder farmers' seed needs are therefore met by the informal sector. With the low level, or even absence, of the involvement of large-scale seed companies, it is important to strengthen the informal sector and use it as a means of providing resource-poor farmers with quality seeds of improved varieties of crops at affordable prices. Concerted efforts are being made to promote the dissemination of seeds of improved cowpea varieties in many SSA countries. Farmer-to-farmer seed diffusion was jointly promoted by IITA, IAR and Kano State Agricultural and Rural Development Authority to disseminate new cowpea varieties (IT90K-277-2 and IT93K-452-1) in the late 1990s. About 8 kg of cowpea seeds were given to each primary farmer selected to establish a 0.4 ha of seed farm. The 300 kg of seeds produced by each of the primary farmers (foundation seeds step 1 = FS 1) were distributed/sold to 12 secondary farmers. Each secondary farmer in turn established a 0.4 ha seed farm (FS 2) and the 300 kg produced by each farmer gave a total of 3600 kg which was enough to plant 1444 × 0.4 ha of commercial crop (Utoh and Ajeigbe 2009).

This strategy was found to be faster and cheaper for seed dissemination than previously used methods. In Nigeria, community seed production was promoted by the National Program for Food Security. Farmers were trained in seed production strategies and linked to seed companies and research institutions for renewal of seed stocks. The role of extension agents is very important in seed production and adoption of new improved varieties. In southern Borno State, the Promoting Sustainable Agriculture in Borno State program implemented by IITA and national partners selected and trained seed producers and assisted them with establishing community-based seed multiplication schemes in 30 communities that covered three agroecological zones (Fatokun et al. 2012b). The TL II project helped to establish an awareness creation system for improved varieties through field days, demonstrations, seed fairs, agricultural shows, dealing with farmers' research groups/farmer field schools, and distribution of small packs of seed samples. The small seed pack strategy, developed in partnership with the private sector, was helpful in getting seeds of improved varieties to many more farmers. Marketing seed in small quantities of 1- or 2-kg packs that are within the reach of smallholder farmers was found to be both profitable to a small private seed company and attractive to farmers (Fatokun et al. 2012b). Over 12,000 farmers were reached with this method over a 3-year period (2010–2012) and this further popularized some improved cowpea varieties.

Improved market linkages have encouraged seed producers to increase seed production to supply a growing market. Market development for cowpea seed resulted in increased production and sales of cowpea, making significant contributions to improving livelihood and poverty reduction. Over 188 MT of seed was sold by seed producers in Nigeria, 31.5 MT in Mali, and 93.7 MT sold in Niger, within the first phase of TL II project (Fatokun et al. 2012b). This market is now established and paying good prices for seeds, a situation likely to be sustained.

Based on our experience with the TL II project, strengthening of community-based organizations, in particular the farmers' groups and associations, through training and support for quality certified and foundation seed production, reinforced with postharvest processing, storage, distribution, and marketing will ensure that quality the seed of newly developed and released varieties will be available to a majority of farmers.

Another major constraint to seed trade in cowpea is the susceptibility of the seeds to the bruchid weevil (*Callosobruchus maculatus*). This insect is capable of destroying all seed stored by farmers within 6 months. No cowpea variety has resistance to this insect pest. In order to protect the seeds from damage by the insect, farmers use a number of methods which are mostly not effective. In recent times, researchers at Purdue University, USA have come up with a simple and cheap chemical-free method for storing cowpea seeds. The technology is commonly referred to as Purdue Improved Cowpea Storage (PICS). With this method farmers and seed retailers are able to store cowpea seeds longer than hitherto and this should encourage them to produce and keep seeds for planting in the following cropping season.

References

Abate T, Alene AD, Bergvinson D et al (2012) Tropical grain legumes in Africa and South Asia: knowledge and opportunities. International Crops Research Institute for the semi-arid tropics (ICRISAT), Nairobi

Agbicodo EM (2009) Genetic analysis of abiotic and biotic resistance in cowpea (*Vigna unguiculata* (L.) Walp.). PhD Dissertation, Wageningen University, Wageningen, The Netherlands

Ahmed FE, Hall AE, DeMason DA (1992) Heat injury during floral development in cowpea (*Vigna unguiculata* (L.) Walp.). Am J Bot 79:784–791

Akella V, Lurquin PF (1993) Expression in cowpea seedlings of chimeric transgenes after electroporation into seed-derived embryos. Plant Cell Rep 12:110–117

Allen DJ, Thottappilly G, Rossel HW (1982) Cowpea mottle virus: field resistance and seed transmission in virus tolerant cowpea *Vigna unguiculata*. Ann Appl Biol 100:331–336

Bashir M (1992) Serological and biological characterization of seed-borne isolates of blackeye cowpea mosaic and cowpea aphid borne mosaic potyviruses in *Vigna unguiculata* (L.) Walp. PhD Dissertation, Oregon State University, USA

Boukar O, Massawe F, Muranaka S et al (2011) Evaluation of cowpea germplasm lines for protein and mineral concentrations in grains. Plant Genet Resour Character Util 9:515–522

Charmet G, Robert N, Perretant MR et al (2001) Marker assisted recurrent selection for cumulating QTLs for bread-making related traits. Euphytica 119:89–93

Citadin CT, Ibrahim AB, Aragão FJL (2011) Genetic engineering in Cowpea (*Vigna unguiculata*): history, status and prospects. GM Crops 2:144–149

Citadin CT, Cruz ARR, Aragão FJL (2013) Development of transgenic imazapyr-tolerant cowpea (*Vigna unguiculata*). Plant Cell Rep 32:537–543

Close TJ, Wanamaker S (2001) HarvEST: EST databases. http://harvest.ucr.edu.

Close TJ, Lucas MR, Muñoz-Amatriain M, Mirebrahim H, Wanamaker S, Barkley NA, Clair SS, Guo YN, Lo S, Huynh BL, Ndeye AD, Santos JRP, Joseph BTB, Jean-Baptiste TDLS, Drabo I, Kusi F, Atokple I, Boukar O, Fatokun C, Cisse N, Xu P, Roberts PA, Lonardi S (2015). A new SNP-genotyping resource for cowpea and its deployment for breeding. Plant and Animal Genome Conference, San Diego January 10–14, 2015. https://pag.confex.com/pag/xxiii/webprogram/Paper14976.html

Diop NN, Ehlers JD, Wanamaker S et al (2012) An improved genetic consensus map of cowpea [*Vigna unguiculata* (L.) Walp]. In: Boukar O Coulibaly O, Fatokun CA et al (eds) Innovative research along the cowpea value chain. Proceedings of the Fifth World Cowpea Conference on Improving livelihoods in the cowpea value chain through advancement in science. Saly, Senegal 27 September–1 October 2010. International Institute of Tropical Agriculture, Ibadan, Nigeria, pp 116–127

Drabo, I (2014) Personal communication to BL Huynh

Duke JA (1981) *Vigna unguiculata* (L.) Walp. ssp. *unguiculata*. In: Handbook of Legumes of world economic importance. Plenum Press, New York

Dumet D, Adeleke R, Faloye B (2008) Regeneration guidelines: cowpea. In: Dulloo ME, Thormann I, Jorge MA et al (eds) Crop specific regeneration guidelines (CD ROM). CGIAR System-wide Genetic Resources Programme, Rome, Italy, p 8

Dumet D, Mishra SK, Boukar O et al (2010) Key access and utilization descriptors for cowpea genetic resources. Dillon S, Kainz W, Bharadwaj C et al (reviewers) Bioversity International, Rome, Italy
Dumet D, Fatokun C, Pasquet R et al (2012) Sharing of responsibilities of cowpea and wild relatives in long term conservation. In: Boukar O, Coulibaly O, Fatokun CA. et al (eds) Innovative research along the cowpea value chain. Proceedings of the Fifth World Cowpea Conference on Improving livelihoods in the cowpea value chain through advancement in science. Saly, Senegal, 27 September–1 October 2010. International Institute of Tropical Agriculture, Ibadan, Nigeria, pp 56–65
Ehlers JD, Hall AE, Patel PN et al (2000) Registrations of cultivars: registration of 'California Blackeye 27' cowpea. Crop Sci 40:849–863
Ehlers JD, Hall AE, Roberts PA et al (2002) Blackeye varietal improvement. In: University of California dry bean research, 2002 Progress report. California Dry Bean Advisory Board, Dinuba, USA, pp 27–47
Ehlers JD, Sanden BL, Frate CA et al (2009) Registration of 'California Blackeye 50' Cowpea. J Crop Regist 3:236–240
Ehlers JD, Diop NN, Boukar O et al (2012) Modern approaches for cowpea breeding. In: Boukar O, Coulibaly O, Fatokun CA et al (eds) Innovative research along the cowpea value chain. Proceedings of the Fifth World Cowpea Conference on Improving livelihoods in the cowpea value chain through advancement in science, Saly, Senegal, 27 September–1 October 2010. International Institute of Tropical Agriculture, Ibadan, Nigeria, pp 3–16
Emechebe AM, Florini DA (1997) Shoot and pod diseases of cowpea induced by fungi and bacteria. In: Singh BB, Mohan Raj DR, Dashiell KE et al (eds) Advances in cowpea research. (Co-publication of International Institute of Tropical Agriculture (IITA) and Japan International Research Center for Agricultural Sciences (JIRCAS)). IITA, Ibadan, pp 176–192
Emechebe AM, Shoyinka SA (1985) Fungal and bacterial diseases of cowpea in Africa. In: Singh SR, Rachie KO (eds) Cowpea research, production and utilization. Wiley, Chichester, pp 173–192
Evans M, Boulter D (1974) Chemical methods suitable for screening for cowpea protein content and quality in cowpea (*Vigna unguiculata*) meals. J Sci Food Agric 25:311–322
Fatokun CA (2002) Breeding cowpea for resistance to insect pests: attempted crosses between cowpea and *V. vexillata*. In: Fatokun CA, Tarawali SA, Singh BB et al (eds). Challenges and opportunities for enhancing sustainable cowpea production. Proceedings of the World Cowpea Research Conference III held at the International Institute of Tropical Agriculture (IITA), Ibadan, Nigeria, 4–8 September 2000, pp 52–61
Fatokun CA, Singh BB (1987) Interspecific hybridization between *Vigna pubescens* Wilcz. and *Vigna unguiculata* (L.) Walp. through embryo culture. Plant Cell Tissue Organ Cult 9:229–233
Fatokun CA, Boukar O, Muranaka S (2012a) Evaluation of cowpea (*Vigna unguiculata* (L.) Walp.) germplasm lines for tolerance to drought. Plant Genet Resour Charact Util 10:171–176
Fatokun CA, Boukar O, Kamara A et al (2012b) Enhancing cowpea productivity and production in drought-prone areas of Sub-Saharan Africa. In: Abate T (ed) Four seasons of learning and engaging smallholder farmers: progress of phase 1. International Crops Research Institute for the Semi-Arid Tropics, Nairobi, pp 81–112
Fery RL, Dukes PD (1994) Genetic analysis of the green cotyledon trait in southernpea (*Vigna unguiculata* (L.) Walp.). J Am Soc Hort Sci 119:1054–1056
Garcia JA, Hille J, Goldbach R (1986) Transformation of cowpea *Vigna unguiculata* cells with an antibiotic resistance gene using a Ti-plasmid-derived vector. Plant Sci 44:37–46
Garcia JA, Hille J, Vos P et al (1987) Transformation of cowpea *Vigna unguiculata* with a full-length DNA copy of cowpea mosaic virus M-RNA. Plant Sci 48:89–98
Hampton RO, Thottappilly G, Rossel HW (1997) Viral diseases of cowpea and their control by resistance-conferring genes. In: Singh BB et al. (eds) Advances in cowpea research. (Co-publication of IITA and JIRCAS). IITA, Ibadan, pp 159–175
Hearne SJ, Franco J, Magembe E (2012) Optimal units of selection: how many plants are representative of diversity in collections of cowpea landrace accessions? In: Boukar O, Coulibaly O,

Fatokun CA et al (eds) Innovative research along the cowpea value chain. Proceedings of the Fifth World Cowpea Conference on Improving livelihoods in the cowpea value chain through advancement in science, Saly, Senegal, 27 September–1 October 2010. International Institute of Tropical Agriculture. Ibadan, Nigeria, pp 103–109

Higgins TJV, Gollasch S, Molvig L et al (2012) Insect-protected cowpeas using gene technology. In: Boukar O, Coulibaly O, Fatokun CA et al (eds) Innovative research along the cowpea value chain. Proceedings of the Fifth World Cowpea Conference on Improving livelihoods in the cowpea value chain through advancement in science, Saly, Senegal, 27 September–1 October 2010. International Institute of Tropical Agriculture, Ibadan, Nigeria, pp 131–137

Huguenot C, Furneaux MT, Hamilton RI (1997) Further characterization of cowpea aphid-borne mosaic and blackeye cowpea mosaic potyviruses. In: Singh BB, Mohan Raj DR, Dashiell KE et al (eds). Advances in cowpea research. (Co-publication of International Institute of Tropical Agriculture (IITA) and Japan International Research Center for Agricultural Sciences (JIR-CAS)). IITA, Ibadan, pp 231–239

Huynh BL, Close TJ, Roberts PA et al (2013a) Gene pools and the genetic architecture of domesticated cowpea. Plant Genome 6:1–8

Huynh BL, Ehlers JD, Close TJ et al (2013b) Enabling tools for modern breeding of cowpea for biotic stress resistance. In: Varshney R, Tuberosa R (eds) Translational genomics for crop breeding, Volume I: biotic stress. Wiley-Blackwell, New York, pp 183–200

Huynh BL, Ehlers JD, Ndeve A et al (2015) Genetic mapping and legume synteny of aphid resistance in African cowpea (*Vigna unguiculata* L. Walp.) grown in California. Mol Breed 35:36. doi:10.1007/s 1 1032-915-0254-0

International Board for Plant Genetic Resources (1983) Cowpea descriptors. IBPGR Secretariat, Rome

Kononowicz AK, Cheah KT, Narasimhan ML et al (1997) Developing a transformation system for cowpea (*Vigna unguiculata* [L.] Walp.). In: Singh BB, Mohan Raj DR, Dashiell KE et al (eds) Advances in cowpea research. (Co-publication of International Institute of Tropical Agriculture (IITA) and Japan International Research Center for Agricultural Sciences (JIRCAS)). IITA, Ibadan, pp 361–371

Lane JA, Moore THM, Child DV et al (1997) Variation in virulence of *Striga gesnerioides* on cowpea: new sources of crop resistance. In: Singh BB, Mohan Raj DR, Dashiell KE et al (eds) Advances in cowpea research. (Co-publication of International Institute of Tropical Agriculture (IITA) and Japan International Research Center for Agricultural Sciences (JIRCAS)). IITA, Ibadan, pp 225–230

Lucas MR, Diop NN, Wanamaker S et al (2011) Cowpea-soybean synteny clarified through an improved genetic map. Plant Genome 4:1–11

Lucas MR, Ehlers JD, Huynh BL, Diop NN et al (2013a) Markers for breeding heat tolerant cowpea. Mol Breed 31:529–536

Lucas MR, Huynh BL, Ehlers JD et al (2013b) High-resolution SNP genotyping reveals a significant problem among breeder resources. Plant Genome 6:1–5

Lush WM, Evans LT, Wien HC (1980) Environmental adaptation of wild and domesticated cowpea (*Vigna unguiculata* (L.) Walp.). Field Crops Res 3:173–187

Mahalakshmi V, Ng Q, Lawson M, Ortiz R (2007) Cowpea [*Vigna unguiculata* (L.) Walp.] core collection defined by geographical, agronomical and botanical descriptors. Plant Genet Resour 5:113–119

Marechal R, Mascherpa JM, Stainer F (1978) Etude taxonomique d'un groupe complexe d'espèces des genres *Phaseolus* et *Vigna* (Papillinionaceae) sur la base des données morphologiques, et polliniques, traits par l'analyse informatique. Bossiera 28:1–273

Muchero W, Ehlers JD, Roberts PA (2008) Seedling stage drought-induced phenotypes and drought-responsive genes in diverse cowpea genotypes. Crop Sci 48:541–552

Muchero W, Diop NN, Bhat PR et al (2009a) A consensus genetic map of cowpea [*Vigna unguiculata* (L.) Walp.] and synteny based on EST-derived SNPs. Proc Natl Acad Sci USA 106:18159–18164

Muchero W, Ehlers JD, Close TJ et al (2009b) Mapping QTL for drought stress-induced premature senescence and maturity in cowpea [*Vigna unguiculata* (L.) Walp.]. Theor Appl Genet 118:849–863

Muchero W, Roberts PA, Diop NN et al (2013) Genetic architecture of delayed senescence, biomass and grain yield under drought stress in cowpea. PLoS ONE 8:1–10

Myers GO (1996) Hand crossing of cowpeas. IITA Research Guide 42. IITA, Ibadan

Nielsen SS, Brandt WE, Singh BB (1993) Genetic variability for nutritional composition and cooking time in improved cowpea lines. Crop Sci 33:469–472

Njoku E (1958) The photoperiodic response of some Nigerian plants. J West Afr Sci Assoc 4:99–112

Ng NQ, Maréchal R (1985) Cowpea taxonomy, origin and germplasm. In: Singh SR, Rachie KO (eds) Cowpea research, production and utilization. Wiley, Chichester, pp 11–21

Ng Q, Singh BB (1997) Cowpea. In: Fuccillo D, Sears L, Stapleton P (eds) Biodiversity in trust: conservation and use of plant genetic resources in CGIAR centres. Cambridge University Press, Cambridge, pp 82–99

Omo-Ikerodah EE, Fawole I, Fatokun C (2008) Genetic mapping of quantitative trait loci (QTLs) with effects on resistance to flower bud thrips (*Megalurothrips sjostedti*) in recombinant inbred lines of cowpea (*Vigna unguiculata* (L.) Walp). African J Biotechnol 7:263–270

Ouedraogo JT, Ouedraogo M, Gowda BS et al (2012) Development of sequence characterized amplified region (SCAR) markers linked to race-specific resistance to *Striga gesnerioides* in cowpea (*Vigna unguiculata* L.). African J Biotech 11(62):12555–12562. http://www.academicjournals.org/AJB. doi:10.5897/AJB12.805

Pasquet RS, Padulosi S (2012) Genus *Vigna* and cowpea (*V. unguiculata* [L.] Walp.) taxonomy: current status and prospects. In: Boukar O, Coulibaly O. Fatokun CA. et al (eds). Innovative research along the cowpea value chain. Proceedings of the Fifth World Cowpea Conference on Improving livelihoods in the cowpea value chain through advancement in science, Saly, Senegal, 27 September–1 October 2010. International Institute of Tropical Agriculture, Ibadan, Nigeria, pp 66–87

Patel PN (1982) Genetics of cowpea reactions to two mosaic virus from Tanzania. Phytopathol 72:460–466

Penza R, Akella V, Lurquin PF (1992) Transient expression and histological localization of a gus chimeric gene after direct transfer to mature cowpea embryos. Biotechniques 13:576–580

Popelka JC, Gollasch S, Moore A et al (2006) Genetic transformation of cowpea (*Vigna unguiculata* L.) and stable transmission of the transgenes to progeny. Plant Cell Rep 25:304–312

Pottorff M, Wanamaker S, Ma YQ et al (2012) Genetic and physical mapping of candidate genes for resistance to *Fusarium oxysporum* f.sp. *tracheiphilum* race 3 in cowpea [*Vigna unguiculata* (L.) Walp]. PLoS ONE 7:1–11

Pottorff M, Li G, Ehlers JD et al (2014) Genetic mapping, synteny, and physical location of two loci for *Fusarium oxysporum* f. sp. *tracheiphilum* race 4 resistance in cowpea [*Vigna unguiculata* (L.) Walp]. Mol Breed 33:779–791

Rachie KO (1985) Introduction. In: Singh SR, Rachie KO (eds) Cowpea research, production and utilization. Wiley, Chichester, pp XXI–XXVIII

Singh SR, Singh BB, Jackai LEN et al (1983) Cowpea research at IITA, Ibadan, Nigeria. Inf Ser 14:1–20

Singh SR, Jackai LEN, Thottappilly G et al (1992) Status of research on constraints to cowpea production. In: Thottappilly G, Monti L, Mohan-Raj DR et al (eds) Biotechnology: enhancing research on tropical crops in Africa. (CTA/IITA co-publication). IITA, Ibadan, pp 21–26

Singh BB, Ehlers JD, Sharma B et al (2002) Recent progress in cowpea breeding. In: Fatokun CA, Tarawali SA, Singh BB et al (eds) Challenges and opportunities for enhancing sustainable cowpea production. Proceedings of the World Cowpea Research Conference III held at the International Institute of Tropical Agriculture (IITA), Ibadan, Nigeria, pp 22–40

Solleti S, Bakshi S, Purkayastha J et al (2008) Transgenic cowpea (*Vigna unguiculata*) seeds expressing a bean α-amylase inhibitor 1 confer resistance to storage pests, bruchid beetles. Plant Cell Rep 27:1841–1850

Taiwo MA, Provvidenti R, Gonsalves D (1982) Inheritance of resistance to blackeye cowpea mosaic virus in *Vigna unguiculata*. J Hered 72:433–434

Terao T, Watanabe I, Matsunaga R et al (1997) Agro-physiological constraints in intercropped cowpea: an analysis. In: Singh BB, Mohan Raj DR, Dashiell KE et al (eds) Advances in cowpea research. (Co-publication of International Institute of Tropical Agriculture (IITA) and Japan International Research Center for Agricultural Sciences (JIRCAS)). IITA, Ibadan, pp 129–140

Utoh NO, Ajeigbe HA (2009) Dissemination of legume and cereal certified seeds using the community seed approach. In: Ajeigbe HA, Abdoulaye T, Chikoye D (eds) Legume and cereal seed production for improved crop yields in Nigeria. Proceedings of the Training Workshop on Production of Legume and Cereal Seeds held on 24 January–10 February 2008 at IITA-Kano Station, Kano, Nigeria, pp 61–63

Warrag MOA, Hall AE (1983) Reproductive responses of cowpea to heat stress: genotypic differences in tolerance to heat at flowering. Crop Sci 23:1088–1092

Watanabe I, Hakoyama S, Terao T et al (1997) Evaluation methods for drought tolerance of cowpea. In: Singh BB, Mohan Raj DR, Dashiell KE et al (eds) Advances in cowpea research. (Co-publication of International Institute of Tropical Agriculture (IITA) and Japan International Research Center for Agricultural Sciences (JIRCAS)). IITA, Ibadan, pp 141–146

Westengen OT, Jeppson S, Guarino L (2013) Global ex-situ crop diversity conservation and the Svalbard Global Seed Vault: assessing the current status. PLoS ONE 8:e64146

Westphal E (1974) Pulses in Ethiopia: their taxonomy and agricultural significance. Agricultural research report, center for agricultural publishing and documentation, Wageningen, The Netherlands, pp 213–232

Chapter 8
Grass Pea

Nuno Felipe Almeida, Diego Rubiales and Maria Carlota Vaz Patto

1 Introduction

Grass pea (*Lathyrus sativus* L.) is a multipurpose robust grain legume crop. It can grow in both drought- and flooding-prone environments and poor soils due to its hardy and penetrating root systems (Campbell 1997; Vaz Patto et al. 2006b). It has a high nutritional value (protein content ranging from 25 to 30 %), being important both for human food and animal feed. In what concerns human consumption, it can be consumed uncooked as a green snack, cooked in a stew, milled into flour or by roasting the seed (Peña-Chocarro and Peña 1999). In addition to its uses as food and feed, symbiosis with rhizobia allows an efficient nitrogen fixation in the soil, lowering the inputs needed in crop rotation and making them suitable to be used as green manure in sustainable farming systems (Hanbury et al. 2000). As an example of its versatility, grass pea is easily introduced in intercropping systems, rotations or used along with paddy rice in relay cropping systems (Abd El Moneim et al. 2001; Campbell et al. 1994; Hillocks and Maruthi 2012).

There is great potential for the expansion in the utilization of grass pea in dry areas or zones which are becoming more drought prone, with increased salinity or increased tendency to suffer from biotic stresses. However, the crop is unpopular

N. F. Almeida (✉)
Instituto de Tecnologia Química e Biológica António Xavier (ITQB),
Av. da República, EAN, Apartado 127, 2781-901 Oeiras, Portugal
e-mail: nalmeida@itqb.unl.pt

D. Rubiales
Institute for Sustainable Agriculture, CSIC, Avenida Menendez Pidal s/n,
14004 Córdoba, Spain
e-mail: diego.rubiales@ias.csic.es

M. C. Vaz Patto
Instituto de Tecnologia Química e Biológica António Xavier (ITQB),
Av. da República, EAN, Apartado 127, 2781-901 Oeiras, Portugal
e-mail: cpatto@itqb.unl.pt

A. M. De Ron (ed.), *Grain Legumes*, Handbook of Plant Breeding 10,
DOI 10.1007/978-1-4939-2797-5_8

with governments and donors because the plant contains small amounts of a toxin, β-*N*-ozalyl-L-α, β-diaminopropanoic acid (ODAP). Although this toxin can cause a neuronal disorder, known as "lathyrism", the condition develops in humans with a 6 % chance only when grass pea is consumed in large quantities, unaccompanied by other foodstuffs in an unbalanced diet and during a long period of time (Lambein et al. 2009). Also, seeds can be partly detoxified by various processing methods (Kumar et al. 2011; Kuo et al. 2000).

Even though this robust crop is rightly considered as a model crop for sustainable agriculture and despite the lathyrism stigma, the development of new breeding technologies and the growing interest in its use in Mediterranean-type environments, all over the world, will provide a bright future to this crop (Vaz Patto et al. 2006b; Vaz Patto and Rubiales 2014).

2 Origin and Systematics

Grass pea belongs to genus *Lathyrus,* within the Fabaceae family (syn. Leguminosae), subfamily Faboideae (syn. Papilionoideae), tribe Fabeae (syn. Vicieae), along with genera *Pisum, Vicia, Lens* and *Vavilovia* (Kenicer et al. 2005; Schaefer et al. 2012; Smýkal et al. 2011; Wojciechowski et al. 2004).

Natural distribution of grass pea has been completely obscured by human cultivation. Its use for food, feed and forage makes it difficult to distinguish between wild and domesticated populations, toughening the task to precisely locate its centre of origin (Kumar et al. 2013). The most probable grass pea centre of origin is believed to have been in the eastern Mediterranean or Fertile Crescent, around 6000 before present (BP). This has been supported by archaeobotanical and recent phylogenetic reports (Kislev 1989; Schaefer et al. 2012), refuting the hypothesis by Smartt (1984) that the centre of origin was located in Southwest or Central Asia. Domestication of grass pea seems to have occurred alongside with other pulses, being normally found with early domesticates of pea (*Pisum sativum* L.), lentils (*Lens culinaris* Medik.) and bitter vetch (*Vicia ervilia* (L.) Willd.; Erskine et al. 1994).

Hopf (1986) hypothesized that *L. sativus* is a derivative from *Lathyrus cicera,* its genetically nearest wild species. In addition, in what concerns domestication in Southern Europe (France and Iberian peninsula), evidences of cultivation of *L. cicera* were found, dating from 4000 or 3000 BP, suggesting that expansion of *L. sativus* farming may have also led to the domestication of the local *L. cicera* (Campbell 1997).

Within the economically important legume crops and model species, *P. sativum* is reported as the closest relation to grass pea, followed by lentil, faba bean (*Vicia faba* L.), barrel medic (*Medicago truncatula* Gaertn.), chickpea (*Cicer arietinum* L.) and *Lotus corniculatus* L. (Asmussen and Liston 1998; Ellison et al. 2006; Wojciechowski et al. 2004).

The infrageneric classification of *Lathyrus* genus has been revised several times, the one reported by Kupicha (1983) being the most accepted one. In this treatment, the genus is organized in 13 clades (Orobus, Lathyrostylis, Lathyrus, Orobon, Pratensis, Aphaca, Clymenum, Orobastrum, Viciopsis, Linearicarpus, Nissolia, Neurolobus and Notolathyrus). This morphological-based classification has been recently supported by molecular phylogenetic studies using sequence data from the internal transcribed spacer (ITS) region and from cpDNA (Kenicer et al. 2005, 2009). Schaefer et al. (2012), using nuclear and chloroplast phylogenetic data, further suggested that the genus *Lathyrus* is not monophyletic and recommended that a more natural classification would be to transfer *Pisum* and *Vavilovia* to a then monophyletic *Lathyrus* genus.

3 Varietal Groups

Great morphological variation is reported in grass pea, especially in vegetative characters such as leaf length, while, for instance, its floral characters are much less variable, showing a clear grouping in flower colour (Fig. 8.1; Jackson and Yunus 1984), as well as its seed and yield traits (Hanbury et al. 1999). Several studies divided grass pea accessions broadly into two groups: those from the Indian subcontinent and those from the Mediterranean region. Jackson and Yunus (1984) reported that all blue-flowered accessions came from Southwest and South Asia, while the white and mixed-coloured accessions had a more western distribution, from the Canary Isles to the western republics of the Soviet Union. These authors also pointed out that white-flowered accessions only had white seeds with no secondary markings on the seed coat. In accordance with this, Hanbury et al. (1999) reported that Mediterranean accessions were characterized by larger and whiter seeds, selected for human consumption, with higher yield potential than the Indian accessions. Grass pea small-seeded accessions are considered more primitive types

Fig. 8.1 *Blue*-flowered *Lathyrus sativus* genotype

and normally associated with hardened seeds like what happens in other Old World grain legumes such as pea, chickpea or lentil (Chowdhury and Slinkard 2000).

A particular case is the germplasm selected for forage, in the Mediterranean region, with landraces with broad leaves and pods, but low seed yield (Chowdhury and Slinkard 2000; Kumar et al. 2013).

4 Genetic Resources and Utilization

Conservation of *Lathyrus* genetic resources has recently attracted more attention because of the potential role of this species under the climate change scenario (Kumar et al. 2013).

Grass pea is mentioned in two conservation programmes for major food legumes. One is the International Treaty on Plant Genetic Resources for Food and Agriculture (ITPGRFA; FAO 2009), which aims at guaranteeing food security through conservation of biodiversity, fair exchange and sustainable use of plant genetic resources. This is being accomplished by establishing a global system to provide farmers, plant breeders and scientists access to plant genetic materials, ensuring that recipients share benefits with the countries where they have been originated and by recognizing the contribution of farmers to the diversity of crops used as food.

The other, a more specific programme developed by the Global Crop Diversity Trust (CGDT) in collaboration with the International Center for Agriculture Research in the Dry Areas (ICARDA), aims for a long-term conservation strategy of *L. sativus*, *L. cicera* and *L. ochrus* (GCDT 2009). This programme is detailing the current status of national collections and identifying gaps in collections of these three species from areas of diversity. Their strategy recommends that documentation on collections should be upgraded and that more work should be carried out on characterizing and evaluating collections for key traits, making this data widely available (Gurung and Pang 2011).

Several ex situ and a few in situ conservation examples exist for grass pea germplasm. The largest *Lathyrus* ex situ collections are maintained at the Conservatoire Botanique National des Pyrénées et de Midi-Pyrénées in France (4.477 accessions; previously at Pau University), by the ICARDA comprising 3.239 accessions and by the National Bureau of Plant Genetic Resources (NBPGR) in India (2.619 accessions). Smaller, but still relevant, collections are maintained by other banks such as the Germplasm Resource Information Network (GRIN) from the US Department of Agriculture (USDA) in the USA, the Leibniz Institute of Plant Genetics and Crop Plant Research (IPK) in Germany and the Centro de Recursos Fitogenéticos (CRF) from the Instituto Nacional de Investigación y Tecnología Agraria y Alimentaria (INIA) in Spain. Backups from 2.134 grass pea accessions, from 44 countries, are deposited at the Svalbard Global Seed Vault (http://www.nordgen.org/sgsv/, accessed June 2014). In what concerns in situ conservation, five genetic reserves for *Lathyrus* diversity conservation have been proposed in Syria and Turkey (Heywood et al. 2007). These authors also stressed the importance of increasing public awareness for the significance of crop wild relatives in agricultural development and the need for their simultaneous conservation.

This conserved germplasm represents a valuable reservoir of diversity, providing access to sources of a wide range of interesting agromorphological traits such as earliness, plant architectural traits, disease and pest tolerance, as well as low ODAP content. Characterization of this diversity through phenotyping and genotyping studies will unveil novel alleles that can be used to improve this crop. Diversity characterization in *Lathyrus* germplasm has focused, for example, on ODAP content (Fikre et al. 2008; Grela et al. 2012; Kumar et al. 2011), phenology and yield (Grela et al 2012; Mera 2010), parasitic weed resistance (Fernández-Aparicio et al. 2012), disease resistance (Gurung et al. 2002; Vaz Patto et al. 2006a; Vaz Patto and Rubiales 2009) or quality traits (Granati et al. 2003). Some of these characterization studies have represented the first step of selection programmes.

5 Major Breeding Achievements

Conventional grass pea-breeding programmes have been established in several countries, including Australia (Hanbury et al. 1995), Bangladesh (Malek 1998), Canada (Campbell and Briggs 1987), China (Yang and Zhang 2005), Chile (Mera et al. 2003), Ethiopia (Tadesse and Bekele 2003), India (Lal et al. 1986; Pandey et al. 1996), Nepal (Yadav 1996) and Syria (Abd El-Moneim et al. 2000). Some of these breeding programmes are still active, but most are small in comparison to other legume crops (Vaz Patto et al. 2011).

Due to the occurrence of lathyrism in humans, major breeding programmes are essentially aimed for low ODAP content, besides productivity and adaptability. This has resulted at present in several *L. sativus* or *L. cicera* breeding lines or released varieties with reduced ODAP content (from 0.5 to 1.5 %, down to 0.01 % or less; Kumar et al. 2011). For instance, low ODAP cultivars have been released in several countries, such as 'Wasie' in Ethiopia, 'Ali-Bar' in Kazakhstan and 'Gurbuz 1' in Turkey (ICARDA 2006, 2007). Similarly, low ODAP, high-yielding cultivars have been released in India such as 'Pusa 24', 'Prateek', 'Ratan' and 'Mahateora' (ICAR 2009). In Bangladesh, examples of the low ODAP and high-yielding varieties are 'BARI Khesari 1', 'BARI Khesari 2' and 'BARI Khesari 3' (Malek 1998) or the 'BINA Khesari 1' (Kumar et al. 2011). In Canada, high yield and low ODAP (0.03 %) 'LS8246' was released for feed and fodder (Campbell and Briggs 1987), and in addition, a high N-fixation variety, 'AC Greenfix' was released specially as green manure (Krause and Krause 2003). In Chile, 'Luanco-INIA', a large-seeded, high-yielding grass pea variety was released, used locally as feed and for export, especially for some European markets where larger seed size is desirable for human consumption (Mera et al. 2003). Finally, in Australia, the variety 'Ceora' was bred to be used as forage, hay or as a green manure crop (Siddique et al. 2006). Also in Australia, a *L. cicera* cultivar, 'Chalus', was selected for high yields and low ODAP levels (Hanbury and Siddique 2000).

6 Specific Goals in Current Breeding

Low ODAP content is still one important goal of many of the current grass pea-breeding programmes. Nevertheless, other traits have always been associated with this.

Increased yield has been a selection criterion for most crop improvement programmes. However, some of the yield components that affect yield such as double podding or increased seeds per pod have received insufficient attention. Also the biomass yield of *L. sativus* has started to receive more attention during the past few years (Campbell 1997; Abd El Moneim et al. 2001; Vaz Patto et al. 2006b). This is a very important area due to the large potential of this crop for forage and straw in the North African and South Asian regions (Campbell 1997). Additionally, undesirable traits such as prostrate plant habit, indeterminate growth, late maturity and pod shattering (Rybinski 2003) are being handled by several breeding programmes.

The concentrated effort on reducing ODAP content resulted in many other areas of evaluation and crop improvement, such as resistance to biotic and abiotic stresses, being neglected. However, with the release of these low ODAP lines, the development of varieties with increased resistance to prevalent pests and diseases has gained new strength. This crop is usually grown by poor farmers and under poor management, where it is difficult to adopt chemical control for diseases and pests. Therefore, the development of varieties having resistance to prevalent biotic stresses is essential, and more efforts are required in this area of improvement of this very hardy pulse crop (Vaz Patto et al. 2006b).

7 Breeding Methods and Specific Techniques

Collection and evaluation of germplasm, local or introduced, is the cornerstone in any breeding programme. Subsequent hybridization and selection of the resulting progeny using different strategies will allow incorporating interesting traits into a more adapted background. This may include backcrossing, recurrent selection, single-seed descent and pedigree/bulk breeding methods. All of these methods can be applied on grass pea improvement.

Grass pea is predominantly a self-pollinated crop, although outcrossing up to 30 % has been reported (Ben Brahim et al. 2001; Chowdhury and Slinkard 1997; Rahman et al. 1995). Large size of flower, bright colour of petals, flower density and nectar production are reported to influence the outcrossing in *Lathyrus* species (Kiyoshi et al. 1985). Entomophilic pollination in grass pea is due especially to bees and bumblebees (Kumar et al. 2011). Due to this observed outcrossing level, in most grass pea-breeding programmes, crosses are done under controlled conditions, in greenhouse or under insect-proof coverings (Vaz Patto et al. 2011).

Conventional grass pea breeding focussed essentially on hybridization of selected accessions, with the screening and evaluation of the resulting progeny. In the particular case of breeding to reduce ODAP content, low ODAP accessions are crossed with high-yield material with good agronomic potential (Campbell 1997).

Intergeneric hybridization, although difficult, is possible with *L. amphicarpos* and *L. cicera* (Yunus and Jackson 1991). Crosses have also been made with other species such as *L. chrysanthus, L. gorgoni, L. marmoratus* and *L. pseudocicera* (Heywood et al. 2007), but only ovules were produced.

Also with the objective of reducing ODAP content, grass pea has been subjected to induced mutagenesis by physical and/or chemical mutagens. Other traits have been affected by mutagenesis such as plant habit, maturity, branching, stem shape, leaf size, stipule shape, flower colour and structure, pod size, seed size and colour and NaCl tolerance (Biswas 2007; Nerkar 1972, 1976; Rybinski 2003; Talukdar 2009a, b, 2011). In vitro culture was also employed, inducing somaclonal variation (Ochatt et al. 2002a; Roy et al. 1993; Zambre et al. 2002). Induced mutagenesis and somaclonal variation created new diversity, allowing the selection of lines with interesting traits.

Ochatt et al. (2002b) developed an in vitro system coupled with in vitro stages in order to shorten regeneration cycles, obtaining up to almost four cycles per year. However, this approach is only applicable when few seeds/plant are intended, as in single-seed descendant breeding schemes.

The advent of various molecular-marker techniques and the ability to transfer genes across different organisms, using transgene technology, have begun to have an impact on plant genome research and breeding. These techniques offer new approaches for improving important agronomic traits in *Lathyrus* species and breaking down transfer barriers to related legume species (Vaz Patto et al. 2006b). This would allow exploring the variability existing in other *Lathyrus* genepools and hopefully transfer the interesting grass pea traits to related legume species.

Genetic transformation of grass pea was attempted with only one successful report obtaining stable transformed plants (Barik et al. 2005). Given that regeneration protocols for grass pea are often genotype specific, it may be necessary either to develop more generally applicable protocols or to adapt the protocol after transformation (Ochatt et al. 2013).

8 Integration of New Biotechnologies in Breeding Programmes

In order to be able to perform marker-assisted selection (MAS), it is necessary to identify molecular markers that are closely linked to the trait of interest. Once a trait is associated with a marker (or more), plants can be selected early on its growth stage, allowing a faster and more efficient breeding process.

Until now, only two linkage maps using molecular markers were developed for *L. sativus*. One developed by Chowdhury and Slinkard (1999) used 11 random amplified polymorphic DNA (RAPD) markers, 1 isozyme marker and 1 morphological trait (flower colour). The other linkage map was constructed by Skiba et al. (2004), using 47 RAPDs, 7 cross-amplified pea microsatellite simple sequence repeats (SSR) markers and 13 cleaved amplified polymorphic sequence (CAPS) markers and was used to study the genetic basis of resistance to Ascochyta blight. Nevertheless, these maps were not informative enough to allow bridging that mapping information between them, as reviewed by Vaz Patto et al. (2006b).

Compared to other grain legumes such as pea, faba bean or chickpea, genomic recourses for grass pea are still scarce. In July 2014, the National Center for Biotechnology Information (NCBI) database had made available the information of 178 EST sequences from a cDNA library of one *L. sativus* accession inoculated with *Mycosphaerella pinodes* (Skiba et al. 2005), 89 nucleotide sequences mainly from the Bowman–Birk (BBI) inhibitor coding sequences (41 accessions), chloroplast sequences (21 accessions) and 216 protein sequences (44 amino acid sequences from BBI inhibitors, 150 sequences from chloroplast proteins).

Specific molecular markers have been developed or adapted for grass pea in order to assist diversity studies and further develop linkage maps. Almeida et al. (2014a) studied the transferability of molecular markers from *M. truncatula, P. sativum, L. culinaris, Lupinus* spp. and *V. faba* to *Lathyrus* spp. and their application in mapping and diversity studies. Cross-genera amplification of molecular markers provided an alternative for the development of new molecular markers on understudied genus, allowing also performing comparative mapping between the sequence donor and the target species. This survey for similar genetic regions among closely related species will contribute to the potential future exchange of interesting traits among them.

Earlier molecular markers, specific or cross amplification studies in grass pea, included the work of Shiferaw et al. (2011) that successfully amplified nine expressed sequence tag-simple sequence repeats (EST-SSRs) developed from the EST sequences of Skiba et al. (2005) and 12 EST-SSRs from *M. truncatula,* which have been previously proven to be transferable to other legume species by Gutierrez et al. (2005).

Lioi et al. (2011) were able to genotype in a grass pea diversity study, ten SSRs developed from nucleotide sequences stored at public databases, being nine from *L. sativus* sequences and one from a *L. japonicus* sequence.

More recently, Yang et al. (2014) employed next-generation sequencing (NGS) to develop 144 specific grass pea SSRs, from which, 74 where polymorphic and therefore useful for diversity studies and genetic mapping.

The first grass pea expression analysis was performed by Skiba et al. (2005), identifying 29 potential defence-related genes differentially expressed in response to *M. pinodes* inoculation. These included genes associated with pathogen recognition, the phenylpropanoid pathway, hypersensitivity, pathogenesis-related and disease resistance response proteins.

In addition, expression analysis using RNA-sequencing was also employed in grass pea to tackle the molecular mechanisms underlying prehaustorial rust resistance (Almeida et al. 2014b). These authors identified several pathogenesis-related proteins as possibly involved in grass pea resistance to rust, that included some regulated by the well-studied mildew resistance locus O (MLO) gene. In this study, several potential rust effectors were also identified. These could be used as probes to identify target grass pea host proteins, as a first step in the development of effector-driven legume breeding, maximizing the durability of resistance against this quickly evolving pathogen (Vleeshouwers et al. 2011). Finally this RNA-sequencing study also identified several polymorphic single-nucleotide polymorphism (SNPs) and EST-SSRs between parental lines of existing grass pea segregating recombinant inbred lines (RILs), allowing its use for linkage mapping.

Expression analysis of the response to infection with *Ascochyta lathyri* in a resistant grass pea accession was performed using deepSuperSAGE. This approach has identified several differentially expressed genes (Almeida et al. 2015), opening the way to a powerful route of identification of candidate resistance genes and more detailed study of resistance gene networks in *L. sativus* (Vaz Patto and Rubiales 2014).

9 Future Prospects

The present paradigm change towards the study of crop species instead of focusing on model species will aid in the development of plant species that have been neglected. Lowering costs in high-throughput sequencing and the development of high-throughput phenotyping have encouraged the development of new molecular tools to boost the genetic characterization and utilization of the rich *Lathyrus* germplasm.

Grass pea research was tied to the persecution of an ODAP-free variety for several decades. This has hampered progress in this crop for the improvement of other traits. As an example and despite its importance, ODAP-related research should not block the understanding of the reasons behind the success of grass pea when dealing with biotic and abiotic stresses, and for which it is considered a survival crop. In an alternative to low ODAP varieties, an option might be improving quality traits that can lower ODAP's negative effects. These include increasing the content in homoarginine, cysteine or methionine. Although this is an old objective, it is still unachieved, due to the presence of technical barriers in the regeneration of transformed tissues (Girma and Korbu 2012) or the high influence of genotype × environment in those traits (Piergiovanni et al. 2011; Piergiovanni and Damascelli 2011).

For quicker progress on these and other quantitative traits improvement via MAS, it would also be useful to have a saturated linkage map, including cross-transferable markers to other related species such as pea, faba bean or the model *M. truncatula*, to apply in quantitative trait loci (QTL) mapping. In this way, comparative mapping would also be possible to other closely related legume species, assisting knowledge transfer among these species and facilitating candidate gene discovery for the detected potential QTL regions.

With the development of high-throughput and dense genotyping, assessment of the correlation between phenotype and genotype, needed for the development of MAS approaches, has shifted from focusing on two parental lines, differing strongly in phenotype, to the analysis of populations of unrelated individuals. Association mapping panels by sampling more genetic diversity can take advantage of many more generations of recombination and avoid the time-consuming generations of crosses (Morrell et al. 2012). High-throughput genotyping associated with a core collection evaluation will facilitate trait dissection and gene discovery through association mapping as well as characterization of the collection genetic structure (Cobb et al. 2013). That is why Vaz Patto and Rubiales (2014) supported the idea of concentrating international evaluation efforts on to a grass pea core collection, representative of all the existing diversity, but of a manageable size. For adaptive traits, core and mini-core collections may not capture the needed diversity (Gepts 2006). As an alternative, the Focused Identification of Germplasm Strategy (FIGS) approach, which is a trait-based approach with high probability of identification of desired genetic material (Khazaei et al. 2013) is being applied at ICARDA to the *Lathyrus* germplasm collection to develop subset collections.

In terms of grass pea plant resources for functional genomics studies, various mapping populations including RILs, near isogenic lines (NILs) and targeting induced local lesions in genomes (TILLING) populations are critically needed for trait–marker association and gene inactivation/deletion studies (Kumar et al. 2013).

10 Seed Production

Several grass pea improved varieties have been originated from various breeding programmes as already described in the section "Major Breeding Achievements" section. As in any plant species with outcrossing frequency rate up to 30 %, special efforts are needed for cultivar conservation. Strategies like an isolation distance or the use of a buffer crop between cultivars when producing seeds are essential to maintain genetic purity and phenological features of the developed cultivars.

Nevertheless, the most common seed available is from landraces or farmers' varieties, inherently heterogeneous. These farm-saved seeds are obtained and traded within an informal seed system where seeds are exchanged among farmers that mainly do not sell the product of the seed, but use it for self-consumption.

In conclusion, presently available genetic resources, established breeding achievements and recent biotechnological progress, associated with a growing international interest on grass pea cultivation, will definitely provide a bright future to this highly potential crop.

References

Abd El-Moneim AM, van Dorrestein B, Baum M et al (2000) Improving the nutritional quality and yield potential of grasspea (*Lathyrus sativus* L.). Food Nut Bull 21:493–496

Abd El Moneim A, van Dorrestein B, Baum M et al (2001) Role of ICARDA in improving the nutritional quality and yield potential of grasspea (*Lathyrus sativus* L.) for subsistence farmers in dry areas. Lathyrus Lathyrism Newsl 2:55–58

Allkin R, Goyder DJ, Bisby FA et al (1986) Names and synonyms of species and subspecies in the Vicieae. Vicieae Database Project Issue 3. Southampton University, Southampton

Almeida NF, Leitão ST, Caminero C et al (2014a) Transferability of molecular markers from major legumes to *Lathyrus* spp. for their application in mapping and diversity studies. Mol Biol Rep 41:269–283

Almeida NF, Krezdorn N, Rotter B et al (2015) *Lathyrus sativus* transcriptome resistance response to *Ascochyta lathyri* investigated by deepSuperSAGE analysis. Front Plant Sci 6:178

Almeida NF, Leitão ST, Krezdorn N et al (2014b) Allelic diversity in the transcriptomes of contrasting rust-infected genotypes of *Lathyrus sativus*, a lasting resource for smart breeding. BMC Plant Biol (14:376)

Asmussen CB, Liston A (1998) Chloroplast DNA characters, phylogeny, and classification of *Lathyrus* (Fabaceae). Am J Bot 85:387–401

Barik DP, Mohapatra U, Chand PK (2005) Transgenic grasspea (*Lathyrus sativus* L.): factors influencing agrobacterium-mediated transformation and regeneration. Plant Cell Rep 24:523–531

Ben Brahim N, Combes D, Marrakchi M (2001) Autogamy and allogamy in genus *Lathyrus*. Lathyrus Lathyrism Newsl 2:21–26

Biswas AK (2007) Induced mutation in grass pea. In: Ochatt SJ, Jain SM (eds) Breeding of neglected and under-utilised crops, spices and herbs. Science, Plymouth, pp 29–39

Campbell CG (1997) Promoting the conservation and use of underutilized and neglected crops. Grass pea. *Lathyrus sativus* L., vol 18. International Plant Genetic Resources Institute, Rome

Campbell CG, Briggs CJ (1987) Registration of low neurotoxin content *Lathyrus* germplasm LS 8246. Crop Sci 27:821

Campbell CG, Mehra RB, Agrawal SK et al (1994) Current status and future strategy in breeding grasspea (*Lathyrus sativus*). Euphytica 73:167–175

Choi HK, Mun JH, Kim DJ et al (2004) Estimating genome conservation between crop and model legume species. Proc Natl Acad Sci U S A 101(43):15289–15294

Chowdhury MA, Slinkard AE (1997) Natural outcrossing in grasspea. J Hered 88:154–156

Chowdhury MA, Slinkard AE (1999) Linkage of random amplified polymorphic DNA, isozyme and morphological markers in grasspea (*Lathyrus sativus*). J Agr Sci 133:389–395

Chowdhury MA, Slinkard AE (2000) Genetic diversity in grasspea (*Lathyrus sativus* L.). Genet Resour Crop Evol 47:163–169

Cobb JN, DeClerck G, Greenberg A et al (2013) Next-generation phenotyping: requirements and strategies for enhancing our understanding of genotype-phenotype relationships and its relevance to crop improvement. Theor Appl Genet 126:867–887

Ellison NW, Liston A, Steiner JJ et al (2006) Molecular phylogenetics of the clover genus (*Trifolium*-Leguminosae). Mol Phylogenet Evol 39:688–705

Erskine W, Smartt J, Muehlbauer F (1994) Mimicry of lentil and the domestication of common vetch and grass pea. Econ Bot 48:326–332

Erskine W, Siddique K, Khan T et al (2001) Utilisation of grain legume diversity. In: Maxted N, Bennett S (eds) Plant genetic resources of legumes in the mediterranean. Current plant science and biotechnology in agriculture, vol 39. Springer, Dordrecht, pp 311–326

FAO (2009) International treaty on plant genetic resources for food and agriculture. Food and Agriculture Organization of the United Nations (FAO), Rome

Fernández-Aparicio M, Flores F, Rubiales D (2012) Escape and true resistance to crenate broomrape (*Orobanche crenata* Forsk.) in grass pea (*Lathyrus sativus* L.) germplasm. Field Crops Res 125:92–97

Fikre A, Korbu L, Kuo YH et al (2008) The contents of the neuro-excitatory amino acid beta-ODAP (beta-N-oxalyl-L-alpha, beta-diaminopropionic acid), and other free and protein amino acids in the seeds of different genotypes of grass pea (*Lathyrus sativus* L.). Food Chem 110:422–427

GCDT (2009) Strategy for the ex situ conservation of *Lathyrus*, with special reference to *Lathyrus sativus*, *L. cicera*, and *L. ochrus*. Global Crop Diversity Trust (GCDT), Rome

Gepts P (2006) Plant genetic resources conservation and utilization: the accomplishments and future of a societal insurance policy. Crop Sci 46:2278–2292

Girma D, Korbu L (2012) Genetic improvement of grass pea (*Lathyrus sativus*) in Ethiopia: an unfulfilled promise. Plant Breed 131:231–236

Granati E, Bisignano V, Chiaretti D et al (2003) Characterization of Italian and exotic *Lathyrus* germplasm for quality traits. Genet Resour Crop Evol 50:273–280

Grela E, Rybiński W, Matras J et al (2012) Variability of phenotypic and morphological characteristics of some *Lathyrus sativus* L. and *Lathyrus cicera* L. accessions and nutritional traits of their seeds. Genet Resour Crop Evol 59:1687–1703

Gurung A, Pang EK (2011) *Lathyrus*. In: Kole C (ed) Wild crop relatives: genomic and breeding resources. Springer, Berlin, pp 117–126

Gurung AM, Pang ECK, Taylor PWJ (2002) Examination of *Pisum* and *Lathyrus* species as sources of ascochyta blight resistance for field pea (*Pisum sativum*). Australas Plant Pathol 31:41–45

Gutierrez MV, Vaz Patto MC, Huguet T et al (2005) Cross-species amplification of *Medicago truncatula* microsatellites across three major pulse crops. Theor Appl Genet 110:1210–1217

Hanbury CL, Siddique KHM (2000) Registration of 'Chalus' *Lathyrus cicera* L. Crop Sci 40:1199

Hanbury CD, Sarker A, Siddique KHM et al (1995) Evaluation of *Lathyrus* germplasm in a Mediterranean type environment in south-western Australia. CLIMA Occasional Publication (Cooperative Research Centre for Legumes in Mediterranean Agriculture) no 8, Perth, p 23

Hanbury CD, Siddique KHM, Galwey NW et al (1999) Genotype-environment interaction for seed yield and ODAP concentration of *Lathyrus sativus* L. and *L. cicera* L. in Mediterranean-type environments. Euphytica 110:45–60

Hanbury CD, White CL, Mullan BP et al (2000) A review of the potential of *Lathyrus sativus* L. and *L. cicera* L. grain for use as animal feed. Anim Feed Sci Technol 87:1–27

Heywood V, Casas A, Ford-Lloyd B et al (2007) Conservation and sustainable use of crop wild relatives. Agr Ecosyst Environ 121:245–255

Hillocks R, Maruthi M (2012) Grass pea (*Lathyrus sativus*): is there a case for further crop improvement? Euphytica 186:647–654

Hopf M (1986) Archaeological evidence of the spread and use of some members of the Leguminosae family. In: Barigozzi C (ed) The Origin and domestication of cultivated plants: symposium. Elsevier, Oxford, New York and Tokyo, UK, USA and Japan pp 35–60

ICAR (2009) Project Coordinator's report of all India coordinated research project on mungbean, urdbean, lentil, lathyrus, rajmash and pea. Indian Council of Agricultural Research (ICAR), New Delhi

ICARDA (2006) ICARDA Annual report 2005. International Center for Agricultural Research in the Dry Areas. Aleppo, Syria, pp 54–55

ICARDA (2007) ICARDA Annual report 2006. International Center for Agricultural Research in the Dry Areas. Aleppo, Syria, pp 57–58

Jackson MT, Yunus AG (1984) Variation in the grass pea (*Lathyrus sativus* L.) and wild-species. Euphytica 33:549–559

Jørgensen IH (1992) Discovery, characterization and exploitation of Mlo powdery mildew resistance in barley. Euphytica 63:141–152

Kenicer GJ, Kajita T, Pennington RT et al (2005) Systematics and biogeography of *Lathyrus* (Leguminosae) based on internal transcribed spacer and cpDNA sequence data. Am J Bot 92:1199–1209

Kenicer G, Nieto-Blásquez EM, Mikić A et al (2009) *Lathyrus* – diversity and phylogeny in the genus. Grain Legumes 54:16–18

Khazaei H, Street K, Bari A et al (2013) The FIGS (Focused Identification of Germplasm Strategy) approach identifies traits related to drought adaptation in *Vicia faba* genetic resources. PLoS ONE 8:e63107

Kislev ME (1989) Origins of the Cultivation of *Lathyrus sativus* and *Lathyrus cicera* (Fabaceae). Econ Bot 43:262–270

Kiyoshi Y, Toshiyuki F, Blumenreich ID (1985) Isozymic variation and interspecific crossability in annual species of genus *Lathyrus* L. In: Kaul AK, Combes D (eds) *Lathyrus* and Lathyrism. Pau, France, pp 118–129

Krause D, Krause I (2003) New green manuring *Lathyrus sativus* variety AC Greenfix available in USA. Lathyrus Lathyrism Newsl 3:13–14

Kumar S, Bejiga G, Ahmed S et al (2011) Genetic improvement of grass pea for low neurotoxin (beta-ODAP) content. Food Chem Toxicol 49:589–600

Kumar S, Gupta P, Barpete S et al (2013) Grass pea. In: Singh M, Upadhyaya HD, Bisht IS (eds) Genetic and genomic resources of grain legume improvement. Elsevier Science, Oxford, pp 269–292

Kuo YH, Bau HM, Rozan P et al (2000) Reduction efficiency of the neurotoxin beta-ODAP in low-toxin varieties of *Lathyrus sativus* seeds by solid state fermentation with *Aspergillus oryzae* and *Rhizopus microsporus* var *chinensis*. J Sci Food Agr 80:2209–2215

Kupicha FK (1983) The infrageneric structure of *Lathyrus*. Notes Roy Bot Gard Edinburgh 41:209–244

Lal MS, Agrawal I, Chitale MW (1986) Genetic improvement of chickling vetch (*Lathyrus sativus* L.) in Madhya Pradesh, India. In: Kaul AK, Combes D (eds) *Lathyrus* and Lathyrism. Third World Medical Research Foundation, New York, pp 146–160

Lambein F, Ngudi DD, Kuo YH (2009) Progress in prevention of toxico-nutritional neurodegenerations. http://wwwatdforumorg/journal/html/2009-34/8. Accessed July 2014

Lioi L, Sparvoli F, Sonnante G et al (2011) Characterization of Italian grasspea (*Lathyrus sativus* L.) germplasm using agronomic traits, biochemical and molecular markers. Genet Resour Crop Evol 58:425–437

Malek MA (1998) Genetic resources of grass pea (*Lathyrus sativus* L.) in Bangladesh. In: Mathur PN, Rao VR (eds) Lathyrus Genetic Resources Network: Proc IPGRI-ICARDA-ICAR Regional Working Group Meeting. National Bureau of Plant Genetic Resources, IPGRI Office for South Asia, New Delhi, India, pp 1–6

Mera M (2010) Inheritance of seed weight in large-seed grass pea *Lathyrus sativus* L. Chil J Agr Res 70:357–364

Mera M, Tay J, France A et al (2003) Luanco-INIA, a large-seeded cultivar of *Lathyrus sativus* released in Chile. Lathyrus Lathyrism Newsl 3:26

Morrell PL, Buckler ES, Ross-Ibarra J (2012) Crop genomics: advances and applications. Nat Rev Genet 13:85–96

Nerkar YS (1972) Induced variation and response to selection for low neurotoxin content in *Lathyrus sativus*. Ind J Genet Plant Breed 32:175–180

Nerkar YS (1976) Mutation studies in *Lathyrus sativus*. Ind J Genet Plant Breed 36:223–229

Ochatt SJ, Muneaux E, Machado C et al (2002a) The hyperhydricity of in vitro regenerants of grass pea (*Lathyrus sativus* L.) is linked with an abnormal DNA content. J Plant Physiol 159:1021–1028

Ochatt SJ, Sangwan RS, Marget P et al (2002b) New approaches towards the shortening of generation cycles for faster breeding of protein legumes. Plant Breed 121:436–440

Ochatt S, Pech C, Grewal R et al (2009) Abiotic stress enhances androgenesis from isolated microspores of some legume species (Fabaceae). J Plant Physiol 166:1314–1328

Ochatt SJ, Conreux C, Jacas L (2013) Flow cytometry distinction between species and between landraces within *Lathyrus* species and assessment of true-to-typeness of in vitro regenerants. Plant Syst Evol 299:75–85

Pandey RL, Chitale MW, Sharma RN et al (1996) Status of *Lathyrus* research in India. In: Arora RK, Mathur PN, Riley KW et al (eds) *Lathyrus* Genetic Resources in Asia: Proc Regional Workshop. Indira Gandhi Agricultural University, Raipur, India. IPGRI Office for South Asia, New Delhi, India, pp 45–52

Peña-Chocarro L, Peña LZ (1999) History and traditional cultivation of *Lathyrus sativus* L. and *Lathyrus cicera* L. in the Iberian peninsula. Veget Hist Archaeobot 8:49–52

Piergiovanni AR, Damascelli A (2011) L-Homoarginine accumulation in grass pea (*Lathyrus sativus* L.) dry seeds. A preliminary survey. Food Nut Sci 2:207–213

Piergiovanni AR, Lupo F, Zaccardelli M (2011) Environmental effect on yield, composition and technological seed traits of some Italian ecotypes of grass pea (*Lathyrus sativus* L.). J Sci Food Agric 91:122–129

Rahman MM, Kumar J, Rahman MA et al (1995) Natural outcrossing in *Lathyrus sativus* L. Ind J Genet Plant Breed 55:204–207

Roy PK, All K, Gupta A et al (1993) β-N-Oxalyl-L-α, β-diaminopropionic acid in somaclones derived from internode explants of *Lathyrus sativus*. J Plant Biochem Biotechnol 2:9–13

Rybinski W (2003) Mutagenesis as a tool for improvement of traits in grass pea (*Lathyrus sativus* L.). Lathyrus Lathyrism Newsl 3:30–34

Schaefer H, Hechenleitner P, Santos-Guerra A et al (2012) Systematics, biogeography, and character evolution of the legume tribe Fabeae with special focus on the middle-Atlantic island lineages. BMC Evol Biol 12:250

Shiferaw E, Pè M, Porceddu E et al (2011) Exploring the genetic diversity of Ethiopian grass pea (*Lathyrus sativus* L.) using EST-SSR markers. Mol Breed 30:789–797

Siddique KHM, Hanbury CL, Sarker A (2006) Registration of 'Ceora' grass pea registration by CSSA. Crop Sci 46:986–986

Skiba B, Ford R, Pang ECK (2004) Construction of a linkage map based on a *Lathyrus sativus* backcross population and preliminary investigation of QTLs associated with resistance to ascochyta blight. Theor Appl Genet 109:1726–1735

Skiba B, Ford R, Pang ECK (2005) Construction of a cDNA library of *Lathyrus sativus* inoculated with *Mycosphaerella pinodes* and the expression of potential defence-related expressed sequence tags (ESTs). Physiol Mol Plant Pathol 66:55–67

Smartt J (1984) Evolution of grain legumes.1. Mediterranean pulses. Exp Agric 20:275–296

Smýkal P, Kenicer G, Flavell AJ et al (2011) Phylogeny, phylogeography and genetic diversity of the *Pisum* genus. Plant Genet Resour 9:4–18

Tadesse W, Bekele E (2003) Phenotypic diversity of Ethiopian grass pea (*Lathyrus sativus* L.) in relation to geographical regions and altitudinal range. Genet Resour Crop Evol 50:497–505

Talukdar D (2009a) Dwarf mutations in grass pea (*Lathyrus sativus* L.): origin, morphology, inheritance and linkage studies. J Genet 88:165–175

Talukdar D (2009b) Recent progress on genetic analysis of novel mutants and aneuploid research in grass pea (*Lathyrus sativus* L.). Afr J Agr Res 4:1549–1559

Talukdar D (2011) Isolation and characterization of NaCl-tolerant mutations in two important legumes, *Clitoria ternatea* L. and *Lathyrus sativus* L.: induced mutagenesis and selection by salt stress. J Med Plant Res 5:3619–3628

Vaz Patto MC, Rubiales D (2009) Identification and characterization of partial resistance to rust in a germplasm collection of *Lathyrus sativus* L. Plant Breed 128:495–500

Vaz Patto MC, Rubiales D (2014) *Lathyrus* diversity: available resources with relevance to crop improvement – *L. sativus* and *L. cicera* as case studies. Ann Bot 113:895–908

Vaz Patto MC, Fernández-Aparicio M, Moral A et al (2006a) Characterization of resistance to powdery mildew (*Erysiphe pisi*) in a germplasm collection of *Lathyrus sativus*. Plant Breed 125:308–310

Vaz Patto MC, Skiba B, Pang ECK et al (2006b) *Lathyrus* improvement for resistance against biotic and abiotic stresses: from classical breeding to marker assisted selection. Euphytica 147:133–147

Vaz Patto MC, Hanbury CD, Moorhem MV et al (2011) Grass pea. In: Kole C (ed) Genetics, genomics and breeding of cool season grain legumes. Genetics, genomics and breeding of crop plants. Science, Enfield, pp 151–204

Vleeshouwers VG, Raffaele S, Vossen JH et al (2011) Understanding and exploiting late blight resistance in the age of effectors. Annu Rev Phytopathol 49:507–531

Wojciechowski MF, Lavin M, Sanderson MJ (2004) A phylogeny of legumes (Leguminosae) based on analysis of the plastid matK gene resolves many well-supported subclades within the family. Am J Bot 91:1846–1862

Yadav CR (1996) Genetic evaluation and varietal improvement of grasspea in Nepal. In: Arora RK, Mathur PN, Riley KW et al (eds) *Lathyrus* genetic resources in Asia: Proc Regional Workshop. Indira Gandhi Agricultural University, Raipur, India. IPGRI Office for South Asia, New Delhi, India, pp 21–28

Yang H-M, Zhang X-Y (2005) Considerations on the reintroduction of grass pea in China. Lathyrus Lathyrism Newsl 4:22–26
Yang T, Jiang J, Burlyaeva M et al (2014) Large-scale microsatellite development in grasspea (*Lathyrus sativus* L.), an orphan legume of the arid areas. BMC Plant Biol 14:65
Yunus AG, Jackson MT (1991) The gene pools of the grasspea (*Lathyrus sativus* L.). Plant Breed 106:319–328
Zambre M, Chowdhury B, Kuo YH et al (2002) Prolific regeneration of fertile plants from green nodular callus induced from meristematic tissues in *Lathyrus sativus* L. (grass pea). Plant Sci 163:1107–1112

Chapter 9
The Legume–Rhizobia Symbiosis

Jean-Jacques Drevon, Nora Alkama, Adnane Bargaz, A. Paula Rodiño, Kiriya Sungthongwises and Mainassara Zaman-Allah

1 Introduction

Rhizobial symbiosis is the term for symbiosis between nitrogen-fixing soil bacteria commonly called rhizobia, and roots of legumes like soya beans, broad beans and stems in some species like *Sesbania* sp. However, nonleguminous plant can also form symbiosis with nitrogen-fixing bacteria, such as actinomycetes, of the genus *Frankia* (Tate 1995). Hellriegel and Wilfarth (1888) isolated the first rhizobia able to induce the nodulation in a controlled experimental environment, thus establishing the basis of the symbiotic relationship between legumes and rhizobial species.

J.-J. Drevon (✉)
Eco&Sols, INRA, 1 Place Viala, 34000 Montpellier, France
e-mail: drevonjj@supagro.INRA.FR

N. Alkama
Department des Sciences Agronomiques, Universite de Mouloud Mammeri, Tizi Ouzou, Algeria
e-mail: alkamanora@yahoo.fr

A. Bargaz
Biosystems and Technology, Swedish University of Agricultural Sciences, Sundsvägen 16, 230 53 Alnarp, Sweden
e-mail: bargazadnane@yahoo.fr

A. P. Rodiño
Biology of Agrosystems, Misión Biológica de Galicia (MBG), Spanish National Research Council (CSIC), El Palacio-Salcedo, 36143 Pontevedra, Spain
e-mail: aprodino@mbg.csic.es

K. Sungthongwises
Faculty of Agriculture, Khon Kaen University, Khon Kaen, Thailand
e-mail: skiriy@kku.ac.th

M. Zaman-Allah
CIMMYT, Southern Africa Regional Office, Harare, Zimbabwe
e-mail: mainassara.za@gmail.com

A. M. De Ron (ed.), *Grain Legumes*, Handbook of Plant Breeding 10,
DOI 10.1007/978-1-4939-2797-5_9

This symbiosis is readily observable since the nodules are numerous and usually 1–3 mm in size. During symbiotic life, rhizobia differentiate into bacteroids encapsulated within the peribacteroid membrane (PBM) inside the infected cells constituting an organite called the symbiosome (Young and Haukka 1996; Sy et al. 2001). Bacteroids are able to alter metabolic and molecular processes in the cells and cause cellular enlargement.

The nodules formed by most of these bacteria are benign and mutually useful structures, conducting the N_2 fixation for the host plants. Generally, symbiosis is characterized by a more or less distinct host preference or even host specificity. Poor nodulation may occur even if good seed inoculation practices were used (Zahran 1999). The large differences of the combination between rhizobial strains and legume genotypes may offer the potential to increase N_2 fixation by both rhizobial and legume selection (Peoples and Craswell 1992). In addition, genetic diversity of legumes and rhizobia strongly affect SNF as a function of the environment (Poehlman 1991). The rhizobia–legume symbiosis is the primary source of biologically fixed nitrogen for agricultural system (Vance 1997; Kinzig and Socolow 1994). Some 40–60 million metric tons (Mt) of N_2 are fixed by agriculturally important legumes annually, with another 3–5 million Mt fixed by legumes in natural ecosystems (Smil 1999). The following revision of the literature focuses on symbiotic nitrogen fixation (SNF) of legumes.

2 Nodule Formation

Many legumes can nodulate with a diversity of rhizobia strains. Thus, most legume species have a specific rhizobia strain that maximizes N_2 fixation. However, some rhizobia strains are effective and some are ineffective for SNF. Ineffective strains will form many small nodules on the legume root but fix little or no nitrogen. Effective rhizobia strains that fix high rates of nitrogen form fewer but larger nodules that have dark pink centres. In soil, rhizobia are mobile and rely on chemical signals like phenolic compounds secreted by the roots, for example, luteolin by *Medicago truncatula,* within the rhizosphere of legumes (Peters et al. 1986).

The interaction between a particular strain of rhizobia and the "appropriate" legume is mediated by a Nod factor (NF), a lipochitin oligosaccharide secreted by the rhizobia and recognized by *transmembrane receptors* on the *root-hair cells* of the legume. NF are produced by rhizobia and act as the main morphogenic molecules regulating nodule organogenesis (Ferguson and Mathesius 2003). They can be recognized by receptors for chitin oligosaccharides (Stacey and Shibuya 1997). Molecules which are related to NF may have a more general role in plant development (Spaink et al. 1993; van der Holst et al. 2001). Benhamou and Asselin (1989) and Spaink et al. (1993) have identified within plants, molecules related in their structure to chitin oligosaccharides which are known to play a role in animal development. More particularly, chitin oligosaccharides are substrates for chitinases, which have been shown to play a role in different aspects of plant development (Collinge et al. 1993). Modifying chitin structures by expression of the bacterial nodA and nodB genes, which modify NF in rhizobia, led to changes in plant development (Schmidt et al. 1993). If the combination is correct, the rhizobia enter the root-hair cell or an epithelial cell if the infection proceeds by root-hair curling or lateral-root cracking, then multiply within the infection thread progressing extracellularly within the cortex (Crespi and Galvez 2000; Stougaard 2001; Kistner and Parniske 2002). This infection thread is constructed by the root cells and is formed only in response to the infection (Fernández-Pascual et al. 2007). When the infection thread reaches a cell in the developing cortex, the rhizobia are engulfed by *endocytosis* into endosomes. At this time, the cell goes through several rounds of mitosis without *cytokinesis* so the cell becomes *polyploid.* The cortex cells then begin to divide rapidly forming a nodule. This organogenesis of nodule starts earlier as a response to the NF and progresses during the progression of the infection thread (Fernández-Pascual et al. 2007). Each nodule becomes connected by the xylem and phloem to the vascular system of the plant.

Thus, the development of nodules, while dependent on rhizobia, is a well-coordinated developmental process of the plant. This response is driven by the translocation of *cytokinins* from epidermal cells to the cells of the cortex (Sturtevant and Taller 1989) with subsequent nodule formation in other regions of the root (Sutherland et al. 1990). Several pieces of evidence suggest that rhizobia induce changes in the cytokinin balance of the root, through cytokinin synthesis, turnover or sensitivity in the roots during nodule initiation (Ferguson and Mathesius 2003). The reactivation of the cell cycle was demonstrated to initiate nodule primordium formation (Foucher and Kondorosi 2000; Goormachtig et al. 1997; Yang et al. 1994). In pea, Newcomb et al. (1976) showed that nodule cytokinin levels were related to the maturity of the nodule. The application of cytokinins induced the formation of pseudo-nodule structures on legumes and nonlegumes, such as *Nicotiana tabacum* (tobacco; Arora et al. 1959), *Alnus glutinosa* (Rodriguez-Barrueco and Bermudez de Castro 1973), *Pisum sativum* (Libbenga and Harkes 1973), *Macroptilium atropurpureum* (siratro; Relic et al. 1994) and *Medicago sativa* (Cooper and Long 1994; Bauer et al. 1996). Moreover, since cytokinins can induce starch formation, they probably also play a role in setting up a carbohydrate sink for the developing nodule (Bauer et al. 1996).

The rhizobia may go through a period of rapid multiplication within the nodule. The term symbiosome refers to the organelle-like structure composed of the PBM, the peribacteroid space which it encloses and the bacteroids within that space. Symbiosomes are formed when rhizobia are released from the infection thread into the infected cells of the root cortex. This occurs by endocytosis of the infection thread membrane and results in encapsulation of the bacteria in a plant membrane. Only a few bacteria infect any one cortical cell, and after endocytosis, the rhizobia proliferate resulting in the cytoplasm of mature infected cells being packed with symbiosomes. The symbiosomes are intercalated with a honeycomb-like network of actin filaments, which may play a role in the organization of the symbiosomes and metabolite movement between them (Whitehead et al. 1998). In the nodules of temperate legumes such as pea, all symbiosomes contain only single bacteria. In tropical associations, such as soya bean, the symbiosomes become multi-bacteroid. Infected root cells may contain up to 20,000 bacteroids.

3 Nodule Function

Generally, effective root nodules are big and have a pink colour by contrast with ineffective root nodules that remain small and with a white colour (Poehlman 1991). N_2 fixation starts between 10 and 21 days after infection (Marschner 1995). Nodules occur at 9 days after sowing (DAS), and N_2 fixation activity occurs at 11 DAS in soya beans and 14–25 DAS in mung beans. The number of nodules will increase until flowering stage but the size and weight, leghaemoglobin and nitrogenase activity will be maximized at 30 DAS. Pawar and Ghulghule (1980) found that nodulation of cowpea occurs at 7 DAS, then the number of nodules and the weight will be maximized at 21 DAS and decrease after 28 DAS. Bushby (1991) studied nodulation pattern of the mung bean variety Satin and found that this variety produces 40 % of the nodule weight before flowering stage which will reach 60 % at flowering stage.

SNF with legumes host is catalyzed by the nitrogenase enzyme with the following reaction:

$$N_2 + 8e^- + 8\,H^+ + 16\,ATP \rightarrow 2NH_3 + H_2 + 16\,ADP + 16\,Pi$$

The overall activity of the whole nitrogenase complex decomposes in two steps: first step consists in ferredoxin transfer of electron to Fe in component II of nitrogenase (dinitrogenase reductase); in second step, component II will transfer electron to component I (dinitrogenase). Two molecules of ATP are used for transferring of each one electron. The steps repeat will load electrons in component II until it reaches a redox potential that will make it possible to reduce N_2 and H^+ into NH_3 and H_2 (Fig. 9.1). Electron source to reduce N_2 comes from leaf photosynthates. Three major factors are involved in the large potential of nodulated legumes for N_2 fixation, namely photosynthate supply, O_2 concentration, fixed-N export.

The first factor is the direct supply of photosynthates to the N_2-fixing nodules. About 30 % of carbohydrates from photosynthesis are transferred to roots in order to support symbiosis (Söderström and Read 1987; Söderström 1992; Jones and Darrah 1996; Farrar and Jones 2000).

During the vegetative growth stage of the legume, the nodules may consume as much as 20 % of the photosynthates in a legume like cowpea (Pate and Herridge 1978) and half of these photosynthates are respired as CO_2. Although, between up to 30 % of the respired CO_2 can be reassimilated by the nodules via phosphoenolpyruvate (PEP) carboxylase providing up to 25 % of the carbon needed for the synthesis of malate and aspartate (Fig. 9.2) for the assimilation of NH_3 and export to the host plant (Deroche and Carrayol 1988). Consequently, SNF will decrease or stop if nodulated-roots lack carbohydrates supply.

Fig. 9.1 Nodule conductance to O_2 diffusion

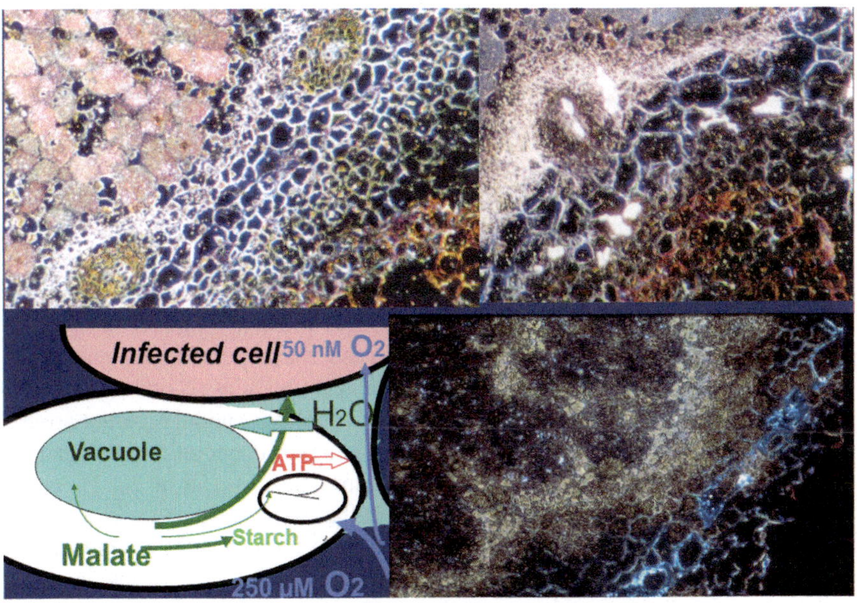

Fig. 9.2 Infection process in the cells of the nodule. *ATP* adenosine triphosphate

The second factor is effective maintenance of O_2 concentrations in the interior of the nodules for protection of the nitrogenase against inactivating O_2. The low O_2 concentrations within nodules are determined by two mechanisms: The existence of an O_2 diffusion barrier in the cortex of the nodules; the high respiration rates of the bacteroids in an O_2-limited environment. Indeed, a short-term decrease in the nodule conductance to oxygen diffusion appears to be linked with the inhibition of nitrogenase activity by water deficiency (Durand et al. 1987; Weisz et al. 1985) or by exposure to acetylene (Witty et al. 1984; Drevon and Hartwig 1997). It has been suggested that the variable component of the nodule conductance involves changes in the distribution of air spaces within the nodule cortex. These changes have been supposed to result from an occlusion of intercellular spaces (Sinclair and Goudriaan 1981; Sheehy et al. 1985; Davis et al. 1987; Witty et al. 1987; Hunt et al. 1988) and/or changes in the volume of some of its cells (Walsh et al. 1989; Parsons and Day 1990; Vance and Heichel 1991). Short-term decreases in *Lupinus albus* conductance, associated with lowering temperature, detopping or darkening plants, or root exposure to nitrate, were correlated with the occlusion of cortical intercellular spaces with a glycoprotein (Iannetta et al. 1993). In the inner cortex, both the "boundary layer" and the more internal "distributing zone" may be implicated in long- and/or short-term changes in the nodule conductance to oxygen diffusion (Parsons and Day 1990). We reported that the inhibition of soya bean nodule nitrogenase activity and/or respiration by a short-term exposure to supra optimal pO_2 or NaCl was linked with a rapid decrease in the nodule O_2 conductance caused by cell collapse in the inner cortex (Drevon et al. 1988; Serraj et al. 1994). This structural variation would involve a change in the size and water content of intercellular spaces thus modifying the nodule conductance.

In addition, the high O_2 consumption in nodules for provision of energy as ATP for nitrogenase creates a great potential for production of toxic oxygen species. Leghaemoglobin is probably the most important of these sources of reactive O_2 because this protein is extremely abundant in nodules and is subject to autoxidation in which O_2^-, H_2O_2 and OH^- are produced (Puppo et al. 1981; Puppo and Halliwell 1988). Oxidative damage in nodules is especially important during nodule senescence when free radical production and lipid peroxidation are enhanced by elevated levels of free iron (Becana and Klucas 1992). For protection against this toxicity, a defence mechanism by the enzymes of the ascorbate–glutathione cycle rapidly responds to changes in O_2 concentrations in legume nodules (Dalton et al. 1991). Controlling nutrient transport and oxygen permeability at nodular cortex are particularly important aspects for the effective function of the nodule, and negatively influenced by nutrient deficiency, salinity and drought stress (Serraj 2002).

Finally, the third factor is the rapid export of the fixed nitrogen (Vance and Gantt 1992). NH_3 produced from SNF is thought to move through PBM to cytoplasm of the host cell. Then NH_3 is incorporated into amino acids by glutamate synthase-glutamine oxoglutarate aminotransferase (GS-GOGAT), and partly transformed into ureides such as allantoin and allantoic acid (Pate et al. 1980). These products are then transported to the above part of the plant in order to produce amino acid, protein and other N compounds for the growth of the plants.

4 Environmental Factors Limiting Nitrogen Fixation

Among environmental factors that influence the quantity of nitrogen fixed, the temperature is essential for nodule formation. In temperate zones, the development of nodules in legumes is maximized between 17 and 22 °C, while in the tropical zone nodules best develop between 19 and 35 °C. When the temperature reaches these extrema, the size of nodules is reduced and only a low bacteroids population can develop. Moreover, temperature extrema also reduce nitrogenase activity significantly. In addition, the competitiveness of the rhizobia in forming nodules and the effectiveness of the rhizobia species–host plant symbiosis to fix N_2 are controlled by a series of edaphic, chemical and biophysical factors (Wangnai 1998).

4.1 Nitrate Inhibition of the Rhizobial Symbiosis

Nitrate (NO_3^-), ammonium (NH_4^+) and urea around 15–20 kg N ha^{-1} stimulate leguminous plants growth at the beginning of the vegetative growth. Nitrogen supply is needed during the initial phase of nodulation, when the plant cannot rely on symbiotic N_2 fixation (Sousanna and Hartwig 1996). The nitrate nutrition of the legume can however inhibit SNF (Arrese-Igor et al. 1997). It has been suggested that there are multiple effects of nitrate inhibition. In some studies, nitrate nutrition was responsible for stopping nodulation and decreasing the number of nodules as well as their weight. A decrease in the activity of nitrogenase (Fujikake et al. 2003) and N_2 fixation activity was moreover recorded (Arrese-Igor et al. 1998). Other studies suggested that the regulation of N_2 fixation has a direct effect on nitrogenase activity in the nodules (Ribet and Drevon 1995b; Drevon and Hartwig 1997; Almeida et al. 2000). This inhibition occurs in areas of legume cultivation, especially in temperate regions where it prevents the maximum exploitation of SNF.

In addition, Gordon and James (1997) have reported that sucrose synthase activity and gene transcription in legume nodules are reduced by nitrate absorption. A decrease in nitrogenase activity inside the nodules was observed within 18 h after exposing soya bean plants to nitrate application and coupled with the disappearance of sucrose synthase mRNA within 24 h. These observations suggest that nitrate feeding may produce a plant-to-nodule signal which could affect both the O_2 diffusion barrier and the expression of the sucrose synthase gene. The existence of mutants that hypernodulate even in the presence of nitrate also support the hypothesis that nitrate is not the inhibiting factor itself, but that it leads to secondary signals that suppress nodulation (Carroll et al. 1985). Although mechanisms of signals remain unclear, Neo and Layzell (1997) have suggested that glutamine and/or asparagine could act as signals. It is possible that, in legumes, at least some of the effects of nitrate are also mediated by auxin. According to the auxin burst hypothesis (Gresshoff 1993), high auxin levels inhibit nodule formation. It is hypothesized that nitrate increases the sensitivity of the root to auxin, thus reducing nodule formation. An effect of nitrate on the auxin response pathway has been found in *Arabidopsis thaliana* (Zhang et al. 1999). The nodular NO_3^- reduction is suggested to be the major cause of this inhibition.

The nodular reduction of NO_3^- could also inhibit bacteroidal nitrogenase activity via NO_2^- generation and/or reducing power competition. In addition, the nitrogenase activity of *ex planta* bacteroids (Stephens and Neyra 1983) and of free-living rhizobia (Keister and Evans 1976) is inhibited by NO_3^- in strains having nitrate reductase (NR) capacity, this inhibition being alleviated in NR mutants (Stephens and Neyra 1983). The mechanism of this *ex planta* inhibition could be related to nitrite, the first product of NO_3^- reduction. Indeed this ion accumulates in isolated bacteroids subjected to NO_3^- in microaerobiotic conditions (Rigaud et al. 1973); it is also detected in bacteroids extracted from nodules of symbiosis receiving NO_3^- (Becana et al. 1985). It inhibits nitrogenase at concentrations as low as 0.1 mM (Trinchant and Rigaud 1982). Finally, the local effect of nitrate inhibition on nodulation and N_2 fixation may be related to the fact that high accumulation of nitrate is restricted in the root parts in direct contact with nitrate (Ohyama et al. 1993). The intensity of inhibition of nodule growth by NO_3^- is linked to the treatment period as well as the tolerance of legume species (Fujikake et al. 2003).

4.2 Salinity and Drought

Abiotic constraints such as salinity and drought (González et al. 2001) strongly reduce SNF. Several hypotheses have been advanced to explain the negative effect of salt on SNF in plant legumes: diminished photosynthate supply to the nodule (Vessey and Waterer 1992; Georgiev and Atkins 1993); reduced supply of respiratory substrates to the bacteroids (Delgado et al. 1994); alterations in the oxygen diffusion barrier which reduces oxygen flux into the nodule and avoids nitrogenase damage (Serraj et al. 1994). The provision of substrates by the legume host in order to support N_2 in the nodules is an important facet of effective symbiosis, which has been studied more extensively in ureide-exporting nodules (Day and Copeland 1991; Gordon and James 1997). Salinity can seriously change the photosynthetic carbon metabolism, leaf-chlorophyll content, as well as photosynthetic efficiency (Seeman and Critchley 1985; Sharkey et al. 1985). However, salinity is known to boost the nodular carbohydrate content, and sucrose is the predominant carbohydrate in legume root nodules (Fougère et al. 1991; Gordon et al. 1993).

Eventually, there is genetic variability among legumes regarding their tolerance to salinity. Some legumes such as *Vicia faba, Phaseolus vulgaris* and *Glycine max* are more salt tolerant than others as *Pisum sativum*. It has been reported that some *Vicia faba* tolerant lines sustained SNF under saline conditions (Cordovilla et al. 1995). Other legumes, such as *Prosopis* spp. (Fagg and Stewart 1994), *Acacia* spp. (Zhang et al. 1991) and *Medicago sativa* (Abdel-Wahab and Zahran 1983) are salt tolerant. However, these legume hosts are less tolerant to salt than are their rhizobia. Salt inhibits the initial steps of rhizobia–legume symbioses. Indeed, soil microbial populations are also negatively affected by increasing salt concentrations as a result of direct toxicity as well as through osmotic constraint (Tate 1995).The hypothesis that the oxygen diffusion barrier is solely responsible for the control of nodule N_2 fixation during osmotic constraint has been challenged (Guérin et al. 1990; Diaz del Castillo et al. 1994; Diaz del Castillo and Layzell 1995) because nodule metabolic potential is also impaired in response to drought. Under drought conditions, only the activity of sucrose synthase, the key sucrose hydrolytic enzyme, declined. It is possible, therefore, that the reduced potential to metabolize sucrose may be an important factor contributing to the overall response of soya bean nodules to drought. Sucrose synthase activity and transcript levels also declined significantly and rapidly in nodules when soya bean plants were subjected to dark-induced constraint (Gordon et al. 1993). Therefore, it is possible that the sensitivity of sucrose synthase gene expression and the in vivo activity of sucrose synthase may be a common feature of the response of soya bean nodules to environmental constraints.

In drought conditions, the reduced import of photosynthates may cause less carbohydrate to be available for nodule function (Durand et al. 1987). Another possibility is that export of carbohydrates from nodules could be blocked because of a lack of water (Walsh 1990). The resulting accumulation of N products somehow causes feedback inhibition of nitrogenase activity or ammonia assimilation. In previous studies, an accumulation of ureides under drought was observed in both shoots (Serraj and Sinclair 1996) and nodules (Serraj et al. 1999; Vadez et al. 2000). Serraj et al. (2001) suggested that ureides accumulation within the plant could result in the inhibition of nodulation either as a result of a direct feedback within the nodule or an indirect feedback originated from shoots. Moreover, supply of external ureides increased ureide concentration in leaves and inhibited nitrogenase activity (Serraj et al. 1999; Vadez et al. 2000). Albrecht et al. (1994) reported that nodule initiation, growth and activity are all more sensitive to water deficit than are general root and shoot metabolism implying that soil moisture deficiency has a pronounced effect on N_2 fixation.

The response of nodulation and N_2 fixation to water deficit depends on the growth stage of the plants. Pena-Cabriales and Castellanos (1993) pointed out that water deficit imposed during vegetative growth was more detrimental to nodulation and SNF than that imposed during the reproduction stage. Sellstedt et al. (1993) found that water deficiency decreased the amount of N derived from N_2 fixation by about 26% when measured by the acetylene reduction assay. Moreover, during the harvest period of soya bean in symbiosis with rhizobia, nodule P concentrations and P use efficiency declined linearly with soil and root water content (Franson et al. 1991).

4.3 Soil Acidity

It is also well documented that legume productivity is limited by soil acidity which is a significant problem for agricultural production in many areas of the world (Graham 1992; Clarke et al. 1993; Bordeleau and Prevost 1994; Correa and Barneix 1997). Most leguminous plants require a neutral or slightly acidic soil for growth, especially when they depend on SNF (Brockwell et al. 1991; Bordeleau and Prevost 1994). Poehlman (1991) showed that the suitable soil pH for rhizobia activity was 6.5. However, legumes and their rhizobia exhibit varied responses to acidity. Some species, like alfalfa (*Medicago sativa*) are extremely sensitive to acidity, while others, such as *Lotus tenuis,* tolerate relatively low soil pH (Correa and Barneix 1997). The failure of legumes to nodulate under acid soil conditions is common, especially in soils of pH less than 5.0. The inability of some rhizobia to persist under such conditions is one cause of nodulation failure (Carter et al 1994; Bayoumi et al. 1995). However, poor nodulation can occur even when a viable rhizobial population can be demonstrated (Graham 1992; Graham et al. 1994). Taylor et al. (1991) reported that acidity had more severe effects on rhizobial multiplication than did Al stress and low P conditions. Low pH reduced the number of *Rizhobium leguminosarum* bv. *trifolii* cells in soils, which resulted in no or ineffective nodulation by clover plants (Ibekwe et al. 1997). The number of nodules, the nitrogenase activity, the nodule ultrastructure and the fresh and dry weights of nodules were affected to a greater extent at a low–medium pH (4.5; Vassileva et al. 1997). Similar results were obtained with *Bradyrhizobium japonicum* in symbiosis with soya bean.

Legume species differ greatly in their response to low pH with regard to growth and nodulation (Tang and Thomson 1996). Recently, it has been found that the amount of N_2 fixed by forage legumes on low-fertility acidic soil is dependent on legume growth and persistence (Thomas et al. 1997). Recent reports indicated the destructive effects of acidic soils on rhizobia–legume symbiosis and N_2 fixation. Low pH reduced the number of *R. leguminosarum* bv. *trifolii* cells in soils, which resulted in no or ineffective nodulation by clover plants (Ibekwe et al. 1997). The number of nodules, the nitrogenase activity, the nodule ultrastructure and the fresh and dry weights of nodules were affected to a greater extent at a low medium pH (4.5) (Vassileva et al. 1997). In acidic soils with pH of 5.0, where heavy metal activity is relevant, the presence of available aluminium inhibits nodulation (Bordeleau and Prevost 1994). Graham (1992) reported that Al is more detrimental to nodulation than it is to plant growth in legumes. At pH 4.5 and with 0.5 mM Ca, nodulation in cowpea was delayed by 13 mM Al and nodule number and dry weight were severely depressed (Alva et al. 1990).

The pH and calcium concentration strongly affect root exudation and the adsorption of the rhizobia at the root surface (Richardson et al. 1988; Carter et al. 1994). Availability of Ca in acidic soils with a high Al content appears very important for nodulation. At the concentration between 0 and 1.5 mM, the adsorption of *Sinorhizobium meliloti* on *Medicago sativa* is linearly related to the concentrations of calcium and magnesium. Higher calcium concentrations are required to compensate for the negative effect of low pH (i.e. high H^+ concentrations) on the adsorption of rhizobia (Caetano-Anollés et al. 1989). Thus, the negative effects of low pH and low calcium concentrations on nodulation of alfalfa are reflected at the first step of nodulation. In addition, similar observations indicated that the nodule number, nitrogenase activity and nodule ultrastructure of the common bean, *Phaseolus vulgaris*, were greatly affected at low Ca concentration (0.13 mM) in acidic soils with a pH of 4.5 units (Vassileva et al. 1997). Calcium moreover is essential in several mechanisms of energy storage and transfer being indispensable in the reduction process of N_2 to NH_4 (Vidor et al. 1983). The critical Ca level for nodule formation in pigeon pea (*Cajanus cajan*) and guar is more than 50 mM, whereas peanut and cowpea nodulated very well in solution culture with 2 mM calcium (Bell et al. 1989). Nodulation and nodule development in cowpea were strongly depressed at low pH (4.5–5.5) and low calcium concentration (0.05–2.5 mM; Alva et al. 1990).

4.4 Phosphorus Requirement of the Symbiosis

Acid-weathered soils of the tropics and subtropics are particularly prone to P deficiency. Worldwide, phosphorus is considered as the principal yield-limiting nutrient along with nitrogen (Zahran 1999). Phosphorus deficiency is a primary constraint to plant growth in many terrestrial ecosystems (Bonser et al. 1996). Under low soil pH, phosphate is adsorbed by clay minerals. Other factors such as low soil moisture affect on the availability of phosphorus (Karmakar et al. 1997; Raychaudhury et al. 2003). P deficiency has two main causes: (i) the low content in total P of some soils poor in organic matter or highly lessivated and (ii) the complexation of P with cations such as Ca, Al or Fe, under the form of insoluble oxides which are unavailable for the plants, like in acid soils. Nowadays, vast areas of potentially good land are still agriculturally poor because of P deficiency (Basu et al. 2008). The availability of P in the soil for plants and microorganisms can moreover be limited by the competition with microorganisms in the rhizosphere and geochemical sinks. In addition, P deficiency in the soils can also be caused by low phosphorus contents in the parent material.

Sanginga et al. (1996) reported that legumes introduced in cereal-based cropping systems with P applications showed an enhanced ability to establish, nodulate and grow. Under P-deficient conditions, P fertilization will usually result in enhanced nodule number and mass and greater N_2 fixation per plant and per gram of nodules (Serraj and Adu-Gyamfi 2004). Besides increasing nodulation and N_2 fixation, P fertilizer may increase grain yields (Mugwira et al. 1997). The increasing common bean grain yields up to 900 kg ha^{-1} were obtained by applied doses of P as high as 352 kg P ha^{-1}. Previous observations in the northern Guinea savanna have shown that legumes require about 30 kg P ha^{-1} for optimal growth and N_2 fixation (Fig. 9.3; Weber et al., personal communication).

Bell et al. (1990) reported that suitable phosphorus concentration for groundnut is around 0.3–0.4 % P of dry weight. Shu-Jie et al. (2007) showed that P fertilization stimulates the growth of nodulated plants and N_2 fixation. This hypothesis has been confirmed with the isotopic method ^{15}N for soya bean at the end of its cycle (Pongsakul and Jensen 1991) and for the pea for which an optimal fertilization in P stimulates tenfolds more the accumulation of N than the growth. Cassman et al. (1993) has shown that the application of P fertilizer increased number of nodules of soya bean grown in acid soils. Moreover, the application of P fertilizer at the correct place and time has positive effects on growth and N_2 fixation of soya bean in acid

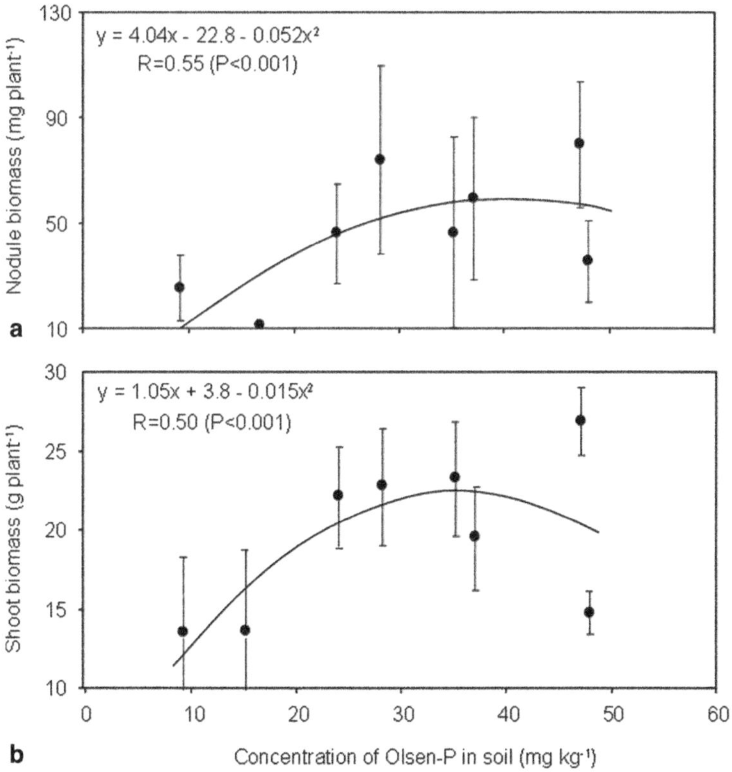

Fig. 9.3 Growth of nodule (**a**) and shoot (**b**) for different P concentration

soils. This may increase the availability of P in the vicinity of the soya bean seedlings and has a positive effect on the survival of introduced rhizobia which stimulate root growth and then promote infection of the plants.

Phosphate is essential for SNF of legumes (Waluyo et al. 2004). The P requirements for N_2 fixation have been investigated in various legume crops such as cowpea, pea, soya bean and *Acacia mangium* (Serraj and Adu-Gyamfi 2004). First et al. (1987) demonstrated that leguminous species differ in their phosphorus requirement for growth from 0.8 to 3.0 mM. Regarding the phosphorus use efficiency (PUE) concept, it was initially defined by Siddiqi et al. (1991) and discussed by Gourley et al. (1994), and it has gained scientific interest (Cure et al. 1988). According to Cassman et al. (1981), efficient P utilization in N_2-fixing symbioses may be closely related to an adequate P partitioning between shoot and nodulated root, and between root and nodules.

ATP is an essential energy provider molecule for the metabolism of organic compounds containing P such as sugar phosphates, phospholipids, nucleic acids, nucleotides and coenzymes which are key molecules for biological metabolisms (Schachtman et al. 1998). Plants dependent on SNF have therefore high ATP requirements for nodule development and function (Ribet and Drevon 1996), and need additional P for signal transduction and membrane biosynthesis. Phosphorus is also particularly important for leguminous plants because of its influence on the activity of the rhizobia species. P supplies in the soil are essential to N_2-fixing soya bean plants compared to N resources in the soil (Serraj and Adu-Gyamfi 2004). Wan Othman et al. (1991) reported that nodulation of cowpea (*Vigna unguiculata*) was impaired by a very low P status of the soil. Therefore, maximum benefits from N_2 fixation depend on soil P availability (Kennedy and Cocking 1997).

The symbiosis between legumes and rhizobia requires additional P uptake for the initial growth of the plant and to establish functioning nodules (Serraj and Adu-Gyamfi 2004). Shu-Jie et al. (2007) suggested that the P deficiency specifically inhibited the nodule development and thereby the total N_2 fixation. Several physiological characteristics of the nodule such as N_2-dependent growth, nodule respiration and control of oxygen diffusion affect SNF under phosphorus deficiency. Al-Niemi et al. (1997) suggest that bacteroids can be P limited even when plants have received otherwise adequate P levels. Cassman et al. (1980) showed that for soya bean, the P requirement for the nodules appeared higher than for root and shoot growth. Phosphorus concentrations in the nodule are often significantly higher than those in shoot or root tissue (Serraj and Adu-Gyamfi 2004). Hart (1989) suggested that nodules are strong sinks for P and range in P content from 0.7 to 1.2% of dry matter. Vadez et al. (1997) concluded that the P concentration in nodules can be threefold higher than in the other organs confirming the sink ability of nodules for P in nodulated legumes.

However, P requirement for N_2 fixation has been shown to vary among genotypes in pigeon pea and mung beans or *Casuarina–Frankia* symbioses (Serraj and Adu-Gyamfi 2004). Differences in N_2 fixation related to the efficiency of utilization of P were also found among soya bean genotype (Guanawardena et al. 1993) and *A. mangium* populations (Vadez et al. 1995). In addition, the growth rate of most rhizobia strains is reduced by low levels of P (Al-Niemi et al. 1997). However, strains of rhizobia differ markedly in tolerance to phosphorus deficiency (Beck and Munns 1985). This P-deficiency response occurred when the medium P concentration decreased below 1 mM. Nodulation and N_2 fixation and survival of rhizobia in soil are particularly affected under low P and acid soil conditions (Graham and Vance 2003).

Nodule biomass has been proven to be highly correlated to the availability of P for the plant. P deficiency decreases the number of nodules per plant (Singleton et al. 1985) and/or the individual mass of nodules (Gates 1974; Jacobsen 1985; Israel 1987; Guanawardena et al. 1992), the mass of the bacteroid in soya bean (Sa and Israel 1991) as well as the specific nitrogenase activity of nodules (Jacobsen 1985; Singleton et al. 1985; Israel 1987; Ribet and Drevon 1995b; Vadez et al. 1996; Drevon and Hartwig 1997; Qiao et al. 2007). Additional effects of P deficiency have been reported in common bean, soya bean, lupin and alfalfa, such as to increase the absorption surface and density of the roots resulting in more exploration of soil volume (Vance 2001), and to acidify the rhizosphere by root exudates (Neumann and Römheld 1999) and H^+efflux (Tang et al. 2001a, b, 2004). Finally, P may increase the nodulation and stimulate the nitrogenase activity by improving the plant growth (Gentili and Huss-Danell 2003; Yang 1995; Reddell et al. 1997).

5 The Diversity of Rhizobial Symbioses in Their N_2 Fixation Under Low P Soil

Previous studies have proven that although N_2 fixation increases P demand of the plant for *Pisum sativum* (Jacobsen 1985) and for *G. max* at the end of its development cycle (Cassman et al. 1981; Israel 1987), it does not increase P demand for *Stylosanthes* (Gates 1974), *Trifolium* (Robson et al. 1981), *Vigna unguiculata* (Cassman et al. 1981) and *Aracia mangium* (Sun et al. 1992).

Among legumes, cowpea is more tolerant to phosphorus deficiency than others like soya bean and common bean (Cassman et al. 1981; Gómez et al. 2002). Moreover, SNF in common beans is affected by P deficiency more than in other legumes (Israel 1987; Ribet and Drevon 1995a, 1996; Vadez et al. 1996, 1997; Drevon and Hartwig 1997). Consequently, SNF is considered to be less efficient in common beans than in other legumes (Pereira and Bliss 1987; Isoi and Yoshida 1991). In works reviewed by Bliss (1993), high SNF of common bean under P deficiency was reported to be related to nodule number (Jebara et al. 2005), nodule mass (Kipe-Nolt and Giller 1993; Kipe-Nolt et al. 1993), late nodule senescence (Hungria and Franco 1988), early nodulation (Chaverra and Graham 1992) and secondary nodulation (Wolyn et al. 1989). Regarding common bean *Phaseolus vulgaris,* there is evidence of genotypic variability for SNF at different levels of available P which show a possibility of selecting bean cultivars able to support biological N_2 fixation under low P soils (Pereira and Bliss 1989; García et al. 1996; Vadez et al. 1999).

Sanginga et al. (2000) studied PUE and N balance of cowpea on 94 lines in a low P soil of the derived savanna zone in West Africa. The cowpea lines fixed on average 22 kg N ha^{-1}, which was 70 % of the plant total N. The N balance based on the difference between the amount of N_2 fixed and N exported through the harvest, ranged between −10.6 kg N ha^{-1} and +7.7 kg N ha^{-1}. Based on its adaptability to grow in low P soils and overall positive N balance, the cowpea line IT81D-715 should be recommended for cultivation when P is the limiting factor.

Most studies of low P tolerance in beans have been undertaken in hydroponics or sand–alumina culture systems in order to optimize the control of P supply (Whitaker et al. 1976; Pereira and Bliss 1987; Vadez et al. 1997, 1999). In these conditions, Vadez et al. (1999) identified tolerant lines with threefold greater N fixed per P supplied than sensitive lines, and higher P use efficiency for SNF, the ratio of N fixed per nodule P concentration. They also found that P deficiency delayed the onset of nitrogenase activity, measured as acetylene reduction activity (ARA) and accelerated the ARA decline during pod filling. They concluded, in agreement with Pena-Cabriales and Castellanos (1993) and Vikman and Vessey (1992), that the time course of ARA may be a valuable trait for screening SNF tolerance to P deficiency.

In addition, the best-adapted genotypes increased the soil P availability by about 50 % after a culture cycle (Ankomah et al. 1995; Rajput and Singh 1996). The later was associated with an increase in SNF covering 89 % of the plant N requirement and an accumulation of 200 kg ha–^{1} N in the soil (Sanginga 2003). Attempts have been made to increase bean productivity by selecting lines able to fix adequate N_2 under conditions of low P availability (Pereira and Bliss 1987; Vadez et al. 1999). Conducting a screening procedure through 600 lines of common bean, Vadez et al. (1999) found a large genotypic variability for SNF tolerance to P deficiency. They reported that SNF tolerance to P deficiency was mostly found among late flowering varieties. The SNF tolerance to P deficiency was correlated with (i) low nodule P concentration, (ii) intense and early nodulation and (iii) large root dry weight and high nodule nitrogenase activity.

The lower values of the ratio SNF/nodule dry weight for tolerant lines than for sensitive lines indicate that the tolerant lines were essentially early N_2-fixing symbioses. The negative correlation between SNF and nodule P concentration suggests higher PUE for N_2 fixation (Fig. 9.4). This conclusion that the tolerance was mostly due to efficient utilization of P for SNF is also suggested by the lower correlation between SNF and P concentration in other organs than in nodules (Vadez et al.

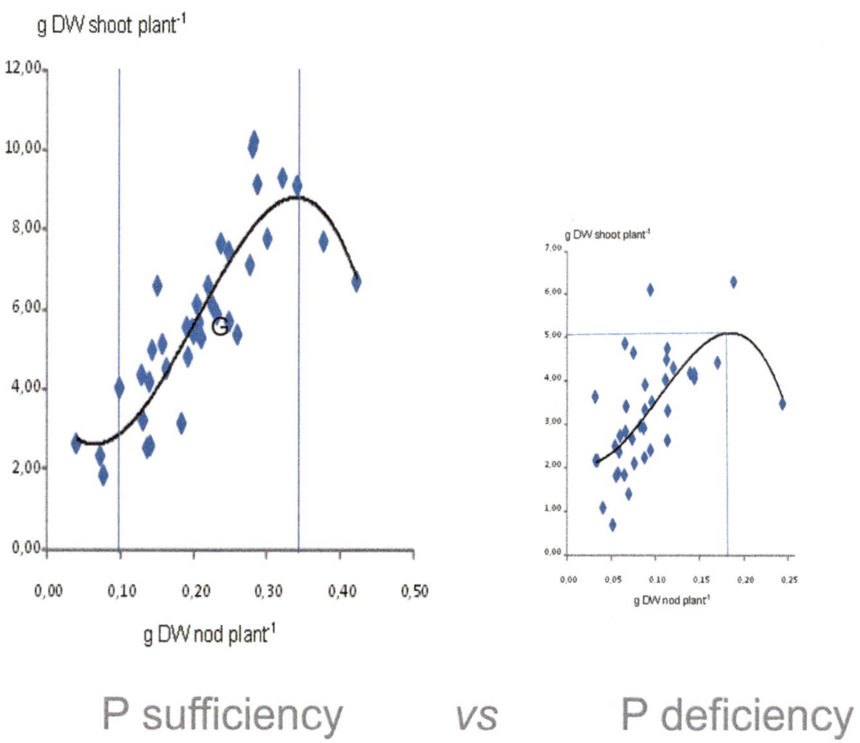

Fig. 9.4 Efficiency in use of the rhizobial symbiosis. *PUE* phosphorus use efficiency, *SNF* symbiotic nitrogen fixation, *DW* dry weight, *EURS* efficiency in use of rhizobial symbiosis

1999). The correlations between SNF and P concentrations in shoot or root would be a consequence of the effect of nodule PUE on the N-determined growth of root and shoot, and the subsequent dilution effect on P partitioning in these organs. The lower P concentration in nodules of tolerant lines compared to sensitive lines may also explain the higher nodule mass of tolerant lines as a result of a lower immobilization of P in nodule structures, allowing more nodules to develop. This observation may also explain the higher nodule persistence in tolerant lines during reproductive stages when P is remobilized for pod and seed development. However, it does not exclude that some tolerant lines have relatively higher nodule efficiency (Vadez et al. 1999).

The lower difference in growth than in N fixed or ARA per tolerant versus sensitive line, indicates limited benefit in growth subsequently to SNF tolerance to P deficiency, though the SNF tolerance to P deficiency is associated with higher PUE for plant growth. Vadez et al. (1999) concluded that the improvement of SNF for low P soils might be feasible through an extension of vegetative growth using climbing indeterminate progenitors in common bean. Although, it may be possible to breed among early flowering lines for enhanced SNF and tolerance to P deficiency with the ratio SNF per plant P concentration in glasshouse hydroponic culture.

References

Abdel-Wahab HH, Zahran HH (1983) The effects of water stress on N_2 (C_2H_2)-fixation and growth of *Medicago sativa* L. Acta Agron Acad Sci Hung 32:114–118

Albrecht SL, Bennett JM, Boote KJ (1994) Relationship of nitrogenase activity to plant water stress in field grown soybeans. Field Crop Res 8:61–71

Almeida JPF, Hartwig UA, Frehner M et al (2000) Evidence that P deficiency induces N feedback regulation of symbiotic N_2 fixation in white clover (*Trifolium repens* L.). J Expt Bot 51:1289–1297

Al-Niemi TS, Kahn ML, McDermott TR (1997) P metabolism in the bean-*Rhizobium tropici* symbiosis. Plant Physiol 113:1233–1242

Alva AK, Assher CJ, Edwards DG (1990) Effect of solution pH, external calcium concentration and aluminum activity on nodulation and early growth of cowpea. Aust J Agr Res 41:359–365

Ankomah AB, Zapata F, Danso SKA et al (1995) Cowpea varietals differences in uptake of phosphorus from Gafsa phosphate rock in a low-P ultisol. Fert Res 41:219–225

Arora N, Skoog F, Alien ON (1959) Kinetin-induced pseudonodules on tobacco roots. Am J Bot 46:610–613

Arrese-Igor C, Minchin FR, Gordon AJ et al (1997) Possible causes of the physiological decline in soybean nitrogen fixation in the presence of nitrate. J Expt Bot 48:905–913

Arrese-Igor C, Gordon AJ, Minchin FR et al (1998) Nitrate entry and nitrite formation in the infected region of soybean nodules. J Expt Bot 49:41–48

Basu M, Bhadoria PBS, Mahapatra SC (2008) Growth, nitrogen fixation, yield and kernel quality of peanut in response to lime, organic and inorganic fertilizer levels. Bioresource Technol 99:4675–4683

Bauer P, Ratet P, Crespi MD et al (1996) Nod-factors and cytokinins induce similar cortical cell divisions, amyloplast deposition and *MsENOD12A* expression patterns in alfalfa roots. Plant J 10:91–105

Bayoumi HEA, Biro B, Balazsy S et al (1995) Effects of some environmental factors on *Rhizobium* and *Bradyrhizobium* strains. Acta Microbiol Immunol Hung 42:61–69

Becana M, Klucas RV (1992) Transition metals in legume root nodules: iron-dependent free radical production increases during nodule senescence. Proc Natl Acad Sci U S A 89:8958–8962

Beck DP, Munns DN (1985) Effect of calcium on the phosphorus nutrition of *Rhizobium meliloti*. Soil Sci Soc Am J 49:334–337

Becana M, Aparicio-Tejo PM, Sanchez-Diaz M (1985) Levels of ammonia, nitrite and nitrate in alfalfa root nodules supplied with nitrate. J Plant Physiol 119:359–367

Bell RW, Edwards DG, Adher CJ (1989) External calcium requirements for growth and nodulation of six tropical food legumes grown in flowing solution culture. Aust J Agr Res 40:85–96

Bell RW et al (1990) Growth and nodulation of tropical food legumes in dilute solution culture. Plant Soil 122:249–258

Benhamou N, Asselin A (1989) Attempted localisation for a substrate for chitinases in plant cells reveals abundant N-acetyl-D-glucosamine residues in secondary walls. Biol Cell 67:341–350

Bliss FA (1993) Breeding common bean for improved biological nitrogen fixation. Plant Soil 152:71–79

Bonser AMYM, Lynch J, Snapp S (1996) Effect of phosphorus deficiency on growth angle of basal roots in *Phaseolus vulgaris*. New Phytol 132:281–288

Bordeleau LM, Prevost D (1994) Nodulation and nitrogen fixation in extreme environments. Plant Soil 161:115–124

Brockwell J, Pilka A, Holliday RA (1991) Soil pH is a major determinant of the numbers of naturally-occurring *Rhizobium meliloti* in non-cultivated soils of New South Wales. Aust J Expt Agr 31:211–219

Bushby HVA (1991) Nodulation and nitrogen fixation of mungbeans in Australia. In: Imrie BC, Lawn RJ (eds) Mungbean: the Australian experience. Proceedings of the first Australian mungbeans workshop, Brisbane

Caetano-Anollés G, Lagares A, avelukes G (1989) Adsorption of *Rhizobium* meliloti to alfalfa roots: dependence on divalent cations and pH. Plant Soil 117:67–74

Carroll BJ, McNEil DL, Gresshoff PM (1985) Isolation and properties of soybean (*Glycine max*) mutants that nodulate in the presence of high nitrate concentrations. Proc Natl Acad Sci U S A 82:4164–4166

Carter JM, Gardner WK, Gibson AH (1994) Improved growth and yield of faba beans (*Vicia faba* cv. Fiord) by inoculation with strains of *Rhizobium leguminosarum* biovar *viciae* in acid soils in south-west Victoria. Aust J Agr Res 45:613–623

Cassman KG, Whitney AS, Stockinger KR (1980) Root growth and dry matter distribution of soybean as affected by phosphorus stress, nodulation, and nitrogen source. Crop Sci 20:239–244

Cassman KG, Whitney AS, Fox RL (1981) Phosphorus requirement of soybean and cowpea as affected by mode of N nutrition. Agron J 73:17–22

Cassman KG, Singleton PW, Linquist BA (1993) Input/output analysis of the cumulative soybean response to phosphorus on an Ultisol. Field Crops Res 34:23–36

Chaverra MH, Graham PH (1992) Cultivar variation affecting early nodulation of common bean. Crop Sci 32:1432–1436

Clarke LM, Dilworth MJ, Glenn AR (1993) Survival of *Rhizobium meliloti* WSM419 in laboratory culture: effect of combined pH shock and carbon. Soil Biol Biochem 25:1289–1291

Collinge DB, Kragh KM, Mikkelsen JD et al (1993) Plant chitinases. Plant J 3:31–40

Cooper JB, Long SR (1994) Morphogenetic rescue of *Rhizobium meliloti* nodulation mutants by trans-zeatin secretion. Plant Cell 6:215–225

Cordovilla MP, Ocana A, Ligero F et al (1995) Salinity effects on growth analysis and nutrient composition in four grain legumes-*Rhizobium* symbiosis. J Plant Nutr 18:1595–1609

Correa OS, Barneix AJ (1997) Cellular mechanisms of pH tolerance in *Rhizobium loti*. World J Microbiol Biotech 13:153–157

Crespi M, Galvez S (2000) Molecular mechanisms in root nodule development. J Plant Growth Regul 19:155–166

Cure JD, Rufty TW, Israel DW (1988) Phosphorus stress effects on growth and seed yield responses of non nodulated soybean to elevated carbon dioxide. Agron J 80:897–902

Dalton DA, Post CJ, Langeberg L (1991) Effects of ambient oxygen and of fixed nitrogen on concentrations of glutathione ascorbate, and associated enzymes in soybean root nodules. Plant Physiol 96:812–818

Davis LC, Erickson LE, Jones GT (1987) Diffusion and reaction in root nodules. Crit Rev Biotech 7:43–95

Day DA, Copeland L (1991) Carbon metabolism and comparmentation in nitrogen-fixing legume nodules. Plant Physiol Biochem 29:185–201

Delgado MJ, Ligero F, Lluch Cl (1994) Effects of salt stress on growth and nitrogen fixation by pea, faba-bean, common bean and soybean plants. Soil Biol Biochem 26:371–376

Deroche ME, Carrayol E (1988) Nodule phosphoenolpyruvate carboxylase: a review. Plant Physiol 74:775–782

Diaz del Castillo L, Layzell DB (1995) Drought stress, permeability to O_2, diffusion and the respiratory kinetics of soybean root nodules. Plant Physiol 107:1187–1194

Diaz del Castillo L, Hunt S, Layzell DB (1994) The role of oxygen in the regulation of nitrogenase activity in drought-stressed soybean nodules. Plant Physiol 106:949–955

Drevon JJ, Hartwig UA (1997) Phosphorous deficiency increases the argon induced decline of nodule nitrogenase activity in soybean and alfalfa. Planta 201:463–469

Drevon JJ, Kalia VC, Heckmann MO et al (1988) In situ open-flow assay of soybean root nodular acetylene reduction activity: influenceof acetylene and oxygen. Plant Physiol Biochem 26:73–78

Durand JL, Sheehy JE, Minchin FR (1987) Nitrogenase activity, photosynthesis and nodule water potential in soybean plants experiencing water deprivation. J Expt Bot 38:311–321

Fagg CW, Stewart JL (1994) The value of *Acacia* and *Prosopis* in arid and semi-arid environments. J Arid Environ 27:3–25

Farrar JF, Jones DL (2000) The control of carbon acquisition by roots. New Phytol 147:43–53

Ferguson BJ, Mathesius U (2003) Signaling interactions during nodule development. J Plant Growth Reg 22:47–72

Fernández-Pascual M, Pueyo JJ, Felipe MR de et al (2007) Singular features of the *Bradyrhizobium-Lupinus* symbiosis. Dynamic soil, dynamic plant. Global Science Books, Springer-Verlag, Berlin, pp 1–16

First AJ, Smith FW, Edwards DG (1987) External phosphorus requirements of five tropical grain legumes grown in flowing-solution culture. Plant Soil 99:75–84

Foucher F, Kondorosi E (2000) Cell cycle regulation in the course of nodule organogenesis in *Medicago*. Plant Mol Biol 43:773–786

Fougère F, Le Rudulier D, Streeter JGl (1991) Effects of salt stress on amino acid, organic acid, and carbohydrate composition of roots bacteroids and cytosol of alfalfa (*Medicago sativa* L.). Plant Physiol 96:1228–1236

Franson RL, Brown MS, Bethlenfalvay GJ (1991) The *Glycine-Glomus-Bradyrhizobium* symbiosis. XI. Nodule gas exchange and efficiency as a function of soil and root water status in mycorrhizal soybean. Plant Physiol 83:476–482

Fujikake H, Yamazaki A, Ohtake N et al (2003) Quick and reversible inhibition of soybean root nodule growth by nitrate involves a decrease in sucrose supply to nodules. J Expt Bot 54:1379–1388

García A, Hernandez G, Nuviola A et al (1996) Efecto del P sobre el rendimiento y extracción del NP de frijol cultivado en tres suelos. Agron Mesoam 7:99–102

Gates CT (1974) Nodule and plant development in *Stylosanthes humilis* HBK: symbiosis response to phosphorus and sulphur. Aust J Bot 22:45–55

Gentili F, Huss-Danell K (2003) Local and systemic effects of phosphorus and nitrogen on nodulation and nodule function in *Alnus incana*. J Expt Bot 54:2757–2767

Georgiev GI, Atkins CA (1993) Effects of salinity on N_2 fixation, nitrogen metabolism and export and diffusive conductance of cowpea root nodules. Symbiosis 15:239–255

Gómez LV, Vadez V, Hernandez G et al (2002) Evaluación de la tolerancia al estrés de fósforo en caupí (*Vigna unguiculata* L. Walp) en Cuba. I. Cultivo en solución nutritiva. Agron Mesoam 13:59–65

González EM, Galvez L, Arrese-Igor C (2001) Abscisic acid induces a decline in nitrogen fixation that involves leghaemoglobin, but is independent of sucrose synthase activity. J Expt Bot 52:285–293

Goormachtig S, Alves-Ferreira M, Montague M et al (1997) Expression of cell cycle genes during *Sesbania rostrata* stem nodule development. Mol Plant-Microbe Interact 10:316–325

Gordon AJ, James CL (1997) Enzymes of carbohydrate and amino acid metabolism in developing and mature nodules of white clover. J Exp Bot 48:895–903

Gordon AJ, Ougham HJ, James CL (1993) Changes in levels of gene transcripts and their corresponding proteins in nodules of soybean plants subjected to dark-induced stress. J Expt Bot 44:1453–1460

Gourley CJP, Allan DL, Ruselle MP (1994) Plant nutrient efficiency: a comparison of definitions and suggested improvement. Plant Soil 158:29–37

Graham PH (1992) Stress tolerance in *Rhizobium* and *Bradyrhizobium*, and nodulation under adverse soil conditions. Can J Microbiol 38:475–484

Graham PH, Vance C (2003) Legumes: importance and constraints to greater use. Plant Physiol 131:872–877

Graham PH, Draeger K, Ferrey ML et al (1994) Acid pH tolerance in strains of *Rhizobium* and *Bradyrhizobium*, and initial studies on the basis for acid tolerance of *Rhizobium tropici* UMR1899. Can J Microbiol 40:198–207

Gresshoff PM (1993) Molecular genetic analysis of nodulation genes in soybean. Plant Breeding Rev 11:275–318

Guanawardena SFBN, Danson SKA, Zapata F (1992) Phosphorus requirement and nitrogen accumulation by three mungbean (*Vigna radiata* L. Welzek) cultivars. Plant Soil 147:267–274

Guanawardena SFBN, Danson SKA, Zapata F (1993) Phosphorus requirement and source of nitrogen in three soybean (*Glycine max*) genotypes, Bragg, nts 382 and Chippewa. Plant Soil 151:1–9

Guérin V, Pladys D, Trinchant JC et al (1990) Nitrogen fixation (C_2H_2 reduction) by broad bean (*Vicia faba* L.) nodules and bacteroids under water-restricted conditions. Plant Physiol 92:595–601

Hart AL (1989) Nodule phosphorus and nodule activity in white clover. NZ J Agr Res 32:145–149

Hellriegel H, Wilfarth H (1888) Untersuchungen uöber die tickstoffnahrung der Gramineen und Leguminosen. Beilageheft zu der Zeitschrift des Vereins fuör die Ruöbenzucker-Industrie des Deutschen Reiches, Buchdruckerei der Post, Kayssler, Berlin

Hungria M, Franco AA (1988) Nodule senescence in *Phaseolus vulgaris* (L.). Trop Agr 65:341–346

Hunt S, Gaito ST, Layzell DB (1988) Model of gas exchange and diffusion in legume nodules. II. Characterisation of the diffusion "barrier" and estimation of the concentration of CO_2, H_2 and N_2 in the infected cells. Planta 173:128–141

Iannetta PPM, De Lorenzo C, James EK et al (1993) Oxygen diffusion in lupin nodules: I. Visualization of diffusion barrier operation. J. Exp Bot 44:1461–1467

Ibekwe AM, Angle JS, Chaney RL et al (1997) Enumeration and nitrogen fixation potential of *Rhizobium leguminosarum* biovar *trifolii* grown in soil with varying pH values and heavy metal concentrations. Agr Ecosyst Environ 61:103–111

Isoi T, Yoshida S (1991) Deficient nitrogen fixation of common-bean (*Phaseolus vulgaris*). Soil Sci Plant Nutr 37:559–563

Israel DW (1987) Investigation of the role of phosphorus in symbiotic nitrogen fixation. Plant Physiol 84:835–840

Jacobsen I (1985) The role of phosphorus in nitrogen fixation by young pea plants (*Pisum sativum*). Physiol Plant 64:190–196

Jebara M, Aouani ME, Payre H et al (2005) Nodule conductance varied among common bean (*Phaseolus vulgaris*) genotypes under phosphorus deficiency. J Plant Physiol 162:309–315

Jones DL, Darrah PR (1996) Re-sorption of organic compounds by roots of *Zea mays* L. and its consequences in the rhizosphere. III. Characteristics of sugar influx and efflux. Plant Soil 178:153–160

Karmakar S, Mittra BN, Gosh BC (1997) Comparative efficiency of organic and industrial wastes in improving productivity of acid lateritic soil under rice-groundnut cropping system. In: First all India peoples technology Congress, Calcutta, 21–23 Feb 1997, pp 24–27

Kennedy IR, Cocking EC (1997) Biological nitrogen fixation: the global challenge and future needs. Rockefeller Foundation Bellagio conference proceedings. SUN Fix Press, University of Sydney, Sydney, pp 1–83

Kinzig AP, Socolow RH (1994) Human impacts on the nitrogen cycle. Physiol Today 47:24–31

Kipe-Nolt JA, Giller KE (1993) A field evaluation using the ^{15}N isotope dilution method of lines of *Phaseolus vulgaris* L. bred for increased nitrogen fixation. Plant Soil 152:107–114

Kipe-Nolt JA, Vargas H, Giller KE (1993) Yield and biological nitrogen fixation of common bean (*Phaseolus vulgaris* L.) in Peru. Plant Soil 152:87–91

Keister DL, Evans WR (1976) Oxygen requirement for acetylene reduction by pure cultures of Rhizobia J Bacteriol 129:149–153

Kistner C, Parniske M (2002) Evolution of signal transduction in intracellular symbiosis. Trends Plant Sci 7:511–518

Libbenga KR, Harkes PAA (1973) Initial proliferation of cortical cells in the formation of root nodules in *Pisum sativum* L. Planta 114:17–28

Marschner H (1995) Mineral nutrition of higher plants, 2nd edn. Academic, New York. pp 1–889

Mugwira LM, HAque I, Lupwayi NZ et al (1997) Evaluation of phosphorus uptake and use efficiency and nitrogen fixation potential by African clovers. Agr Ecos Environ 65:169–175

Neo HH, Layzell DB (1997) Phloem glutamine and the regulation of O_2 diffusion in legume nodules. Physiol Planta 113:259–267

Neumann G, Römheld V (1999) Root excretion of carboxylic acids and protons in phosphorus-deficient plants. Plant Soil 211:121–130

Newcomb W, Syono K, Torrey JG (1976) Development of an ineffective pea root nodule: morphogenesis, fine structure, and cytokinin biosynthesis. Can J Bot 55:1891–1907

Ohyama T, Mizukoshi K, Nishiwaki T (1993) Distribution of ureide originated from nodules attached to the upper roots and nitrate derived from lower roots in soybean plants cultivated by double piled pots. Bull Fac Agric (Niigata University, Niigata, Japan) 45:107–116

Parsons R, Day DA (1990) Mechanism of soybean nodule adaptation to different oxygen pressures. Plant Cell Environ 13:501–512

Pate JS, Herridge DF (1978) Partitioning and utilization of net photosynthates in nodulated annual legumes. J Expt Bot 29:401–412

Pate JS, Craig AA, Shane TW et al (1980) Nitrogen nutrition and xylem transport of nitrogen in ureides-producing grain legumes. Plant Physiol 65:961–965

Pawar NB, Ghulghule JN (1980) Preliminary observation on occurrence and senescence of root nodules in some pluses. Trop Grain Leg Bull 18:3–5

Pena-Cabriales JJ, Castellanos JZ (1993) Effect of water stress on N_2 fixation and grain yield of *Phaseolus vulgaris* L. Plant Soil 152:151–155

People MB, Craswell ET (1992) Biological nitrogen fixation: investments, expectations and actual contributions to agriculture. Plant Soil 141:13–39

Pereira PA, Bliss FA (1987) Nitrogen fixation and plant growth of common bean (*Phaseolus vulgaris* L.) at different levels of phosphorus availability. Plant Soil 104:84–104

Pereira P, Bliss FA (1989) Selection of common bean (*Phaseolus vulgarus* L.) for N_2 fixation at different levels of available phosphorus under field and environmentally controlled conditions. Plant Soil 115:75–82

Peters NK, Frost JW, Long SR (1986) A plant flavone, luteolin, induces expression of *Rhizobium meliloti* nodulation genes. Science 233:977–980

Poehlman JM (1991) The mungbeans. Mohan Primlani for Oxford and IBH, New Delhi, pp 1–343

Pongsakul P, Jensen ES (1991) Dinitrogen fixation and soil N uptake by soybean as affected by phosphorus availability. Plant Physiol 14:809–823

Puppo A, Halliwell B (1988) Generation of hydroxyl radicals by soybean nodule leghaemoglobin. Planta 173:405–410

Puppo A, Rigaud J, Job D (1981) Role of superoxide anion in leghemoglobin autoxidation. Plant Sci Lett 22:353–360

Qiao YF, Tang C, Hang XZ et al (2007) Phosphorus deficiency delays the onset of nodule function in soybean (*Glycine max* Murr.). J Plant Nutr 30:1341–1353

Rajput AL, Singh TP (1996) Response of nitrogen and phosphorus with and without *Rhizobium* inoculation on fodder production of cowpea (*Vigna unguiculata*). Indian J Agron 41:91–94

Raychaudhury M, Ngachan SV, Raychaudhury S et al (2003) Yield response of groundnut to dual inoculation and liming of an acid hill Ultisol of Manipur. Indian J Agr Sci 73:86–88

Reddell P, Yang Y, Shipton WA (1997) Do *Casuarina cunninghamiana* seedlings dependent on symbiotic N_2 fixation have higher phosphorus requirement than those supplied with adequate fertilizer nitrogen? Plant Soil 189:213–219

Relic B, Perret X, Estrada-Garcia MT et al (1994) Nod factors of *Rhizobium* are the key to the legume door. Mol Microbiol 13:171–178

Ribet J, Drevon JJ (1995a) Increase in permeability to oxygen and in oxygen uptake of soybean nodules under limiting phosphorus nutrition. Plant Physiol 94:298–304

Ribet J, Drevon JJ (1995b) Phosphorus deficiency increases the acetylene-induced decline in nitrogenase activity in soybean (*Glycine max* (L.) Merr.). J Expt Bot 46:1479–1486

Ribet J, Drevon JJ (1996) The phosphorus requirement of N_2-fixing and urea-fed *Acacia mangium*. New Phytol 132:383–390

Richardson AE, Djordjevic MA, Rolfe BG et al (1988) Effect of pH, Ca and Al on the exudation from clover seedlings of compounds that induce the expression of nodulation genes in *Rhizobium trifolii*. Plant Soil 109:37–47

Rigaud J, Bergersen FJ, Turner GL et al (1973) Nitrate dependent aerobic acetylene reduction and nitrogen fixation by soybean bacteroids. J Gen Microbiol 77:137–144

Robson AD, O'Hara GW, Abbott LK (1981) Involvement of phosphorus in nitrogen fixation by subterranean clover (*Trifolium subterraneum* L.). Aust J Plant Physiol 8:427–436

Rodrigez-Barrueco C, Bermudez de Castro F (1973) Cytokinin-induced pseudonodules on *Alnus glutinosa*. Plant Physiol 29:277–280

Sa TM, Israel DW (1991) Energy status and functioning of phosphorus-deficient soybean nodules. Plant Physiol 97:928–935

Sanginga N (2003) Role of biological nitrogen fixation in legume based cropping systems; a case study of West Africa farming systems. Plant Soil 252:25–39

Sanginga N, Okogun JA, Akobundu IO et al (1996) Phosphorus requirement and nodulation of herbaceous and shrub legumes in low P soils of a Guinean savanna in Nigeria. App Soil Eco 3:247–255

Sanginga N, Thottappilly G, Dashiell K (2000) Effectiveness of rhizobia nodulating recent promiscuous soyabean selections in the moist savanna of Nigeria. Soil Biol Biochem 32:127–133

Schachtman DP, Reid RJ, Ayling SM (1998) Phosphorus uptake by plants: from soil to cell. Plant Physiol 116:447–453

Schmidt J, Röhrig H, John M, Wienecke U, Stacey G, Koncz C, Schell J (1993) Alteration of plant growth and development by Rhizobium nodA and nodB genes involved in the synthesis of oligosaccharide signal molecules. Plant J 4:651–658

Seeman JR, Critchley C (1985) Effects of salt stress on the growth, ion content, stomatal behaviour and photosynthetic capacity of salt-sensitive species, *Phaseolus vulgaris* (L). Planta 164:151–162

Sellstedt A, Staahl L, Mattsson M et al (1993) Can the ^{15}N dilution technique be used to study N_2 fixation in tropical tree symbioses as affected by water deficit? J Expt Bot 44:1749–1755

Serraj R (2002) Response of symbiotic nitrogen fixation to drought and salinity stresses. Physiol Mol Biol Plants 8:77–86

Serraj R, Adu-Gyamfi J (2004) Role of symbiotic nitrogen fixation in the improvement of legume productivity under stressed environements. West Afr J App Eco 6:95–109

Serraj R, Sinclair TR (1996) Processes contributing to N_2-fixation insensitivity to drought in the soybean cultivar Jackson. Crop Sci 36:961–968

Serraj R, Roy G, Drevon JJ (1994) Salt stress induces a decrease in the oxygen uptake of soybean nodules and in their permeability to oxygen diffusion. Plant Physiol 91:161–168
Serraj R, Vadez V, Denison RF et al (1999) Involvement of ureides in nitrogen fixation inhibition in soybean. Plant Physiol 119:289–296
Serraj R, Vadez V, Sinclair TR (2001) Feedback regulation of symbiotic N_2 fixation under drought stress. Agron 21:621–626
Sharkey TD, Seeman JR, Berry JA (1985) Photosynthesis in intact leaves of C_3 plants: physics, physiology and rate limitations. Bot Rev 51:53–105
Sheehy JE, Minchin FR, Witty JF (1985) Control of nitrogen fixation in a legume nodule: an analysis of the role of oxygen diffusion in relation to nodule structure. Ann Bot 55:549–562
Shu-Jie M, Yun-Fa Q, Xiao-Zeng H et al (2007) Nodule formation and development in soybeans (*Glycine max* L.) in response to phosphorus supply in solution culture. Pedosphere 17:36–43
Siddiqi MY, Glass ADM, Ruth TJ (1991) Studies of the uptake of nitrate in barley. 11. Compartmentation of NO_3^-. J Expt Bot 42:1455–1463
Sinclair TR, Goudriaan J (1981) Physical and morphological constraints on transport in nodules. Plant Physiol 67:143–145
Singleton PW, Abdel-Magid HM, Tavares JW (1985) Effect of phosphorus on the effectiveness of strains of *Rhizobium japonicum*. Soil Sci Soc Am J 49:613–616
Smil V (1999) Nitrogen in crop production: an account of global flows. Global Biogeochemical Cycles 13:647–662
Söderström B (1992) The ecological potential of the ectomycorrhizal mycelium. In: Read DJ, Lewis DH, Fitter AH et al (eds) Mycorrhiza functioning. CAB International, Wallingford, pp 77–83
Söderström BE, Read DJ (1987) Respiratory activity of intact and excised ectomycorrhizal mycelial systems growing in unsterilized soil. Soil Biol Biochem 11:231–237
Sousanna JF, Hartwig UA (1996) The effects of elevated CO_2 on symbiotic N_2 fixation: a link between the carbon and nitrogen cycles in grassland ecosystems. Plant Soil 187:321–332
Spaink HP, Wijfjes AHM, Vanvliet TB et al (1993) Rhizobial lipo-oligosaccharide signals and their role in plant morphogenesis. Are analogous lipophilic chitin derivatives produced by the plant? Aust J Plant Physiol 20:381–392
Stacey G, Shibuya N (1997) Chitin recognition in rice and legumes. Plant Soil 194:161–169
Stephens BD, Neyra CA (1983) Nitrate and nitrite reduction in relation to nitrogenase activity in soybean nodules and *Rhizobium japonicum* bacteroids. Plant Physiol 71:731–735
Stougaard J (2001) Genetics and genomics of root symbiosis. Curr Opin Plant Biol 4:328–335
Sturtevant DB, Taller BJ (1989) Cytokinin production by *Bradyrhizobium japonicum*. Plant Physiol 89:1247–1252
Sun JS, Simpson RJ, Sands R (1992) Nitrogenase activity of two genotypes of *Acacia mangium* as affected by phosphorus nutrition. Plant Soil 144:51–58
Sutherland TD, Bassam BJ, Schuller LJ et al (1990) Early nodulation signals of the wild type and symbiotic mutants of soybean (*Glycine max*). Mol Plant Microbe Interact 3:122–128
Sy A, Giraud E, Jourand P et al (2001) Methylotrophic *Methylobacterium* bacteria nodulate and fix nitrogen in symbiosis with legumes. J Bacteriol 183:214–220
Tang C, Thomson BD (1996) Effects of solution pH and bicarbonate on the growth and nodulation of a range of grain legumes. Plant Soil 186:321–330
Tang C, Hinsinger P, Jaillard B et al (2001a) Effect of phosphorus deficiency on the growth, symbiotic N_2 fixation and proton release by two bean (*Phaseolus vulgaris*) genotypes. Agron 21:683–689
Tang C, Hinsinger, P, Drevon JJ et al (2001b) Phosphorus deficiency impairs early nodule functioning and enhances proton release in roots of *Medicago truncatula* L. Ann Bot 88:131–138
Tang C, Drevon JJ, Jaillard B et al (2004) Proton release of two genotypes of bean (*Phaseolus vulgaris* L.) as affected by N nutrition and P deficiency. Plant Soil 260:59–68
Tate RL (1995) Soil microbiology (symbiotic nitrogen fixation). Wiley, New York, pp 307–333
Taylor RW, Williams ML, Sistani KR (1991) Nitrogen fixation by soybean-*Bradyrhizobium* combinations under acidity, low P and high Al stresses. Plant Soil 131:293–300

Thomas RJ, Askawa NM, Rondon MA et al (1997) Nitrogen fixation by three tropical forage legumes in an acid soil savanna of Colombia. Soil Biol Biochem 29:801–808

Trinchant JC, Rigaud J (1982) Nitrite and nitric oxide as inhibitors of nitrogenase from soybean bacteroids. Appl Environ Microbiol 44:1385–1388

Vadez V, Lim G, Durand P et al (1995) Comparative growth and symbiotic performance of four *Acacia mangium* provenances from Papua New Guinea in response to the supply of phosphorus at various concentrations. Biol Fert Soils 19:60–64

Vadez V, Rodier F, Payre H et al (1996) Nodule permeability to O_2 and nitrogenase linked respiration in bean landraces varying in the tolerance of N_2 fixation to P deficiency. Plant Physiol Biochem 34:871–878

Vadez V, Beck DP, Drevon JJ (1997) Utilization of the acetylene reduction assay to screen for tolerance of symbiotic N_2 fixation to limiting P nutrition in common bean. Plant Physiol 99:227–232

Vadez V, Lasso JH, Beck DP et al (1999) Variability of N_2 fixation in common bean (*Phaseolus vulgaris* L.) under P deficiency is related to puse efficiency. Euphytica 106:231–242

Vadez V, Sinclair TR, Serraj R (2000) Asparagine and ureide accumulation in nodules and shoots as feedback inhibitors of N_2 fixation in soybean. Plant Physiol 110:215–223

Vance CP (1997) Biological fixation of nitrogen for ecology and sustainable agriculture (Legocki A, Bothe H, Pühler A (eds)). Springer-Verlag, Berlin, pp 179–186

Vance CP (2001) Symbiotic nitrogen fixation and phosphorus acquisition, plant nutrition in world of declining renewable resources. Plant Physiol 127:390–397

Vance CP, Gantt JS (1992) Control of nitrogen and carbon metabolism in root nodules. Plant Physiol 85:266–274

Vance CP, Heichel GH (1991) Carbon in N_2 fixation: limitation or exquisite adaptation. Ann Rev Plant Physiol Plant Mol Bio 42:373–392

Van der Holst PPG, Schlaman HRM, Spaink HP (2001) Proteins involved in the production and perception of oligosaccharides in relation to plant and animal development. Curr Opin Struct Biol 11:608–616

Vassileva V, Milanov G, Ignatov G et al (1997) Effect of low pH on nitrogen fixation of common bean grown at various calcium and nitrate levels. J Plant Nutr 20:279–294

Vessey JK, Waterer J (1992) In search of the mechanism of nitrate inhibition of nitrogenase activity in legume nodules: recent developments. Physiol Planta 84:171–176

Vidor CJ, Kolling JR, Freire J, Scholles D, Brose E, Pedrose MHT (1983) Fixação biológica do nitrogênio pela simbiose Rhizobium e leguminosas. Porto Alegre. IPAGRO. p 52 (Bol. Tec., 11)

Vikman PA, Vessey JK (1992) Gas-exchange activity, carbohydrate status, and protein turnover in root nodule subpopulations of field pea (*Pisum sativum* L. cv. Century). Plant Soil 151:31–38

Walsh KB (1990) Vascular transport and soybean nodule function. 111. Implications of a continual phloem supply of carbon and water. Plant Cell Environ 13:893–901

Walsh KB, Canny MJ, Layzell DB (1989) Vascular transport and soybean nodule function. II. A role for phloem supply in product export. Plant Cell Environ 12:721–723

Waluyo SH, Lie TA, Mannetje L't (2004) Effect of phosphate on nodule primordia of soybean (*Glycine max* Merrill) in acid soils in rhizotron experiments in acid soil in rhizotron experiments. Indonesian J Agr Sci 5:37–44

Wangnai S (1998) Nitrogen fixation: *Rhizobium*-leguminous. Soil Sciences Department, Faculty of Agriculture, Kasetsart University, Kasetsart

Wan Othman WM, Lie TA, 't Mannetje L et al (1991) Low level phosphorus supply affecting nodulation, N_2 fixation and growth of cowpea (*Vigna unguiculata* L. (Walp)). Plant Soil 135:67–74

Weisz PR, Denison RF, Sinclair TR (1985) Response to drought stress of nitrogen fixation (acetylene reduction) rates by field-grown soybean. Plant Physiol 78:525–530

Whiteaker GG, Gerloff GC, Gabelman WH et al (1976) Intraespecific differences in growth of beans at stress levels of phosphorus. J Am Soc Hort Sci 101:472–475

Whitehead LF, Day DA, Hardham AR (1998) Cytoskeletal arrays in the cells of soybean root nodules: the role of actin microfilaments in the organisation of symbiosomes. Protoplasma 203:194–205

Witty JF, Minchin FR, Sheehy JE et al (1984) Acetylene induced changes in the oxygen diffusion conductance and nitrogenase activity of legume root nodules. Ann Bot 53:13–20
Witty JF, Skot L, Revsbech NP (1987) Direct evidence for changes in the resistance of legume root nodules to O_2 diffusion. J Exp Bot 38: 1129–1140
Wolyn DJ, Attewell J, Ludden PW et al (1989) Indirect measures of N_2 fixation in common bean (*Phaseolus vulgaris* L.) under field conditions: the role of lateral root nodules. Plant Soil 113:181–187
Yang Y (1995) The effects of phosphorus on nodule formation in the *Casuarina-Frankia* symbiosis. Plant Soil 176:161–169
Yang WC, de Blank C, Maskiene I et al (1994) *Rhizobium* Nod factors reactivate the cell cycle during infection and nodule primordium formation, but the cell cycle is only completed in primordium formation. Plant Cell 6:1415–1426
Young JPW, Haukka KE (1996) Diversity and phylogeny of rhizobia. New Phytol 133:87–94
Zahran HH (1999) *Rhizobium*-legume symbiosis and nitrogen fixation under severe conditions and in an arid climate. Microbiol Mol Bio Rev 63:968–989
Zhang XP, Karsisto M, Harper R et al (1991) Diversity of *Rhizobium* bacteria isolated from the root nodules of leguminous trees. J Syst Bacteriol 41:104–113
Zhang H, Jennings A, Barlow PW et al (1999) Dual pathways for regulation of root branching by nitrate. Proc Natl Acad Sci U S A 96:6529–6534

Chapter 10
Nutritional Value

Francesca Sparvoli, Roberto Bollini and Eleonora Cominelli

1 Introduction

Consumption of legumes is associated with physiological and health benefits, such as prevention of cardiovascular disease, obesity, diabetes mellitus and cancer, as indicated by an increasing number of studies. The growing body of research on these health benefits has stimulated interest in developing innovative technologies to expand the use of pulses in food products. Nevertheless, growing global food security challenges and protein malnutrition remain critical in many countries around the world.

Currently, nearly 870 million people are suffering from chronic undernourishment. Moreover, 100 million children under five are underweight, and this childhood malnutrition is a cause of death for more than 2.5 million children every year. Most hungry and undernourished people live on diets based on very high amounts of staple foods but few micronutrient-rich foods such as fruits, vegetables and animal and fish products. As a consequence, they lack the required protein, fat, vitamin A, iodine, zinc and iron. A particular type of malnutrition is the so-called hidden hunger, a chronic lack of vitamins and minerals as a consequence of poor dietary quality, which negatively impacts on health, cognition, function, survival and economic development. Hidden hunger mainly affects women and children from the lower-income groups in developing countries; however, it also claims victims in the developed world, where people apparently do not look hungry. Indeed, obesity or being overweight can be a sign that bodies are still hungry for crucial micronutrients. The impact of hidden hunger is serious: Globally stunted growth and anaemia

F. Sparvoli (✉) · R. Bollini · E. Cominelli
CNR Institute of Agricultural Biology and Biotechnology, Via Bassini15, 20133 Milan, Italy
e-mail: sparvoli@ibba.cnr.it

R. Bollini
e-mail: bollini@ibba.cnr.it

E. Cominelli
e-mail: cominelli@ibba.cnr.it

A. M. De Ron (ed.), *Grain Legumes*, Handbook of Plant Breeding 10,
DOI 10.1007/978-1-4939-2797-5_10

in children become a major cause of health problems in later life, particularly the increasing prevalence of overweight/obesity and chronic diseases. This results in a double burden for the health systems, with associated massive health costs and negative impact on economic productivity.

Grain legumes (pulses) are considered an essential source of nutrients and are also recognized as a poor man's meat, showing their importance for people of developing countries, where the consumption of animal protein is limited by non-availability or is self-imposed because of religious or cultural habits. Furthermore, legume seeds contain many bioactive and/or antinutritional compounds, such as phytate, oligosaccharides, phenolic compounds, nonprotein amino acids, lectins and enzyme inhibitors that play metabolic roles in humans or animals that frequently assume these seeds. These effects may be regarded as positive, negative or both (Champ 2002).

Considerable genetic variation has been reported in the chemical composition of legume seeds, both among and within species. In addition, chemical composition can be modified by environmental factors during plant development, since many of the bioactive compounds are secondary metabolites produced during seed development and maturation. Existing data show that the balance between deleterious and beneficial effects of these compounds depends on their chemical structure, concentration, time of exposure and interaction with other dietary components. Therefore, it is important to know not only the amounts but also the types of compounds in the food and how they affect human body. The scientific understanding of how these bioactive molecules act on organisms is an important challenge for the future research and a special attention should be paid to the potential synergistic effects between/among the different classes of bioactive compounds (Rochfort and Panozzo 2007).

In this chapter, main classes of bioactive compounds, together with some species-specific ones, are described in relation to their biological activities, abundance in legume crops and role in nutrition and health.

2 Major Seed Proteins: Storage Proteins and Enzyme Inhibitors

Legume seeds contain large amounts of proteins, mostly with a storage role, ranging from about 16 % (dry weight) in cowpea, pigeon pea and chickpea to as much as about 50 % in lupin and soybean, according to species, genotypes within species and environments (Table 10.1). Storage proteins are synthesized during seed development, stored in specific subcellular compartments, the storage vacuoles or protein bodies and then hydrolyzed during germination to provide nitrogen and carbon skeletons for the developing seedling. The major storage proteins of legume seeds are oligomeric globulins and albumins, which usually account for about 70 and 20 % of the total protein, respectively.

Table 10.1 Range of variation (% of seed weight) of principal constituents of grain legume seeds

		Carbohydrates			
Species	Protein	Total	Fibres	Starch	Oil
Common bean	20.9–30.1[a]	54–64[a]	10[c]	41.5[c]	1.3–2.5[a]
Pea	21.9–31[a]	52–62[a]	5.9–12.7[b]	18.6–54.5[b]	1.3–3[a]
Chickpea	16–28[a]	54–66[a]	2.7–9[b]	42–54.9[b]	3.1–7[a]
Faba bean	24.3–32.2[a]	57–60[a]	7.5–13.1[b]	37–51.5[b]	1.1–4
Cowpea	16–36[b]	56–68[a]	6.3[b]	46.84–53.63[e]	1–1.3[a]
Lentil	20.6–32[a]	54–58[a]	12–14.7[b]	46–49.7[b]	1–2.1[a]
Pigeon pea	15.9–24.1[b]	57.3–58.7[a]	10[c]	44.3[c]	1.2–1.6[b]
Mung bean	23.3–27.7[a]	61–62[a]	7–12.9[b]	45[c]	0.7–2.4
Lupinus spp.	28–47[c]	26–47[c]	3[d]	0.4[b]	4–15.5[c]
Soybean	26.5–55.2[b]	30.2–35[c]	20[c]	1.5[c]	6.5–28.7[b]

[a] Chibbar et al. 2010
[b] Burstin et al. 2011
[c] Hedley 2001
[d] Reddy et al. 1984
[e] Sreerama et al. 2012

2.1 Globulins

Globulins are classified as 7S and 11S proteins, according to their sedimentation coefficients (S), and are collectively named vicilins and legumins, respectively. Legumins are compact hexamers of about 350–400 kDa, and each monomer is made up of two disulphide-bonded subunits derived from posttranslational proteolysis of a single precursor polypeptide. Vicilins are typically trimeric proteins of 150–190 kDa that lack cysteine residues and hence cannot form disulphide bonds. Their subunit compositions vary considerably, mainly because of differences in the extent of posttranslational processing (proteolysis and glycosylation). Vicilins can be divided into two groups: In the first one, the precursor polypeptides are extensively fragmented to give rise to mature subunits in the range of 12–34 kDa. Conversely, the precursor polypeptides belonging to the second group undergo little or no posttranslational cleavage, and mature polypeptides are about 40–76 kDa. Pea vicilin is a typical example of group one vicilins, while soybean β-conglycinin, and common bean (*Phaseolus vulgaris* L.) are among the best characterized of the second group. Although legumins and vicilins are both present in most legumes, their relative abundance is highly variable, and some species are virtually devoid of either one or the other. Vicilins are usually less abundant then legumins, but a remarkable exception exists in *P. vulgaris* and *P. lunatus,* in which normally vicilin is the most abundant storage protein. In *P. coccineus,* the ratio between vicilin and legumin is highly variable, and seeds virtually depleted of legumin or phaseolin have been reported (Durante et al. 1989).

From a nutritional point of view, the amino acid profile of legume storage proteins reveals low amounts of the essential sulphur-containing amino acids (i.e. methionine and cysteine) and tryptophan, while lysine, another essential amino acid, is quite abundant. Legume proteins complement very well those of cereals, which are normally rich in sulphur amino acids and poor in lysine and threonine. Besides the composition in essential amino acids, the nutritional quality of seed proteins is also largely determined by their digestibility. In fact, amino acids composition only represents the potential nutritional quality of a protein, being their bioavailability critical for the supply of amino acids in the diet. A number of data obtained with experimental approaches devised to assess the bioavailability of amino acids in foods concurrently demonstrated that seed proteins have a lower overall nutritional quality than animal proteins. This can be related to their low content in sulphur amino acids, the compact proteolysis-resistant structure of the native protein and the presence of antinutritional compounds in the seed, which may affect digestibility of proteins themselves as well as of other components. For example, the nutritive value of common bean phaseolin, the 7S globulin of this species which normally accounts for about 40–50 % of total seed proteins, is limited by a low content in sulphur amino acids and by resistance to enzymatic hydrolysis, even after heat treatment (Montoya et al. 2006). The three-dimensional structures of 7S vicilins and 11S legumins have been resolved and confirmed that the two globulins are structurally related (Lawrence et al. 1994). Vicilin trimers are arranged in a disk-shaped fashion with each monomer arranged around a threefold symmetry axis and are rich of β-sheet and β-turn structures which have been proposed to be responsible of the low digestibility of globulins (Carbonaro et al. 2012; Lawrence et al. 1994).

Although the nutritional value and digestibility of globulins are not optimal, there are data showing that soybean β-conglycinin, the 7S globulin, is responsible for cholesterol/triglyceride-lowering activity, and it seems that the N-terminal extension domain of the α′ chain is responsible for inducing this biological response (Consonni et al. 2011). A similar positive role in reduction of hypercholesterolemia (and prevention of cardiovascular risk) has been shown for lupin proteins, the major role being played by γ-conglutin, an unusual basic 7S-type globulin specific of white lupin (Sirtori et al. 2004). Interestingly, γ-conglutin is also able to counteract the plasma glucose increase and improve insulin sensitivity when administered to rats, thus suggesting a potential use in the control of glycaemia in type 2 diabetes (Lovati et al. 2012).

Different approaches have been undertaken to improve the nutritional quality of legume storage proteins, and many have been directed towards increasing the content of sulphur amino acids and/or change in relative abundance/type of storage proteins. An example strategy, applied in common bean, consisted in the selection and breeding for highly digestible phaseolin types (Montoya et al. 2010). In fact, comparison of the degree of hydrolysis of 43 different phaseolin types showed variability ranging from 11 to 27 % for uncooked phaseolin and from 57 to 96 % for heat-treated one (Montoya et al. 2008). An alternative approach is the manipulation of seed protein compositions by decreasing the percentage of those types with low or limiting amino acid content. Using three common bean lines differing for major storage proteins content (devoid of phaseolin and/or major lectins), Taylor et al. (2008) evaluated the impact of storage protein deficiency on protein accumulation and amino acid composition, especially those containing sulphur, in mature seeds.

They found that deficiency of phaseolin and major lectins was associated with a progressive and compensatory increase of the content of other proteins, mainly legumin, α-amylase inhibitor (αAI) and mannose lectin FLT-3 receptor interacting lectin (FRIL; Marsolais et al. 2010). However, the most interesting finding was that the deficiency of some classes of storage proteins caused a modulation of sulphur amino acid content. The deficient lines showed a decrease of *S*-methyl-Cys and *γ*-Glu-*S*-methyl-Cys (both nonprotein amino acids) that were compensated with an increase of the Cys (70 %) and Met (10 %) pools, and the combined content raised from 18.9 to 26.8 mg/g protein, a value slightly above Food and Agriculture Organization (FAO) guidelines of 25 mg/g protein for human nutrition.

Albumins are the second most abundant class of legume storage proteins. They are water-soluble proteins comprising most of the bioactive polypeptides, such as lectins, protease inhibitors and αAI. The abundance of these bioactive molecules is quite variable in the different legumes. Lectins are widespread in many legume seeds, while αAI activity has been detected only in few legume species. Remarkably, the majority of these proteins have evolved within the seed as a protective mechanism against insects, fungi, predators and a number of stress conditions (Chrispeeels and Raikhel 1991). On the other hand, very often, the biological activity of these proteins is also responsible for the nutraceutical and health properties of legumes, thus the interest for the potential uses of these molecules has increased in recent years.

2.2 Lectins

Lectins are a family of highly homologous glycoproteins that exhibit specific and reversible carbohydrate-binding properties. As a result, lectins can bind to specific sugars and glycoproteins on the surface of cells in the gut wall, thereby interfering with nutrient breakdown and absorption.

Many lectins are able to agglutinate red blood cells, thus their presence is traditionally measured by their haemagglutinating activity (HA). Lectins' abundance and their biological activity in legume seeds vary among species as well as among genotypes of the same species (Table 10.2). Null or very-low lectin activity/presence has been reported for chickpea, lupin and *Vigna* genus, on the contrary, seeds of *Phaseolus* species have the highest content, although common bean genotypes devoid of lectins have been identified (Campion et al. 2009a; Confalonieri et al. 1992).

Growth suppression, diarrhoea and bloating are the most common effects of raw lectin ingestion by humans and livestock (Vasconcelos and Oliveira 2004). The toxicity of lectins is very often due to their high resistance to proteolysis and stability over a large range of pH. Even though some lectins are heat sensitive, they are not always completely destroyed by cooking because of the use of gentle cooking methods, such as dry heat and short cooking times. Lectin activity can, to various degrees, be removed from foods by different technological processes. For

Table 10.2 Comparison of haemagglutinating activity of different legume seed protein extracts towards rabbit or human erythrocytes. One unit of haemagglutinating activity *(HU)* is defined as the amount of seed extract per ml (range from 12,500 to 0.1 μg/ml) in the last serial dilution giving 50 % of agglutination (the lowest the HU value, the highest is the haemagglutinating activity of the sample). (Data adapted from Grant et al. 1983)

	Haemagglutinating activity range	
Species	Rabbit erythrocytes	Human erythrocytes (AB type)
Phaseolus vulgaris (white kidney beans)	6–24	12–390
Phaseolus vulgaris (pinto beans)	6250–12,500	12,500
Phaseolus coccineus	1.5–12	98–390
Phaseolus acutifolius	1.5–12	24
Phaseolus lunatus	12,500	98
Lentil	49–780	1560–6250
Pea	49–195	3120
Chickpea	12,500	12,500
Cowpea	12,500	12,500
Pigeon pea	12,500	12,500
Mung bean	12,500	12,500
Faba bean	49–3120	3120–12,500
Soybean	24–390	12,500

example, soaking, autoclaving and toasting completely destroyed the lectin in *P. lunatus* (Adeparusi 2001). Apart from common bean, microwave heating adequately destroys haemagglutinins and trypsin inhibitors in legume seeds without affecting protein quality, and irreversible lectin denaturation is achieved using boiling water (Hernandez-Infante et al. 1998).

Active lectins, that survived cooking and/or passage in the gastrointestinal tract, may induce changes in some or all of the digestive, absorptive, protective or secretory functions of the whole digestive system and affect cellular proliferation and turnover. For example, phytohaemagglutinin (PHA), the common bean lectin, binds to the gastric mucosal and parietal cells inhibiting the gastric acid secretion in conscious rats (Kordas et al. 2001), while in pigs it causes an increase in stomach weights and thickness (Radberg et al. 2001).

Although lectins are considered antinutrients, positive and beneficial roles for human health and nutrition have also been reported. Studies have revealed that oral administration of low doses of lectins can produce beneficial effects on the digestive/absorptive efficiency of the gut, its immune system and bacterial ecology and that, by modulating the secretion of gut hormones, some lectins can influence the body's endocrine system with beneficial consequences for general metabolism (Pusztai and Bardocz 1996). Some lectins may play a key role in preventing certain cancers (De Mejia and Prisecaru 2005) or can be used as therapeutic agents for preventing or controlling obesity (Bardocz et al. 1996).

2.3 Protease Inhibitors

Many legume seeds also contain inhibitors of proteolytic enzymes. These are considered antinutritional molecules interfering with protein digestion, due to their ability to irreversibly inhibit, if not properly inactivated, the action of digestive enzymes, such as trypsin, chymotrypsin, carboxypeptidases and elastase. However, once inactivated, protein inhibitors may even play a positive nutritional role, due to their high content of sulphur-containing amino acids compared to the majority of the seed proteins.

Most protease inhibitors belong to two major classes, the Kunitz trypsin inhibitors, particularly abundant in soybean seeds, and the Bowman–Birk inhibitors, that are widely found in the other legume seeds (Clemente et al. 2011; Oliva et al. 2011). Kunitz-type inhibitors have a molecular mass of about 20 kDa, with two disulphide bridges, and act specifically against trypsin. The Bowman–Birk inhibitors are double-headed inhibitors of 8–9 kDa with a high proportion of disulphide bonds. They usually comprise two distinct binding loops responsible for the inhibition of two identical or different proteases (chymotrypsin and/or trypsin) per inhibitor molecule. Differences in inhibitor concentrations and activity have been reported among legume species as well as varieties (Guillamon et al. 2008) and may be affected by environmental conditions during seed maturation (Piergiovanni and Pignone 2003). Trypsin inhibition (measured as trypsin inhibitor units per mg, TIU/mg) can range from negligible, as in the *Lupinus* spp., to very abundant in soybean (43–84 TIU/mg) and common bean (17–51 TIU/mg). The TIU content of different *Lathyrus* cultivars is in the range 19–30 TIU/mg sample, and this is higher than in chickpea (15–19 TIU/mg) and pea (6–15 TIU/mg). Most lentil and faba bean cultivars have lower values (3–8 and 5–10 TIU/mg sample, respectively) (Guillamon et al. 2008).

Kunitz and Bowman–Birk inhibitors' antinutritional effects are not only a consequence of inhibition of intestinal protein digestion for their presence in a diet consisting of free amino acids still has adverse effects resulting in decreased growth (Lajolo et al. 2004). It has been proposed that these inhibitors act suppressing the negative feedback regulation of pancreatic secretions through the release of hormone cholecystokinin from the intestinal mucosa (Liener 1994). The consequence is the stimulation of pancreas enlargement and hypersecretion of digestive enzymes (sulphur-rich proteins), causing a loss of sulphur-rich endogenous proteins. This would depress growth, as legume seed proteins are deficient in sulphur amino acids. On the other hand, the presence of trypsin inhibitors in the diet has been linked also to health-promoting properties. Bowman–Birk inhibitors are effective in preventing or suppressing carcinogen-induced transformation in vitro and carcinogenesis in animal assays. Anti-inflammatory properties of protease inhibitors have also been demonstrated (Ware et al. 1999). A number of patents on the use of Bowman–Birk inhibitors to combat obesity (Defreitas et al. 2003), degenerative and autoimmune diseases (Kennedy and Rostami 2004) and, in general, skeletal muscle atrophy (Sweeney et al. 2005) have been released.

2.4 α-Amylase Inhibitors

Presence of αAI differs greatly among legume species and the best described and most abundant are those found in *Phaseolus* species. Grant et al. (1995), analyzing the αAI levels in seeds of a number of legumes found in Europe, detected the highest contents of αAI in *Phaseolus* species (2–4 g/kg). Much lower levels (0.1–0.2 g/kg) were found for lima bean, cowpea, chickpea, faba bean and sweet lupin, while no αAI activity was found in seeds of adzuki bean, lentil, mung bean, pea, soybean and winged bean. In common bean, αAIs belong to a group of evolutionarily related seed proteins, comprising lectins and arcelins, whose presence has been frequently associated to resistance against phytophagous insects (Pueyo and Delgado Salinas 1997; Zaugg et al. 2013).

αAIs do not inhibit plant amylases, while they are active against a different type of amylases, such as human (saliva and pancreatic), porcine, fungal and, most interesting, insect amylases (Ishimoto et al. 1995). Hence, it is not surprising that most of the studies looking for αAI presence in seeds have been motivated by the protective role of this molecule against predatory insects. The best characterized αAI is that of common bean; it is a quite stable molecule being relatively heat resistant (it is still active after 30 min at 80 °C). In native conditions, it can resist tryptic digestion up to 24 h at 37 °C (Adeparusi 2001); however, proper thermal treatment, such as heating for at least 10 min at 100 °C, is sufficient for complete inactivation and loss of resistance to trypsin (Sparvoli et al. 1999). These properties could in part explain the biological effects of αAI. In clinical studies, purified αAI inhibited intraduodenal amylase. High dietary intakes of αAI can cause a number of potentially deleterious alterations in the metabolism of experimental animals. Starch digestion in the rat small intestine was also inhibited, with occasional blockage of the caecum, particularly at daily intakes of αAI higher than 20 mg, leading to losses of body nitrogen, lipids and carbohydrates and growth depression (Pusztai et al. 1995). In humans, αAI consumption decreases postprandial plasma glucose and insulin levels (Jain et al. 1989; Layer et al. 1986), thus the starch-blocking properties of αAI have prompted several studies to exploit the use of this molecule to control obesity as well in the prevention and treatment of diabetes (Obiro et al. 2008).

3 Starch, Fibres and Oligosaccharides

Starch and fibres, along with carbohydrate derivatives, such as oligosaccharides, constitute the major components of seed carbohydrates, making up to 30–40 % of seed dry matter in those species with higher protein content, such as lupin and soybean, and up to 50–65 % in less protein-rich legumes species (Table 10.1). There is increasing evidence that the addition of fermentable fibre to the diet alters the function and structure of the gut, modifies the production of gut-derived hormones, and is associated with improved whole-body glucose homeostasis even in the absence of disease.

3.1 Starch

Generally, starch constitutes the largest part of carbohydrate fraction, accounting for 35–45 % of seed dry weight, the only exceptions being soybean and lupin in which oil replaces starch as the main energy storage source (Table 10.1). Amylose and amylopectin are the two basic components of starch; amylose is a linear molecule with molecular weight ranging between 70 and 200 kDa, whereas amylopectin is a highly branched molecule consisting of main chains of (1–4)- α-D-glucose with short chains of (1–6)- α-D-glucose-linked branches with molecular weight greater than 2×10^4 kDa.

Starch composition is one of the determinants that define legume seed nutritional value and health effects. On the basis of its susceptibility to amylases and consequent digestibility profile, starch is classified as rapidly digestible starch (RDS), slowly digestible starch (SDS) and resistant starch (RS). RDS is rapidly digested and absorbed in the duodenum and proximal regions of the small intestine leading to a rapid elevation of blood glucose and usually a subsequent episode of hyperglycaemia. SDS is slowly digested in the small intestine to provide sustained glucose release with a low initial glycaemia and subsequently a slow and prolonged release of glucose. On the contrary, RS is not hydrolyzed in the small intestine and is fermented by the colonic microflora in the large intestine, producing short-chain fatty acids (SCFA) that provide additional energy to the body along with butyrate that is beneficial to colonic health (Aller et al. 2011; Fig. 10.1). Due to the different method used to evaluate the content of starch fractions, it is difficult to make a meaningful comparison of the levels of SDS, RDS and RS starches among legume seeds. Factors such as amylose content, crystallinity and amylopectin structur have been shown to influence SDS and RS levels (Hoover et al. 2010). Most starches from grain legumes have a relatively high amylose content (30–40 %) compared to those from cereals or tubers. These characteristics may lead to increase in RS content after processing, hence having important effects on human physiology (Guillon and Champ 2002), such as promoting slow and moderate postprandial glucose and insulin responses (Sievenpiper et al. 2009).

3.2 Dietary Fibre (DF)

DFs consist of chemically heterogeneous molecules such as cellulose, and noncellulosic polysaccharides like hemicellulose, pectins, oligosaccharides and lignin derived from structural carbohydrates of the plant cell walls. DFs resist digestion and absorption in the small intestine and are partially or completely fermented in the large intestine, thus exerting various physiological effects with health implications (Tharanathan and Mahadevamma 2003). Depending on their water solubility, total DF (TDF) are classified into insoluble DF (IDF) and soluble DF (SDF). The first class (IDF) is made up by cellulose, hemicellulose and lignin, and its consumption reduces intestinal transit time, thereby improving laxation. The second class

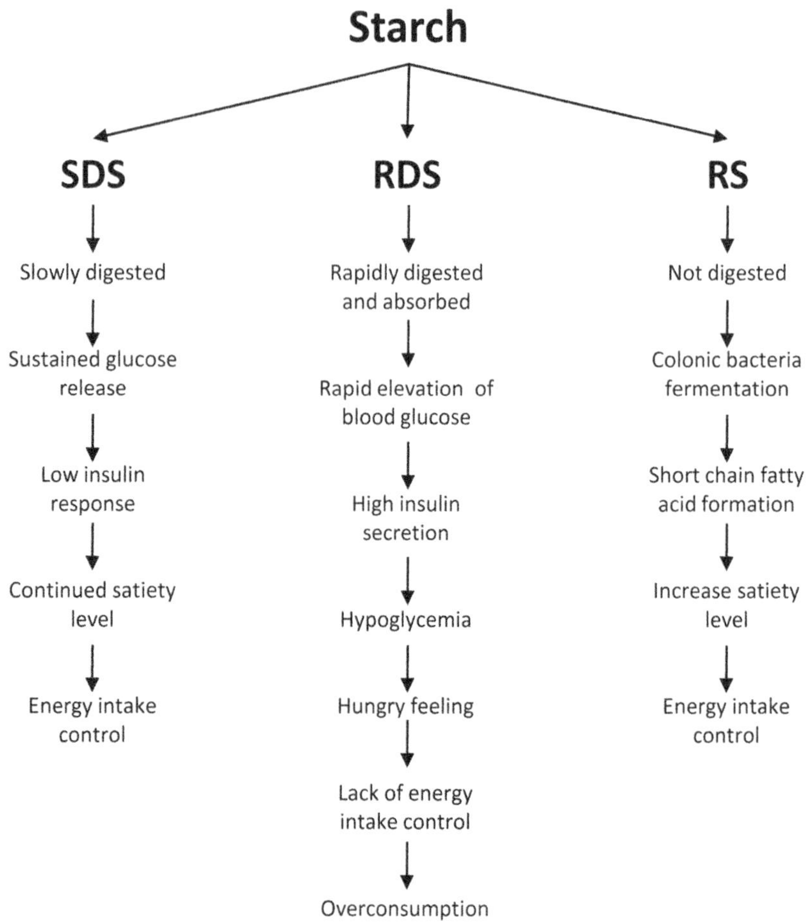

Fig. 10.1 Scheme representing starch classification and its main postprandial effects. *SDS* slowly digestible starch, *RDS* rapidly digestible starch, *RS* resistant starch. (Adapted from Aller et al. 2011)

(SDF) consists of oligosaccharides, glucans, gums and pectins, and its action is mainly in helping lowering blood cholesterol and regulating glucose. The composition and concentration of DF in legume seeds vary depending on their localization: the seed coat or the cotyledons, with the first having higher DF concentrations. DFs from seed coats contain large amounts of cellulose (35–57 %) and less amounts of hemicellulose and pectins, while the major polysaccharides in cotyledons are pectin compounds (about 55 %), cellulose (9 %) and non-starchy glycans (ranging between 6 and 12 %; Guillon and Champ 2002). Cellulose has been found to be the major constituent of crude fibre in pea and common bean, while hemicellulose is more abundant in lentil, faba bean, pigeon pea and mung bean (Reddy et al. 1984).

3.3 Oligosaccharides

The most common oligosaccharides in legume seeds are α-galactosides, which are soluble low-molecular-weight sugars mostly represented by the raffinose family oligosaccharides. They are α-(1→6)-galactosides linked to carbon C-6 of the glucose moiety of sucrose and include raffinose (trisaccharide), stachyose (tetrasaccharide) and verbascose (pentasaccharide). Their relative and total abundance vary among species and cultivars. Appreciable levels of these oligosaccharides, ranging between 0.4 and 16.1 % of dry matter, are accumulated in the seeds of lentil, chickpea, lupin, pea and faba bean. Lupins contain higher levels of stachyose and raffinose compared to peas and faba beans, while verbascose is more abundant in peas and faba beans compared to lupin, chickpea and common bean (Muzquiz et al. 2012); stachyose is the most abundant α-galactoside in common bean (Diaz-Batalla et al. 2006). Interestingly, in chickpea there is a marked difference in oligosaccharides content between desi-type and kabuli-type chickpeas with the last having 16.8 % higher content than the former (Saini and Knights 1984).

From a nutritional point of view, α-galactosides are considered antinutritional factors as they are not hydrolyzed by mucosal enzymes in the small intestine of monogastric animals and are then fermented in the lower gut by resident anaerobic bacteria with the consequent production of carbon dioxide and hydrogen gases that are responsible for digestive discomfort (Rochfort and Panozzo 2007). On the other hand, α-galactosides have also proven to exert prebiotic effects by promoting the beneficial activity of specific members of the intestinal microflora, thus improving gut health by suppressing intestinal putrefaction, reducing constipation and diarrhoea, stimulating the immune system and increasing resistance to infection (Buddington et al. 2002).

4 Minerals and Phytic Acid

Legume seeds are an excellent source of essential minerals, particularly iron, zinc and calcium (Campos-Vega et al. 2010). The highest levels of iron can be found in seeds of common bean, faba bean, mung bean and lentil. High zinc contents have been reported for *Lupinus* spp., lentil and chickpea, while the highest calcium content is found in seeds of common bean, lupin, faba bean and chickpea (Table 10.3).

The increasing concerns about food security, together with a widespread "hidden hunger", have stimulated research on crop biofortification, that, in many cases, has been translated into wide screenings of natural variability to identify donor genotypes with high mineral content. These genetic materials have been further used for breeding high Fe and high Zn varieties as well as for the identification of useful molecular markers to assist breeding (Amarakoon et al. 2012; Blair et al. 2013; DellaValle et al. 2013; Nair et al. 2013).

Table 10.3 Range of variation of essential minerals (μg/g seed dry weight) and phytic acid content (mg/g seed dry weight) in different legume species

Species	Fe	Zn	Ca	PA	Reference
Mung bean	44.5–107	23.3–48	273	1.8–5.8	Sompong et al. 2010; Taunk et al. 2012
Pea	46–73	39–63	622–1219	3.1–7.1	Amarakoon et al. 2012; Muzquiz et al. 2012; Trinidad et al. 2010
Lentil	64.6–90	44–73.1	480–1280	2.5–12.2	Cabrera et al. 2003; Karakoy et al. 2012; Muzquiz et al. 2012; Thavarajah et al. 2009
Chickpea	46–77	37–74	517–1974	2.8–13.6	Grant et al. 2003; Konietzny and Greiner 2003; Muzquiz et al. 2012; Thavarajah and Thavarajah 2012; Wang et al. 2003
Pigeon pea	54	61	514	3.5–17.5	Chitra et al. 1995; Sompong et al. 2010; Trinidad et al. 2010
Lupinus spp.	24–108	29–176	1350–2225	6–8.9	Muzquiz et al. 2012; Porres et al. 2007; Trugo et al. 1993
Cowpea	106	65	209	5.4	Chitra et al. 1996; Grant et al. 2003; Konietzny and Greiner 2003
Common bean	62–280	10–42	562–4065	3.4–28.7	Cabrera et al. 2003; Doria et al. 2012; Grant et al. 2003; Guzman-Maldonado et al. 2000; Konietzny and Greiner 2003; Muzquiz et al. 2012; Trinidad et al. 2010
Faba bean	55–110	20.5–58	610–1973	5.9–15	Cabrera et al. 2003; Campos-Vega et al. 2010; Konietzny and Greiner 2003; Muzquiz et al. 2012; Oomah et al. 2011; Uzun et al. 2011
Soybean	161	66	1502	4.8–20.1	Muzquiz et al. 2012; Trinidad et al. 2010

Although legume seeds have a good content in essential minerals, they also accumulate significant amounts of compounds that lower their nutritional value by lowering nutrient bioavailability. Phytic acid (*myo*-inositol-1,2,3,4,5,6-hexa-*kis*-phosphate, $InsP_6$, PA) and its lower phosphorylated derivatives ($InsP_5$ and $InsP_4$) are some of such compounds. Phytic acid is the main phosphorous storage form in seed and is stored as a mineral complex (phytate salts) in specific subcellular structures, called globoids, in the protein vacuole of embryo and cotyledonary cells. It accounts for an average of 75 % of total seed P and constitutes 1–3 % of dry weight. However, PA and its less abundant derivatives, $InsP_5$ and $InsP_4$, are well recognized antinutrients, as, during gastrointestinal passage, they bind trace elements (e.g. Fe, Zn, Ca and Mg) and reduce their absorption leading, under certain dietary circumstances, to mineral (mostly Fe, Zn, Ca) deficiencies (Gibson et al. 2006). Phytic acid also interferes with other nutrient absorption: Its ability to complex with proteins decreases their solubility, therefore, impacting on digestive enzyme activity (Urbano et al. 2000). Recent studies have also identified PA as an antioxidant and have demonstrated that it possesses anticarcinogenic/antineoplastic properties, can reduce or prevent kidney stone formation, and plays important roles in many physiological processes (Raboy 2003; Vucenik and Shamsuddin 2006). The amount of PA accumulated in seeds varies among species, varieties and soil P availability; however, accumulation of very low PA levels has been detected only in induced mutants. Among legumes, mung bean, pea, lentil and chickpea have relatively lower levels of PA compared to common bean, faba bean and soybean (Table 10.3).

Mutations that reduce the level of seed PA (low phytic acid, *lpa*) have been identified in major crops such as maize, rice, wheat, common bean, pea and soybean (Campion et al. 2009b; Rasmussen et al. 2010; Warkentin et al. 2012; Wilcox et al. 2000). Decreased accumulation of seed PA varies depending on the type of mutation and generally ranges between 30 and 90 %. The highest reductions are found associated to mutations affecting a multidrug resistance-associated protein (MRP)-type adenosine triphosphate (ATP)-binding cassette (ABC) transporter, which is a high-affinity $InsP_6$ transporter (ABCC5) necessary for PA vacuolar storage (Maroof et al. 2009; Panzeri et al. 2011). Since PA and inositol phosphate derivatives play key roles in plant and cell functions, reduction in plant agronomic performance and fitness has been reported for a number of *lpa* mutants. Moreover, the extent of the negative pleiotropic effects of the mutation appears to correlate to the level of PA reduction (Panzeri et al. 2011). Despite this, there are indications that in some cases low PA levels in the seed are compatible with good plant performance and seed viability, as shown in common bean (Campion et al. 2013), or there is a potential for breeding to obtain acceptable performances, as shown for other crops (Israel et al. 2007). From a nutritional point of view, recent papers demonstrated that *lpa* mutants are effectively biofortified, being able to provide more micronutrients to humans than their wild type (WT) counterparts (Petry et al. 2013).

5 Phenolic Compounds: Tannins and Anthocyanins

Phenolic compounds are mainly represented by tannins and flavonoids and are mostly accumulated in the seed coats where they contribute to the determination of the color. Total phenolic compounds vary in composition and contents across different species, tissues, stages of development and in response to environmental factors (Caldas and Blair 2009; Diaz-Batalla et al. 2006; Marles et al. 2010; Oomah et al. 2011). A survey on phenolic compounds content in legume species can be found in the US Department of Agriculture (USDA) flavonoid, isoflavone and proanthocyanidin databases (USDA, 2013).

In nutritional terms, the major effect of tannins is the reduction of protein digestibility by inhibition of proteolytic activity and/or formation of indigestible complexes with dietary protein. Tannins also form complexes with polysaccharides and iron in the gastrointestinal lumen; therefore, they reduce the efficiency of carbohydrate absorption and the bioavailability of the minerals in the grain. Despite these concerns, tannins may function as anticarcinogenic compounds and antioxidants (Serrano et al. 2009), thus the balance between health benefits and antinutritional effects is important when planning breeding work for this trait. Moreover, since some legume seeds, such as faba bean and peas, are also used as a protein source for feeding monogastric animals, tannin-free varieties are considered superior to tannin-containing ones (Crepon et al. 2010).

Flavonoids have been shown to exert many beneficial roles on human health since they possess diverse biological activities such as antioxidation, antiageing, anticancer, antiinflammation, antiatherosclerosis, cardiovascular protection, improvement of endothelial function, as well as inhibition of angiogenesis and cell proliferation activities. In legumes, the highest polyphenolic contents are found in dark, highly pigmented seed varieties, mostly belonging to *Phaseolus* and *Vigna* species. Xu and Chang (2007) made a comparative analysis of phenolic composition in a number of widely cultivated legume species and showed that lentil, black and red varieties of common bean and black soybean have the highest total phenolic content (TPC), total flavonoid content (TFC) and condensed tannins content (CTC). These high phenolic contents correlate with the highest antioxidant activities, as assessed with different evaluation methods (22,2-diphenyl-1-picrylhydrazyl radical scavenging assay DPPH; ferric reducing antioxidant power, FRAP; and oxygen radical absorbance capacity, ORAC; Table 10.4). In a recent work, black chickpea genotypes with high TPC and TFC content as well as FRAP levels have been described (Segev et al. 2010).

Common beans, and in general *Phaseolus* species, exhibit a wide variety of seed coat colours and patterns, thus they have been the subject of most of the published studies regarding composition and abundance of the different classes of phenolic compounds in legume species. Extensive genetic analyses have identified specific loci, controlling seed coat colour (*P, C, R, J, D, G, B, V* and *Rk*, that regulate flavonol and anthocyanin synthesis) and pattern (*T, Z, L, J, Bip* and *Ana*), and 12 quantitative trait loci (QTL) controlling condensed tannin concentration. Among them,

Table 10.4 Phenolic contents in seeds of different legume species. (Adapted from Xu et al. 2007)

Species	Market class	TPC (mg GAE/g)	TFC (mg CAE/g)	CTC (mg GAE/g)	DPPH (μmol Trolox eq/g)	FRAP (mmol Fe^{2+}eq/100 g)	ORAC (μmol Trolox eq/g)
Pea	Yellow pea	0.86–1.14	0.09–0.17	0.22–0.58	0.57–2.65	0.62–0.82	3.26–12.8
	Green pea	0.65–0.99	0.05–0.15	0.23–0.61	0.98–2.25	0.43–0.86	1.73–9.95
Lentil		4.86–9.6	3.04–4.54	3.73–10.2	19.07–19.87	8.75–12.44	59.55–95.19
Common bean	Black bean	3.37–6.99	2.51–3.3	4.09–5.73	14.49–18.95	6.05–9.70	48.91–92.73
	Navy bean	0.57	0.92	0.47	1.48	1.27	13.3
	Small red bean	5.76	4.24	5.16	17.9	4.53	70.58
Soybean	Black soybean	5.57	4.04	1.96	18.44	9.43	131.34
	Yellow soybean	1.74–1.82	1.06–1.24	0.37–0.79	0.92–1.83	1.09–1.49	35.1–44.23
Chickpea		0.98	0.72	0.52	1.26	0.8	9.26

TPC total phenolic content, *TFC* total flavonoid content, *CTC* condensed tannins content, *GAE* gallic acid equivalent, *CAE* catechin equivalents, *DPPH* 2,2-diphenyl-1-picrylhydrazyl radical scavenging assay, *FRAP* ferric reducing antioxidant power, *ORAC* oxygen radical absorbance capacity the *P* gene plays a key role in the regulation of seed colour, since it is epistatic on the expression of *C, D* and *J,* and it is considered the controlling factor for the presence or absence of flavonoids in the seed coat (Caldas and Blair 2009). An example of the variability of phenolic compounds in common bean has been reported by Diaz-Batalla et al. (2006) who quantified the concentrations of flavonoids (kaempferol, quercitin) and phenolic acids (p-hydoxybenzoic acid, vanillic acid, p-coumaric acid and ferulic acid) in a collection of ten cultivated and four wild varieties of Mexican bean seeds. A similar analysis was performed on a collection of Italian common bean landraces (Doria et al. 2012) and the content of genistein and daidzein isoflavones was also assessed. Only some genotypes contained this class of compounds and the highest values were 101 μg/g for daidzein (average value 36.7 μg/g) and 21.6 μg/g for genistein (average value 9.14 μg/g). However, these values are very far from those reported in soybean, which is a well-known good font of daidzein (470 μg/g) and genistein (740 g/g; Rochfort and Panozzo 2007).

6 Saponins

Saponins are naturally occurring compounds widely distributed and particularly abundant in legume seeds. Saponins consist of a lipid-soluble nucleus, having either a steroid or a triterpenoid aglycone structure, with one or more side chains of water-soluble carbohydrates. Based on their aglycone structure, saponins are generally categorized into three main groups as groups A, B and E. The main saponin components in legumes are the group B saponins, which contain the aglycone, soyasapogenol B. Many saponins are bitter and reduce the palatability of livestock feeds and have long been considered antinutrients due to toxicity and their haemolytic activity (Khalil and El-Adawy 1994). However, only a few are toxic since an enormous structural diversity within this chemical class exists, depending on the aglycone structure, the attachment of the glycosidic moieties and the nature of the glycosides. Saponins are attracting considerable interest as a result of their diverse properties. Clinical studies have suggested that they are health-promoting components that affect the immune system in ways that help to protect the human body against cancer, and also lower cholesterol levels. Saponins decrease blood lipids, lower cancer risks and lower blood glucose response. A high saponin diet can be used in the inhibition of dental caries and platelet aggregation, in the treatment of hypercalciuria in humans, and as an antidote against acute lead poisoning. In epidemiological studies, saponins have been shown to have an inverse relationship with the incidence of renal stones (Shi et al. 2004).

Presence of saponins has been reported in many edible legumes such as lupin, lentil, chickpea, faba bean, as well as soybean, bean and pea. Contents vary from 10 μg/g in pea up to 5000–6000 μg/g in soybean and chickpea (Campos-Vega et al. 2010). A study on a group of Italian landraces of common bean has reported an average soyasapogenol B content of 304 μg/g, and values were ranging from 105 to 454 μg/g (Doria et al. 2012). Another study on Spanish varieties showed a variation from 890 up to 2050 μg/g and a third study on navy bean seeds reported even higher values up to 7620 μg/g (Burbano et al. 1999; Shi et al. 2009).

7 Other Minor Antinutritional/Bioactive Compounds: Vicine, Convicine, L-DOPA and ODAP

The nutritional quality of some legume seeds may be affected also by other more species-specific molecules of different chemical origin. The seeds of faba bean accumulate vicine and convicine, two pyrimidine glycosides, whose aglycone forms, divicine and isouramil, respectively, are the causative agent of favism. This is a haemolytic anaemia that affects male individuals carrying a specific genetic defect in the gene coding for erythrocyte-located glucose-6-phosphate dehydrogenase. Vicine and convicine have antinutritional effects also in the diet of monogastric animals, and several efforts, related to breeding as well as to processing treatments, have been undertaken to reduce their amounts in seeds (Crepon et al. 2010).

Faba bean, together with *Mucuna pruriens* (a tropical legume also known as velvet bean), is one of the best plant sources of L-3,4-dihydroxyphenylalanine (L-DOPA), a naturally occurring nonprotein isomer of the amino acid 3,4-dihydroxyphenylalanine, which is potentially toxic if ingested in large amounts. L-DOPA has been reported to cause serious hallucinations in addition to gastrointestinal disturbances, such as nausea, vomiting and anorexia. Despite this, a lot of interest exists for this compound, since it is the major ingredient in medicines used to treat Parkinson disease (PD) patients. In fact, L-DOPA is a substrate of L-DOPA decarboxylase, which converts L-DOPA to the biologically active catecholamine dopamine, a compound that is depleted in the brain of people affected by PD. In faba bean, L-DOPA accumulates mostly in embryo axis of germinating seeds, and levels of around 75 mg/g dry weight have been detected after 9 days of germination, while much lower amounts (0.34 mg/g dry weight) have been found in the seed (Goyoaga et al. 2008). On the contrary, L-DOPA levels in the seeds of *Mucuna* species are around 3.1–6.7 % dry weight, and can reach up to 9 % (Pras et al. 1993).

Another nonprotein amino acid, the neuroexcitatory, β-*N*-oxalyl-L-α, β-diaminopropionic acid (ODAP), is found in *Lathyrus sativus* seeds. It is responsible for neurolathyrism, a disease associated to prolonged overconsumption of this protein-rich seed in a monotonous diet and consisting in the degeneration of upper motor neurons and the irreversible paralyzing of the legs in up to 6 % of the affected individuals. Assessment of ODAP content in *L. sativus* has shown that germplasm from South Asia contained relatively high amounts of ODAP (0.7–2.4 % dry weight), whereas those from North Africa, Syria, Turkey and Cyprus had significantly lower quantities of ODAP (0.02–1.2 %). No accessions were found to be free of the toxin (Hillocks and Maruthi 2012)

8 Health Benefits of Grain Legumes

Legumes have been consumed for thousands of years for their nutritional qualities, but only during the past few decades the potential impact of pulses on human health has been revived. Many different studies have reported that the consumption of pulses have beneficial physiological effects in the prevention and control of a broad range of chronic and degenerative diseases such as obesity, cardiovascular diseases (CVD), diabetes and cancer which are typical of industrialized societies (Bazzano et al. 2011). Pulses could potentially be considered as "functional foods" in addition to their accepted role of providing proteins and fibres. The consumption of pulses is in fact recommended as part of healthy eating by governments and health organizations globally. A high consumption of pulses is also one of the eight components of the highly lauded Mediterranean diet. However, according to FAO, the current consumption ratio of cereal grains to pulses in the diet is 8:1, which is considerably different from the ideal consumption ratio of 2:1. Particularly, the consumption of pulses in the Western world remains quite low, at less than 3.5 kg/capita per year, while in other parts of the world annual pulse consumption can range from 10 kg/capita (South America and India) to 40 kg/capita (Burundi). The role of pulses in the prevention and control of different pathologies is summarized below.

8.1 Metabolic Syndrome

The metabolic syndrome includes different risk factors of chronic diseases such as CVDs and diabetes, abdominal obesity, atherogenic dyslipidemia (high level of serum triglycerides and LDL cholesterol and low blood concentrations of HDL cholesterol), raised blood pressure, insulin resistance, proinflammatory state and prothrombotic state.

Results of two meta-analyses showed the long-term benefits of pulse consumption (2–5 cups per week for 3–12 weeks) on risk factors of the metabolic syndrome (Bazzano et al. 2011; Sievenpiper et al. 2009). Another study demonstrated that frequent consumption (5 cups/week over 8 weeks) of different legumes (yellow peas, chickpeas, navy beans and lentils) in an ad libitum diet reduced risk factors of metabolic syndrome in overweight and obese adults. These effects were similar or even stronger, depending on the different parameters analyzed, to the ones obtained with an energy-restricted diet (by 2093 kJ/day, corresponding to 500 kcal/day) implemented by counselling (Mollard et al. 2012).

The effects of consumption of grain legumes in the reduction of CVDs, obesity and diabetes mellitus are very strictly correlated and dependent on the different nutritional and nutraceutical components present in pulses, as recently reviewed (Hayat et al. 2014) and shown in the model reported in Fig. 10.2. Particularly, due to their slow release of carbohydrates, pulses are considered as low glycemic index foods (Atkinson et al. 2008), contributing to a reduction in the insulinemic responses. On the other hand, the consumption of pulses, through different mechanisms, reduces serum total cholesterol and LDL cholesterol (Anderson and Major 2002).

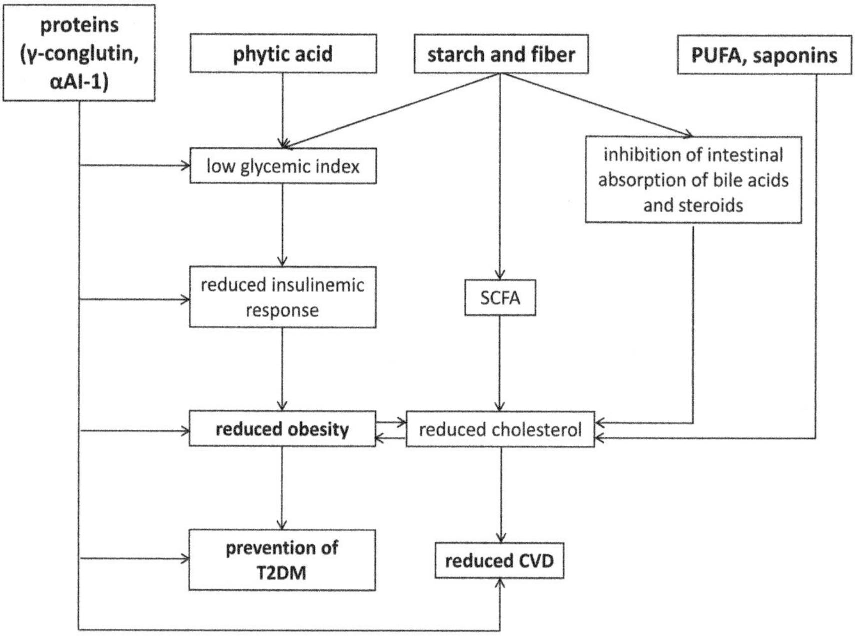

Fig. 10.2 A simplified model representing how different pulse compounds can protect from diseases associated with metabolic syndrome. *PUFA* polyunsaturated fatty acids, *SCFA* short-chain fatty acid, T2DM type 2 diabetes mellitus, *CVD* cardiovascular disease (Adapted from Hayat et al. 2014)

8.2 *Diabetes Mellitus*

Type 2 diabetes mellitus (T2DM) is a chronic metabolic disorder, resulting from insulin resistance, a condition in which cells fail to use insulin properly. This disorder leads to several secondary complications, including cardiovascular disease, chronic renal failure, diabetic retinopathy, hypertension, atherosclerosis, coronary artery disease and hyperlipidaemia. Approximately 150 million people worldwide are affected by T2DM, with a projection of 300 million people being affected by 2025. Diabetes has become a serious public health problem, particularly in developed countries.

The glycemic index (GI) of a food is defined as the incremental area under the blood glucose curve following ingestion of a test food, expressed as a percentage of the corresponding area following an equivalent load of a reference carbohydrate, either glucose or white wheat bread. After consumption of high GI foods, there is a large, rapid increase in blood sugar levels (glycemic response) and in response a rapid increase in insulin levels, while the consumption of low GI foods is correlated to a reduction in postprandial glucose and insulin elevations. Pulses are foods with low GI (Atkinson et al. 2008). Epidemiological studies have suggested that the consumption of foods with low GI protects against the development of T2DM and is useful also in the management of T2DM patients (Campos-Vega et al. 2010; Venn and Mann 2004). Particularly, people consuming about three portions per week of whole grain foods have lower probability (risk reduction of 20–30%) to develop T2DM than low consumers (<3 servings per week; Venn and Mann 2004). Different short-term studies have shown that consumption of seeds of common beans and other pulses typically reduces postprandial glucose elevations in nondiabetic and diabetic individuals compared with most starchy foods. Moreover, pulses combined with a high GI food produce a glycaemic response that is intermediate between the high- and low GI foods, but it is not clear if the nature of the effect is additive or linear (Tappy et al. 1986; Thompson et al. 2012). A clear association between a higher intake of legumes and a reduced risk of T2DM was particularly evident from results of a large, prospective, population-based study of middle-aged Chinese women. In this study, 64,227 women with no history of T2DM, cancer or cardiovascular disease at study recruitment, were followed up for an average of 4.6 years. An inverse association between quintiles of total legume intake of three mutually exclusive legume groups (peanuts, soybeans and other legumes) and T2DM incidence was observed (Villegas et al. 2008).

Different mechanisms of action have been proposed to explain the low glycemic response to legumes consumption (Hutchins et al. 2012). The high content of viscous fibres in pulses contributes to the low glycemic response, as they form a gel-like substance along the digestive tract, slowing down absorption rate of nutrients. The inclusion of a viscous fibre with a test meal is able to reduce the blood glucose response by an average of 44% (Wolever and Jenkins 2001). However, the viscous fibre component of legumes is not sufficient to determine the low glycemic and insulin response to legumes, as the addition of bean fibre to a potato meal is significantly less effective in lowering glucose and insulin response than a bean meal alone (Tappy et al. 1986).

Legumes are particularly rich in amylose; its lower molecular weight, smaller surface area and linear structure make it subject to slower digestion than amylopectin. The presence of RS in legumes results in the lower availability of glucose with the consequent slow entry of glucose in the bloodstream, the reduction of the demand for insulin, the lowering of the GI and the insulinemic postprandial response (Tappy et al. 1986). The protein fraction of pulses may also reduce starch digestibility and consequently glycemic response by directly interacting with starch (Alli and Baker 1980).

Moreover, proteins from pulses with a specific role in the prevention and management of diabetes have been extensively studied, such as γ-conglutin and α-amylase inhibitor isoform 1 (αAI-1; Barrett and Udani 2011; Lovati et al. 2012). The use of seeds decoctions of white lupin as "antidiabetic" is well known in the old pharmacopoeia. In the past years, γ-conglutin has been identified as the molecule conferring hypoglycemic properties to lupin. It was demonstrated that it has a glucose-lowering effect in normal rats upon glucose overload trials. This effect is very similar to that of metformin, a well-known antidiabetic drug. Moreover, the chronic oral γ-conglutin treatment in rats, in which hyperglycaemia had been induced, attenuated the rise in plasma glucose and insulin (Lovati et al. 2012). γ-conglutin is able to interact in vitro with mammalian insulin (Magni et al. 2004). To explain its antidiabetic effect, it has been hypothesized that it acts as an insulin-like agent. These data suggest the potential use of this protein in the control of glycaemia in patients with manifest or preclinical diabetes as well as for applications as functional foods and dietary supplements (Lovati et al. 2012; Magni et al. 2004). Another pulse protein with a well-characterized role in lowering the postprandial increases in blood sugar level is αAI-1 from common bean, also referred to as phaseolamin in starch blockers preparations (Barrett and Udani 2011). Starches are digested into sugars by α-amylase secreted in the saliva and by the pancreas and consequently absorbed in the small intestine. The use of purified forms of αAI-1 as a dietary supplement reduces the postprandial spikes of glucose and insulin following a high-GI meal (Obiro et al. 2008).

The glycemic response is influenced also by phytic acid. It was demonstrated that the consumption of unleavened bread made from navy bean flour, containing phytic acid, significantly reduced blood glucose area compared with that of bread made with wheat flour, while the opposite effect was obtained removing phytic acid. Moreover, phytic acid is also able to directly interact with starch and inhibit in this way starch digestibility (Thompson et al. 1987). As phytic acid binds cations such as Ca^{2+}, its presence may also reduce the stability of α-amylase, dependent on Ca^{2+} (Yoon et al. 1983).

Another possible mechanism of action responsible for the low glycemic response to pulses consumption is independent on components but depends on the fact that beans are commonly consumed in their whole form or minimally processed with little or no grinding. The integrity of the cell wall is then maintained after eating. Moreover, the cell wall of pulses is generally more resistant to digestion than the cell wall of cereal grains. These aspects may contribute to slow digestion and consequent low glycemic response (Noah et al. 1998).

8.3 Obesity and Overweight

Based on data from the World Health Organization (WHO), in 2008 approximately 1.4 billion adults worldwide were overweight and at least 500 million were obese. Increased consumption of foods rich in DFs, such as pulses, is associated with a lower body mass index (BMI), defined as the individual's body mass divided by the square of their height. Moreover, intake of foods with a high-fibre content helps in reaching satiety faster, and this effect is maintained longer as fibre-rich foods require a longer time to chew and digest in the intestinal system (Marlett et al. 2002). Different epidemiological studies demonstrated the efficacy of combined diets containing wholegrains and pulses in conferring a lower average BMI, a smaller waist circumference and demonstrated the negative correlation of this kind of diet with waist-to-hip ratio (Koh-Banerjee and Rimm 2003). Only a few published studies have specifically measured the effects of pulses consumption on body weight and satiety. Papanikolaou and Fulgoni (2008) reported on the association of consumption of beans with dietary quality and obesity risk in >8000 adult participants in the

US National Health and Nutrition Examination Survey (1999–2002). Compared to nonconsumers, bean consumers ($N = 1475$) had a lower body weight and a smaller waist size. Additionally, consumers of beans had a 23 % reduced risk of increased waist size and a 22 % reduced risk of being obese (Papanikolaou and Fulgoni 2008). Experimental studies in humans to evaluate the effects of pulse consumption on weight loss were performed during intentional energy restriction or without energy restriction. The first group of studies showed that, when pulse consumption is coupled with energy restriction, there is a beneficial effect on weight loss and on other parameters important to evaluate obesity risk. One of these trials was performed on 30 overweight and obese participants consuming a reduced energy intake diet (30 % energy restriction based on initial energy requirement) for 8 weeks eating either four servings/week of pulses or none. Results showed significantly greater decreases in BMI and body weight, expressed as percent of initial value, in the pulse-consuming group compared with the control group (22.0 vs 20.9 kg/m^2 and 27.8 vs 25.3 %, respectively). However, percentage body fat and waist circumference decreases did not differ significantly between groups (Hermsdorff et al. 2011). Similar results were obtained from other studies (Abete et al. 2009; Karlström et al. 1987). On the other hand, randomized controlled trials performed without energy restriction did not support a beneficial effect of pulses on weight loss, as reviewed by McCroryet al. (2010).

There is available evidence in support of pulse grains' ability to induce satiety. Subjects consuming at least 1200 g/week canned chickpeas for 12 weeks reported a significant increase in satiety compared with subject consuming their habitual diet (Murty et al. 2010). Similar effects on reducing appetite were described for navy beans (Wong et al. 2009). In both cases, despite an increase in DF intake, no differences in total energy intake were observed.

RS and DF are mainly responsible of the pulse effects on the control of appetite through increased satiety for their low digestibility. The fermentation of fibre and RS by bacteria in the large intestine generates specific SCFA, mainly butyric acid (Marinangeli and Jones 2012). This compound is the main product of indigestible fractions of black beans, lentils and chickpea by human microbiota, as demonstrated by in vitro fermentation (Hernández-Salazar et al. 2010). Rats fed high diets containing 25 % adzuki beans or two varieties of common beans significantly increased cecal butyric acid concentrations, compared with rats fed control cornstarch diet (Han et al. 2003). Butyrate was shown to increase hepatic and muscle expenditure as well as fat oxidation, mitochondrial oxidation and biogenesis when supplemented to mice diet (Gao et al. 2009). Thus, increased production of SCFA by fermentation of RS and fibres is an underlying reason for the protective benefits by the consumption of pulses (Finley et al. 2007). Moreover, high-fibre foods are believed to stimulate and prolong cholecystokinin secretion, a gastrointestinal peptide acting as hunger suppressant (Holt et al. 1992). Therefore, it is reasonable to hypothesize that appropriate dosages of pulse grain fibres can stimulate cholecystokinin release (de Graaf et al. 2004).

Protein component of pulses also plays an important role in weight management. Proteins, compared to carbohydrates, produce the highest thermic effect of food, which depends on the energetic costs associated with dietary peptide catabolism, protein synthesis and gluconeogenesis (Robinson et al. 1990). Amino acid composition could facilitate increase in energy expenditure (Marinangeli and Jones 2012). Arginine and glutamine, present at high level in pulses, have been shown to possess thermogenic properties (Iwashita et al. 2006; McKnight et al. 2010).

A specific antiobesity role for some proteins has also been described. Extracts of the already-cited phaseolamin have an antiobesity effect, as shown by different studies, although some conflicting results have been reported (Barrett and Udani 2011; Obiro et al. 2008). The αAI-1 inhibitory action results in the mobilisation of body fat reserves, due to energy restriction. In different studies, the efficacy of a commercial αAI-1 extract, referred to as Phase 2® (Pharmachem Laboratories, Inc., Kearny, NJ, USA), in reducing obesity was reported. Celleno et al. examined the effects of a dietary supplement containing 445 mg of Phase 2® on body composition of overweight human subjects in a 30-day study. They found greater reduction of body weight, BMI, fat mass, adipose tissue thickness and waist/hip/thigh circumferences, while maintaining lean body mass in subjects treated with Phase 2® compared to subjects receiving placebo (Celleno et al. 2007). Similar results were obtained in other studies, as reviewed by Barrett and Udani 2011. On the other hand, other reports did not confirm the efficacy of this starch blocker (Chokshi 2006). The effect of the extracts depends on a given manufacturer's methods of extraction, as regards the maintenance of high anti-amylase activity and purity (Obiro et al. 2008). As raw beans contain the lectin PHA, a highly toxic protein if consumed in native conditions, the protocol for the preparation of starch blockers requires a specialized process to inactivate haemagglutinin activity. To overcome this problem, common bean genotypes not able to accumulate PHA in the seed have been developed (Bollini et al. 1999). On the other hand, it was suggested that PHA as a dietary adjunct or therapeutic agent may be efficacious to stimulate gut function and ameliorate obesity if a safe and effective dose range can be established. The effects of inclusion of different levels of raw kidney bean of high lectin content (27 g/kg meal) in the diet of obese Zucker rats and their lean littermates in comparison with pair-fed controls were tested. It was shown that the growth of both obese and lean rats on bean diets was retarded by the daily bean intake in a dose-dependent manner, and most of this decrease was because bean-fed rats contained less body fat than the controls after 10 days (Pusztai et al. 1998). Consumption of bean-derived lectins was shown to increase cholecystokinin secretion, compared with controls fed lactalbumin, contributing to induce satiety (Herzig et al. 1997).

8.4 Cardiovascular Disease

CVDs are the number one cause of death, globally accounting for 30 % of all deaths, and are projected to remain the single leading cause of death by 2030.

In general, increased consumption of soluble fibre from foods results in reduced serum total cholesterol and LDL-cholesterol and has an inverse correlation with coronary heart disease (CHD) mortality (Noakes et al. 1999). Legume consumption has been associated with lower risks of CVD and CHD in observational epidemiologic studies (Bazzano et al. 2001; Kushi et al. 1999). For example, a study involving a total of 9632 men and women revealed a significant inverse relationship between legume intake and risk of CHD and CVD. In fact, legume consumption four times or more per week, compared with less than once a week, was associated with a 22 % lower risk of CHD and an 11 % lower risk of CVD (Bazzano et al. 2001).

Among the different controlled trials that have examined the potential hypocholesterolaemic effects of a diet rich in non-soy legumes, such as peas, lentils, different market classes of common beans, lima beans, chickpeas and faba beans, the majority identified positive effects, particularly in some cases (Anderson et al. 1984; Nervi et al. 1989; Sowmya and Rajyalakshmi 1999), while in a very few studies no effect was identified (Cobiac et al. 1990; Mackay and Ball 1992; Winham et al. 2007). A meta-analysis of randomized controlled trials was conducted to quantify the direction and magnitude of the potential effect that consumption of non-soy legumes may have on serum cholesterol concentrations (Bazzano et al. 2011). From 140 reports on the subject, the authors selected ten publications including studies performed on a total of 268 participants, and in which a comparison between a non-soy and a control diet was carried out for at least 3 weeks. This meta-analysis study provided a strong evidence that non-soy legume consumption lowers serum total cholesterol (Bazzano et al. 2011). Very recently, another study confirmed the efficacy of a diet rich in pulses (two servings daily of beans, chickpeas, peas or lentils, about 150 g/day) for reducing CVD risk factors in individuals 50 years or older, an age group who are at increased risk of this disease and on which a few studies had focused before (Abeysekara et al. 2012).

Low glycaemic index of pulses is important for lowering the risk of CVD (Duranti 2006). Moreover, several components of pulses are likely to contribute to their cholesterol-lowering effects. Soluble fibre is thought to bind to bile acids in the intestines and prevent reabsorption into the body. Consequently, an increase in the production of bile acids reduces the liver pool of cholesterol, triggering uptake of serum cholesterol by the liver, thereby lowering circulating cholesterol in the blood (Galisteo et al. 2008).

Another mechanism for the reduction of serum cholesterol depends on the activity of SCFA, particularly propionate, which alters metabolic pathways resulting in reduced serum cholesterol (Venter et al. 1990).

Chickpea contains a higher amount of fat than other pulses, and it is a relatively good source of nutritionally important polyunsaturated fatty acids (PUFAs), oleic acid and linoleic acid, constituting almost about 50–60 % of chickpea fat (Jukanti et al. 2012). It was shown that the intake of PUFA such as linoleic acid has a beneficial effect on serum lipids, insulin sensitivity and haemostatic factors, thereby it could be helpful in lowering the risk of CHD (Hu et al. 2001).

Recent evidence suggests that legume saponins, in addition to their anticancer activity may also be beneficial for hyperlipidemia (Shi et al. 2004) and in reducing the risk of heart diseases in humans (Geil and Anderson 1994).

8.5 Cancer

Cancer is a leading cause of death, mainly in industrialized countries, in the USA, for example, it is second only to CVD. In many developing countries cancer incidence appears much lower, but it is expected to raise due to increased control over infectious diseases and control of childhood diseases, leading to rise in life expectancy and in proportion of elderly people (Khatib and Aljurf 2008).

Different epidemiological analyses indicated a decreased risk of death associated with colon, breast and prostate cancer in countries with higher consumption of pulses (Mathers 2002). Moreover, experiments performed on laboratory animals have confirmed these results. One such study showed that feeding black or navy beans to rats exposed to the carcinogen azoxymethane reduced both the incidence and number of colon tumours by 50 % (Hangen and Bennink 2002). Most of these studies were focused on common beans, while a few investigations have been performed on other pulses. For example, a 64 % suppression of azoxymethane-induced aberrant cryptic foci was shown in mice fed with 10 % chickpea flour (Murillo et al. 2004). It was reported a systematic comparative study on antiproliferation effects of hydrophilic extracts from 13 commonly consumed food legumes (green pea, yellow pea, chickpea, lentil, yellow soybean, black soybean, pinto bean, black bean, small red bean, red kidney bean, mung bean, adzuki bean and black-eyed pea), using nine in vitro cultured human cancer cell lines (Xu and Chang 2012). Among the legume tested, lentil, the four common beans, mung bean and adzuki bean exhibited dose-dependent inhibitory effects against cell proliferation of all cancer cell lines. In particular, adzuki bean exhibited the strongest antiproliferative properties in a dose-dependent manner against seven of the nine cancer cell lines. Moreover, other legumes tested showed antiproliferative effects only against some cancer cell lines (Xu and Chang 2012). These results indicate that commonly consumed legumes may serve as an excellent dietary source for cancer prevention and further studies are warranted to characterize the potentiality of different legumes in cancer protection.

There are several bioactive food components of pulses that could be responsible for the cancer-preventive effect, as shown in the model reported in Fig. 10.3.

It is recognized that a major role in the anticancer effects of food is played by phenolic components. Beneficial effects of isoflavonoids in preventing breast and prostate cancers have been extensively studied (McCue and Shetty 2004). The already-mentioned paper from Xu and Chang (2012) showed that the total phenolic content of 13 food legumes exhibited significant linear correlation with the overall antioxidant activities. Coloured common beans, black soybean, lentil, adzuki bean and mung bean exhibited stronger antioxidant capacities and cancer cell proliferation inhibitory effects on different human cancer cell lines, as compared to green and yellow peas, chickpea, yellow soybean and black-eyed pea. Although the antioxidant properties of polyphenols have been extensively investigated, more recently their real impact as antioxidants has been reconsidered and questioned. In fact, there is an emerging view that these molecules may act not only by scavenging reactive

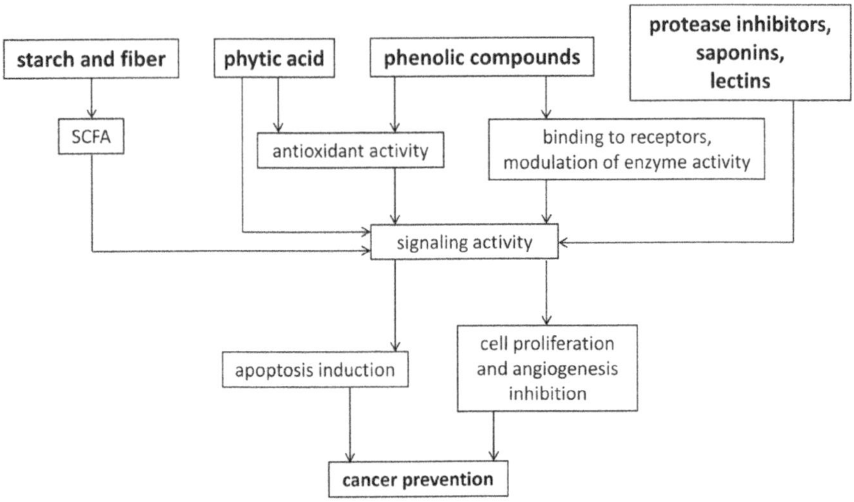

Fig. 10.3 A simplified model representing how different pulse compounds can protect from cancer. *SCFA* short-chain fatty acids. (Adapted from Hayat et al. 2014)

oxygen and nitrogen species or suppressing their production but also by enhancing the endogenous antioxidant capacity of cells/tissues (e.g. glutathione synthesis) or by influencing signalling pathways through interaction with proteins, enzymes and nuclear receptors, as recently reviewed (Martin et al. 2013). The anticancer activity of polyphenols has been associated with lower leukocyte immobilization, apoptosis induction, cell proliferation and angiogenesis inhibition (Garbisa et al. 2001; Nijveldt et al. 2001).

Phytic acid is a broad-spectrum antineoplastic agent, playing an important role in cancer prevention as well as in control of experimental tumour growth, progression and metastasis. Phytic acid seems to be responsible for the epidemiological link between high-fibre diets (high phytic acid content) and low incidence of some cancers. Phytic acid, after its rapid intake and dephosphorylation, enters the pool of inositol phosphates and acts as a strong antioxidant, enhances immune function, elicits anti-inflammatory activity, modifies phase I and II metabolizing enzymes, modulates oncogene expression, normalizes abnormal cell proliferation, induces cell differentiation, induces apoptosis and inhibits angiogenesis (Vucenik and Shamsuddin 2006).

Saponins are another class of non-nutrient bioactive compounds for which epidemiological studies suggest anticancer activity (Shi et al. 2004). Soyasaponin I properties have been mainly investigated, and its molecular activity was identified. This compound is able to inhibit the transfer of sialic acids to the nonreducing terminal positions on sugar chains of glycoconjugates, a process correlated with oncogenic transformation and metastatic potential (Chang et al. 2006). Moreover, saponins are able to regulate the apoptosis pathway enzymes, leading to programmed cell death of cancer cells (Zhu et al. 2005).

Also, lectins present in legumes may play a key role in preventing certain cancers (Campos-Vega et al. 2010). In vitro studies demonstrated, for example, that *Vicia faba* agglutinin (VFA) stimulated the morphological differentiation and reduced the malignant phenotype of colon cancer cells (Jordinson et al. 1999). The inclusion of PHA from raw kidney bean in the diet of a murine model for non-Hodgkin lymphoma tumour greatly reduced, in a dose-dependent manner, the growth rate of the tumour, either as an intraperitoneal ascites tumour or as a solid subcutaneous one (Pryme and Bardocz 2001). The number of Krebs II lymphosarcoma tumour cells in the ascitic fluid of mice fed a PHA diet for 8 days was three times lower than in mice fed a control diet (Bardocz et al. 1997). There is scientific evidence for different anticarcinogenic mechanisms of action of lectins, including binding to tumoural cell membranes, cytotoxic effects of lectins on tumour cells (decrease in protein synthesis and induction of apoptosis), reduction of cell proliferation and stimulation of the immune system (De Mejia and Prisecaru 2005).

Different in vitro and in vivo experiments have shown that protease inhibitors have anticarcinogenic properties. Although the majority of these studies were performed with soybean, as reviewed by Roy et al. (2010), more recently, the antiproliferative effects on human colon cancer cells of two recombinant wild-type Bowman–Birk inhibitors from pea seeds has been reported (Clemente et al. 2005).

All the molecules present in pulses having anticancer properties described so far are soluble in aqueous–alcohol extracts, while RSs, present in high amount in pulses, together with nonstarch polysaccharides, are primarily insoluble residues from aqueous–alcohol extracts. Colon carcinogenesis was induced by azoxymethane treatment in obese *ob/ob* mice fed with diet containing cooked navy beans (whole beans), the insoluble or soluble fraction of aqueous–alcohol extracts or a standard diet (Bobe et al. 2008). In comparison to control-fed mice, the incidence rates of various types of colon lesions were detected in fewer mice on either bean fraction diet. Moreover, no significant differences in incidence rate of various types of colon lesions between mice-fed diets containing bean residue or bean extract were observed. These results suggest that the insoluble fraction, containing RS, contributes in a similar way of the soluble fraction to the cancer-protective effect of cooked navy beans (Bobe et al. 2008). The cancer-preventive effect of RSs and nonstarch polysaccharides was previously shown (Bauer-Marinovic et al. 2006). As already discussed, SCFA, mainly butyric acid, are the products of the bacterial fermentation of resistant starches. They have protective effect against colon cancer as against the majority of tumours developed in the distal colon. Butyrate was reported to induce apoptosis, growth arrest and differentiation in colon cancer cell lines (Barnard and Warwick 1993); its effect is due to its activity in histone hyperacetylation and downregulation of epidermal growth factor receptor (Archer et al. 1998). The low GI of legumes attenuates the postprandial insulin response contributing to their cancer-preventive effect as hyperinsulinaemia and hyperglycaemia are reported to increase the risk of colon cancer (Jenkins et al. 2002).

Acknowledgment This work was partially supported by the Programme FILAGRO "Strategie innovative e sostenibili per la filiera agroalimentare", as part of the activities defined within the Accordo Quadro Consiglio Nazionale delle Ricerche and Regione Lombardia.

References

Abete I, Parra D, Martinez JA (2009) Legume-, fish-, or high-protein-based hypocaloric diets: effects on weight loss and mitochondrial oxidation in obese men. J Med Food 12:100–108

Abeysekara S, Chilibeck PD, Vatanparast H et al (2012) A pulse-based diet is effective for reducing total and LDL-cholesterol in older adults. Br J Nutr 108(Suppl 1):S103–S110

Adeparusi E (2001) Effect of processing on the nutrients and anti-nutrients of lima bean (*Phaseolus lunatus* L.) flour. Nahrung-Food 45:94–96

Aller E, Abete I, Astrup A et al (2011) Starches, sugars and obesity. Nutrients 3:341–369

Alli I, Baker BE (1980) Constitution of leguminous seeds: the microscopic structure of proteins isolated from *Phaseolus* beans. J Sci Food Agric 31:1316–1322

Amarakoon D, Thavarajah D, McPhee K et al (2012) Iron-, zinc-, and magnesium-rich field peas (*Pisum sativum* L.) with naturally low phytic acid: a potential food-based solution to global micronutrient malnutrition. J Food Compos Anal 27:8–13

Anderson JW, Major AW (2002) Pulses and lipaemia, short- and long-term effect: potential in the prevention of cardiovascular disease. Br J Nutr 88(Suppl 3):S263–271

Anderson JW, Story L, Sieling B et al (1984) Hypocholesterolemic effects of oat-bran or bean intake for hypercholesterolemic men. Am J Clin Nutr 40:1146–1155

Archer S, Meng S, Wu J et al (1998) Butyrate inhibits colon carcinoma cell growth through two distinct pathways. Surgery 124:248–253

Atkinson F, Foster-Powell K, Brand-Miller J (2008) International tables of glycemic index and glycemic load values: 2008. Diabetes Care 31:2281–2283

Bardocz S, Grant G, Pusztai A (1996) The effect of phytohaemagglutinin at different dietary concentrations on the growth, body composition and plasma insulin of the rat. Br J Nutr 76:613–626

Bardocz S, Grant G, Duguid TJ et al (1997) Intracellular levels of polyamines in Krebs II lymphosarcoma cells in mice fed phytohaemagglutinin-containing diets are coupled with altered tumour growth. Cancer Lett 121:25–29

Barnard JA, Warwick G (1993) Butyrate rapidly induces growth inhibition and differentiation in HT-29 cells. Cell Growth Differ 4:495–501

Barrett ML, Udani JK (2011) A proprietary alpha-amylase inhibitor from white bean (*Phaseolus vulgaris*): a review of clinical studies on weight loss and glycemic control. Nutr J 10:24

Bauer-Marinovic M, Florian S, Müller-Schmehl K (2006) Dietary resistant starch type 3 prevents tumor induction by 1,2-dimethylhydrazine and alters proliferation, apoptosis and dedifferentiation in rat colon. Carcinogenesis 27:1849–1859

Bazzano LA, He J, Ogden LG et al (2001) Legume consumption and risk of coronary heart disease in US men and women: NHANES I epidemiologic follow-up study. Arch Intern Med 161:2573–2578

Bazzano LA, Thompson AM, Tees MT et al (2011) Non-soy legume consumption lowers cholesterol levels: a meta-analysis of randomized controlled trials. Nutr Metab Cardiovasc Dis 21:94–103

Blair M, Izquierdo P, Astudillo C et al (2013) A legume biofortification quandary: variability and genetic control of seed coat micronutrient accumulation in common beans. Front Plant Sci 4:275

Bobe G, Barrett KG, Mentor-Marcel RA et al (2008) Dietary cooked navy beans and their fractions attenuate colon carcinogenesis in azoxymethane-induced ob/ob mice. Nutr Cancer 60:373–381

Bollini R, Carnovale E, Campion B (1999) Removal of antinutritional factors from bean *Phaseolus vulgaris* seeds. Biotechnol Agron Soc Environ 3:217–219

Buddington R, Kelly-Quagliana K, Buddington K et al (2002) Non-digestible oligosaccharides and defense functions: lessons learned from animal models. Br J Nutr 87:S231–S239
Burbano C, Muzquiz M, Ayet G et al (1999) Evaluation of antinutritional factors of selected varieties of *Phaseolus vulgaris*. J Sci Food Agric 79:1468–1472
Burstin J, Gallardo K, Mir R et al (2011) Improving protein content and nutrition quality. In: Pratrap A, Kumar J (eds) Biology and breeding of food legumes. CABI, New Delhi, pp 314–328
Cabrera C, Lloris F, Gimenez R et al (2003) Mineral content in legumes and nuts: contribution to the Spanish dietary intake. Sci Total Environ 308:1–14
Caldas G, Blair M (2009) Inheritance of seed condensed tannins and their relationship with seedcoat color and pattern genes in common bean (*Phaseolus vulgaris* L.). Theor Appl Genet 119:131–142
Campion B, Perrone D, Galasso I et al (2009a) Common bean (*Phaseolus vulgaris* L.) lines devoid of major lectin proteins. Plant Breeding 128:199–204
Campion B, Sparvoli F, Doria E et al (2009b) Isolation and characterisation of an *lpa* (low phytic acid) mutant in common bean (*Phaseolus vulgaris* L.). Theor Appl Genet 118:1211–1221
Campion B, Glahn R, Tava A et al (2013) Genetic reduction of antinutrients in common bean (*Phaseolus vulgaris* L.) seed, increases nutrients and in vitro iron bioavailability without depressing main agronomic traits. Field Crops Res 141:27–37
Campos-Vega R, Loarca-Pina G, Oomah B (2010) Minor components of pulses and their potential impact on human health. Food Res Int 43:461–482
Carbonaro M, Maselli P, Nucara A (2012) Relationship between digestibility and secondary structure of raw and thermally treated legume proteins: a Fourier transform infrared (FT-IR) spectroscopic study. Amino Acids 43:911–921
Celleno L, Tolaini MV, D'Amore A et al (2007) A dietary supplement containing standardized *Phaseolus vulgaris* extract influences body composition of overweight men and women. Int J Med Sci 4:45–52
Champ MM (2002) Non-nutrient bioactive substances of pulses. Br J Nutr 88:S307–S319
Chang WW, Yu CY, Lin TW et al (2006) Soyasaponin I decreases the expression of alpha2,3-linked sialic acid on the cell surface and suppresses the metastatic potential of B16F10 melanoma cells. Biochem Biophys Res Commun 341:614–619
Chibbar R, Ambigaipalan P, Hoover R. (2010) Molecular diversity in pulse seed starch and complex carbohydrates and its role in human nutrition and health. Cereal Chem 7:342–52
Chitra U, Vimala V, Singh U et al (1995) Variability in phytic acid content and protein digestibility of grain legumes. Plant Foods Hum Nutr 47:163–172
Chitra U, Singh U, Rao P (1996) Phytic acid, in vitro protein digestibility, dietary fiber, and minerals of pulses as influenced by processing methods. Plant Foods Hum Nutr 49:307–316
Chokshi D (2006) Toxicity studies of blockal, a dietary supplement containing phase 2 starch neutralizer (phase 2), a standardized extract of the common white kidney bean (*Phaseolus vulgaris*). Int J Toxicol 25:361–371
Chrispeeels M, Raikhel N (1991) Lectins, lectin genes, and their role in plant defense. Plant Cell 3:1–9
Clemente A, Gee JM, Johnson IT et al (2005) Pea (*Pisum sativum* L.) protease inhibitors from the Bowman-Birk class influence the growth of human colorectal adenocarcinoma HT29 cells in vitro. J Agr Food Chem 53:8979–8986
Clemente A, Sonnante G, Domoney C (2011) Bowman-Birk inhibitors from legumes and human gastrointestinal health: current status and perspectives. Curr Protein Pept Sci 12:358–373
Cobiac L, McArthur R, Nestel PJ (1990) Can eating baked beans lower plasma cholesterol? Eur J Clin Nutr 44:819–822
Confalonieri M, Bollini R, Berardo N et al (1992) Influence of phytohaemagglutinin
on the agronomic performance of beans *(Phaseolus vulgaris* L). Plant Breeding 109:329–334
Consonni A, Lovati M, Parolari A et al (2011) Heterologous expression and purification of the soybean 7S globulin alpha' subunit extension region: in vitro evidence of its involvement in cell cholesterol homeostasis. Protein Expr Purif 80:125–129
Crepon K, Marget P, Peyronnet C et al (2010) Nutritional value of faba bean (*Vicia faba* L.) seeds for feed and food. Field Crops Res 115:329–339

de Graaf C, Blom WA, Smeets PA et al (2004) Biomarkers of satiation and satiety. Am J Clin Nutr 79:946–961
De Mejia E, Prisecaru V (2005) Lectins as bioactive plant proteins: a potential in cancer treatment. Crit Rev Food Sci Nutr 45:425–445
Defreitas Z, Ausich R, Newman J et al (2003) Composition and method for reducing post-prandial blood glucose. Canadian Patent CA 2483633, 13 Nov 2003
DellaValle D, Thavarajah D, Thavarajah P et al (2013) Lentil (*Lens culinaris* L.) as a candidate crop for iron biofortification: is there genetic potential for iron bioavailability? Field Crops Res 144:119–125
Diaz-Batalla L, Widholm J, Fahey G et al (2006) Chemical components with health implications in wild and cultivated Mexican common bean seeds (*Phaseolus vulgaris* L.). J Agr Food Chem 54:2045–2052
Doria E, Campion B, Sparvoli F et al (2012) Anti-nutrient components and metabolites with health implications in seeds of 10 common bean (*Phaseolus vulgaris* L. and *Phaseolus lunatus* L.) landraces cultivated in southern Italy. J Food Compos Anal 26:72–80
Durante M, Bernardi R, Lupi M et al (1989) *Phaseolus coccineus* storage proteins. 2. Electrophoretic analysis and erythroagglutinating activity in various cultivars. Plant Breed 102:58–65
Duranti M (2006) Grain legume proteins and nutraceutical properties. Fitoterapia 77:67–82
Finley JW, Burrell JB, Reeves PG (2007) Pinto bean consumption changes SCFA profiles in fecal fermentations, bacterial populations of the lower bowel, and lipid profiles in blood of humans. J Nutr 137:2391–2398
Galisteo M, Duarte J, Zarzuelo A (2008) Effects of dietary fibers on disturbances clustered in the metabolic syndrome. J Nutr Biochem 19:71–84
Gao Z, Yin J, Zhang J et al (2009) Butyrate improves insulin sensitivity and increases energy expenditure in mice. Diabetes 58:1509–1517
Garbisa S, Sartor L, Biggin S et al (2001) Tumor gelatinases and invasion inhibited by the green tea flavanol epigallocatechin-3-gallate. Cancer 91:822–832
Geil PB, Anderson JW (1994) Nutrition and health implications of dry beans: a review. J Am Coll Nutr 13:549–558
Gibson R, Perlas L, Hotz C (2006) Improving the bioavailability of nutrients in plant foods at the household level. Proc Nutr Soc 65:160–168
Goyoaga C, Burbano C, Cuadrado C et al (2008) Content and distribution of vicine, convicine and L-DOPA during germination and seedling growth of two *Vicia faba* L. varieties. Eur Food Res Technol 227:1537–1542
Grant G, More L, Mckenzie N et al (1983) A survey of the nutritional and hemagglutination properties of legume seeds generally available in the UK. Br J Nutr 50:207–214
Grant G, Edwards J, Pusztai A (1995) Alpha-amylase inhibitor levels in seeds generally available in Europe. J Sci Food Agr 67:235–238
Grant G, Duncan M, Alonso R et al (2003) Peas and lentils. In: Caballero B (ed) Encyclopedia of food sciences and nutrition, 2nd edn. Elsevier, Amsterdam, pp 4433–4440
Guillamon E, Pedrosa M, Burbano C et al (2008) The trypsin inhibitors present in seed of different grain legume species and cultivar. Food Chem 107:68–74
Guillon F, Champ MM (2002) Carbohydrate fractions of legumes: uses in human nutrition and potential for health. Br J Nutr 88(Suppl 3):S293–S306
Guzman-Maldonado S, Acosta-Gallegos J, Paredes-Lopez O (2000) Protein and mineral content of a novel collection of wild and weedy common bean (*Phaseolus vulgaris* L.). J Sci Food Agr 80:1874–1881
Han KH, Fukushima M, Shimizu K et al (2003) Resistant starches of beans reduce the serum cholesterol concentration in rats. J Nutr Sci Vitaminol (Tokyo) 49:281–286
Hangen L, Bennink MR (2002) Consumption of black beans and navy beans (*Phaseolus vulgaris*) reduced azoxymethane-induced colon cancer in rats. Nutr Cancer 44:60–65
Hayat I, Ahmad A, Masud T eta l (2014) Nutritional and health perspectives of beans (*Phaseolus vulgaris* L.): an overview. Crit Rev Food Sci Nutr 54:580–592

Hedley CL (2001) Carbohydrates in grain legume seeds: improving nutritional quality and agronomic characteristics. CABI, New York
Hermsdorff HH, Zulet M, Abete I et al (2011) A legume-based hypocaloric diet reduces proinflammatory status and improves metabolic features in overweight/obese subjects. Eur J Nutr 50:61–69
Hernandez-Infante M, Sousa V, Montalvo I et al (1998) Impact of microwave heating on hemagglutinins, trypsin inhibitors and protein quality of selected legume seeds. Plant Hum Nutr 52:199–208
Hernández-Salazar M, Osorio-Diaz P, Loarca-Piña G et al (2010) In vitro fermentability and antioxidant capacity of the indigestible fraction of cooked black beans (*Phaseolus vulgaris* L.), lentils (*Lens culinaris* L.) and chickpeas (*Cicer arietinum* L.). J Sci Food Agr 90:1417–1422
Herzig KH, Bardocz S, Grant G et al (1997) Red kidney bean lectin is a potent cholecystokinin releasing stimulus in the rat inducing pancreatic growth. Gut 41:333–338
Hillocks R, Maruthi M (2012) Grass pea (*Lathyrus sativus*): is there a case for further crop improvement? Euphytica 186:647–654
Holt S, Brand J, Soveny C et al (1992) Relationship of satiety to postprandial glycaemic, insulin and cholecystokinin responses. Appetite 18:129–141
Hoover R, Hughes T, Chung H et al (2010) Composition, molecular structure, properties, and modification of pulse starches: a review. Food Res Int 43:399–413
Hu FB, Manson JE, Willett WC (2001) Types of dietary fat and risk of coronary heart disease: a critical review. J Am Coll Nutr 20:5–19
Hutchins AM, Winham DM, Thompson SV (2012) *Phaseolus* beans: impact on glycaemic response and chronic disease risk in human subjects. Br J Nutr 108 (Suppl 1):S52–S65
Ishimoto M, Suzuki K, Iwanaga M et al (1995) Variation of seed alpha-amylase inhibitors in the common bean. Theor Appl Genet 90:425–429
Israel D, Kwanyuen P, Burton J et al (2007) Response of low seed phytic acid soybeans to increases in external phosphorus supply. Crop Sci 47:2036–2046
Iwashita S, Mikus C, Baier S et al (2006) Glutamine supplementation increases postprandial energy expenditure and fat oxidation in humans. J Parenter Enteral Nutr 30:76–80
Jain N, Boivin M, Zinsmeister A et al (1989) Effect of ileal perfusion of carbohydrates and amylase inhibitor on gastrointestinal hormones and emptying. Gastroenterology 96:377–387
Jenkins DJ, Kendall CW, Augustin LS et al (2002) Glycemic index: overview of implications in health and disease. Am J Clin Nutr 76:266S–273S
Jordinson M, El-Hariry I, Calnan Det al (1999) *Vicia faba* agglutinin, the lectin present in broad beans, stimulates differentiation of undifferentiated colon cancer cells. Gut 44:709–714
Jukanti AK, Gaur PM, Gowda CLet al (2012) Nutritional quality and health benefits of chickpea (*Cicer arietinum* L.): a review. Br J Nutr 108(Suppl 1):S11–S26
Karakoy T, Erdem H, Baloch F et al (2012) Diversity of macro- and micronutrients in the seeds of lentil landraces. Sci World J Article ID 710412, 9 pages, doi:10.1100/2012/710412
Karlström B, Vessby B, Asp NG et al (1987) Effects of leguminous seeds in a mixed diet in non-insulin-dependent diabetic patients. Diabetes Res 5:199–205
Kennedy A, Rostami A (2004) Use of bowman birk inhibitor for the treatment of multiple sclerosis and other autoimmune diseases. US Patent 20040142050, 22 Jul 2004
Khalil A, El-Adawy T (1994) Isolation, identification and toxicity of saponin from different legumes. Food Chem 50:197–201
Khatib O, Aljurf M (2008) Cancer prevention and control in the Eastern Mediterranean region: the need for a public health approach. Hematol Oncol Stem Cell Ther 1:44–52
Koh-Banerjee P, Rimm EB (2003) Whole grain consumption and weight gain: a review of the epidemiological evidence, potential mechanisms and opportunities for future research. Proc Nutr Soc 62:25–29
Konietzny U, Greiner R (2003) Phytic acid: nutritional impact. In: Caballero B (ed) Encyclopedia of food sciences and nutrition, 2nd edn. Elsevier, Amsterdam, pp 4555–4563
Kordas K, Szalmay G, Bardocz S et al (2001) Phytohaemagglutinin inhibits gastric acid but not pepsin secretion in conscious rats. J Physiol 95:309–314

Kushi LH, Meyer KA, Jacobs DR (1999) Cereals, legumes, and chronic disease risk reduction: evidence from epidemiologic studies. Am J Clin Nutr 70:451S–458S
Lajolo F, Genovese M, Pryme I et al (2004) Beneficial (antiproliferative) effects of different substances. In: Muzquiz M (ed) Recent advances of research in antinutritional factors in legume seeds and oilseeds. Wageningen Academic, Wageningen, pp 123–135
Lawrence M, Izard T, Beuchat M et al (1994) Structure of phaseolin at 2-center-dot-2 Angstrom resolution—implications for a common vicilin/legumin structure and the genetic-engineering of seed storage proteins. J Mol Biol 238:748–776
Layer P, Zinsmeister A, Dimagno E (1986) Effects of decreasing intraluminal amylase activity on starch digestion and postprandial gastrointestinal function in humans. Gastroenterology 91:41–48
Liener I (1994) Implications of antinutritional components in soybean foods. Crit Rev Food Sci Nutr 34:31–67
Lovati MR, Manzoni C, Castiglioni S et al (2012) Lupin seed γ-conglutin lowers blood glucose in hyperglycaemic rats and increases glucose consumption of HepG2 cells. Br J Nutr 107:67–73
Mackay S, Ball MJ (1992) Do beans and oat bran add to the effectiveness of a low-fat diet? Eur J Clin Nutr 46:641–648
Magni C, Sessa F, Accardo E et al (2004) Conglutin gamma, a lupin seed protein, binds insulin in vitro and reduces plasma glucose levels of hyperglycemic rats. J Nutr Biochem 15:646–650
Marinangeli CP, Jones PJ (2012) Pulse grain consumption and obesity: effects on energy expenditure, substrate oxidation, body composition, fat deposition and satiety. Br J Nutr 108(Suppl 1):S46–S51
Marles M, Balasubramanian P, Bett K (2010) Differential accumulation of polyphenolics in black bean genotypes grown in four environments. J Agr Food Chem 58:7001–7006
Marlett JA, McBurney MI, Slavin JL et al (2002) Position of the American dietetic association: health implications of dietary fiber. J Am Diet Assoc 102:993–1000
Maroof M, Glover N, Biyashev R (2009) Genetic basis of the low-phytate trait in the soybean line CX1834. Crop Sci 49:69–76
Marsolais F, Pajak A, Yin F et al (2010) Proteomic analysis of common bean seed with storage protein deficiency reveals up-regulation of sulfur-rich proteins and starch and raffinose metabolic enzymes, and down-regulation of the secretory pathway. J Proteomics 73:1587–1600
Martin C, Zhang Y, Tonelli C et al (2013) Plants, diet, and health. Annu Rev Plant Biol 64:19–46
Mathers JC (2002) Pulses and carcinogenesis: potential for the prevention of colon, breast and other cancers. Br J Nutr 88(Suppl 3):S273–S279
McCrory MA, Hamaker BR, Lovejoy JC et al (2010) Pulse consumption, satiety, and weight management. Adv Nutr 1:17–30
McCue P, Shetty K (2004) Health benefits of soy isoflavonoids and strategies for enhancement: a review. Crit Rev Food Sci Nutr 44:361–367
McKnight JR, Satterfield MC, Jobgen WS et al (2010) Beneficial effects of L-arginine on reducing obesity: potential mechanisms and important implications for human health. Amino Acids 39:349–357
Mollard RC, Luhovyy BL, Panahi S et al (2012) Regular consumption of pulses for 8 weeks reduces metabolic syndrome risk factors in overweight and obese adults. Br J Nutr 108 (Suppl 1):S111–S122
Montoya C, Lalles J, Beebe S et al (2006) Influence of the *Phaseolus vulgaris* phaseolin level of incorporation, type and thermal treatment on gut characteristics in rats. Br J Nutr 95:116–123
Montoya C, Leterme P, Victoria N et al (2008) Susceptibility of phaseolin to in vitro proteolysis is highly variable across common bean varieties (*Phaseolus vulgaris*). J Agr Food Chem 56:2183–2191
Montoya C, Lalles J, Beebe S et al (2010) Phaseolin diversity as a possible strategy to improve the nutritional value of common beans (*Phaseolus vulgaris*). Food Res Int 43:443–449
Murillo G, Choi JK, Pan O et al (2004) Efficacy of garbanzo and soybean flour in suppression of aberrant crypt foci in the colons of CF-1 mice. Anticancer Res 24:3049–3055

Murty CM, Pittaway JK, Ball MJ (2010) Chickpea supplementation in an Australian diet affects food choice, satiety and bowel health. Appetite 54:282–288
Muzquiz M, Varela A, Burbano C et al (2012) Bioactive compounds in legumes: pronutritive and antinutritive actions. Implications for nutrition and health. Phytochem Rev 11:227–244
Nair R, Yang R, Easdown W et al (2013) Biofortification of mungbean (*Vigna radiata*) as a whole food to enhance human health. J Sci Food Agr 93:1805–1813
Nervi F, Covarrubias C, Bravo P et al (1989) Influence of legume intake on biliary lipids and cholesterol saturation in young Chilean men. Identification of a dietary risk factor for cholesterol gallstone formation in a highly prevalent area. Gastroenterology 96:825–830
Nijveldt RJ, van Nood E, van Hoorn DE et al (2001) Flavonoids: a review of probable mechanisms of action and potential applications. Am J Clin Nutr 74:418–425
Noah L, Guillon F, Bouchet B et al (1998) Digestion of carbohydrate from white beans (*Phaseolus vulgaris* L.) in healthy humans. J Nutr 128:977–985
Noakes M, Clifton P, McMurchie T (1999) The role of diet in cardiovascular health. A review of the evidence. Aust J Nutr Diet 56:S3–S22
Obiro WC, Zhang T, Jiang B (2008) The nutraceutical role of the *Phaseolus vulgaris* alpha-amylase inhibitor. Br J Nutr 100:1–12
Oliva ML, Ferreira R da S, Ferreira JG et al (2011) Structural and functional properties of kunitz proteinase inhibitors from Leguminosae: a mini review. Curr Protein Pept Sci 12:348–357
Oomah B, Luc G, Leprelle C et al (2011) Phenolics, phytic acid, and phytase in canadian-grown low-tannin faba bean (*Vicia faba* L.) genotypes. J Agr Food Chem 59:3763–3771
Panzeri D, Cassani E, Doria E et al (2011) A defective ABC transporter of the MRP family, responsible for the bean *lpa1* mutation, affects the regulation of the phytic acid pathway, reduces seed *myo*-inositol and alters ABA sensitivity. New Phytol 191:70–83
Papanikolaou Y, Fulgoni VL (2008) Bean consumption is associated with greater nutrient intake, reduced systolic blood pressure, lower body weight, and a smaller waist circumference in adults: results from the national health and nutrition examination survey 1999–2002. J Am Coll Nutr 27:569–576
Petry N, Egli I, Campion B et al (2013) Genetic reduction of phytate in common bean (*Phaseolus vulgaris* L.) seeds increases iron absorption in young women. J Nutr 143:1219–1224
Piergiovanni A, Pignone D (2003) Effect of year-to-year variation and genotype on trypsin inhibitor level in common bean (*Phaseolus vulgaris* L.) seeds. J Sci Food Agr 83:473–476
Porres J, Aranda P, Lopez-Jurado M et al (2007) Nitrogen fractions and mineral content in different lupin species (*Lupinus albus*, *Lupinus angustifolius*, and *Lupinus luteus*). Changes induced by the alpha-galactoside extraction process. J Agr Food Chem 55:7445–7452
Pras N, Woerdenbag H, Batterman S et al (1993) *Mucuna pruriens* – improvement of the biotechnological production of the anti-Parkinson drug L-DOPA by plant-cell selection. Pharmacy World Sci 15:263–268
Pryme IF, Bardocz S (2001) Anti-cancer therapy: diversion of polyamines in the gut. Eur J Gastroenterol Hepatol 13:1041–1046
Pueyo J, Delgado Salinas A (1997) Presence of alpha-amylase inhibitor in some members of the subtribe phaseolinae (Phaseoleae: Fabaceae). Am J Bot 84:79–84
Pusztai A, Bardocz S (1996) Biological effects of plant lectins on the gastrointestinal tract: metabolic consequences and applications. Trends Glycosci Glycotechnol 8:149–165
Pusztai A, Grant G, Duguid T et al (1995) Inhibition of starch digestion by alpha-amylase inhibitor reduces the efficiency of utilization of dietary proteins and lipids and retards the growth of rats. J Nutr 125:1554–1562
Pusztai A, Grant G, Buchan WC et al (1998) Lipid accumulation in obese Zucker rats is reduced by inclusion of raw kidney bean (*Phaseolus vulgaris*) in the diet. Br J Nutr 79:213–221
Raboy V (2003) Myo-Inositol-1,2,3,4,5,6-hexakisphosphate. Phytochemistry 64:1033–1043
Radberg K, Biernat M, Linderoth A et al (2001) Enteral exposure to crude red kidney bean lectin induces maturation of the gut in suckling pigs. J Anim Sci 79:2669–2678
Rasmussen S, Ingvardsen C, Torp A (2010) Mutations in genes controlling the biosynthesis and accumulation of inositol phosphates in seeds. Biochem Soc Trans 38:689–694

Reddy N, Pierson M, Sathe S et al (1984) Chemical, nutritional and physiological-aspects of dry bean carbohydrates – a review. Food Chem 13:25–68
Robinson SM, Jaccard C, Persaud C et al (1990) Protein turnover and thermogenesis in response to high-protein and high-carbohydrate feeding in men. Am J Clin Nutr 52:72–80
Rochfort S, Panozzo J (2007) Phytochemicals for health, the role of pulses. J Agr Food Chem 55:7981–7994
Roy F, Boye JI, Simpson BK (2010) Bioactive proteins and peptides in pulse crops: pea, chickpea and lentil. Food Res Int 43:432–442
Saini H, Knights E (1984) Chemical constitution of starch and oligosaccharide components of desi and kabuli chickpea (*Cicer arietinum*) seed types. J Agr Food Chem 32:940–944
Segev A, Badani H, Kapulnik Y et al (2010) Determination of polyphenols, flavonoids, and antioxidant capacity in colored chickpea (*Cicer arietinum* L.). J Food Sci 75:S115–S119
Serrano J, Puupponen-Pimia R, Dauer A et al (2009) Tannins: current knowledge of food sources, intake, bioavailability and biological effects. Mol Nutr Food Res 53:S310–S329
Shi J, Arunasalam K, Yeung D et al (2004) Saponins from edible legumes: chemistry, processing, and health benefits. J Med Food 7:67–78
Shi J, Xue S, Ma Y et al (2009) Kinetic study of saponins B stability in navy beans under different processing conditions. J Food Eng 93:59–65
Sievenpiper JL, Kendall CWC, Esfahani A et al (2009) Effect of non-oil-seed pulses on glycaemic control: a systematic review and meta-analysis of randomised controlled experimental trials in people with and without diabetes. Diabetologia 52:1479–1495
Sirtori CR, Lovati MR, Manzoni C et al (2004) Proteins of white lupin seed, a naturally isoflavone-poor legume, reduce cholesterolemia in rats and increase LDL receptor activity in HepG2 cells. J Nutr 134:18–23
Sompong U, Kaewprasit C, Nakasathien S et al (2010) Inheritance of seed phytate in mungbean (*Vigna radiata* (L.) Wilczek). Euphytica 171:389–396
Sowmya P, Rajyalakshmi P (1999) Hypocholesterolemic effect of germinated fenugreek seeds in human subjects. Plant Foods Hum Nutr 53:359–365
Sparvoli F, Daminati M, Cantoni R et al (1999) Possible role(s) of glycosylation in the biological properties of lectins and related proteins. In: Bardocz S, Hajos G, Pusztai A (eds) COST 98: effects of antinutrients on the nutritional value of legume diets, vol 6. EU, Luxemburg, pp 43–49
Sreerama Y, Sashikala V, Pratape V, Singh V (2012) Nutrients and antinutrients in cowpea and horse gram flours in comparison to chickpea flour: Evaluation of their flour functionality. Food Chem. 131:462–468
Sweeney H, Morris C, Kennedy A (2005) Bowman-birk inhibitor compositions for treatment of muscular atrophy and degenerative muscle disease. Patent Publication WO2005011596, 10 Feb 2005
Tappy L, Wursch P, Randin J et al (1986) Metabolic effect of precooked instant preparations of bean and potato in normal and in diabetic subjects. Am J Cl Nutr 43:30–36
Taunk J, Yadav N, Yadav R et al (2012) Genetic diversity among greengram (*Vigna radiata* (L.) Wilczek) genotypes varying in micronutrient (Fe and Zn) content using RAPD markers. Indian J Biotech 11:48–53
Taylor M, Chapman R, Beyaert R et al (2008) Seed storage protein deficiency improves sulfur amino acid content in common bean (*Phaseolus vulgaris* L.): redirection of sulfur from gamma-glutamyl-S-methyl-cysteine. J Agr Food Chem 56:5647–5654
Tharanathan R, Mahadevamma S (2003) Grain legumes – a boon to human nutrition. Trends Food Sci Thecnol 14:507–518
Thavarajah D, Thavarajah P (2012) Evaluation of chickpea (*Cicer arietinum* L.) micronutrient composition: biofortification opportunities to combat global micronutrient malnutrition. Food Res Int 49:99–104
Thavarajah D, Thavarajah P, Sarker A et al (2009) Lentils (*Lens culinaris* Medikus subspecies *culinaris*): a whole food for increased iron and zinc intake. J Agr Food Chem 57:5413–5419
Thompson LU, Button CL, Jenkins DJ (1987) Phytic acid and calcium affect the in vitro rate of navy bean starch digestion and blood glucose response in humans. Am J Clin Nutr 46:467–473

Thompson S, Winham D, Hutchins A (2012) Bean and rice meals reduce postprandial glycemic response in adults with type 2 diabetes: a cross-over study. Nutr J 11:23
Trinidad T, Mallillin A, Loyola A et al (2010) The potential health benefits of legumes as a good source of dietary fibre. Br J Nutr 103:569–574
Trugo L, Donangelo C, Duarte Y et al (1993) Phytic acid and selected mineral-composition of seed from wild-species and cultivated varieties of lupin. Food Chem 47:391–394
Urbano G, Lopez-Jurado M, Aranda P et al (2000) The role of phytic acid in legumes: antinutrient or beneficial function? J Physiol Biochem 56:283–294
USDA, Special Interest Databases on Flavonoids. Release 3.1. http://www.ars.usda.gov/Services/docs.htm?docid=24953.
Uzun A, Gucer S, Acikgoz E (2011) Common vetch (*Vicia sativa* L.) germplasm: correlations of crude protein and mineral content to seed traits. Plant Foods Hum Nutr 66:254–260
Vasconcelos I, Oliveira J (2004) Antinutritional properties of plant lectins. Toxicon 44:385–403
Venn B, Mann J (2004) Cereal grains, legumes and diabetes. Eur J Cl Nutr 58:1443–1461
Venter CS, Vorster HH, Cummings JH (1990) Effects of dietary propionate on carbohydrate and lipid metabolism in healthy volunteers. Am J Gastroenterol 85:549–553
Villegas R, Gao YT, Yang G et al (2008) Legume and soy food intake and the incidence of type 2 diabetes in the Shanghai women's health study. Am J Clin Nutr 87:162–167
Vucenik I, Shamsuddin AM (2006) Protection against cancer by dietary IP_6 and inositol. Nutr Cancer 55:109–125
Wang TL, Domoney C, Hedley CL et al (2003) Can we improve the nutritional quality of legume seeds? Plant Physiol 131:886–891
Ware J, Wan X, Newberne P et al (1999) Bowman-Birk inhibitor concentrate reduces colon inflammation in mice with dextran sulfate sodium-induced ulcerative colitis. Dig Dis Sci 44:986–990
Warkentin T, Delgerjav O, Arganosa G et al (2012) Development and characterization of low-phytate pea. Crop Sci 52:74–78
Wilcox J, Premachandra G, Young K et al (2000) Isolation of high seed inorganic P, low-phytate soybean mutants. Crop Sci 40:1601–1605
Winham DM, Hutchins AM, Johnston CS (2007) Pinto bean consumption reduces biomarkers for heart disease risk. J Am Coll Nutr 26:243–249
Wolever T, Jenkins D (2001) Effect of dietary fiber and foods on carbohydrate metabolism. In: Spiller G (ed) CRC Handbook of dietary fiber in human nutrition, 3rd edn. CRC, Boca Raton, pp 321–362
Wong CL, Mollard RC, Zafar TA et al (2009) Food intake and satiety following a serving of pulses in young men: effect of processing, recipe, and pulse variety. J Amer Coll Nutr 28:543–552
Xu B, Chang S (2007) A comparative study on phenolic profiles and antioxidant activities of legumes as affected by extraction solvents. J Food Sci 72:S159–S166
Xu B, Chang SKC (2012) Comparative study on antiproliferation properties and cellular antioxidant activities of commonly consumed food legumes against nine human cancer cell lines. Food Chem 134:1287–1296
Xu B, Yuan S, Chang S (2007) Comparative analyses of phenolic composition, antioxidant capacity, and color of cool season legumes and other selected food legumes. J Food Sci 72:S167–S177
Yoon JH, Thompson LU, Jenkins DJ (1983) The effect of phytic acid on in vitro rate of starch digestibility and blood glucose response. Am J Clin Nutr 38:835–842
Zaugg I, Magni C, Panzeri D et al (2013) QUES, a new *Phaseolus vulgaris* genotype resistant to common bean weevils, contains the Arcelin-8 allele coding for new lectin-related variants. Theor Appl Genet 126:647–661
Zhu J, Xiong L, Yu B et al (2005) Apoptosis induced by a new member of saponin family is mediated through caspase-8-dependent cleavage of Bcl-2. Mol Pharmacol 68:1831–1838

Chapter 11
Seed Physiology and Germination of Grain Legumes

Jaime Kigel, Leah Rosental and Aaron Fait

1 Introduction

Grain legumes belong to the Fabaceae (=Leguminosae) family in which the structure and anatomy of the seeds, as well as their main mechanism of dormancy and germination control based on seed-coat impermeability to water, have been remarkably conserved, despite the prolonged time span since the origin of the family during the early Tertiary, ca. 60 million years (Myr) ago (Cronk et al. 2006). The major crop legumes are included in the Hologalegina and the Millettoid/Phaseoloid sister clades of the Papilionoideae (=Faboideae) subfamily (Bruneau et al. 2013). Temperate herbaceous legumes, such as the pulses chickpea *(Cicer arietinum),* lentil *(Lens culinaris),* garden pea *(Pisum sativum),* faba bean *(Vicia faba),* sweet pea (*Lathyrus* spp.), and pasture and fodder species (*Trifolium* spp., *Medicago* spp.) belong to the Hologalegina clade. In contrast, tropical and subtropical legumes such as common bean *(Phaseolus vulgaris),* lima bean *(Phaseolus lunatus),* runner bean *(Phaseolus coccineus),* tepary bean *(Phaseolus acutifolius),* cowpea *(Vigna unguiculata)*, mungbean *(Vigna radiata),* pigeon pea *(Cajanus cajan),* and soybean *(Glycine max)* belong to the Millettoid/Phaseoloid clade. Lupins (*Lupinus* spp.) are included in the minor Genistoid clade (Wojciechowski et al. 2004). Germination responses of the different legume crops to environmental conditions are strongly associated with taxonomic affiliation and climatic regions of origin.

J. Kigel (✉)
Hebrew University of Jerusalem, POB 10, 76100 Rehovot, Israel
e-mail: kigel@agri.huji.ac.il

L. Rosental
French Associates Institute for Agriculture & Biotechnology of Drylands, Jacob Blaustein Institutes for Desert Research, Ben-Gurion University of the Negev, 48990 Beer Sheva, Israel
e-mail: leah.zucker@mail.huji.ac.il

A. Fait
Blaustein Institutes for Desert Research, Ben Gurion University, 84990 Midreshet Ben-Gurion, Israel
e-mail: fait@bgu.ac.il

A. M. De Ron (ed.), *Grain Legumes,* Handbook of Plant Breeding 10,
DOI 10.1007/978-1-4939-2797-5_11

2 Seed Changes During Domestication of Grain Legumes

The legume family has 41 species with domesticated crops, the greatest number compared to any other plant family (Harlan 1992). Grain legumes are grown for their highly nutritious seeds, already an important component of the human diet in early civilizations (Zohary et al. 2012). As in other seed crops, the "domestication syndrome" in legumes involved a set of traits that differentiate between domesticated crops and their wild ancestors (Fuller 2007; Weedin 2007; Zohary et al. 2012; Abbo et al. 2014). Regarding seed traits, the syndrome included loss of seed dispersal, an increase in seed size, and loss of seed dormancy. Loss of seed dispersal through selection for non-dehiscent pods was a primary step in legume domestication. Together with synchronous seed maturation, it allows higher yields, since losses due to pod dehiscence are prevented and harvesting can be carried out when most seeds on a plant have matured. As to seed size, many wild legume species produce relatively large seeds rich in proteins, presumably as a consequence of symbiotic nitrogen fixation, making them an attractive food source. Large seeds rich in nitrogen facilitate seedling establishment, particularly in nutrient-poor tropical soils. Furthermore, larger seeds generally produce bigger and more competitive seedlings (Nakamura 1988; Al-Karaki 1998; Bond et al. 1999; Gan et al. 2006), with higher stress tolerance (Westoby et al. 1992; Gholami et al. 2009) and able to emerge from deeper burial due to greater seed reserves (Lush and Wien 1980; Kluyver et al. 2013). Therefore, large seeds are also more suitable for sowing and cultivation practices. Indeed, it has been proposed that increased seed size was selected for during domestication by tillage and cultivation (Harlan et al. 1973). Seed mass of cultivated grain legumes is often up to tenfold larger compared to wild relatives, as in common bean (Gepts and Debouck 1991; Santalla et al. 2004), mungbean and azuki bean (Isemura et al. 2007), rice bean (*Vigna umbellata;* Isemura et al. 2010), lablab bean (*Lablab purpureus;* Maass and Usongo 2007), and soybean (Kluyver et al. 2013). The increase in seed mass of grain legumes with domestication is usually linked to a reduction in seed-coat thickness and increased seed-coat permeability (Lush and Evans 1980).

Regarding seed germination, domestication resulted in the reduction and loss of seed dormancy. Plant species usually possess dormancy mechanisms that disperse germination in time, thus reducing the risk of catastrophic germination and population extinction due to soil seed-bank depletion, as well as preventing sibling competition, particularly in annual species (Kigel 1995; Baskin and Baskin 2014). In domesticated crops, in contrast, fast and uniform germination is necessary to ensure early plant establishment before severe weed competition occurs and to allow synchronous seed set and maturation before harvest. Loss of dormancy leads to simultaneous germination after sowing once water availability and temperature allow it. In domesticated crops, lack of dormancy was most probably selected for by sowing from harvested seeds, because seeds that do not germinate will not contribute to the harvested seed population (Fuller and Allaby 2009). Moreover, in several grain legumes, early domestication occurred independently at different locations, implying that in these species loss of dormancy took place several times in historical times. The fact that seed-coat permeability in legumes is controlled by a few genes (see Sect. 3.10) probably facilitated loss of dormancy in the different domestication events (Zohary et al. 2012). This is the case for common bean that was domesticated in South America and Meso America in different times and locations (Brucher 1988; Gepts and Debouck 1991; Kaplan and Lynch 1999; Gepts et al. 2005). Seeds of *Phaseolus vulgaris* ssp. *aborigineous,* a wild relative of common bean, are dormant due to seed-coat impermeability (Kaplan 1965).

3 Seed Structure and Germination Control in Legume Seeds

3.1 Seed Dormancy and Hardseededness

A comprehensive classification consisting of five classes of seed dormancy has been proposed by Baskin and Baskin (2014, 2004): physiological (PD), morphological (MD), morpho-physiological (MPD), physical (PY), and combinational (PY+PD). MD refers to seeds that have an underdeveloped embryo that requires time to grow and germinate. Physiological embryo dormancy, the most prevalent type of dormancy, is regulated by hormonal action involving abscisic acid (ABA) and gibberellins (GA). PY is due to the presence of water impermeable seed covers that prevent water uptake by the seed, and can be broken by mechanical or chemical scarification. PY occurs in 18 evolutionarily advanced angiosperm families (Gamma-Arachchige et al. 2013), sometimes combined with physiological embryo dormancy, namely PY+PD (Baskin et al. 2000). PY is widespread in the legume family and has been extensively studied. It usually occurs in the wild ancestors of cultivated legumes and is one of the key traits modified through domestication (Gepts and Debouck 1991; Abbo et al. 2011; Zohary et al. 2012). In recent years, PY+PD has been found in an increasing number of papilionoid legumes—species of *Vicia, Trifolium, Medicago, Lathyrus* (van Assche and Vandelook 2010; Hu et al. 2013). The presence of physiological dormancy is probably more common than presently known in papilionoid legumes and in the wild ancestors of legume crops.

Seeds with PY that are unable to imbibe and germinate when in contact with free water are defined as "hard (impermeable) seeds," in contrast to "soft seeds" which swell during imbibition. Thus, "hardseededness" is the inability of seeds to absorb water and germinate. Legumes' hard seeds may remain impervious to water even after prolonged imbibition (Rolston 1978; Tran and Cavanagh 1984). The biological and ecological advantages of the hard-seeded character are a longer seed life span due to protection against seed decay under humid conditions and postponement of germination, thus building up seed reserves in the soil (Baskin and Baskin 2014; Taylor 2005). For instance, wild azuki bean is a summer annual with impermeable seeds that during winter in wet soil at low temperatures inhibiting germination. Permeable seeds that imbibe under these conditions are lost since they fail to germinate in the cold soil and decay due to the action of soil pathogens (Kaga et al. 2008). In wild legumes, degree of hardseededness varies within and between cohorts of seeds produced by different plants in the natural population. Variation in the fraction of impermeable seeds and in the duration of hardseededness arises from genetic differences, environmental effects during plant and seed development and from intra-plant somatic effects related to nodal position of the pods and to position of the seeds within the pods (Nakamura 1988; Kigel 1995). Thus, wild legumes produce seed cohorts that vary in their propensity to imbibe water, spreading germination of their offspring over time. However, hardseededness is an undesirable trait for cultivation of grain legumes and for the food processing industry (Ross et al. 2010). Seed soaking is the initial step in the thermal processing of raw legume seeds in species with toxic compounds and antinutritional factors in their seeds. Furthermore, impermeable seed coats delay seed imbibition, increasing cooking time and fuel use and may lower the quality of the food product. On the other hand, the association between increased seed-coat permeability and thinner seed coats make seeds more susceptible to mechanical damage during harvest and seed processing. Hence, past selection and current breeding programs aim at creating lines with seed coats that are fairly permeable to water and yet reasonably strong to reduce mechanical damage to the seeds.

3.2 Legume Seed Structure and Seed-Coat Anatomy

Seed development has been thoroughly studied in grain legumes, particularly in pea (van Dongen et al. 2003; Nadeau et al. 2011). Legume seeds develop from campylotropous ovules covered by two integuments, have well-developed embryos with storage cotyledons and a seed coat with a distinctive anatomy (Boesewinkel and Bouman 1995; Werker 1997). The more massive, outer integument differentiates into the seed testa that is composed by several cell layers with different functions (Watson 1948; Chamberlin et al. 1994; Ma et al. 2004). The outermost layer, the epidermis, develops into a uniseriate palisade layer of elongated and narrow macrosclereids, the Malpighian cells oriented perpendicularly to the seed surface. The palisade layer is covered by a thin cuticle and lacks intercellular spaces. The macrosclereids are tightly packed, with thick and often lignified cell walls that partially occlude the cell lumen. Their outermost cell wall is usually thicker and often suberized, forming the macrosclereid cap. Cells of the subepidermal layer differentiate into a continuous uniseriate sheath of osteosclereids (bone shaped, hourglass cells), with uneven cell wall thickenings and an extensive network of intercellular spaces. The macrosclereid and osteosclereid layers contribute to the mechanical strength of the seed coat. Deeper layers, called the nutritional tissue, consist of branched parenchymatic cells with intercellular spaces and collapses during seed maturation due to the pressure exerted by the expanding cotyledons. The thinner, two-layered inner integument is obliterated during seed ontogeny, but may produce a cuticle on the surface facing the endosperm (Werker 1997). In species that retain an endospermic layer in the mature seed, this internal cuticle adheres to the aleurone layer when the inner integument degenerates.

A main function of the seed coat during seed development is to supply nutrients arriving via the funiculus and distribute them to the growing embryo through a vascular net embedded in its parenchyma tissue (Patrick and Offler 2001). The vasculature of the seed coat differs among papilionoid legumes—in soybean and common bean (Phaseoleae tribe) the seed coats have an extensive vascular network, whereas in pea and faba bean (Vicieae tribe), as well as in *Medicago truncatula,* it consists of a single chalazal vein with two lateral branches (van Dongen et al. 2003; Wang and Grusak 2005). The parenchyma tissue is responsible for the post-phloem symplastic transport of nutrients from the seed coat to the embryo cotyledons. Solutes imported by the phloem, such as sucrose, amino acids, and minerals, are unloaded first into the parenchyma cells, then released into the apoplast and conveyed thereafter through the endosperm to the growing embryo and storage cotyledons (Patrick and Offler 2001; van Dongen et al. 2003).

In contrast to the relatively uniform structure of the testa, the presence, amount, and function of the endosperm in the mature seed vary among papilionoid legumes (Kirkbridge et al. 2003). About 66% of the 452 genera in the subfamily have endosperm in the mature seeds. In the other 154 genera, the endosperm is fully consumed and obliterated during seed ontogeny and is characteristically absent in mature seeds of the Phaseolae tribe (e.g., *Phaseolus, Vigna, Cicer;* Yeung and Cavey 1990), but may remain as a uniseriate aleurone layer of unknown function, as in soybean (Ma et al. 2004). In genera with endospermic seeds, the endosperm typically encloses the embryo. Of these, in 97 genera, the endosperm is thick and functions as a storage tissue with cell walls rich in galactomannans, thus complementing storage in the cotyledons, as in fenugreek *(Trigonella foenum-graecum),* crimson clover *(Trifolium incarnatum)* and lucerne (*Medicago sativa;* Reid and Meier 1972). In these seeds, the testa is separated from the embryo by a well-developed endosperm with an outermost aleurone layer, responsible for synthesis and secretion of the enzymes involved in degradation of the galactomannans stored in the cell walls of the inner, larger endosperm cells. This type of endosperm is well developed in seeds of the legume *Cyamopsis tetragonoloba,* an important industrial source of guar gum (McClendon et al. 1976). In addition to its storage role, swelling of the strongly hydrophilic galactomannans in the endosperm during seed imbibition causes seed-coat rupture, thus facilitating radicle protrusion (Reid and Bewley 1979). Most interestingly, in 177 genera, the endosperm remains as a continuous thin layer of living cells of unknown function. In some species, the endosperm is reduced to an aleurone layer lacking intercellular spaces, while in others it includes a few additional layers of endosperm cells (Watson 1948; Ma et al. 2004). In the mature seeds of these species, the endosperm surrounds the entire embryo, enclosing the radicle within a sheath. In *Medicago truncatula* (Bolingue et al. 2010) and *Trifolium repens* (Jakobsen et al. 1994), this endospermic envelope is not fused to the testa and is composed of several layers of living cells, similarly to the endosperm cover involved in the physiological dormancy of Solanaceae and Asteraceae

(Finch-Savage and Leubner-Metzger 2006). However, its function in papilionoid legumes has not been elucidated till now.

Altogether, development and differentiation of the seed coat in legumes is a complex and highly regulated process. Cell layers differentiate rapidly according to the changing role of the seed coat—from nutrient transport and metabolism, during the embryo growth phase, to the structural and chemical characteristics required for protection and control of germination of mature seeds after dispersal (Miller et al. 1999). This complexity is reflected in the transcriptome analysis of the seed coat in *Medicago truncatula,* which contains more than 30,000 genes (Verdier et al. 2013).

3.3 Anatomical and Chemical Bases of Seed-Coat Impermeability

Seed-coat impermeability in papilionoid legumes is due to structural and chemical changes that take place during the last stages of seed maturation. Seed dehydration and subsequent contraction of cells in the seed coat result in mechanical compression and a closer cell packing that drastically reduce rate of water uptake by the seeds. Complete impermeability requires, however, the presence of hydrophobic compounds in the cells and closure of all water gaps in the seed coat (Werker 1997; Gamma-Arachchige et al. 2013). Yet, there is no consensus regarding the nature of these hydrophobic compounds. The palisade layer is generally considered as the main permeability barrier in legume seeds. The waxy layer and thin cuticle covering the palisade layer are not accountable for seed impermeability in many legumes, since abrasion or organic solvents did not render the seed coat permeable to water (Rolston 1978; Werker 1997). Nevertheless, cracks in the cuticle lead to permeable seeds in some legumes, such as soybean (Ma et al. 2004). A subcuticular mucilaginous layer is present in the outer cell wall of the palisade cells. It is composed of pectic insoluble materials that contract upon desiccation and undergo chemical changes that make them hard and hydrophobic, with a very low swelling capacity (Werker 1997).

In *Pisum,* the mucilaginous layer is conspicuous in impermeable seeds of the wild species *Pisum elatius, Pisum fulvum,* and *Pisum humile,* but barely distinguished in cultivated *Pisum sativum* with a permeable seed coat (Werker et al. 1979). In contrast, in *Lupinus palaestinus,* the mucilaginous layer remains soft, but seeds are impermeable (Werker 1997). Impermeability of the palisade layer has been attributed to the deposition of callose and suberin in the macrosclereids (*Melilotus alba*—Riggio-Bevilaqua et al. 1989), to suberized caps on the outer walls of the macrosclereids (*Trifolium subterraneum*—Aitken 1939), and to the presence of phenolic compounds in the cell walls and lumen of macrosclereids and osteosclereids (*Pisum* spp., Werker et al. (1979)). In *Pisum elatius* and *Pisum fulvum* with impermeable seeds, catechol oxidase activity increases sharply during seed maturation and is associated with a higher content of phenols in the seed coat. These changes did not occur in cultivated *Pisum sativum* with permeable seed coat (Marbach and

Mayer 1974). When seeds of *Pisum elatius* were dried in the absence of O_2 the seed coat became totally permeable to water, thus suggesting that oxidation of phenolic compounds renders the seed-coat impermeability (Marbach and Mayer 1975). In black common bean, phenolic compounds such as tannins are associated to reduced water uptake (Sievwright and Shipe 1986). Similarly, in permeable lines of common bean, seeds have less phenolic compounds in the osteosclereids at the site of initial water entry in the raphe and chalazal region, compared to semi-hard lines (Holubowicz et al. 1988). In navy lines of common bean, faster water uptake was associated with seed coats containing lower levels of phenolics, tannins, and unsaturated fatty acids, compared to pinto bean lines with slower water uptake (Ross et al. 2010). Other studies, however, indicate that phenolic compounds and callose do not play a role in seed-coat impermeability of other papilionoid seeds (Serrato-Valenti et al. 1989).

The role of the internal cuticle of the inner integument in the control of seed imbibition by legume seeds has been questioned and apparently does not impair water uptake (Ma et al. 2004). Localization of the impermeability barrier has been attempted by piercing the seed coat to different depth layers and correlating with their anatomy and composition. Seeds became permeable after piercing the whole palisade layer in *Lupinus palaestinum* (Werker 1997) and up to the endosperm in *Coronilla varia* (McKee et al. 1977), while in *Medicago rotata* an intact osteosclereid layer prevents water uptake (Russi et al. 1992). Thus, different tissues can contribute to seed-coat impermeability in papilionoid legumes, even though the palisade layer is generally viewed as the main barrier to water uptake. Moreover, the fact that seed-coat impermeability can be achieved by different combinations of structures and hydrophobic compounds in the testa shows the potential plasticity of testa development and mechanisms of germination control in grain legumes, despite their uniform morphology.

3.4 The Roles of the Micropyle, Hilum, and Lens in Legume Seeds

The seed coat of papilionoid legumes has several specialized structures—the hilum, micropyle, and lens (or strophiole)(Fig. 11.1; Kirkbridge et al. 2003), which have different roles in the regulation of both seed dehydration and water uptake by the dry seed. The micropyle is a narrow gap in the integuments through which the pollen tube enters the ovule during the fertilization process. In some species and cultivars, it remains open in the mature seed, while in others is closed by shrinkage and compression of surrounding cells upon seed desiccation, or occluded by hydrophobic compounds thus preventing water entry (Werker 1997). The radicle tip is oriented towards the micropyle area and protrudes through it during germination. The hilum is a specialized structure that differentiates in the seed coat at the place where the seed detaches from the funiculus, the structure connecting the developing seed to the fruit. An abscission layer differentiates that facilitates seed detachment.

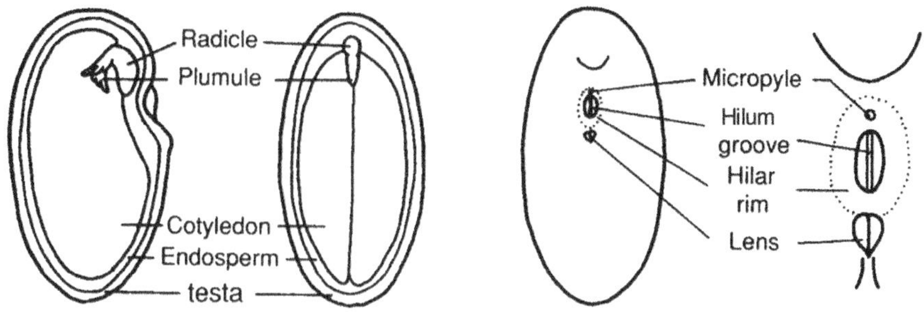

Fig. 11.1 Structure of papilionoid legume seed and position of micropyle, hilum, and lens. (Kirkbridge et al. 2003)

In legume seeds, the funiculus is fused to the outer integument forming the raphe, a ridge on the seed coat of the mature seed. The hilum of papilionoid legumes has a complex structure in which the palisade layer is interrupted by a hilar groove with a longitudinal slit—the hilum aperture (Fig. 11.2). In addition, a counter-palisade layer of macrosclereid cells derived from the funiculus epidermis is fused to the palisade layer of the hilum. A cuticle develops between the fused palisade layers. Importantly, chemical composition of the cell wall differs in the palisade compared to the counter-palisade macrosclereids, with a higher content of hydrophilic compounds in the counter-palisade layer (Lush and Evans 1980; Werker 1997). This leads to differential swelling and contraction of the two layers, causing opening and closure of the hilum aperture in response to changes in humidity outside the seed. The osteosclereids layer is absent under the hilum. Instead, a tracheid bar is present and runs underneath the hilum groove. This is a unique structure found only in papilionoid seeds (Lersten 1982). It has tracheid-like cells, larger than tracheids in the vascular bundles of the seed. They are oriented perpendicularly to the hilar groove and have cell walls densely pitted, with the pit membrane absent in many pits. It probably plays a role as an avenue for water vapor loss during seed dehydration and later for water diffusion through the hilum during seed imbibition. The lens

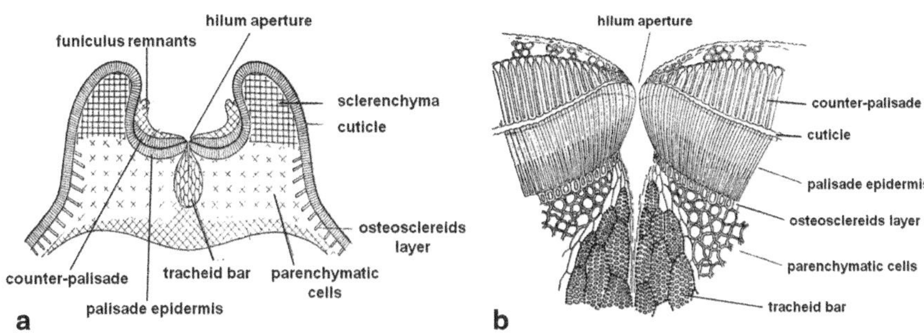

Fig. 11.2 Median section of the hilum (**a**) and enlarged view of hilum structure (**b**) in the lupin seed. (Adapted from Hyde (1954))

(or strophiole) is an area of modified seed coat, which often appears externally as a lens-shaped structure on the raphe near the hilum, opposite to the micropyle. In the lens, the macrosclereids of the palisade layer are longer and the osteosclereid layer is absent (Werker 1997). The long macrosclereids can split apart more easily, producing a small fissure—the strophiolar cleft, through which water can enter into the seed (Hagon and Ballard 1970).

The function of the hilum in papilionoid seeds was described and experimentally confirmed in a seminal paper by Hyde (1954). In legume seeds with impermeable seed coats, the hilum operates as a hygroscopically activated valve controlling seed dehydration after seed dispersal: the hilum slit opens when the seed is in dry air and closes when the outside air is moist (*Trifolium*—Hyde (1954), common bean—Hanma and Denna (1962), wild cowpea—Lush and Evans (1980)). Under dry conditions, the slit opens by the contraction of the counter-palisade layer, allowing water vapor loss to the dryer atmosphere. Under humid conditions, the counter-palisade expands, closing the hilum aperture and preventing entry of moisture. Empty cells of the tracheid bar and the large intercellular spaces in the osteosclereid layer most likely facilitate water vapor diffusion through the open hilum to the atmosphere. Thus, papilionoid seeds can reach very low water content despite wide fluctuations in external relative humidity and also prevent water entry due to hilum closure. The lowest water content in legume seeds with a functional hilum is established during equilibrium with the driest conditions to which the seeds were exposed, thus prolonging seed longevity (Hyde 1954; Ellis and Roberts 1982). In domesticated grain legumes in which the hilum is not functional, seed water content fluctuates according to external humidity and seed longevity is shorter. Full hilum closure can be prevented by incomplete development of the palisade layer or by structures interfering with closure, such as a more outward position of the tracheid bar as in cultivated pea (Werker 1997).

3.5 Seed Water Content and Hardseededness

Seed water content is a major factor affecting water-absorption characteristics of legume seeds. Impermeability of legume seeds develops as they lose moisture during maturation and after dispersal. Seed-coat permeability already decreases at 20–25 % water content, reaching full impermeability once water content falls below a threshold, for example, 7 % in *Trifolium subterraneum* (Fairbrother 1991), 9 % in *Lupinus varius* (Quinlivan 1968), 10 % in wild cowpea (Lush and Evans 1980), and 12 % in lentil (Tang et al. 1994). Thereafter, water diffusion between seed and atmosphere takes place through the hilum functioning as a hygroscopically activated valve in the impermeable seed coat. In general, impermeability is reversible in grain legumes if water content is above 10–12 %, but becomes irreversible below 6–7 % (Ellis et al. 1985).

Time course of water absortion depends on initial seed water content. In lentil, seeds had high rates of absorption at 16–24 % water content, with water entering through the whole seed surface. At 10–12 %, permeability is reduced by four orders of magnitude as a result of cell shrinkage and decreased capillarity of the seed coat, and the closure of the lens and micropyle. Below 10 % the hilum is closed and seeds became impermeable to water (Tang et al. 1994). In wild Lima bean, seeds are dispersed with 12–14 % water content near the end of the dry season; they lack dormancy and germinate readily if water is available. Yet, since in their natural environment no rainfall occurs at the time of seed dispersal and the seeds remain on the soil surface exposed to high temperatures (60–65 °C) and low humidity, seed water content is further reduced to 7–8 %, seeds became impermeable and dormant and are incorporated into the soil seed bank thus persisting for several years (Degreef et al. 2002). In semi-hard lines of common bean, seed-coat permeability is reversible and varies according to seed water content. Dickson and Boettger (1982) defined semi-hard lines as lines in which seeds do not imbibe during the first 24 h when their water content is lower than 8 %, but imbibe readily if higher than 10 %. In these lines, water entry is localized at the raphe and chalazal region of the seeds. Below 6 % water content palisade cells at the raphe become wider and shorter providing better sealing and seeds become impermeable. Increasing the water content to 12 % reversed the changes in shape and length of the palisade cells, rendering the seed coat permeable to water (Holubowicz et al. 1988).

3.6 Physical Dormancy and Disruption of Seed-Coat Impermeability

Under natural conditions seed-coat impermeability can be disrupted by exposure to extreme diurnal changes in temperature and humidity, cycles of wetting and drying such as after rainfall events or dew condensation, and freezing and thawing (Tran and Cavanagh 1984). These changes cause expansion and contraction of cells and eventually produce cracks in the seed coat or disrupt the testa at specific weak points, such as the lens. Direct damage to the seed coat as by fire, abrasion by soil particles, the action of soil microorganism and passage through the digestive tract also lead to increased seed permeability and germination in legume seeds. Such processes may take years to complete and are of ecological significance, since they affect seed-bank density and the spread of germination over time (Baskin and Baskin 2014; Taylor 2005).

PY can be artificially removed by mechanical and chemical scarification (Rolston 1978; Tran and Cavanagh 1984; Argel and Parton 1999). Priming treatments are applied to commercial legume seeds in order to improve germination rate and uniformity (soybean—Marwat et al. (2008), *Medicago sativa*—Zhang et al. (2007)). The enhanced permeability due to environmental factors or priming treatments leads to oxygen uptake and activation of metabolic processes (Bai et al. 2012) which enhance energy production, redox balance, and the biosynthesis of compounds for the full resumption of metabolism in preparation for germination (Benamara et al. 2003; Fait et al. 2006).

In regions with Mediterranean climate, breaking of PY has been attributed to the wide amplitude of daily temperature fluctuation at or near the soil surface during the warm dry summer, in the range of 30–60 °C (Taylor 2005). Rate of softening decreases with increasing depth of burial in the soil, concomitantly with the decreasing amplitude of the diurnal temperature cycle. Less known are the factors involved in the disruption of PY of papilionoid legumes in regions with temperate climate, with much smaller daily fluctuations in soil temperature. In an intriguing study, van Assche et al. (2003) showed that exposure of legume seeds to chilling renders the seed coats sensitive to daily temperature alternation, with their combined action breaking seed-coat impermeability. The effect of chilling is reversible and disappears by exposing the seeds to higher temperatures during the warmer season. This reversal results in seasonal cycling in the ability of legume seeds to respond to the alternating temperatures necessary to break seed-coat permeability and, therefore, in seasonal germination (Baskin 2003).

Softening of hard seeds is a very slow process at low temperatures, and probably contributes to persistent legume seed banks in temperate regions. In *Medicago arabica,* rate of seed softening increases exponentially with increasing temperatures, with a Q_{10} higher than two (van Assche and Vandelook 2010). This suggests possible involvement of a chemical reaction in the breakdown of impermeability in the seed coat, in the lens or in the hilum. In Mediterranean annual pasture legumes, thermal degradation and loss of lipids in the seed coats of seeds exposed to high soil temperatures during the summer caused seed-coat fractures, thus increasing permeability (Zeng et al. 2005). Thus, changes in the quality and quantity of the lipid components may modulate seed-coat permeability.

3.7 *Physiological and Combinational Dormancy in Legumes*

In legumes with combinational dormancy (PY+PD), PD prevents accidental germination in case of failure of seed-coat impermeability (Baskin and Baskin 2014). In hard-seed species, PD can be detected in fresh seeds by seed-coat scarification, since a period of after-ripening in dry conditions is required for germination of the scarified seeds. Accumulating evidence shows that PD is quite common in papilionoid legumes, particularly in winter annuals and in species from semiarid and Mediterranean climates (Thomson 1965; van Assche and Vandelook 2010). In *Medicago truncatula,* imbibition at 4 °C of scarified seeds releases PD, allowing earlier germination (Faria et al. 2005). In *Trifolium subterraneum,* PD is a heritable characteristic and its expression is temperature dependent (Morley 1958; Ballard 1961). Scarified imbibed seeds do not germinate at 30 °C, but germinate readily at 15 °C, or after transfer from 30 to 15 °C (Katznelson and Carpenter 1972). High temperatures (25–30 °C) also inhibit germination of several Mediterranean annual legumes of semiarid rangelands (*Onobrychis* spp., *Astragalus* spp., *Medicago* spp.) (Young et al. 1970; Katznelson and Carpenter 1972) and temperate regions (van Assche and Vandelook 2010) that, otherwise, are able to germinate at low temperatures

(5–10 °C). van Assche and Vandelook (2010) proposed that PY+PD of papilionoid legumes evolved in Mediterranean climates, where seedling survival is highest during the wet and cool winters, and was maintained in species that migrated to temperate regions. However, PY+PD has been found also in annual and perennial species of *Vicia* in the Tibetan Plateau, in which PD is lost after 1 year of dry storage (Hu et al. 2013). The presence of PY+PD can differ in closely related taxa—in *Vicia sativa* it is present in subspecies *macrocarpa* but absent in subspecies *nigra* (Uzun et al. 2013).

Synthesis of ABA during seed imbibition appears to play a key role in maintaining PD in *Medicago truncatula* (Bolingue et al. 2010) and in *Vicia angustifolia* (Hu et al. 2013). ABA does not inhibit testa rupture, but inhibits subsequent radicle growth by hindering cell-wall loosening and cell expansion by water uptake (Finch-Savage and Leubner-Metzger 2006; Gimeno-Gilles et al. 2009). Application of fluridone, an inhibitor of carotenoid and ABA synthesis, to fresh scarified seeds increased the rate of germination, suggesting that ABA synthesis maintains PD. Fluridone also reduced the inhibitory effect of continuous light on germination of *Medicago truncatula* (Bolingue et al. 2010). In contrast, application of paclobutrazol, an inhibitor of ent-kaurene oxidase and GA synthesis, reduced the rate of germination of fresh seeds, indicating that endogenous GA's promote germination of *Medicago truncatula* seeds. However external application of GA_3 did not increase germination, implying that endogenous GA's are present in imbibed seeds of *Medicago truncatula* at levels enabling germination. Yet, in *Vicia angustifolia* GA_3 application increased rate of seed germination (Hu et al. 2013).

Hardseededness and physiological dormancy are affected by high temperature in opposite directions—it accelerates impermeability breakdown, but keeps embryos in a dormant state, that is, thermodormancy. Both effects are required to prevent summer germination in species inhabiting regions with frequent summer rains. Under these conditions, strong selection for hardseededness may occur if PD is lacking (Katznelson and Carpenter 1972). Thus, presence of physiological dormancy may act as a buffer against selection for hardseededness. Thermodormancy can be expressed as inhibition of germination during exposure to supraoptimal temperature or by induction of secondary dormancy preventing subsequent germination at lower temperatures (Baskin and Baskin 2014). Thermodormancy is associated with increased levels of ABA and increased embryo sensitivity to ABA (Toh et al. 2008; Leymarie et al. 2009). High-temperature inhibition of germination can be alleviated by suppression of ABA synthesis with fluridone and by exogenous GA (Toh et al. 2008).

In chickpea, inhibition of seed germination by high temperature (30 °C) and ABA was associated to lack of transcription of specific expansin genes needed for cell-wall loosening and subsequent water uptake by the embryo radicle (Hernandez-Nistal et al. 2010). Furthermore, high temperature (30–35 °C) also inhibited ethylene production and germination in chickpea, and its inhibitory action was alleviated by ethylene (Gallardo et al. 1991). In this case, thermoinhibition was due to increased conjugation of the ethylene precursor 1-aminocyclopropane-1-carboxylic acid (ACC) to 1-(malonylamino) cyclopropane-1-carboxylic acid (malonyl-CoA) and inhibition of the ethylene-forming enzyme ACC synthase. Similarly, ethylene increases germination of thermoinhibited seeds of *Medicago truncatula* and *Trifolium subterraneum* (Globerson 1978). Inhibition of ethylene biosynthesis by 2,5-norbonadiene strongly hinders seed germination in pea, and this effect was counteracted by ethylene (Petruzelli et al. 1995).

Combinatorial dormancy in hard-seeded species acts as a double safety mechanism preventing germination of seeds dispersed with incomplete impermeable seed coats or of early softening seeds (Hu et al. 2013). Furthermore, hydrothermal model analysis showed that seeds of *Vicia* species possessing PD had higher median water potential for germination, implying that PD can prevent germination of permeable seeds under low water availability (Hu et al. 2013). But, since PD is transitory and is lowered by winter temperatures, maintenance of a permanent seed bank in legumes is achieved mainly by hardseededness. In regions lacking summer rains, hardseededness is probably enough to control germination. However, germination control based on seed-coat permeability is usually an irreversible process. Once the impermeability barrier is lost, seeds may readily germinate under unfavorable conditions, thus increasing the risk of seedling death (Kigel 1995). Alternative mechanisms to prevent germination of buried seeds during unsuitable seasons are the induction of secondary dormancy by unfavorable conditions such as high temperature, or by cyclic changes in PD coupled with seasonal variation in climatic conditions (Baskin and Baskin 2014). Yet, the induction of secondary dormancy and cyclic dormancy has not been reported for legumes.

3.8 Ports of Water Uptake and Pathways of Hydration in Legume Seeds

Germination begins with the uptake of water by the dry seed, followed by metabolic activation and embryo expansion. Water uptake is a triphasic process, with a fast initial uptake (phase I, i.e., imbibition), followed by a plateau phase (phase II) and a further increase in water uptake (phase III) taking place concomitantly to embryo expansion, splitting of the seed coat and root protrusion (i.e., germination) (Finch-Savage and Leubner-Metzger 2006; Bewley et al. 2013). Papilionoid legumes differ in the location of main ports of initial water uptake through the seed coat after seed softening. Detection of ports of water entry and evaluation of the relative importance of the hilum, micropyle, and lens compared to the whole seed-coat surface have been studied using magnetic resonance imaging (MRI), fluorescent tracer dyes, or by occluding different ports of water entry. The hilum functions as the main port in lentil (Tang et al. 1994), *Vicia* spp. (Aswathaiah 1988), and in wild Lima bean (Degreef et al. 2002). In cultivars of Lima bean with permeable seeds, the micropyle is open and hilum closure is prevented under humid conditions due to an incomplete counter-palisade layer. In contrast, in lines with impermeable seeds the micropyle is occluded and a well-developed counter-palisade layer allows tight closure of the hilum (Stienswat et al. 1971).

In other grain legumes, the primary region of water entry is the lens, as in *Lupinus albus* (Perisse and Planchuelo 2004) and azuki bean (*Vigna angularis;* Isemura et al. 2007), or the hilum and micropyle as in *Lupinus luteus* (Garnczarska et al. 2007). In soybean, the lens is the main port of water entry (Koizumi et al. 2008), but in cultivars with nonfunctional lens water uptake takes place through cuticular cracks in the seed coat (Ma et al. 2004). Thus, cultivars within the same species might differ in the seed-coat ports involved in initial water uptake. In *Lupinus angustifolius* either the micropyle (Serrato-Valenti et al. 1989) or the lens (Perisse and Planchuelo 2004) were reported as ports of water entry. In common bean, 80 % of the water uptake in Great Northern lines was through the micropyle, compared to 18 and 2 % through the hilum and the testa, respectively. In contrast, in Red Mexican lines, the major venues of water uptake were the raphe and hilum, with the micropyle playing a minor role (Kyle and Randall 1963). In other cultivars of common bean, the main ports of water entry were the hilum (Varriano-Marston and Jackson 1981; Heil et al. 1992), the hilum and the lens (Kikuchi et al. 2006), or the hilum and micropyle (Deshpande and Cheryan 1986). This intraspecific variation in the location of initial water entry reported in different studies can be partly due to the diverse methods and time scales used in these studies.

Pathways of seed hydration are structurally organized in legume seeds and are based on changes in the seed coat during early imbibition that directs subsequent water distribution within the legume seed. In permeable seeds, an important role of the seed coat is to slow down the rate of water entry and, at the same time, facilitate water movement within the imbibing seed. In common bean, autoradiographic studies using $^{3}H_2O$ showed that water enters the seed through the hilum and diffuses to the periphery of the cotyledons via the parenchyma cells of the testa, thus allowing uniform hydration of the whole embryo and preventing imbibition damage (Varriano-Marston and Jackson 1981). Micro-MRI showed that in common bean and adzuki bean, water entered through the lens as the sole water port was first distributed throughout the testa and then delivered to the radicle followed by hydration and swelling of the cotyledons (Kikuchi et al. 2006). In common bean water uptake was immediate, with rapid hydration of the whole testa and cotyledon swelling. In adzuki bean, in contrast, water entry was delayed by c. 7 h, and the slow hydration was required to activate the lens. In soybean, water entered near the raphe, diffused first towards the dorsal side of the seed and then to the hilar region, followed by expansion of the radicle and hypocotyl and later by swelling of the cotyledons (Koizumi et al. 2008). In soybean cultivars in which seed-coat permeability is due to cracks in the seed cuticle, permeability of isolated seed coats was five times lower in hard-seeded compared to soft-seeded cultivars, as measured by pressure probe (Meyer et al. 2007). The critical difference between these cultivars was the continuity of a 0.2-μm-thick cuticle covering the palisade layer—small cracks 1–5 μm wide and 20–200 μm long occur in permeable cultivars, but not in impermeable ones (Arechavaleta-Medina and Snyder 1981; Ma et al. 2004). Once water entry starts in permeable seeds of soybean, the dry seed coat increased in volume causing local wrinkles that detach the coat from the cotyledons. These wrinkles are initially small and perpendicular to the long axis of the seed. Further imbibition enlarges the wrinkles that reach the ventral side of the seed, thus channeling water around the seed. Lateral water movement also occurs through the intercellular spaces of the osteosclereid layer and is more extensive at the hilar side due to larger spaces between its cells (McDonald et al. 1988). As the seed coat becomes wet it absorbs water by c. four times its weight, partly due to the large water-holding capacity of the osteosclereid layer. This structural arrangement of the seed coat allows lateral water distribution in the seed and uniform wetting of the embryo, thus preventing imbibition damage due to fast uneven hydration that might cause embryo fracture and solute leakage.

Soybean lines with permeable and hard seeds differ in the strength of the outer cuticle secreted by the palisade cells. In hard-seed lines, the cuticle is more resistant to cracking and richer in hydroxylated fatty acids, while in cultivars with permeable seeds the cuticle lacks mid-chain hydroxylated fatty acids and is prone to cracking (Ma et al. 2004; Shao et al. 2007). These compounds probably modify either the degree of cross-linking between components of the cuticle or its integration with the underlying cell wall. The critical stage at which hardseededness is established during seed development in these soybean lines apparently occurs when cuticle deposition was completed, but the embryo is still expanding (Ranathunge et al. 2010). In cultivars with soft permeable seeds, continued embryo expansion after cessation of cutin secretion causes microscopic cracks in the seed cuticle, while in cultivars with hard non-permeable seeds, these cracks do not appear despite the same pattern of cuticle deposition and seed expansion. Thus, seed permeability depends on whether the external cuticle can withstand the internal forces generated by rapid embryo expansion in the absence of cuticle deposition.

3.9 Imbibition Damage in Grain Legumes

Slow and controlled hydration of the dry seed is essential as the first step in the reconstitution of cell membranes and organelles and reactivation of metabolic processes in the seed (Bewley et al. 2013). Seed imbibition damage is generally attributed to rapid water uptake by the embryo, leading to disruption of cell membranes, solute leakage, and even cell death, thus negatively affecting germination and embryo growth (Powell and Matthews 1978; Duke and Kakefuda 1981). Imbibition damage takes place in the early stages of water uptake (Parrish and Leopold 1977) and affects both the cotyledons and the embryonic axis (Ashworth and Obendorf 1980; McDonald et al. 1988). In the relatively large legume seeds, as water enters and hydrates the outer embryo tissues they begin to swell while internal tissues stay dry, resulting in physical strains that can rupture the embryo.

One of the functions of the seed coat is to restrict water flow during seed imbibition, so that hydration does not proceed too rapidly and imbibition damage is prevented (McDonald et al. 1988; Bewley et al. 2013). Damage due to seed imbibition at low temperature is compounded by chilling damage causing impaired membrane reorganization and solute leakage (Bewley et al. 2013). A particular outcome of damage upon water uptake can be seen in the reorganization of organellar functions. During imbibitions mitochondria gain proteins, lipids, and important active components such as cytochrome oxidase and malate dehydrogenase (Nawa and Asahi 1971). Pea mitochondria were already able to oxidize reduced nicotinamide adenine dinucleotide (NADH) and tricarboxylic acid (TCA) cycle intermediates 12 h after imbibition. At this stage, however, phosphorylation was still not efficient because of membrane damage (Benamara et al. 2003). After 22 h, the outer mitochondria membranes were reconstituted and phosphorylation was improved. Studies of mitochondrial changes during seed germination in diverse plant species—among them pea and mungbean—revealed similar phenomena (Botha et al. 1992; Howell et al. 2006).

A relationship between seed-coat permeability and imbibition damage is often reported in grain legumes. Cultivars with high seed-coat permeability and high incidence of testa injury imbibe more rapidly and may show extensive imbibition damage, as in chickpea (Legesse and Powell 1996), common bean (Powell et al. 1986; Legesse and Powell 1996; Chachalis and Smith 2000), faba bean (Powell and Matthews 1978; Chachalis and Smith 2000), cowpea (Legesse and Powell 1992, 1996; Asiedu and Powell 1998), longbean (*Vigna sesquipedalis;* Abdullah et al. 1991), pea (Chachalis and Smith 2000), and soybean (Oliveira et al. 1984; Toledo et al. 2010). However, imbibition damage is seldom observed in seeds with initial water content above 15–20% (Toledo et al. 2010), in which respiration and metabolic activity rapidly increase with the increase of water content upon imbibition (Bewley et al. 2013). Light-colored seeds of grain legumes generally have faster water uptake and higher propensity to imbibition damage, as found in common bean (Powell et al. 1986), cowpea (Legesse and Powell 1996), pea (Powell 1989), and soybean (Chachalis and Smith 2001). The relationship between seed-coat color and seed water uptake has been studied in isogenic lines of common bean (Wyatt 1977), soybean (Tully et al. 1981), Lima bean (Kannenberg and Allard 1964), and pea (Powell 1989). Absence of pigment in the seed coat is often associated to a syndrome of traits, as shown by comparing pigmented and white seeds: (1) thinner seed coats; (2) shorter and broader macrosclereids; (3) greater seed-coat permeability; (4) the testa is more easily damaged; and (5) seeds germinate more rapidly.

The way by which pigmentation in the testa reduces the rate of seed imbibition is not clear. Pigmented testae have higher levels of lignins and tannins (Morrison et al. 1995), waterproofing substances that may reduce permeability. On the other hand, lack of pigmentation is frequently associated with less adherence of the testa to the cotyledons, thus creating a void between them during initial imbibition (Wyatt 1977; Powell 1989; Nakayama and Komatsu 2008). This void is filled with water that is in direct contact with the embryo, leading to faster water uptake and increasing the risk of imbibition damage. In pea lines with white seeds, the seed coat loosened rapidly in the imbibing seeds facilitating diffusion of water between the testa and cotyledons, while in lines with pigmented seeds the seed coat remained closely attached to the cotyledons (Powell 1989). Similarly, faster imbibition in lines of yellow-seeded soybean was not due to differences in the permeability of the seed coat, but rather to the increased movement of water between the seed coat and the cotyledons (Nakayama and Komatsu 2008). Adherence of the testa to the cotyledons is also the major factor contributing to the slow imbibition of Desi-type chickpea (Legesse and Powell 1996). However, since several lines of legumes with white and light-colored seeds also have impermeable seed coats, seed-coat pigmentation is not a condition necessary for hardseededness. These contrasting observations are not necessarily contradictory, since genetic linkage between loci controlling pigmentation and permeability might be present in different genotypes. Genes affecting pigmentation may also have pleiotropic effects on permeability, due to possible involvement in biosynthesis of polyphenols that impregnate the cell wall.

3.10 Genetics of Seed-Coat Impermeability in Grain Legumes

The seed coat normally develops from maternal tissues, namely from the integuments of the fertilized ovule and from the endosperm (Esau 1977). In legume seeds, the inner integument is obliterated by the expanding embryo and the endosperm is entirely consumed by the embryo in many species (Werker 1997). Thus, permeability of the seed coat is generally controlled by maternal genes expressed during the differentiation of the outer integument into the testa of the mature seed. Evidence supporting embryo effects on seed-coat permeability in legumes is scarce, even though coordinated growth of the embryo and seed-coat tissues is necessary to prevent seed-coat rupture and gaps enabling water entry. Pressure exerted by fast embryo expansion can cause cracks in the outer cuticle of the seed or even in the testa, if their rate of expansion is slower compared to embryo expansion (Qutob et al. 2008).

Few genes are involved in the control of seed-coat permeability in grain legumes, a fact that allowed fast selection for dormancy loss during domestication (Zohary et al. 2012). The site of action of these genes varies among species and cultivars. It can be localized at potential ports of water entry, namely the hilum, micropyle, or lens, or might affect the whole testa. However, the nature of the structural and chemical changes controlled by these genes is less clear. In several papilionoid legumes the hard-seed trait is controlled just by one or two genes. For instance, crosses between cultivated *Lens culinaris* and two hardseed wild species showed that seed-coat impermeability is controlled by a single recessive gene in the homozygous condition in *Lens orientalis* and by a single dominant gene in *Lens ervoides* (Ladizinski 1985). In *Vicia faba* seed testa-imposed dormancy is a monogenic trait (Ramsay 1997), while in *Vicia sativa* it is controlled by a two gene system (Donnelly et al. 1972). One gene (A) is dominant for hardseededness, a second gene (B) is dominant for softseededness when the *A* locus is homozygous recessive *(aa),* while the double recessive genotype *(aabb)* is hardseeded. In *Lupinus hispanicus* (Arrieta et al. 1994), *Lupinus angustifolius* (Forbes and Wells 1968), and *Lupinus luteus* (Mikolajczyk 1966), seed-coat permeability is controlled by one recessive gene. However, Serrato-Valenti et al. (1989) found that in *Lupinus angustifolius,* where the micropyle is the port of water entry in soft seeds, permeability is controlled by two genes. In common bean, seed-coat permeability is controlled by several genes and the soft-seed trait is incompletely dominant (Dickson and Boettger 1982). Location and genetic control of permeability differs among cultivars of common bean (Kyle and Randall 1963). In Great Northern lines, the micropyle was the port of water entry, while in Red Mexican lines the raphe and hilum were the main ports and their permeability is controlled by a single recessive gene pair. In mungbean, the number of genes controlling hardseededness varied from one to a few, depending on the genotypes used in the crosses (James et al. 1999; Humphrey et al. 2005). In soybean, crosses with *Glycine ussuriensis* showed that permeability is controlled by one gene (Marjüshkin et al. 1987), while in crosses with *Glycine formosana* two or four genes controlled permeability (Shahi and Pandey 1982; Verma and Ram 1987), and in both cases genes for permeability are recessive. Mapping of quantitative trait locus (QTL) for hardseededness in soybean using a cross between a cultivar of *Glycine max* ssp. *max* with permeable seeds and wild *Glycine max* ssp. *soja* with impermeable seeds, showed that few major QTL account for most of the variation in this trait (Sakamoto et al. 2004; Liu et al. 2007). Similarly, QTL analyses of a recombinant inbred population of mungbean *(Vigna radiata)* derived from a cross between a soft-seeded and a hard-seeded line revealed four loci for hardseededness (Humphrey et al. 2005). In azuki bean (*Vigna angularis;* Kaga et al. 2008) and yard-long bean (*Vigna unguiculata* ssp. *unguiculata;* Kongjaimun et al. 2012), five and six QTLs, respectively, were identified for seed dormancy-related traits and occur in the same linkage group. In common bean, four unlinked QTLs were identified for seed dormancy in a cross between a cultivar and a wild accession (Koinange et al. 1996).

Altogether, hardseededness of grain legumes should be considered a quantitative trait, with a relatively few genes affecting seed-coat permeability, acting at different locations in the seed coat according to species and cultivars. Analysis of the available information suggests that several processes under genetic control, acting at the biochemical and structural level during seed-coat development, modulate seed-coat permeability, and henceforth seed dormancy, as well as rate of seed hydration. The challenge now is the identification of the genes and the specific processes under their control that take place during seed-coat development, modulating seed permeability by acting at specific locations in the seed coat.

4 Effects of the Parental Environmental on Seed Size and Germination

Seed dormancy and seed size are determined by the genotypes of the maternal plant and the embryo interacting with parental environmental conditions during seed development and maturation (Roach and Wulff 1987; Baskin and Baskin 2014; Donohue 2009). Temperature, day length, light intensity, and quality as well as availability of water and nutrients can change the proportion of dormant seeds (Fenner 1991; Baskin and Baskin 2014) and composition and weight distribution of offspring seeds (Fenner 1992). Yet, some effects can be indirect, such as through pod and seed abortion that may increase seed mass of the remaining seeds by compensatory mechanisms activated under stress conditions (Stephenson 1981; Gross and Kigel 1994; Gusmao et al. 2012).

Despite numerous agronomic studies related to the effects of climatic conditions on yield components, including individual seed mass, few studies focused on the germination behavior of seeds produced under different environmental conditions. Seed size of grain legumes is often modified by temperature and water availability during the seed filling stage. Seed mass is the combined result of rate of seed growth and duration of seed growth. Notably, increasing temperature has opposite effects on these processes: It enhances seed growth rate but shortens the seed-filling period, thus buffering temperature effect on seed mass (pea—Poggio et al. (2005), soybean—Egli et al. (2005)). Chilling (< 10–15 °C) during seed development reduces accumulation of seed reserves such as starch, proteins, and minerals, resulting in smaller seed mass (chickpea—Kaur et al. (2008), soybean—Egli et al. (2005)). Processes underlying these chilling effects may involve loss of chlorophyll and decreased photosynthesis, restricted availability and/or mobilization of assimilates as well as inhibition of enzymes related to the biosynthesis of storage compounds. Low water availability during the seed-filling stage reduce pod and seed number in chickpea (Behboudian et al. 2001) and in common bean (Boutra and Sanders 2001) due to pod abortion, but did not affect seed mass and even increased protein content in chickpea. In pea, in contrast, seed mass was reduced by water scarcity during seed filling, but germination remained high (Fougereux et al. 1997).

5 Environmental Control of Seed Germination

Area of cultivation of different grain legumes is continuously expanding beyond their original climatic regions of adaptation, thus exposing these crops to suboptimal climatic conditions for germination, growth, and yield. This trend, together with climate change effects, is leading to earlier sowings after winter in temperate regions or after summer in regions with Mediterranean climate, in order to avoid drought and heath stress during reproductive and seed-filling stages. Warm as well as cold weather can have negative impacts on seed germination and emergence, particularly in grain legumes grown continuously throughout the year, which may be sown during periods of high or low temperature. In grain legumes in which seed dormancy and seed-coat impermeability have been lost through domestication and breeding, germination is mainly controlled by water availability and soil temperature. Light has generally no effect on their germination, even though osmotic stress may induce a light requirement, as in common bean (Lopes and Takai 1987). The interactive effects of temperature and water availability in the seedbed are of cardinal importance for seed germination. It is well known that each plant species has its own specific temperature requirements: minimum (i.e., base-*Tb*) and maximum temperature *(Tmax)* below and above which no germination occurs, and an optimum temperature *(Topt)* at which germination rate is fastest. Germination rate is often linearly related to temperature, allowing temperature and time to be combined into thermal time. Regarding water availability, both the rate of germination and the proportion of germinating seeds decrease with decreasing soil water potential, but each species has its own threshold—base water potential *(Ψb)*—for germination. However, since water and temperature interact the threshold water potential depends largely on the temperature optimal for germination (Bradford 2002). Variation in *Ψb, To, Topt, Tmax,* and thermal time to 50 % germination *(TT50%)* among grain legumes of temperate, tropical, and subtropical origin are presented in Table 11.1.

Temperate and tropical clades of grain legumes did not differ in threshold water potential for germination (*Ψb* -1–2 MPa). In contrast, clades differ in germination responses to temperature. The Phaseoloid clade with tropical and subtropical species has a much higher *Tb* for germination compared to the Hologalegina and Genistoid clades with temperate species. Seeds of pea are able to germinate even on ice (Macherel et al. 2007). Furthermore, at lower temperatures rates of germination where faster in the Halogalegina clade compared to the Phaseloid clade *(Topt)*. Thus, below 25 °C pea germinates more rapidly than common bean, while above 25 °C common bean germinates faster than pea. Accordingly, thermal time to 50 % germination *(TT50%)* was shorter in the Phaseoloid clade. On the other hand, *Tmax* did not differ among the clades. Thus, differences in cardinal temperature parameters *To, Topt,* and *TT50%* reflect the different climatic and ecogeographic regions of origin as well as the phylogenetic relationships among papilionoid legumes.

Domestication and breeding broadened the range of temperatures that allow germination of grain legumes. For instance, cultivars of cowpea (Lush et al. 1980), common bean, and tepary bean (Scully and Waines 1987) germinate at lower temperatures than their wild relatives. In common bean, a warm-season legume originating in Central and South America, some commercial lines show some level of germination down to 7–10 °C (Kooistra 1971; Dickson and Boettger 1982; Scully and Waines 1987; White and Montes 1993; Zaiter et al. 1994), while no germination below 10 °C occurred in *Phaseolus aborigineus* (Kooistra 1971). Lower *Tb* predicts germination at low soil temperatures in the field. However, cold tolerance at germination is not necessarily associated with cold tolerance at later stages of plant development. Germination base temperatures frequently differ from base temperatures for vegetative and reproductive development (White and Montes 1993). In common bean and pea seedling, elongation requires 2–3 °C higher minimal temperature than germination (Raveneau et al. 2011).

Importance of salinity effects on germination of grain legumes is increasing due to expansion of their cultivation into more marginal land. Salinity can affect seed germination through osmotic effects (Welbaum et al. 1990) or ion toxicity (Munns 2002). Even though it is difficult to separate between the two effects, low water potential is the main limiting factor at low and moderate salinity levels. Grain legume crops are adversely affected by relatively low salt levels (Maas and Hoffman 1977) and possess limited genetic variability for salinity tolerance (Johansen et al. 1988). Common bean is salt sensitive (Maas and Hoffman 1977), but has wild relatives from arid regions possessing higher salinity tolerance—*Phaseolus angustissimus,*

Table 11.1 Relationships between taxonomic affiliation, temperature, and water potential requirements for seed germination of grain legumes. (*Tb* base temperature, *Topt* temperature of highest rate of germination, *Tmax* temperature inhibitory for germination (< empe, *TT50%* thermal time (°C day), *Ψb* base water potential)

Clades	Common Name	Species	Temperature				Water potential	References
			Tb germ (°C)	*T opt* (°C)	*T max* (°C)	*TT50%* (°C day)	*Ψb* MPa	
Hologalegina	Field pea	*Pisum sativum*	−1.8–0.6	25–30	40	42	–	Olivier and Annandale (1998)
			−1.9–0	21–27	29–40	22–40	− 1.7 – − 2.5	Raveneau et al. (2011)
	Chickpea	*Cicer arietinum*	0	32–33	48	42	–	Covell et al. (1986)
			0	31	–	32–43	–	Ellis et al. (1986)
			5.5	–	–	–	−2	Finch-Savage et al. (2005)
	Lentil	*Lens culinaris*	2.5	24	32–34	21	–	Covell et al. (1986)
			1.5	–	–	25	−1.7 – − 2.5	Ellis and Barrett (1994)
	Barrel medic	*Medicago truncatula*	2–3	20–25	30	13–16	−0.7 – − 1.3	Brunel et al. (2009)
	Faba bean	*Vicia faba*	−4 – − 7.5	20–25	42	13	–	Ellis et al. (1987)
			0.4	25	37	–	–	Dumur et al. (1990)

Table 11.1 (continued)

Clades	Common Name	Species	Temperature				Water potential	References
			Tb germ (°C)	*T opt* (°C)	*T max* (°C)	*TT50%* (°C day)	*Ψb* MPa	
Phaseoloid	Common bean	*Phaseolus vulgaris*	5.1–9.6	31–36	46–50	11–15	−1.9 – −2.4	Raveneau et al. (2011)
			7–14	27–35	31–39	16–26	–	Machado-Neto et al. (2006)
			7.3	27	–	–	–	White and Montes (1993)
			15	30	40	–	–	Scully and Waines (1987)
			12	–	–	–	−0.8	Hucl (1993)
			10–12	–	–	20–30	–	Nleya et al. (2005)
			12	–	–	–	–	Otubo et al. (1996)
			–	–	45	–	–	Pena-Valdivia et al. (2002)
	Tepary bean	*Phaseolus acutifolius*	8.3	35	40	–	–	White and Montes (1993)
			10	35	40	–	–	Scully and Waines (1987)
	Cowpea	*Vigna unguiculata*	8.5	40	>45	29	–	Covell et al. (1986)
			6.1–10.5	30–36	40–44	–	–	Craufurd et al. (1996)
	Mungbean	*Vigna radiata*	10	40	>45	15	−2.2	Fyfield and Gregory (1989)
	Soybean	*Glycine max*	4	34	47–55	32	–	Covell et al. (1986)
Genistoid	Narrow-leafed lupin	*Lupinus angustifolius*	−0.8–0.7	20	30	–	−1 – −1.5	Dracup et al. (1993)
			–	–	–	–	−0.8	Perisse et al. (2002)
	Lupin	*Lupinus albus*	–	–	–	–	−1	Perisse et al. (2002)

Phaseolus filiformis, Phaseolus leptostachyus, and *Phaseolus microcarpus.* Accessions of these species had relatively fast germination under high salinity (120 mM NaCl), and were able to germinate even at 180 mM NaCl. These wild *Phaseolus* species may represent a genetic resource for improvement of salinity tolerance in common bean (Bayuelo-Jimenez et al. 2002).

6 Metabolic Aspects of Seed Germination in Legumes

6.1 Metabolic Reorganization upon Imbibition Is Challenged by Slow Oxygen Diffusion

Metabolism dominates cellular activity throughout the process of germination (Bewley et al. 2013), with some pathways activated from the beginning of imbibition while other processes are initiated later on. The major function of cellular metabolism at the onset of imbibition is to generate a redox state that promotes the activity of essential enzymes and produces energy to support germination (Weitbrecht et al. 2011). In addition, change in the cellular redox level also enables the reduction of protein disulfide bonds, through the activity of thioredoxin (Alkhalfioui et al. 2007). This protein modification contributes to starch and protein degradation in *Medicago truncatula* by promoting the susceptibility of protease and amylase inhibitors to proteolysis. The metabolic events upon imbibition are tightly linked to previous events at the end of seed development, that is, seed desiccation, when molecular and physiological processes enter a quiescent state. Storage polymers synthesized during seed development are stored in the dry seed and constitute most of the seed dry weight. The most abundant storage polymers in legume seeds are starch and storage proteins vicillin, legumin, and convicillin (Gallardo et al. 2003). Their accumulation is influenced by the availability of nitrogen during seed development (Weber et al. 2005). Also enzymes, gene transcripts and free metabolites to be used during early stages of germination are present in the dry seeds (Angelovici et al. 2010) and have likely the function to initiate metabolic processes minimizing energy input. For example, mitochondria isolated from dry pea seeds were partially functional and were able to oxidize succinate (Botha et al. 1992). Moreover, ribosomal proteins, RNA-binding proteins, and translation initiation factors were also found in dry pea seeds, in preparation for essential protein synthesis during germination (Wang et al. 2012).

The first hours of imbibition are characterized by increasing oxygen uptake and parallel increase of ATP levels (Botha et al. 1992). Full aerobic respiration including the TCA cycle and the electron transport chain potentially provides 38 mol of ATP per mol of glucose. As mitochondria activity improves during imbibition, the TCA cycle enzyme malate dehydrogenase (MDH) highly increases in activity during early germination (Morohashi et al. 1981; Soeda et al. 2005). In several species, including chickpea, soybean, and mungbean, the role of cytochrome oxidase in mitochondrial respiration during germination was demonstrated by inhibition of total respiration by cyanide (Botha et al. 1992). Nevertheless, in slowly imbibing dry seeds where oxygen is supplied through water uptake, an environment far from being fully aerobic, ATP can be produced by alternative pathways. For instance, cytosolic glycolysis followed by fermentation produces readily available but low levels of ATP—only two mol ATP per mol of glucose. Fermentation is activated in imbibed seeds when utilization of pyruvate by other pathways is limited (Al-Ani et al. 1985).

In soybean as well as in other high-oil seeds, germination was correlated to respiration and energy charge at relatively higher oxygen pressures, with less consistent relations at lower oxygen levels, suggesting the presence of additional limiting factors acting at low oxygen (Al-Ani et al. 1985). In pea seeds, in contrast, lower threshold of oxygen was needed to promote germination. Furthermore, its rate was not correlated to respiration levels (measured as oxygen uptake) nor to adenylate energy charge under low oxygen (based on ATP/ADP and ATP/AMP ratios), emphasizing the relevance of other pathways, such as fermentation, providing energy for germination to proceed. During germination, ethanol fermentation increases in many seeds (Botha et al. 1992), possibly due to an excess of glycolysis over mitochondrial respiration during early germination (Morohashi and Shimokoriyama 1975). Soybean and pea have higher rates of fermentation and overall ATP production compared to other nonlegume seeds under aerobic conditions, which could be related to their larger seed size associated with lower oxygen diffusion (Raymond et al. 1985). Also under anaerobic conditions, fermentation levels were relatively higher in these two species, with ethanol production increasing in soybean but not in pea and lactate dropping very low. Pea displayed the highest tolerance to anoxia and was able to produce 47% of ATP generated under aerobic conditions. Under aerobic conditions alcohol fermentation accounted for 43% of the ATP regeneration in pea and 5% in soybean (Raymond et al. 1985).

6.2 Amino Acids Are Substrates for Energy Buildup During Imbibition

Several studies showed that amino acids can provide an alternative substrate for energy production (Galili 2011). Manipulation of lysine metabolism had a major effect on the TCA cycle metabolism, revealing a strong interaction between lysine metabolism and cellular energy metabolism (Angelovici et al. 2011; Kirma et al. 2012). Also the carbon skeleton from glutamate released from storage proteins in yellow lupine is utilized for respiration, and the amine group is transferred to form asparagine, which accumulates to high levels in germinating seeds (Lehmann and Ratajczak 2008). Pools of free amino acids stored in the dry seeds are readily used during the first hours of imbibition. In *Arabidopsis,* the nonprotein amino acid γ-aminobutyrate (GABA) was shown to provide a substrate of energy metabolism during seed imbibition via the GABA shunt (Fait et al. 2008). A significant role of this amino acid in metabolic reactivation during germination was found also in faba bean. The activity and transcript levels of glutamate decarboxylase (GAD)—the entry enzyme of the GABA shunt—were shown to increase in germinating seeds (Yang et al. 2013). Hypoxic conditions contribute to GABA accumulation in early stages of germinating faba bean (Yang et al. 2013). At first, it was thought that GABA per se could improve the tolerance of the seed to low oxygen conditions by an unknown mechanism. However, it can be safely suggested that the activation of the GABA shunt provides a metabolic continuity to an impaired TCA cycle. The combined effect of high sensitivity to redox balance of 2-oxoglutarate dehydrogenase—the enzyme responsible for glucose incorporation into the TCA cycle—and the long-known induction of GAD during early germination, likely by Ca^{2+} released during the reorganization of the cellular components of the seed, promotes the GABA-shunt activity, maintaining the TCA cycle functional integrity (Fait et al. 2008). Under hypoxia, faba bean also increase polyamine and glutamate content and diamine oxidase (DAO) activity, all of which are lowered after relief of hypoxia. Polyamine degradation supplies about 30% of GABA in hypoxia conditions, as shown by DAO inhibition (Yang et al. 2013). Taken together the GABA pool can play an important role for the initiation and maintenance of TCA cycle functionality in imbibed seeds, when energy and oxygen levels are limited and storage reserves are not yet readily available.

Pathways in which amino acids provide alternative substrates to sugars for energy production in germinating seeds are regulated by a sugar-sensing mechanism (Lehmann and Ratajczak 2008). For example, in germinating yellow lupine *(Lupinus luteus)* seeds, arginine is catabolized by amino transferase into glutamate, which can enter the TCA cycle (Ratajczak et al. 1996). Arginase and urinase activities, the enzymes of arginine catabolism, increase during germination of seeds of several plant species, and the pathway is induced by sugar starvation (Borek et al. 2001). Under sugar starvation, higher levels of urea are also measured in the seeds, a byproduct of elevated rate of nitrogen compounds degradation for use as energy and carbon substrates.

De novo protein synthesis is required for germination. However, the level of free amino acids in dry seeds is not sufficient for the needs of protein synthesis during germination (Bewley et al. 2013) and storage protein degradation in the embryo starts already during the first hours of imbibition. Approaching radicle protrusion, degradation of storage proteins increasingly provides amino acids for de novo protein synthesis and cellular metabolism in the growing embryo (Wang et al. 2012). In legumes, storage proteins are concentrated in protein storage vacuoles (PSVs) in the embryonic axis and the cotyledon cells. A number of enzymes with different roles are involved in the PSV breakdown at different stages during germination (Bewley et al. 2013). The peptidases, which initiate this process during the first hours, are stored in dry seeds while additional proteases accumulate during imbibition (Yomo and Varner 1973). In black gram *(Vigna mungo)*, cysteine endopeptidase SH-EP, which is involved in the mobilization of storage globulin, reaches its peak 4 days after imbibition and decreases thereafter (Okamoto and Minamikawa 1998). Proteasome subunits identified in the pea embryonic axes increased during germination, and probably have a role in storage proteins degradation (Wang et al. 2012). Since amino acids from storage protein are also used as substrates for respiration, degradation of storage proteins to amino acids is enhanced under sugar depletion conditions (Borek et al. 2001, 2012).

On the other hand, amino acid mobilization and usage are inhibited by application of exogenous sucrose (Okamoto and Minamikawa 1998; Lehmann and Ratajczak 2008). As their incorporation into energy metabolism or protein de novo biosynthesis decreases and free amino acids accumulate, the degradation of protein reserves is inhibited (Yomo and Varner 1973).

6.3 Raffinose and Other Sugars Provide Early Carbon Resources for Pre-Germination Energy

Major mobilization of starch reserves starts after radicle protrusion in most seeds. In pea, amylase activity increases during imbibition (Yomo and Varner 1973). Energy carbon resources at imbibition include the raffinose family oligosaccharides (RFOs), which are present as soluble storage reserves in seeds of diverse species (Peterbauer et al. 2001). Their degradation requires α-galactosidase and invertases. During seed development, raffinoses accumulate in the cytoplasm of starchy (Bewley et al. 2013) and oily seeds alike (Baud et al. 2002; Fait et al. 2006), while the degrading enzymes are partitioned in the protein storage vacuole (pea- Blöchl et al. (2008)). They come together during germination. The small and soluble nature of RFOs makes them a more accessible storage reserve than starch. Half of raffinoses in pea seeds are degraded by the time of radicle emergence (Blöchl et al. 2008). Impaired raffinose breakdown substantially lowers germination rate in pea (Blöchl et al. 2007), while it is not essential for germination in soybean (Dierking and Bilyeu 2009).

In legumes the production of mannose from manno-oligosaccharides during imbibition by the activity of β-mannanases and β-mannosidases secreted by the aleurone layer is long known (Reid and Meier 1972, 1973; McCleary and Matheson 1975). Nonogaki et al. (1995, 1998) investigated their role at the tissue level during imbibition in tomato seeds. A galactomannan-hydrolyzing enzyme was localized specifically to the micropylar region of the endosperm of tomato seed prior to germination. The enzyme was endo-β-mannanase and it hydrolyzed galactomannan into oligosaccharides with no release of galactose and mannose. This pre-germination enzyme was identified 18 h after imbibition and increased up to the time immediately before radicle protrusion, 6 h later. The profile of activity of this enzyme corroborates the hypothesis that the physiological role of pre-germination galactomannan-hydrolyzing enzyme(s) is to weaken the endosperm cell wall allowing the radicle to protrude. Later studies (Nonogaki et al. 2000) showed that distinct mannanases are involved in germination and post-germination processes, with LeMAN2 being associated with endosperm cap weakening prior to radicle emergence, whereas LeMAN1 mobilizes galactomannan reserves in the lateral endosperm. Similar studies on the role of the endosperm in the control of germination of endospermic legume seeds have not been performed yet.

6.4 *Lipids and Early Fatty Acid Metabolism*

Storage lipids of yellow lupine seeds in the form of triacylglycerol are first hydrolyzed to glycerol and free fatty acids by lipases, which are inhibited by application of exogenous sucrose to the germination media (Borek and Nuc 2011). Fatty acids are then fed to the β-oxidation and glyoxylate pathways to produce acetyl CoA, succinate, and malate (Pritchard et al. 2002). The β-oxidation and glyoxylate pathways take place mostly in glyoxysomes, which are a specialized form of peroxisome found in seeds. In dry seeds, they are present in a small underdeveloped form, and then during germination, grow and accumulate essential enzymes, such as malate synthase (MLS) and isocitrate lyase (ICL), which are unique to the glyoxylate cycle. The glyoxylate cycle enzymes cytosolic aconitase and ICL are active in both the embryo and the cotyledons of germinating yellow lupine seeds (Borek and Nuc 2011). Their activity is enhanced by exogenous sucrose by effecting gene expression. The products of fatty acid degradation via these pathways can be converted to sugars, amino acids or can be utilized for energy via the TCA cycle (Eastmond and Graham 2001; Borek et al. 2003), or for amino acid biosynthesis. Asparagine, glutamine, and glutamate are also synthesized from lipid-derived carbon skeletons during germination of yellow lupine (Borek et al. 2003).

7 Conclusions

Altogether, better information is required on the effects of stress conditions during seed development of grain legumes on subsequent seed germination as well as on the genetic and physiological control of germination responses to soil water potential and temperature. This knowledge is needed to successfully confront the challenge of climate change in regions of legume cultivation and for further expansion of cultivation into regions with more extreme climatic conditions. The integrated view of seed germination combining seed structural morphology, physiology of dormancy and germination metabolism remains fragmented, mainly because of barriers between disciplines. In the post-genomics era, with the development of marker-assisted breeding strategies, a deeper understanding of the links between seed-coat permeability, seed metabolism (e.g., energy-related or modulating permeability), seed dormancy, and seedling vigor will greatly contribute to the production of improved varieties. Concomitantly, these studies will lead to the identification of genes functionally related to key processes in metabolic resumption upon water imbibition and in the control of germination.

References

Abbo S, Rachamim E, Zehavi Y et al (2011) Experimental growing of wild pea in Israel and its bearing on Near Eastern plant domestication. Ann Bot 107:1399–1404

Abbo S, Pinhasi van-Oss R, Gopher A et al (2014) Plant domestication versus crop evolution: a conceptual framework for cereals and grain legumes. Trends Plant Sci 19:351–360

Abdullah WD, Powell AA, Matthews S (1991) Association of differences in seed vigour in long bean (*Vigna sesquipedalis* L. Fruhw) with testa colour and imbibition damage. J Agr Sci 116:259–264

Aitken Y (1939) The problem of hard sees in subterranean clover. Proc Royal Soc Victoria 51:187–213

Al-Ani A, Bruzau F, Raymond P et al (1985) Germination, respiration, and adenylate energy charge of seeds at various oxygen partial pressures. Plant Physiol 79:885–90

Al-Karaki GN (1998) Seed size and water potential effects on water uptake, germination and growth of lentil. J Agron Crop Sci 181:237–242

Alkhalfioui F, Renard M, Vensel WH et al (2007) Thioredoxin-linked proteins are reduced during germination of *Medicago truncatula* seeds. Plant Physiol 144:1559–1579

Angelovici R, Fernie AR, Galili G et al (2010) Seed desiccation: a bridge between maturation and germination. Trends Plant Sci 15:211–218

Arechavaleta-Medina F, Snyder HE (1981) Water imbibition by normal and hard soybeans. J Am Oil Chem Soc. 58:976–979

Angelovici R, Fait A, Fernie AR, et al (2011) A seed high-lysine trait is negatively associated with the TCA cycle and slows down Arabidopsis seed germination. New Phytol 189:148–159

Argel PJ, Parton CJ (1999) Overcoming legume hardseededness. In: Loch DS, Ferguson JE (eds) Forage seed production: tropical and sub-tropical species. CAB International, Wallingford, pp 247–267

Arrieta V, Besga G, Cordero S (1994) Seed coat permeability and its inheritance in a forage lupin (*Lupinus hispanicus*). Euphytica 75:173–177

Ashworth EN, Obendorf RL (1980) Imbibitional chilling injury in soybean axes: relationship to stelar lesions and seasonal environments. Agron J 72:923–928

Asiedu EA, Powell AA (1998) Comparisons of cultivars of cowpea (*Vigna unguiculata*) differing in seed coat pigmentation. Seed Sci Technol 26:211–222

Aswathaiah B (1988) Structures involved in softening of hard seeds in common vetch. Seed Res 16:63–67

Bai B, Sikron N, Gendler T et al (2012) Ecotypic variability in the metabolic response of seeds to diurnal hydration-dehydration cycles and its relationship to seed vigor. Plant Cell Physiol 53:38–52

Ballard LAT (1961) Studies on dormancy in the seeds of subterranean clover (*Trifolium subterraneum* L.). II. The interaction of time, temperature and carbondioxide during passage out of dormancy. Austral J Biol Sci 4:173–186

Baskin CC (2003) Breaking physical dormancy in seeds—focusing on the lens. New Phytol 158:229–232

Baskin CC, Baskin JM (2014) Seeds. Ecology, biogeography and evolution of dormancy and germination, 2nd edn. Academic Press, p 1600

Baskin JM, Baskin CC (2004) A classification system for seed dormancy. Seed Sci Res 14:1–16

Baskin JM, Baskin CC, Li X (2000) Taxonomy, anatomy and evolution of physical dormancy in seeds. Plant Spec Biol 15:139–152

Baud S, Boutin JP, Miquel M et al (2002) An integrated overview of seed development in *Arabidopsis thaliana* ecotype WS. Plant Physiol Biochem 406:151–160

Bayuelo-Jimenez JS, Craig R, Lynch JP (2002) Salinity tolerance of *Phaseolus* species during germination and early seedling growth. Crop Sci 42:1584–1594

Behboudian HM, Ma Q, Turner NC et al (2001) Reactions of chickpea to water stress: yield and seed composition. J Sci Food Agr 81:1288–1291

Benamara A, Tallon C, Macherel D (2003) Membrane integrity and oxidative properties of mitochondria isolated from imbibing pea seeds after priming or accelerated ageing. Seed Sci Res 13:35–45

Bewley JD, Bradford KJ, Hilhorst HWM et al (2013) Seeds: physiology of development, germination and dormancy, 3rd edn. Plenum, New York

Blöchl A, Peterbauer T, Richter A (2007) Inhibition of raffinose oligosaccharide breakdown delays germination of pea seeds. J Plant Physiol 164:1093–1096

Blöchl A, Peterbauer T, Hofmann J et al (2008) Enzymatic breakdown of raffinose oligosaccharides in pea seeds. Planta 228:99–110

Boesewinkel, FD, Bouman F. (1995) The seed: structure and function. In: Kigel J, Galili G (eds) Seed development and germination. CRC, Boca Raton, pp 1–24

Bolingue W, Vu BL, Leprince O et al (2010) Characterization of dormancy behaviour in seeds of the model legume *Medicago truncatula*. Seed Sci Res 20:97–107

Bond WJ, Honig M, Maze KE (1999) Seed size and seedling emergence: an allometric relationship and some ecological implications. Oecologia 120:132–136

Borek S, Nuc K (2011) Sucrose controls storage lipid breakdown on gene expression level in germinating yellow lupine (*Lupinus luteus* L.) seeds. J Plant Physiol 168:179–803

Borek S, Morkunas I, Ratajczak W et al (2001) Metabolism of amino acids in germinating yellow lupin (*Lupinus luteus* L.) seeds. III. Breakdown of arginine in sugar-starved organs cultivated in vitro. Acta Physiol Plant 23:141–148

Borek S, Ratajczak W, Ratajczak L (2003) A transfer of carbon atoms from fatty acids to sugars and amino acids in yellow lupine (*Lupinus luteus* L.) seedlings. J Plant Physiol 160:539–545

Borek S, Kubala S, Kubala S (2012) Regulation by sucrose of storage compounds breakdown in germinating seeds of yellow lupine (*Lupinus luteus* L.), white lupine (*Lupinus albus* L.) and Andean lupine (*Lupinus mutabilis* Sweet): I. Mobilization of storage protein. Acta Physiol Plant 34:701–711

Botha FC, Potgieter GP, Botha AM (1992) Respiratory metabolism and gene expression during seed germination. Plant Growth Regul 11:211–224

Boutra T, Sanders FE (2001) Influence of water stress on grain yield and vegetative growth of two cultivars of *Phaseolus vulgaris* L. J Agron Crop Sci 187:251–257

Bradford KJ (2002). Applications of hydrothermal time to quantifying and modeling seed germination and dormancy. Weed Sci 50:248–260

Brucher H (1988) The wild ancestors of *Phaseolus vulgaris* in South America. In: Gepts P (ed) Genetic resources of *Phaseolus* beans. Kluwer, Dordrecht, pp 185–214

Bruneau A, Doyle JJ, Herendeen P et al (2013) Legume phylogeny and classification in the 21st century: progress, prospects and lessons for other species-rich clades. Taxon 62:217–248

Brunel S, Teulat-Merah B, Wagner H et al (2009) Using a model-based framework for analyzing genetic diversity during germination and heterotrophic growth of *Medicago truncatula*. Ann Bot 103:1103–1117

Chachalis D, Smith ML (2000) Imbibition behavior of soybean (*Glycine max* (L.) Merrill accessions with different testa characteristics. Seed Sci Technol 28:321–331

Chachalis D, Smith ML (2001) Seed coat regulation of water uptake during imbibition in soybeans (*Glycine max* (L.) Merrill). Seed Sci Technol 29:401–412

Chamberlin MA, Horner HT, Palmer RG (1994) Early endosperm, embryo, and ovule development in *Glycine max* (L.) Merrill. Intl J Plant Sci 155:421–436

Covell S, Ellis RH, Roberts EH et al (1986) The influence of temperature on seed germination rate in grain legumes. I. A comparison of chickpea, lentil, soybean and cowpea at constant temperatures. J Expt Bot 37:705–715

Craufurd PQ, Ellis RH, Summerfield RRJ et al (1996) Development in cowpea (*Vigna unguiculata*) I. the influence of temperature on seed germination and seedling emergence. Exp Agr 32:1–12

Cronk Q, Ojeda I, Pennington RT (2006) Legume comparative genomics: progress in phylogenetics and phylogenomics. Curr Opin Plant Biol 9:99–103

Degreef J, Rocha OJ, Vanderborght T et al (2002) Soil seed bank and seed dormancy in wild populations of Lima bean (Fabaceae): considerations for in situ and ex situ conservation. Am J Bot 89:1644–1650

Deshpande SS, Cheryan M (1986) Microstructure and water uptake of *Phaseolus* and winged beans. J Food Sci 51:1218–1223

Dickson MH, Boettger MA (1982) Heritability of semi-hard seed induced by low seed moisture in beans (*Phaseolus vulgaris* L.). J Am Soc Hortic Sci 107:69–71

Dierking EC, Bilyeu KD (2009) Raffinose and stachyose metabolism are not required for efficient soybean seed germination. J Plant Physiol 166:1329–1335

Donnelly ED, Watson JE, McGuire JA (1972) Inheritance of hard seed in *Vicia*. J Hered 63:361–365

Donohue K (2009) Completing the cycle: maternal effects as the missing link in plant life histories. Phil Trans R Soc B 364:21059–1074

Dracup M, Davies C, Tapscott H (1993) Temperature and water requirements for germination and emergence of lupin. Aust J Exp Agr 33:759–766

Duke SH, Kakefuda G (1981) Role of the testa in preventing cellular rupture during imbibition of legume seeds. Plant Phys 67:449–456

Dumur D, Pilbeam CJ, Craigon J (1990) Use of the Weibull function to calculate cardinal temperatures in faba bean. J Expt Bot 41:1423–1430

Eastmond PJ, Graham IA (2001) Re-examining the role of the glyoxylate cycle in oilseeds. Trends Plant Sci 6:72–78

Egli DB, TeKrony DM, Heitholt JJ et al (2005) Air temperature during seed filling and soybean seed germination and vigor. Crop Sci 45:1329–1335

Ellis RH, Roberts EH (1982) Desiccation, rehydration, germination, imbibition injury and longevity of pea seeds (*Pisum sativum*). Seed Sci Technol 10:501–508

Ellis RH, Barrett S (1994) Alternating temperatures and rate of seed germination in lentil. Ann Bot 74:519–524

Ellis RH, Hong TD, Roberts EH (1985) Handbook of seed technology for genebanks. Volume II. Compendium of specific germination information and test recommendations. IBPGR, Rome, pp. 211–667

Ellis RH, Covell S, Roberts EH et al (1986) The influence of temperature on seed germination rate of grain legumes. II. Intraspecific variation in chickpea at constant temperatures. J Expt Bot 37:1503–1515

Ellis RH, Simon G, Covell S (1987) The influence of temperature on seed germination rate of grain legumes.III. A comparison of five faba bean genotypes at constant temperatures using a new screening method. J Expt Bot 38:1033–1043

Esau C (1977) Anatomy of seed plants, 2nd edn. Wiley, New York, p 576

Fairbrother TE (1991) Effect of fluctuating temperatures and humidity on the softening rate of hard seed of subterranean clover (*Trifolium subterraneum* L.). Seed Sci Technol 19:93–105

Fait A, Angelovici R, Less H et al (2006) Arabidopsis seed development and germination is associated with temporally distinct metabolic switches. Plant Physiol 142:839–854

Fait A, Fromm H, Walter D et al (2008) Highway or byway: the metabolic role of the GABA shunt in plants. Trends Plant Sci 13:14–19

Faria JMR, Buitink J, van Lammeren AAM et al (2005) Changes in DNA and microtubules during loss and re-establishment of desiccation tolerance in germinating *Medicago truncatula* seeds. J Expt Bot 56:2119–2130

Fenner M (1991) The effects of the parent environment on seed germinability. Seed Sci Res 1:75–84

Fenner M (1992) Environmental influences of seed size and composition. Hort Rev 13:183–213

Finch-Savage AE, Leubner-Metzger G (2006) Seed dormancy and the control of germination. New Phytol 171:501–523

Finch-Savage AE, Rowse HR, Dent KC (2005) Development of combined imbibitions and hydrothermal threshold models to stimulate maize and chickpea seed germination in variable environments. New Phytol 165:825–838

Forbes I, Wells HD (1968) Hard and soft seededness in blue lupine (*Lupinus angustifolius* L.): inheritance and phenotype classification. Crop Sci 8:195–197

Fougereux JA, Dore T, Ladonne F et al (1997) Water stress during reproductive stages affects seed quality and yield of pea *(Pisum sativum* L.). Crop Sci 37:1247–1252

Fuller DQ (2007) Contrasting patterns in crop domestication and domestication rates: recent archeobotanical insights from the Old World. Ann Bot 100:903–924

Fuller DQ, Allaby R (2009) Seed dispersal and crop domestication: shattering, germination and seasonality in evolution under cultivation. Ann Plant Rev 38:238–295

Fyfield TP, Gregory PJ (1989) Effects of temperature and water potential on germination, radicle elongation and emergence in mungbean. J Expt Bot 40:667–674

Galili G (2011) The aspartate-family pathway of plants: linking production of essential amino acids with energy and stress regulation. Plant Signal Behav 6:192–195

Gallardo M, Delgado M, Sanchez-Call IM et al (1991) Ethylene production and 1-aminocyclopropane-1-carboxylic acid conjugation in thermoinhibited *Cicer arietinum* L. seeds. Plant Phys 97:122–127

Gallardo K, Signor CL, Thompson RD et al (2003) Proteomics of *Medicago truncatula* seed development establishes the time frame of diverse metabolic processes related to reserve accumulation. Plant Physiol 133:664–682

Gamma-Arachchige NS, Baskin JM, Geneve RL et al (2013). Identification and characterization often new water gaps in seeds and fruits with physical dormancy and classification of water-gap complexes. Ann Bot 112:69–84

Gan Y, Jayakumar R, Zener RP et al (2006) Selection for seed size and its impact on grain yield and quality in kabuli chickpea. Can J Plant Sci 86:345–352

Garnczarska M, Zalewski T, Kempka M (2007) Water uptake and distribution in germinating lupine seeds studied by magnetic resonance imaging and NMR spectroscopy. Physiol Plant 130:23–32

Gepts P, Debouck D (1991) Origin, domestication, and evolution of the common bean (*Phaseolus vulgaris* L.). In: Schoonhoven A, Voysest O (eds) Common beans: research for crop improvement. CIAT, CABI, Wallingford, pp 1–53

Gepts P, Beavis WD, Brummer EC et al (2005) Legumes as a model plant family. Genomics for food and feed. Report of the cross-legume advances through genomics conference. Plant Physiol 137:1228–1235

Gholami A, Sharafi S, Sharafi A et al (2009) Germination of different seed size of pinto bean cultivars as affected by salinity and drought stress. J Food Agr Environ 7:555–558

Gimeno-Gilles C, Lelièvre E, Viau L et al (2009) ABA-mediated inhibition of germination is related to the inhibition of genes encoding cell-wall biosynthetic and architecture: modifying enzymes and structural proteins in *Medicago truncatula* embryo axis. Mol Plant 2:108–119

Globerson D (1978) Germination and dormancy breaking by ethephon in mature and immature seeds of *Medicago truncatula* and *Trifolium subterraneum*. Austral J Agr Res 29:43–49

Gross Y, Kigel J (1994) Differential sensitivity to high temperature of stages in the reproductive development of common bean (*Phaseolus vulgaris*). Field Crop Res 36:201–212

Gusmao M, Siddique KHM, Flower K, Nesbitt et al (2012) Water deficit during the reproductive period of grass pea (*Lathyrus sativus* L.) reduced grain yield but maintained seed size. J Agron Crop Sci 198:430–441

Hagon MW, Ballard LAT (1970) Reversibility of strophiolar permeability to water in seeds of subterranean clover (*Trifolium subterraneum.*). Austral J Biol Sci 23:519–528

Hanma A, Denna D (1962) Moisture movement in impermeable snap bean seed. Quart Bull Mich Agr Exp Sta 44:726–730

Harlan JR (1992) Crops and man, 2nd edn. ASA USDA, Madison, p 284

Harlan JR, de Wet JMJ, Price EG (1973) Comparative evolution of cereals. Evolution 27:311–325

Heil JR, McCarthy MJ, Ozilgen M (1992) Magnetic resonance imaging and modeling of water uptake into dry beans. Lebensmittel-Wissenschaft & Technologie 25:280–285

Hernandez-Nistal J, Martin I, Esteban R et al (2010) Abscisic acid delays chickpea germination by inhibiting water uptake and down-regulating genes encoding cell wall remodeling proteins. Plant Growth Regul 61:175–183
Holubowicz R, Taylor AG, Goffinet MC et al (1988) Nature of the semihard seed characteristic in snap beans. J Am Soc Hort Soc 113:248–252
Howell KA, Millar AH, Whelan J (2006) Ordered assembly of mitochondria during rice germination begins with pro-mitochondrial structures rich in components of the protein import apparatus. Plant Mol Biol 60:201–223
Hu X, Li T, Wang J et al (2013) Seed dormancy in four Tibetan Plateau Vicia species and characterization of physiological changes in response of seeds to environmental factors. Seed Sci Res 23:133–140
Hucl P (1993) Effects of temperature and moisture stress on the germination of diverse common bean genotypes. Can J Plant Sci 73:697–702
Humphrey ME, Lambrides CJ, Chapman SC et al (2005) Relationships between hard-seededness and seed weight in mungbean (*Vigna radiata*) assessed by QTL analysis. Plant Breed 124:292–298
Hyde EOC (1954) The function of the hilum in some Papilonaceae in relation to the ripening of the seed and the permeability of the testa. Ann Bot 18:241–256
Isemura T, Kaga A, Konishi S et al (2007) Genome dissection of traits related to domestication in Azuki Bean (*Vigna angularis*) and comparison with other warm-season legumes. Ann Bot 100:1053–1071
Isemura T, Kaga A, Tomooka N et al (2010) The genetics of domestication of rice bean (*Vigna umbellata*). Ann Bot 106:927–944
Jakobsen HB, Martens H, Lyshede OB (1994) Accumulation of metabolites during seed development in *Trifolium repens* L. Ann Bot 74:409–415
James AT, Lawn RJ, Williams RW et al (1999) Cross fertility of Australian accessions of wild mungbean (*Vigna radiata* ssp. *sublobata*) with green gram (V. *radiata* ssp. *radiata*) and black gram *(V. mungo)*. Austral J Bot 47:601–610
Johansen C, Sexena NP, Chauhan YS et al (1988) Genotypic variation in salinity response of chickpea and pigeonpea. In: Sinha SK (ed) Proceedings of international congress of plant physiology, vol. 2. New Delhi
Kaga A, Isemura T, Tomooka N et al (2008) The genetics of domestication of the azuki bean (*Vigna angularis*). Genet 178:1013–1036
Kannenberg L W, Allard R W (1964) An association between pigment and lignin formation in the seed coat of the Lima bean. Crop Sci 4:621–622
Kaplan L (1965) Archeology and domestication of americam *Phaseolus* (beans). Econ Bot 19:358–368
Kaplan L, Lynch TF (1999) Phaseolus (Fabaceae) in archaeology: AMS radiocarbon dates and their significance for pre-Columbian archaeology. Econ Bot 53:261–267
Katznelson J, Carpenter JA (1972) Germination of subterranean clover in Mediterranean summer. Israel J Bot 21:228–242
Kaur G, Kumar S, Nayyar H et al (2008) Cold stress injury during the pod-filling phase in chickpea (*Cicer arietinum* L.): effects on quantitative and qualitative components of seeds. J Agron Crop Sci 194:457–464
Kigel J (1995) Seed germination in arid and semi-arid regions. In: Kigel J, Galili G (eds) Seed development and germination. Marcel Dekker, New York, pp 645–700
Kikuchi K, Koizumi M, Ishida N et al (2006) Water uptake by dry beans observed by micro-magnetic resonance imaging. Ann Bot 98:545–553
Kirkbridge JH, Gunn Cr, Weitzman A (2003) Fruits and seeds of genera in the subfamily Faboideae (Fabaceae). Technical Bulletin 1890, USDA
Kirma M, Araújo WL, Fernie AR, Galili G (2012) The multifaceted role of aspartate-family amino acids in plant metabolism. J. Expt. Bot 63:4995–5001
Kluyver TA, Charles M, Jones G et al (2013) Did greater burial depth increase the seed size of domesticated legumes? J Expt Bot 64:4101–4108

Koinange EMK, Singh SP, Gepts P (1996) Genetic control of the domestication syndrome in common bean. Crop Sci 36:1037–1045
Kooistra E (1971) Germinability of beans (*Phaseolus vulgaris* L.) at low temperatures. Euphytica 20:208–213
Koizumi M, Kikuchi K, Isobe S et al (2008) Role of seed coat in imbibing soybean seeds observed by micro-magnetic resonance imaging. Ann Bot 102:343–352
Kongjaimun A, Kaga A, Tomooka N et al (2012) The genetics of domestication of yardlong bean *Vigna unguiculata* (L.) Walp. Ann Bot 109:1185–1200
Kyle JH, Randall TE (1963) A new concept of the hard seed character in *Phaseolus vulgaris* L. and its use in breeding and inheritance studies. Proc Am Soc Hort Sci 83:461–475
Ladizinski G (1985) The genetics of hard seeds in *Lens*. Euphytica 34:539–543
Legesse N, Powell AA (1992) Comparisons of water uptake and imbibition damage in eleven cowpea cultivars. Seed Sci Technol 20:173–180
Legesse N, Powell AA (1996) The association between development of seed coat pigmentation during maturation of grain legumes and reduced rates of imbibition. Seed Sci Technol 24:23–32
Lehmann T, Ratajczak L (2008) The pivotal role of glutamate dehydrogenase (GDH) in the mobilization of N and C from storage material to asparagine in germinating seeds of yellow lupine. J Plant Physiol 165:149–58
Lersten NR (1982) Tracheid bar and vestured pits in legume seeds (Leguminosae: Papilionoideae). Am J Bot 69:98–107
Leymarie J, Benech-Arnold R L, Farrant JM et al (2009) Thermodormancy and ABA metabolism in barley grains. Plant Signal Behav 4:205–207
Liu B, Fujita T, Yan ZE et al (2007) QTL mapping of domestication-related traits in soybean (*Glycine max*). Ann Bot 100:1027–1038
Lopes VB, Takai M (1987) Seed germination in *Phaseolus vulgaris* L. I. Osmotic effect on light sensitivity. Arq Biol Techn 30:641–647
Lush WM, Evans LT (1980) The seed coats of cowpeas and other grain legumes: structure in relation to function. Field Crops Res 3:267–286
Lush WM, Wien HC (1980) The importance of seed size in early growth of wild and domesticated cowpeas. J Agr Sci 94:177–182
Lush WM, Evans LT, Wien HC (1980) Environmental adaptation of wild and domesticated cowpeas (Vigna unguiculata). Field Crops Res 3:173–187
Ma F, Cholewa E, Mohamed T et al (2004) Cracks in the palisade cuticle of soybean seed coats correlate with their permeability to water. Ann Bot 94:213–228
Maas EV, Hoffman GJ (1977) Crop salt tolerance. J Irrig Drainage 103:115–134
Maass BL, Usongo MF (2007) Changes in seed characteristics during the domestication of the lablab bean (*Lablab purpureus* (L.) Sweet). Austral J Agr Res 58:9–19
McCleary BV, Matheson NK (1975) Galactomannan structure and β-mannanase and β-mannosidase activity in germinating legume seeds. Phytochemistry 14:1187–1194
McClendon JH, Nolan WG, Wenzler HF (1976) The role of the endosperm in the germination of legumes: galactomannan, nitrogen, and phosphorus changes in the germination of guar (*Cyamopsis tetragonoloba*). Am J Bot 63:790–797
McDonald MB, Vertucci CW, Roos EE (1988) Seed coat regulation of soybean seed imbibition. Crop Sci 28:987–992
McKee GW, Peiffer RA, Mohsenin NN (1977) Seedcoat structure in *Coronilla varia* L. and its relations to hard seed. Agron J 69:53–58
Machado Neto NB, Prioli MR, Gatti AB et al (2006) Temperature effects on seed germination in races of common bean. Acta Sci Agron 28:155–167
Macherel D, Benamar A, Avelange-Macherel M et al (2007) Function and stress tolerance of seed mitochondria. Physiol Plant 129:233–241
Marbach I, Mayer AM (1974) Permeability of seed coats to water as related to drying conditions and metabolism of phenolics. Plant Physiol 54:817–820
Marbach I, Mayer AM (1975) Changes in catechol oxidase and permeability to water in seed coats of *Pisum elatius* during seed development and maturation. Plant Physiol 56:93–96

Marjüshkin VF, Sichkar VI, Michailov VG et al (1987) Inheritance of hardseededness in soybean. Soybean Genet Newsl 14:294–297
Marwat KB, Khan MA, Arif M et al (2008) Seed priming improves emergence and yield of soybean. Pakistan J Bot 40:1169–1177
Meyer CJ, Steudle E, Peterson CA (2007) Patterns and kinetics of water uptake by soybean seeds. J Expt Bot 58:717–732
Mikolajczyk J (1966) Genetic studies in *Lupinus angustifolius*. III. Inheritance of the alkaloid content, seed hardness and length of the growing season in blue lupin. Genet Pol 7:181–196
Miller SS, Bowman LA, Gijzen M et al (1999) Early development of the seed coat of soybean (*Glycine max*). Ann Bot 86:297–304
Morley FHW (1958) The inheritance and ecological significance of seed dormancy in subterranean clover (*Trifolium subterraneum* L.). Aust J Biol Sci 11:261–274
Morohashi M, Shimokoriyama Y (1975) Development of glycolytic and mitochondrial activities in the early phase of germination of *Phaseolus mungo* seeds. J Expt Bot 26:932–938
Morohashi Y, Bewley J, Yeung E (1981) Biogenesis of mitochondria in imbibed peanut cotyledons: influence of the axis. J Expt Bot 32:605–613
Morrison RM, Asiedu EA, Stuchbury T et al (1995) Determination of lignin and tannin content of cowpea seeds coats. Ann Bot 76:287–290
Munns R (2002) Comparative physiology of salt and water stress. Plant Cell Environ 25:239–250
Nadeau CD, Ozga JA, Kurepin LV et al (2011) Tissue specific regulation of gibberellins biosynthesis in developing pea seeds. Plant Physiol 156:897–912
Nakamura R (1988) Seed abortion and seed size variation within fruits of *Phaseolus vugaris*: pollen donor and resource limitation effects. Am J Bot 75:1003–1010
Nakayama N, Komatsu S (2008) Water uptake by seeds in yellow-seeded soybean cultivars with contrasting imbibition behaviors. Plant Prod Sci 11:415–422
Nawa Y, Asahi T (1971) Rapid development of mitochondria in pea cotyledons during the early stage of germination. Plant Physiol 48:671–674
Nleya T, Ball RA, Vandenberg A (2005) Germination of common bean under constant and alternating cool temperatures. Can J Plant Sci 85:577–588
Nonogaki H, Nomaguchi M, Morohashi Y (1995) Endo-β-mannanases in the endosperm of germinated tomato seeds. Physiol Plant 94:328–334
Nonogaki H, Nomaguchi M, Okumoto N et al (1998) Temporal and spatial pattern of the biochemical activation of the endosperm during and following imbibition of tomato seeds. Physiol Plant 102:236–242
Nonogaki H, Gee OH, Bradford KJ (2000) A germination-specific endo-β-mannanase gene is expressed in the micropylar endosperm cap of tomato seeds. Plant Physiol 123:1235–1245
Okamoto T, Minamikawa T (1998) A vacuolar cysteine endopeptidase (SH-EP) that digests seed storage globulin. Characterization, regulation of gene expression, and post-translational processing. J Plant Physiol 152:675–682
Olivier FC, Annandale JG (1998) Thermal time requirements for the development of green pea (*Pisum sativum* L.). Field Crop Res 56:301–307
Oliveira MA, Matthews S, Powell AA (1984) The role of split seed coats in determining seed vigour in commercial seed lots of soybean, as measured by the electrical conductivity test. Seed Sci Technol 12:659–668
Otubo ST, Ramalho MGP, Abreu AFB et al (1996) Genetic control of low temperature tolerance in germination of common bean. Euphytica 89:313–317
Parrish DJ, Leopold AC (1977) Transient changes during soybean imbibition. Plant Physiol 59:1111–1115
Patrick JW, Offler CE (2001) Compartmentation of transport and transfer events in developing seeds. J Expt Bot 52:551–564
Pena-Valdivia CB, Garcia R, Aguirre JR et al (2002) The effects of high temperature on dormancy and hypocotyls-root growth of wild common bean (*Phaseolus vulgaris*). Seed Sci Technol 30:231–248

Perissé P, Planchuelo AM (2004) Seed coat morphology of *Lupinus albus* L. and *Lupinus angustifolius* L. in relation to water uptake. Seed Sci Technol 32:69–77

Perisse P, Aiazzi MT, Planchuelo AM (2002) Water uptake and germination of *Lupinus albus* and *Lupinus angustifolius* under water stress. Seed Sci Technol 30:289–298

Peterbauer T, Lahuta LB, Blöchl A et al (2001) Analysis of the raffinose family oligosaccharide pathway in pea seeds with contrasting carbohydrate composition. Plant Physiol 127:1764–1772

Petruzelli L, Harren F, Perrone C et al (1995) On the role of ethylene in seed germination and early root growth of *Pisum sativum*. J Plant Physiol 145:83–86

Poggio SL, Satorre EH, Dethiou S et al (2005) Pod and seed numbers as a function of photothermal quotient during the seed set period of field pea (*Pisum sativum*) crops. Eur J Agron 22:55–69

Powell AA (1989) The importance of genetically determined seed coat characteristics to seed quality in grain legumes. Ann Bot 63:169–175

Powell AA, Matthews S (1978) The damaging effect of water on dry pea embryos during imbibition. J Expt Bot 29:1215–1229

Powell AA, Oliveira MA, Matthews S (1986) Seed vigour in cultivars of dwarf French bean (*Phaseolus vulgaris*) in relation to testa colour. J Agr Sci 106:419–425

Pritchard SL, Charlton WL, Baker A et al (2002) Germination and storage reserve mobilization are regulated independently in *Arabidopsis*. Plant J 31:639–647

Quinlivan BJ (1968) The softening of hard seeds of sand-plain lupin (*Lupinus varius* L.). Austral J Agr Res 19:507–515

Qutob D, Ma F, Peterson CA et al (2008) Structural and permeability properties of the soybean seed coat. Botany 86:219–227

Ramsay G (1997) Inheritance and linkage of a gene for testa-imposed seed dormancy in faba bean (*Vicia faba* L.). Plant Breed 16:287–289

Ranathunge K, Shao S, Qutob D et al (2010) Properties of the soybean seed coat cuticle change during development. Planta 231:1171–1188

Ratajczak W, Lehmann T, Polcyn W et al (1996) Metabolism of amino acids in germination yellow lupine seeds. I. The decomposition of 14C-aspartate and 14C-glutamate during the imbibition. Acta Physiol Plant 18:13–18

Raveneau MP, Coste F, Moreau-Valancogne P et al (2011) Pea and bean germination and seedling responses to temperature and water potential. Seed Sci Res 2:205–213

Raymond P, Al-Ani A, Pradet A (1985) ATP production by respiration and fermentation, and energy charge during aerobiosis and anaerobiosis in twelve fatty and starchy germinating seeds. Plant Physiol 79:879–84

Reid JSG, Bewley JD (1979) A dual role for the endosperm and its galactomannan reserves in the germinative physiology of Fenugreek *(Trigonella foenum-graecum* L.), an endospermic leguminous seed. Planta 147:145–150

Reid JSG, Meier H (1972) Function of the aleurone layer during galactomannan mobilization in germinating seeds of fenugreek, crimson clover and lucerne: a correlative biochemical and ultrastructural study. Planta 106:44–60

Reid JSG, Meier H (1973) Enzymic activities and galactomannan mobilization in germinating seeds of fenugreek. Secretion of α-galactosidase and β-mannosidase by the aleurone layer. Planta 112:301–308

Riggio-Bevilaqua LR, Roti-Michelozzi G, Modenesi P (1989) The watertight dormancy of *Melilotus alba* seeds. Can J Bot 67:3453–3456

Roach DA, Wulff RD (1987) Maternal effects in plants. Ann Rev Ecol Syst. 18:209–235

Rolston MP (1978) Water impermeable seed dormancy. Bot Rev 44:365–395

Ross KA, Zhang L, Arntfield SD (2010) Understanding water uptake from the induced changes occurred during processing: chemistry of Pinto bean and Navy bean seed coats. Int J Food Prop 13:631–647

Russi L, Cocks PS, Roberts EH (1992) Coat thickness and hard-seededness in some *Medicago* and *Trifolium* species. Seed Sci Res 2:243–249

Sakamoto S, Abe J, Kanazawa A et al (2004) Marker-assisted analysis for soybean hard seededness with isozyme and simple sequence repeat loci. Breed Sci 54:133–139

Santalla M, Menendez-Sevillano MC, Monteagudo AB et al (2004) Genetic diversity of Argentinean common bean and its evolution during domestication. Euphytica 135:75–87
Scully B, Waines JG (1987) Germination and emergence response of common and tepary beans to controlled temperature. Agron J 79:287–291
Serrato-Valenti G, Melone L, Ferro M et al (1989) Comparative studies on testa structure of 'hard-seeded' and 'soft-seeded' varieties of *Lupinus angustifolius* L. (Leguminosae) and on mechanisms of water entry. Seed Sci Technol 17:563–581
Shahi JP, Pandey MP (1982) Inheritance of seed permeability in soybean. Indian J Gen Pl Br. 42:196–199
Shao S, Meyer CJ, Ma F et al (2007) The outermost cuticle of soybean seeds: chemical composition and function during imbibitions. J Expt Bot 58:1071–1082
Sievwright CA, Shipe WF (1986) Effect of storage conditions and chemical treatments on firmness, in vitro protein digestibility, condensed tannins, phytic acid and divalent cations of cooked Black Beans (*Phaseolus vulgaris*). J Food Sci 51:982–987
Soeda Y, Konings MCJM, Vorst O et al (2005) Gene expression programs during *Brassica oleracea* seed maturation, osmopriming and germination are indicators of progression of the germination process and the stress tolerance level. Plant Physiol 137:354–368
Stephenson AG (1981) Flower and fruit abortion: proximate causes and ultimate functions. Ann Rev Ecol Syst 12:253–279
Stienswat W, Pollard LH, Campbell WF (1971) Nature of hardseededness in Lima bean (*Phaseolus lanatus* L.). J Am Soc Hort Sci 96:312–315
Tang J, Sokhansanj S, Sosulski FW (1994) Moisture absorption characteristics of laird lentils and hardshell seeds. Cereal Chem 71:423–428
Taylor GB (2005) Hardseededness in Mediterranean annual pasture legumes in Australia: a review. Aust J Agr Res 56:645–661
Thomson JR (1965) Breaking dormancy in germination tests of Trifolium spp. Proc Int Seed Testing Assoc. 30:905–909
Toh S, Imamura A, Watanabe A, et al (2008) High temperature induced abscisic acid biosynthesis and its role in the inhibition of gibberellin action in Arabidopsis seeds. Plant Physiol 146:1368–1385
Toledo MZ, Cavariani C, França-Neto JB et al (2010) Imbibition damage in soybean seeds as affected by initial moisture content, cultivar and production location. Seed Sci Technol 38:399–408
Tran VN, Cavanagh AK (1984) Structural aspects of dormancy. In: Murray DR (ed) Seed physiology. Vol. II. Academic, Sydney, pp 1–44
Tully RE, Musgrave ME, Leopold AC (1981) The seed coat as a control of imbibitional chilling injury. Crop Sci 21:312–7
Uzun A, Sozen E, Acikgoz E (2013) Seed dormancy and germination of *Vicia sativa* subsp. *nigra* and *Vicia sativa* subsp. *macrocarpa*. Seed Sci Technol 41:137–142
van Assche JA, Vandelook FEA (2010) Combinational dormancy in winter annual Fabaceae. Seed Sci Res 20:237–242
van Assche JA, Debucquoy KLA, Rommens WAF (2003) Seasonal cycles in the germination capacity of buried seeds of some Leguminosae. New Phytol 158:315–323
van Dongen JT, Ammerlaan AMH, Wooterlood M et al (2003) Structure of developing pea seed coat and the post phloem transport pathway of nutrients. Ann Bot 91:729–737
Varriano-Marston E, Jackson GM (1981) Hard to cook phenomenon in beans: structural changes during storage and imbibition. J Food Sci 46:1379–1385
Verdier J, Dessaint F, Schneider C et al (2013) A combined histology and transcriptome analysis unravels novel questions on *Medicago truncatula* seed coat. J Expt Bot 64:459–470
Verma VD, Ram HH (1987) Genetics of seed impermeability in soyabean. J Agr Sci 108:305–310
Wang HL, Grusak MA (2005). Structure and development of *Medicago truncatula* pod wall and seed coat. Ann Bot 95:737–747

Wang WQ, Møller IM, Song SQ (2012) Proteomic analysis of embryonic axis of *Pisum sativum* seeds during germination and identification of proteins associated with loss of desiccation tolerance. J Proteomics 77:68–86
Watson DP (1948) Structure of the testa and its relation to germinationin the Papilionaceae tribes *Trifoliae* and *Loteae*. Ann Bot 12:385–409
Weber H, Borisjuk L, Wobus U (2005) Molecular physiology of legume seed development. Ann Rev Plant Biol 56:253–79
Weedin N (2007) Genetic changes accompanying the domestication of *Pisum sativum*: is there a common genetic basis to the 'Domestication Syndrome' for legumes? Ann Bot 100:1017–1025
Weitbrecht K, Müller K, Leubner-Metzger G (2011) First off the mark: early seed germination. J Expt Bot 62:3289–3309
Welbaum, GE, Tissaoui T, Bradford KJ (1990) Water relations of seed development and germination in muskmelon. III. Sensitivity of germination to water potential and early seedling growth. Plant Physiol 92:1029–1037
Werker E. (1997) Seed anatomy. Gebrüder Borntraeger, Stuttgart
Werker E, Marbach I, Mayer AM (1979) Relation between the anatomy of the testa, water permeability and the presence of phenolics in the genus *Pisum*. Ann Bot 43:765–771
Westoby M, Jurado E, Leishman M (1992) Comparative evolutionary ecology of seed size. Trends Ecol Evol 7:368–372
White JW, Montes R C (1993) The influence of temperature on seed germination in cultivars of common bean. J Expt Bot 44:1795–1800
Wojciechowski MF, Lavin M, Sanderson MJ (2004) A phylogeny of legumes (Leguminosae) based on analysis of the plastid MATK gene resolves many well-supported subclades within the family. Am J Bot 91:1846–1862
Wyatt JE (1977) Seed coat and water absorption properties of seed of near-isogenic snap bean lines differing in seed coat color. J Am Soc Hort Sci 102:478–480
Yang R, Guo Q, Gu Z (2013) GABA shunt and polyamine degradation pathway on gamma-aminobutyric acid accumulation in germinating faba bean (*Vicia faba* L.) under hypoxia. Food Chem 136:152–159
Yeung CE, Cavey MJ (1990) Cellular endosperm formation in *Phaseolus vulgaris*. I. Light and scanning electron microscopy. Can J Bot 66:1209–1216
Yomo H, Varner JE (1973) Control of the formation of amylases and proteases in the cotyledons of germinating peas. Plant Physiol 51:708–713
Young JA, Evans RA, Kay BL (1970) Germination characteristics of range legumes. J Range Manag 23:98–103
Zaiter H, Baydoun E, Sayyed-Hallak M (1994) Genotypic variation in the germination of common bean in response to cold temperature stress. Plant Soil 163:95–101
Zeng LW, Cocks PS, Kailis SG et al (2005) The role of fractures and lipids in the seed coat in the loss of hardseededness of six Mediterranean legume species. J Agr Sci 143:43–55
Zhang S, Hu J, Zhang Y, Xie XJ KA (2007) Seed priming with brassinolide improves lucerne (*Medicago sativa* L.) seed germination and seedling growth in relation to physiological changes under salinity stress. Austral J Agr Res 58:811–815
Zohary D, Hopf M, Weiss E (2012) Domestication of plants in the old world: the origin and spread of domesticated plants in Southwest Asia, Europe, and the Mediterranean basin, 4th edn. Oxford University Press, Oxford, p 264

Chapter 12
Reproductive Biology of Grain Legumes

María José Suso, Penelope J. Bebeli and Reid G. Palmer

1 Introduction

Regulations and consumer attitudes lead agricultural practices towards an ever more environmentally friendly production. There is an increasing interest of legume breeders in using bee pollinators for insect-aided outcrossing for heterosis-mediated exploitation in hybrid seed production and in heterotic open-pollinated populations. The global widespread decline of pollinator populations and diversity continues to cause international concern in commercial agriculture (Burkle et al. 2013). Also, this approach should preserve bee pollinator fauna by providing suitable floral resources within the legume crops themselves. This requires in-depth knowledge of the reproductive biology of legumes.

The reproductive biology of legumes has never been an easy topic. It has been an old and traditional topic never solved. This chapter is not a comprehensive review on legume reproductive biology; it was not possible nor was the aim. Valuable review materials including estimations on pollen-mediated gene flow (PMGF) from genetically modified (GM) crops to wild species are compiled by Andersson and Vicente (2010). Their focus was on the following legume species: chickpea, common bean, cowpea, pigeon pea, peanut and soybean. Also, there has been an excellent

Reid G. Palmer is deceased.

M. J. Suso (✉)
Plant Breeding. Instituto de Agricultura Sostenible (CSIC), Alameda del Obispo s/n, 14080 Córdoba, Spain
e-mail: ge1susom@uco.es

P. J. Bebeli
Department of Crop Science, Laboratory of Plant Breeding and Biometry, Agricultural University of Athens, Iera Odos 75, 11855 Athens, Greece
e-mail: bebeli@aua.gr

R. G. Palmer
Department of Agronomy, Iowa State University, Ames, IA 50010, USA
e-mail: rpalmer@iastate.edu

A. M. De Ron (ed.), *Grain Legumes*, Handbook of Plant Breeding 10,
DOI 10.1007/978-1-4939-2797-5_12

section on the reproductive biology of legumes in the chapter entitled "Biology of Food Legumes" in the book *Biology and Breeding of Food Legumes* (Chaturvedi et al. 2011). Readers, however, do not get core material on pollination and floral biology for breeding strategies. Nowadays, most of the new concepts and research on reproductive biology tends to be confined to ecologists and biologists. The application of their knowledge is necessary to both make progress in breeding and to improve the ecological services of legumes. We have tried to provide important references on significant works published to date relevant to different aspects of legume improvement. Bearing in mind the scope of the chapter, slight overlapping is possible both because of the paucity of the data and because the same topic is viewed from several perspectives.

2 Flower Attributes and Functions

2.1 Flower Morphology and Structure and Why It Matters. Mechanical Fit to Pollinators

Taking into account that bee pollinators are able to distinguish between complex floral traits, discriminate flower visits depending on floral design and display, and reward and attraction traits, the flower attributes and functions should play an essential role in legume breeding. In this section, we provide clues and analyse questions of relevance in relation to how the plant might best manage pollinators to modify (increase or reduce) the level of outcrossing, how to design a crop to sustain pollinators and the effectiveness of different types of flowers to support crop-pollinating wild bees and honeybees and at the same time discuss breeding strategies.

Grain legumes have a typical papilionate type of flower. Flowers have a pentamerous ground plan. The calyx consists of five sepals and the corolla comprises a standard two wings and two lower petals to form a keel. There are ten stamens surrounding the pistil, which is superior in position and differentiates into a gynoecium with stigma, style and ovary (Chaturvedi et al. 2011). According to Tucker (2003), the successive and overlapping order of organ initiation in some legume flowers is intriguing developmentally because of its conflict with the prevailing interpretation of hypotheses concerning timing of determination of organ identity. The ABC model hypothesis of floral organ identity applies to flowers in which all organs of a whorl initiate simultaneously; the order of initiation is sepals, petals, stamens and carpels, and where the whorls do not overlap in time of initiation (Tucker 2003). It does not satisfactorily explain a system in which more than one type of organ is being initiated at the same time, such as that in papilionoid legumes that have overlapping whorls. The ABC model also does not explain the concurrent initiation of the carpel at the same time as petals or stamens, which is typical in legumes. Thus, Tucker (2003) concluded that legume flowers fail to conform to the ABC model. The basic branching pattern in legume inflorescences is racemose, but systems with pseudoracemes and cymose branching do occur (reviewed by Prenner (2013)).

Fig. 12.1 *Vicia faba* flower showing the nectar guides. (Photo: JL Ubera)

The typical papilionated flower has a zygomorphic symmetry. Zygomorphic flowers have one plane of symmetry where one half mirrors the other half, also referred to as monosymmetric or bilaterally symmetric (Kalisz et al. 2006). Floral zygomorphy (flowers with bilateral symmetry) typically manifests two kinds of asymmetries, dorsoventral (DV) and organ internal (IN) asymmetries in floral and organ planes. Zygomorphic development involves the establishment of different body planes with distinct developmental axes, where various asymmetries are generated and superimposed under the control of different regulators. The pea (*Pisum sativum* L.) possesses a conspicuous zygomorphic flower, with an abundant mutant collection, and provides an ideal experimental system to analyse the key regulators in control of floral symmetry. Wang et al. (2008) investigated the loci affecting these asymmetries during the development of floral zygomorphy in pea. Two genes, *LOBED STANDARD 1 (LST1)* and *KEELED WINGS (K)* constitute the DV asymmetry, and *SYMMETRIC PETALS 1 (SYP1)* control IN asymmetry. Genetic analysis demonstrates that DV and IN asymmetries could be controlled independently by two kinds of regulators in pea, and their interactions help to specify the type of zygomorphy. Based on the genetic analysis in pea, they suggested that variation in both the functions and interactions of these regulators could give rise to a wide spectrum of floral symmetries among legume species.

Floral symmetry is a conspicuous cue. It has a particular signal value that is evidently perceived by insect pollinators. Apart from the external symmetry (i.e. symmetry of the outlines of the flower), flowers also display a symmetry of nectar guides which are usually situated near the centre of the flower (Fig. 12.1). This "internal" symmetry is perceived when the insect is very close to the flower. Thus, the insect has two opportunities for detecting the symmetry of the flower: The first opportunity arises when the insect is still at some distance, so that it can perceive the image of the flower as a whole, and the second opportunity arises when the insect comes close enough to resolve the pattern of the nectar guides (Giurfa et al. 1999).

The papilionate flowers also termed keel or flag flowers are very distinctive and commonly involve specializations for triggered pollen release (Fig. 12.2; Willmer 2011). When a visitor lands on such a flower (Fig. 12.3), it will normally grip onto

Fig. 12.2 Floral morphology of *Vicia faba* (*side view*). (Photo: JL Ubera)

Fig. 12.3 Floral visitors of *Vicia faba*. *Eucera numida* the most important pollinator of *Vicia faba* in southern Spain. (Photo: JL Ubera)

the wing petals and insert its tongue (proboscis) between the more or less erect standard petal and the upper edges of the keel, sliding the tongue inward to reach nectar in the staminal tube. There is a secondary pollen presentation system where the style is sharply bent upward terminally, with a brush of fine hairs near the tip and just below the stigmatic surfaces. Normally, the stamens dehisce inside the closed bud, and therefore shed their pollen onto this brush or on the inside of the end of the keel, where the brush picks it up. When the flower opens, the stamens are already withered, but the stigma brush is exposed by the first visitor and carries the pollen onto the visitor's underside. In these flowers, only quite large visitors with long tongues (bees, some hoverflies) can reach the nectar, and their weight is enough to depress the keel and ensure that the brush is exposed. Secondary pollen presentation occurs independently if the flowers are chasmogamous or cleistogamous, in different legumes such as the genera *Phaseolus, Vicia, Lathyrus, Lens* and *Pisum* (Lavin and Delgado 1990).

The morphology of legume flowers is complex, with each component serving a specific function. According to Westerkamp and Weber (1999), the papilionate flowers possess three elementary functional structures: the flag, the keel and the wings. (1) The flag is for visual attraction. Besides its visual role as a semaphore, the flag has at least two other important roles: (a) formation of a tongue guide and (b) function as an abutment. In certain species, nectar guides mark the pattern of the flag (Fig. 12.1). (2) The keel also has two essential functions: (a) it hides pollen from the collecting bees and (b) it has to provide structures that help to release the hidden pollen and deposit it onto a pollinating bee. (3) The wings facilitate the landing of a pollinator as well as the required active handling (lowering and opening) of the keel. The hiding of pollen within a keel necessitates subsequent pollen presentation. In secondary pollen presentation, it is usually the style which acts as the pollen presenter.

2.2 Functional Flower: Advertisement, Discovery and Rewards

2.2.1 General Approach, Advertisement and Discovery

Palmer et al. (2009) summarized the relative importance of floral functional morphology and structure in relation to pollination aspects that are applicable to plant breeding programmes. The significant question is how do floral traits influence bee behaviour? The reason why an insect forages on a particular flower can be partly attributed to differences in floral design and floral display. Flower shape and size can be considered as a function of one flower or of an inflorescence. Thus, it is helpful to deal with the following terminology. Floral design: describes the characteristics of individual flowers including their size, structure, sex condition, colour, scent, nectar production and degree of herkogamy and dichogamy. Floral display: the number of open flowers on a plant and their arrangement within and among inflorescences. The important functional unit for pollination is usually daily inflorescence size (Barrett 1998).

A flower's design and display will affect not only which animals can feed there but also which animals visit and make return visits. Hence, the morphological design and display features are only part of the story (Willmer 2011). Other floral traits are related to the distant and local recruitment by advertising the flower and to management by payment of reward to ensure that the visitor gets sufficient benefit to encourage further visitation. To receive the service of pollen transfer, plants often offer rewards to flower-visiting animals, such as nectar, oil, resin, pollen, breeding sites, etc. Flowers attract pollinators via various stimuli, whether they are olfactory or visual cues acting from a distance or tactile cues at close vicinity to guide pollinators to rewarding resources. Floral traits, resource distribution and cognitive and learning abilities of pollinators influence their behaviour, which in turn is strongly linked to plant mating patterns and gene flow within and among plant populations (Mayer et al. 2011). One aspect of flowers to be considered is the nectar guides. These colour-based guides may help to highlight the architecture of the flower during the approach, making foraging more efficient. Lines converging towards the corolla entrance are common in Fabaceae, effectively directing pollinators towards the rewards (Fig. 12.1; Delgado-Salinas et al. 2011).

Texture in petals and other floral parts may offer both visual and tactile cues to visitors (Kevan and Lane 1985; Whitney et al. 2011; Alcorn et al. 2012). Whitney et al. (2011), reviewed the importance of conical petal epidermal cells to enhance pollination success and concluded that conical epidermal cells significantly increased tactile handling of the flower by pollinators and hence their preference. The production of conical cells is controlled by an MYB-related transcription factor and the mutant *mixta* that has a null allele of this factor was found in *Antirrhinum majus*. In legumes, conical petal epidermal cells have been used as a marker for petal identity but there is no evidence of *mixta* homologues that play a role in differentiation of petal conical cells (Ojeda et al. 2009; Çildir et al. 2012).

It is not appropriate to look only at a flower colour and shape as attractants, given the ability of most flower visitors to detect and respond to scents or odours as well as other cues (Farré-Armengol et al. 2013). Floral scents mostly result from the production of small amounts of simple volatile organic compounds. Farré-Armengol et al. (2013) reviewed the emission of diverse biogenic volatile organic compounds (BVOCs; such as terpenoids, phenylpropanoids/benzenoids, fatty acid derivatives and amino acid derivatives) that plants produce and emit from different floral parts such as petals, sepals, pollen and nectar. The blends of the BVOCs are usually produced in petal conical cells (Whitney et al. 2011) and in scent glands or osmophores during anthesis (Farré-Armengol et al. 2013). Marinho et al. (2014) localized the sites of fragrance production and release in legume flowers in the perianth and particularly the petals, in which scent plays an important role in pollination. An analysis of cut flower scent in *Lathyrus odoratus* showed that the aroma was almost exclusively produced by the standard and the wings while the keel petals and other floral parts emitted very little (Sexton et al. 2005).

2.2.2 Reward: Pollen and Nectar

Regarding the reward traits, pollen, because of its inherent traits (containing small, easily managed nutrient packages), makes it a useful resource to exploit as food. Pollen is a crucial reward for pollen-eating and pollen-gathering visitors including virtually all bees. But for many visitors, it is functionally secondary; an extra reward for animals that primarily forage for nectar sources and pick up pollen incidentally (Willmer 2011).

Nectar is the reward commonly presented by plants to attract potential pollinators, thereby ensuring outcrossing and/or efficient pollination. Floral nectar is a sugar-rich fluid dominated by the hexoses glucose and fructose and the disaccharide sucrose (Brandenburg et al. 2009 for more details). Nectar allows flowers to "outsource" the pollination business to animal vectors, which assure a directional, accurate and efficient transfer of pollen compared to wind pollination. From the plant's perspective, in an ideal scenario, pollinators carry the maximum amount of pollen from one plant to the stigma of a conspecific while consuming minimal nectar. Limitation of nectar availability persuades pollinators to forage on a larger number of flowers and enhance pollen distribution which in turn may increase outcrossing, as showed in faba bean (Suso et al. 2005). It is also worthy to point out that flower depth also affects the pollinator foraging and that deeper flowers often contain more nectar (Harder 1985). Floral tube length is of prime importance to faba bean because longer tube length increases seed and fruit set (Davis 2001; Suso et al. 2005). Pierre et al. (1996) studied two spring-type faba beans, self-fertile and nonself-fertile, in open field conditions for their nectar amount, sugar composition and their attractiveness to bees. Nectar secretion of the self-fertile line was higher than that of the nonself-fertile line. However, the nonself-fertile line was as attractive as the self-fertile line to honeybees; in contrast, *Bombus terrestris* preferred self-fertile line. In soybean, differences in nectar production and composition, and floral morphology among soybean cultivars are affected by the environment. However, a genetic component probably exists as shown by insect-mediated cross-pollination with phenotypic recurrent selection (Ortiz-Perez et al. 2008). Recurrent selection in soybean with male-sterile, female-fertile mutants was reviewed by Lewers and Palmer (1997). These genetic differences in soybean should be amenable to plant breeding manipulation similar to what has been with alfalfa (Teuber et al. 1983, 1990).

Floral nectar is produced by the nectary. The term nectary does not indicate a well-defined anatomical structure, because there are various types of nectaries with different anatomical origins and positions. The term has ecological significance, in that nectaries are the place where liquid substances involved in interactions with animals are produced and offered (Pacini et al. 2003).

A detailed description of the vascular and ultrastruture of the floral and stipular nectaries of *Vicia faba* can be found in Davis et al. (1988) and Davis and Gunning (1992). In addition, mechanisms of nectar secretion in the context of selection for high nectar production, including *Vicia faba,* are summarized by Davis (2001). Plants also offer a nectar reward for protection. Recent studies (Nepi et al. 2012) demonstrated that nectar may have other functions in addition to attracting pollinators: defence against microbial invasion. Nectar has been considered a major floral reward for animals because it is predominantly composed of sugar, but also protein and nonprotein amino acids and essential and nonessential amino acids have been detected. It was discovered that floral nectar contains a large, heterogeneous assemblage of defence proteins. Thus, nectar is much more than the main floral food reward for pollinators. It not only attracts insects but also must defend itself and the gynoecium against invading microorganisms and eventually pathogens.

Nectar is produced in the flowers as a reward for pollinators, and thus plant pollination is assured. Extrafloral nectaries (EFNs) are common (Narbona and Dirzo 2010) and are offered as a food source to predators that defend plants against herbivores (Wäckers et al. 2013). In particular, EFNs are attractive to ants because of the sugary reward, and many plants are visited and patrolled by ants that feed from their extrafloral nectaries. For instance, lima bean bears EFNs on the stipules of its leaves, three pairs of stipules per leaf (Heil 2004). Apart from the leaves, lima beans also have EFN in the inflorescences (Hernández Cumplido et al. 2010). In the case of faba bean, phenotypic plasticity enables many damaged plants to produce additional EFNs or to increase nectar secretion rates to attract natural enemies of herbivores.

Fig. 12.4 View of *Vicia faba* extrafloral nectary. (Photo: JL Ubera)

Ants often increase plant survival and fitness by deterring herbivore damage (Heil et al. 2001). As EFNs of *Vicia faba* are visually conspicuous (Fig. 12.4), additional nectaries may present a greater visual stimulus for ant attraction (Mondor and Addicott 2003). It is important to note that *Vicia faba* plants never produce more than two EFNs per leaf, but rather produce multiple EFN-bearing stipules on the apical meristem prior to the leaves' unfolding (Mondor et al. 2006).

Most floral traits (attraction and mating system) that have been investigated have significant heritability and also a tendency for heritability to vary with mating system (Ashman and Majetic 2006). On the other hand, although floral and extrafloral nectar traits are important for plant reproduction, protection and for breeding strategies, we know little about their genetic basis (Mitchell 2004). Brandenburg et al. (2009) reviewed progress on nectar research. It is typical for nectar to vary substantially in the environment in concentration, composition and volume between populations, plants, also genders and even interfloral and intrafloral variability from day to day. Studies demonstrate that, in addition to strong environmental variation, there is also abundant genetic variation and thus a substantial opportunity for a response to selection on these traits. Nectar production could be increased by breeding (Delaplane and Mayer 2000). Breeding pollinator-friendly varieties requires us to unravel the genetic component of floral characteristics that are associated with pollinators. Genetic control of flower–pollinator specificity has been reviewed by Yuan et al. (2013).

3 Patterns of Mating System and Breeding System

3.1 General Approach

The terms "mating system" and "breeding system" have long been used without clear definition. According to Neal and Anderson (2005), a strong consensus has been reached for the following uses: (1) "mating system" should be used when treating the genetic relatedness and pairings between individuals (e.g. levels of selfing vs outcrossing); (2) "breeding system" should be used with the anatomical/morphological (e.g. gynodioecy and heterostyly) and physiological (e.g. self-compatibility) aspects of individuals and populations; and (3) when addressing general aspects of the reproductive biology of plants, a more general term such as "reproductive system" should be used.

Reproductive characteristics are particularly important in affecting micro-evolutionary processes and patterns because they influence genetic transmission, population genetic structure, selection response and patterns of evolutionary variability (Charlesworth 2006; Barrett 2008). Knowledge about the reproductive system of legume species is essential for choosing appropriate procedures to be used for cultivar development in intensive conventional breeding but also to develop strategies for low-input farming systems, including participatory plant breeding and evolutionary populations.

Agricultural practices affect natural biotic pollination. Agricultural management practices in low-input or organic systems, such as prohibition or reduced use of chemical pesticides and inorganic fertilizers, protection of noncropped habitats and preservation of farming, are particularly beneficial for farmland wildlife. Low-input and organic (LI/O) systems that harbour a high diversity and abundance of flowering plants have a greater floral attractiveness and resource availability, and therefore attract more pollinators compared to conventional fields and are beneficial for population build-up of native bees (Morandin and Winston 2005; Greenleaf and Kremen 2006; Power and Stout 2011). There is clear benefit delivered by pollinators on yield quantity for field bean, but it is not maximized under current agricultural intensification (Bartomeus et al. 2014).

Thus, according to Richards (2001), seed crops with flowers visited by animals have been classified as follows: (1) those not requiring animal visitation. Autogamous plants. Flowers automatically self-pollinate (sometimes known as autofertility), so that fruits and seed set in the absence of pollinators (e.g. lentil, groundnut, pea). However, some autogamous crops do set more seed or better quality seed after insect visits; (2) those requiring an animal visit to set seed or maximize fruit set. Autogamous–allogamous plants with self-fertile flowers which can set some seed and fruit in the absence of animal visitation. However, animal visitation increases either: (a) the proportion of fruit set or (b) the quality of fruit set. This group contains crops such as soybean and broad bean; and (3) allogamous but self-fertile plants. Hermaphrodite flowers which require animal visits for fruit and seed to set, but seed can set following insect-mediated self-pollination, that is, an isolated individual plant can set good seed after visitation (e.g. runner beans, *Phaseolus coccineus*). Crops that are completely dependent on animal pollination are the minority.

Being aware of the significance of selfing in legumes, the question is to understand the mechanics of the pollination process by determining when, and how much self and outcross pollen are transported to stigmas. Lloyd and Schoen (1992) and Barrett and Harder (1996) classified the modes of self-pollination. These differ in whether or not they utilize specialized flowers, whether they involve the transfer of pollen within or between flowers, whether they are autonomous or mediated by vectors, and their timing relative to opportunities for outcrossing. Cleistogamy: specialized closed flowers and pollinator not involved. Geitonogamy involves the transfer of pollen between flowers of the same plant. Facilitated self-pollination occurs when visitors transfer self-pollen within a flower. It is important to point out that low level of autofertility (seed set of isolated plant/artificial cross-fertilization) did not necessarily imply lack of self-pollination under field natural conditions. Prior, competing and delayed self-pollination are similar in being autonomous modes of selfing that occur without the participation of a pollinator. The three modes differ in their timing. Respectively, they occur before, during and after opportunities for outcrossing a flower. Prior selfing occurs when anthers dehisce and stigmas are receptive before anthesis and there is contact between them in unopened buds. Competing selfing resembles facilitated selfing in that it occurs during the same interval as cross-pollination, but it differs in being achieved autonomously and, hence, it is more easily selected. Delayed selfing occurs when the movements of flower parts at the end of anthesis lead to pollen stigma contacts and the fertilization of ovules that have not been previously cross-fertilized. Results in *Vicia faba* (Suso and Maalouf 2010) suggested that self-fertilization is the result of all three major modes: autonomous and facilitated autogamy as well as geitonogamy. Also delayed autogamy is frequent when pollinators are absent at the beginning of flowering. Delayed autogamy may be selected because it provides reproductive assurance. Manipulating the different modes of selfing (prior and delayed), Saxena et al. (2013) obtained good hybrid yields in pigeon pea.

Recognition of the significance of the patterns of mating systems prompted the development of specific tools for measuring the relative frequency of selfing and outcrossing. Palmer et al. (2009) examined empirical methods to study mating systems. Mating events are usually classified according to whether seeds originate from outcrossing or selfing. A common descriptor of the mating system is the estimated outcrossing rate (the proportion of offspring fathered by genetic individuals other than their seed parent). Historically and conventionally, knowledge of mating systems was based on controlled pollination and statistical approaches based on the classical tools of visual markers, such as heritable differences in floral pigmentation. However, procedures used to estimate the degree of cross-pollination in classical marker studies refer to the frequency of hybrids that would have resulted from crossing between pollen donor plants with the dominant marker and the pollen recipients with the recessive marker; that is inter-genotype crossing. Estimations of the cross-pollination take the form of pollen flow source and pollen "sinks" trials. Small groups of plants were planted and pollen receipt was measured and monitored for heterozygote plants. However, the source and sink trials cannot necessarily detect the shortest pollen movement. Source and sink trials typically examine pollen flow between spatially clustered groups of plants, usually separated by at least one meter (Kouam et al. 2012). It has been demonstrated, more than 25 years ago, the utility of allozyme markers to estimate mating parameters, proportion of offspring produced by selfing, or its complement, the outcrossing rate ($t = 1\text{-s}$) (Barrett and Harder 1996). Among outcrossed progeny, it is possible to estimate the degree of biparental inbreeding, the correlation of paternity and the correlation of selfing among families (Ritland 2002). DNA markers with high allelic variation, such as microsatellite loci and more elaborate biometrical models facilitated the development of multilocus approaches (Ritland 1990; Ritland 2002, http://genetics.forestry.ubc.ca/ritland/programs.html web page). Multilocus approaches can use information from all genotypic categories and numerous loci, and thus, more accurately reflect the total amount of outcrossing in open-pollinated populations.

However, mating-system studies in legumes to date have adopted at largely an intercrossing perspective instead of a population-level perspective. Most reports (see below) in the level of cross-pollination are simplistically based on the common intercrossing methodology. This is likely to change because of the development of evolutionary breeding populations for LI/O farming systems.

3.2 Case Studies

Next, relevant studies on mating systems for different legume crops are described and detailed information is provided.

3.2.1 Lentil *(Lens culinaris)*

Erskine and Muehlbauer (1991) by using allozyme markers and the multilocus estimator of Shaw et al. (1981) reported a rate of outcrossing varying from 2.2 and 2.9% in Turkish and Greek germplasm and up to 6.6% in genetic resources originating from Chile. A total of 6.6% of cross-pollination is an order of magnitude above previous estimates. As the lentil flower is normally cleistogamous, a search for the causal vector insect is required to effect cross-pollination.

Horneburg (2006), using the complete dominance of orange cotyledons over yellow cotyledons as genetic marker, investigated the degree of intercrossing among three varieties. The degree of cross-pollination ranged from 0.06 to 5.12%. Results were strongly influenced by cultivar, year and location. The outcrossing rate of individuals also varied within cultivars, the extremes being 0 and 22.2%. They considered that the differences in flower size and colour may have led to different behaviour of pollinating agents. The highest degree of outcrossing was observed in the cultivars with the largest flowers. They explained that only a small number of insects have been observed on flowering lentils, most of them were honeybees (*Apis mellifera* L.) followed by bumblebees (*Bombus* species) and a few hover flies (Syrphidae).

3.2.2 Cowpea *(Vigna unguiculata)*

Cowpea is a self-pollinated crop, however, many insects visit cowpea flowers, which have EFNs and in the process facilitate both self- and cross-pollination. Anthocyanin pigment, a dominant trait, was used to measure the frequency of intercrossing. Results obtained showed that intercrossing occurs in cowpea albeit at a low frequency of less than 1.0% and that cross-pollination occurred up to 31 m distance. The insects most likely to be involved in pollen movement in cowpea are bumblebees and honeybees (Fatokun and Ng 2007).

Asiwe (2009) intended to identify insect pollinators of cowpea and to determine the level of cross-pollination in the crop. To estimate cross-pollination, morphological markers, purple pigments versus no pigmentation at the base of the petioles and leaf shape were used. Results obtained showed that level of intercrossing was higher (0.5–0.85%) when cowpea was planted in alternate rows than in concentric rows (0.01–0.13%). Insects associated with pollen movement (pollinators) were carpenter bees (*Xylocopa virginica* L.), digger bees (*Anthophora occidentalis* L.), honeybees (*A. mellifera* L.), bumblebees (*Bombus grieocollis* and *Bombus pennyslvanicus*) and leaf-cutting bees *(Megachile latimanus)*. Among the insects observed, only honey and bumblebees were found with cowpea pollen dusts on their legs and abdomens and were responsible for the observed level of outcrossing. This was because only heavy insects such as honey and bumblebees with powerful vibrations from their wings could depress the wings of cowpea flowers and expose their stamens and stigmas for pollination.

Kouam et al. (2012) examined the multilocus outcrossing rates in 35 wild cowpea (*Vigna unguiculata* ssp. *unguiculata* var. *spontanea*) populations from West Africa. Multilocus outcrossing rates ranged from 1 to 9.5% (mean 3.4%). These outcrossing rates are markedly higher than previous studies based on pollen flow source and sink trials. Cowpea pollinators either belong to the genus *Xylocopa* or the family Megachilidae.

3.2.3 Vetch (*Vicia* spp.)

Zhang and Mosjidis (1998) used the degree of polymorphism and the variability distribution of seven enzyme systems in 31 accessions of 12 *Vicia* species to infer their mating system. The results demonstrated that most of the *Vicia* species are autogamous, but *Vicia villosa* ssp. *varia* is predominantly a cross-fertilizing species.

3.2.4 Faba Bean *(Vicia faba)*

Vicia faba has been considered a partially allogamous species. The level of allogamy is variable, ranging from 4 to 84% with a mean around 30–60% (Bond and Poulsen 1983; Link 1990; Link et al. 1994; Suso and Moreno 1999; Suso et al. 2001; Gasim et al. 2004). Outcrossing may vary widely geographically (Link et al. 1994; Suso and Moreno 1999) depending on local environmental conditions, particularly the composition of the pollinator fauna (Bond and Kirby 1999, 2001; Pierre et al. 1996, 1999). Growing faba bean enhances the diversity of flowering resources and may help to maintain wild bee pollinators' abundance and diversity. Thus, a key environmental benefit of faba bean is its ability to facilitate diversification of the agro-ecosystem by indirectly enhancing associated diversity of wild fauna that may affect the sustainability of agricultural systems (Köpke and Nemecek 2010).

Suso et al. (2001) found variation in plant daily available flowers, pollinator abundance and pollinator foraging behaviour rates among the same populations of *Vicia faba* growing in different regions (southern Spain and northern France). They reported variation in outcrossing levels, showing that it was lowest in northern France, the region with the lowest pollinator service. Pierre et al. (1999) reported that honeybees *(A. mellifera)* and bumblebees (mainly *B. terrestris*) were the most frequent pollinators in northern France while solitary bees (mainly *Eucera numida*) were the most frequent in southern Spain. Moreover, the pollinator populations differed in their pollinating behaviour: in northern France only 40 % of the pollinators were positive (i.e. entered the flower and induced tripping), because some *B. terrestris* behaved as "robbers" by collecting nectar through a hole they made at the base of the corolla. In southern Spain, 99 ± 6 % of the pollinators were positive.

Benachour et al. (2007) and Aouar-Sadli et al. (2008) studied the pollinating insects of *Vicia faba* in different regions of Algeria. Species of the *Eucera* genus were the most abundant pollinators. Cunningham and Le Feuvre (2013) showed that more stable crop yield of high quality through better pollinator management could be achieved in *Vicia faba*. Nayak et al. (2015) found that the benefits of insect pollination to the production of winter field bean variety 'Clipper', both through increased pod set and seed set were clear. However, the dependence on pollinators could vary between cultivars as has been shown (Suso and Río 2015). It is important to understand varietal effects on the pollination ecology of faba beans and more research into cultivar difference is needed.

3.2.5 Beans (*Phaseolus* spp.)

The five cultivated species in the genus *Phaseolus* are: *Phaseolus vulgaris* L., common bean, snap bean; *Phaseolus lunatus* L., lima bean; *Phaseolus coccineus* L., scarlet runner bean; *Phaseolus polyanthus* Greenmann, the year bean, and *Phaseolus acutifolius* A. Gray, the tepary bean (Debouck and Smartt 1995).

3.2.5.1 Common Bean *(Phaseolus vulgaris)*

Among the cultivated species in the genus *Phaseolus,* the most economically important is the common bean or snap bean *(Phaseolus vulgaris),* which is normally self-pollinated. The outcrossing rates and the gene flow due to natural hybridization have been the objectives of many studies with various results demonstrating low values in a range of 0–1 % and also high values in a range of 6–10 %. These differences have been attributed to the variability of the pollinator presence both in frequency and pollinator species present among the experimental locations (Bliss 1980).

Efficient pollinators like carpenter bees (*Xylocopa brasilianorum* L.) were shown to raise the outcrossing level to 20 % in Puerto Rico (Bliss 1980). Other species of *Xylocopa* (*X. valga, X. violacea, X. iris* and *X. olivieri*) considered as valuable pollinators of some cultivated beans (*Phaseolus* sp.) in Turkey (Özbek 2013). Thrips have been reported as pollinators, but it has not been confirmed (Park et al. 1996).

The benefit of foraging and tripping of bees and bumblebees on the yield of four common bean cultivars in California was estimated under open visitation by insects, in insect proof cages, and controlled visitation with bumblebees. Although the results varied in the different years and lines, two of them responded positively to bee tripping and showed increased seed yield by up to one third (Ibarra-Perez et al. 1999). In a similar recent work, Kingha et al. (2012) studied foraging and pollinating activities in a region in Cameroon. *Xylocopa olivacea* was the most common and nectar was the reward. The effect of the presence of *X. olivacea* on plant yield components was evaluated and showed increasing pod and seed yields as well as seed quality in open pollinated flowers.

Hypocotyl pigmentation, flower or seed colour, and protein markers have been used as genetic markers for the detection of natural crosses in common bean. Mean intercrossing rate of 6.9 % was estimated from the hypocotyls anthocyanin in their progenies (Ibarra-Perez et al. 1997).

Ferreira et al. (2007) used bean cultivars with violet and white flowers as pollen donor and receiver respectively. The heterozygote violet flowers derived from seeds collected from different rows in the four cardinal directions were used to estimate the intercrossing rate that showed the highest level (0.136 %) at a distance of 0.5 m.

In another experiment, also conducted in Brazil, the level of the outcrossing rate was cultivar dependent and practically 0 % at a distance of less than 2.5 m. Two pairs of transgenic cultivars resistant to the herbicide glufosinate ammonium were sown surrounded by their non-transgenic counterparts as recipients of the transgene (Faria et al. 2010).

The research carried so far in *Phaseolus vulgaris* shows that common bean can present variable natural hybridization that could be used for increasing variability in populations. The presence and the density of bees visiting beans, such as honeybee *A. mellifera* and the two solitary bees, *Xylocopa calens* and *Xylocopa incostans* varied in a farmland in Kenya depending on the management of surrounding area (Kasina et al. 2009b).

Seed protein markers detected almost 0 % outcrossing in an experiment at Asturias, Spain (Ferreira et al. 2000). Studies applying molecular markers in diverse bean germplasm indicated that natural hybridization happened with variable frequencies and led to introgression events between Andean and Mesoamerican bean genepools (Angioi et al. 2011).

3.2.5.2 Lima Bean *(Phaseolus lunatus)*

The lima bean *(Phaseolus lunatus)* is a self-compatible annual with a mixed mating system (self-pollination and cross-pollination), mainly autogamous with a low allogamy rate (mean outcrossing rate less than 10 %), qualified as facultative allogamy (Baudoin et al. 2004).

Cross-pollination mechanism, described by Webster et al. (1979) cited by Baudoin et al. (2004), is mediated by honeybees, bumblebees and possibly thrips with major pollinator *A. mellifera* (Delaplane and Mayer 2000; Hardy et al. 1997).

Depending on genotype, growth conditions, plant spacing, prevailing wind direction and native insect populations the outcrossing rate can vary significantly and may reach 48 % (Baudoin et al. 1998 cited by Martínez-Castillo et al. 2007).

Using isozyme markers, Zoro Bi et al. (2005) estimated that the outcrossing rate ranged from 0.027 to 0.268 % across nine wild populations of lima bean *(Phaseolus lunatus)* grown in Costa Rica.

3.2.5.3 Runner Bean *(Phaseolus coccineus)*

The scarlet runner bean *(Phaseolus coccineus)* is considered an allogamous species. In an experiment carried out in Cameroon for two seasons, Pando et al. (2011) studied insects' visitation in inflorescences of plants growing in an open field. Among the 16 species that visited the flowers in the 2-year study, *X. calens* was the most frequent. Nectar was the reward for *X. calens,* whose foraging activities increased the fructification rate, the number of seeds per pod and the percentage of normal seeds.

The biology of flower, nectar secretion and foraging by insects were studied in four runner bean varieties in Poland. The flowers developed exclusively in the morning and the nectar started to be secreted as soon as the bud stage. Runner bean was visited mainly by honeybees and bumblebees for nectar. If the bumblebees populations are short tongued, then their workers open holes and steal the nectar and the honeybees will also steal from the holes, rather than pollinate (Koltowski 2004).

3.2.6 Grass Pea (*Lathyrus* spp.)

There are many *Lathyrus* species cultivated around the world and used as forage/fodder and/or as pulses for human consumption depending on the region and the people's eating habits. The following species are cultivated as pulses in the regions or countries given in parentheses; *Lathyrus annuus* (Europe, N. Africa), *Lathyrus blepharicarpus* (Near East), *Lathyrus cicera* (S. Europe, N. Africa), *Lathyrus clymenum* (Greece), *Lathyrus ochrus* (Greece, Middle East), *Lathyrus sativus* (India, S. Europe, N. Africa) (Kearney and Smartt 1995). The most economically important and widely cultivated *Lathyrus* species is *Lathyrus sativus*. Other important species are *Lathyrus cicera, Lathyrus tingitanus, Lathyrus hirsutus* and *Lathyrus sylvestris*. *Lathyrus odoratus* and *Lathyrus sylvestris* are ornamentals.

The reproductive mode of 14 annual and perennial *Lathyrus* species was studied in upper semiarid zone in Tunisia. The studied species did not flower at the same period. Selfing, natural pollination, and natural pollination following emasculation were applied on 75 flowers per each species, and the frequency of flowers giving pods and the mean number of seeds per pod was measured. The results showed that *Lathyrus latifolius, Lathyrus sylvestris* and *Lathyrus tuberosus,* perennial species, are strictly outcrossing and the main pollinators are bees and bumblebees. These three species are characterized by large bright-coloured flowers. Selfing by bagging flowers showed that *Lathyrus cicera, Lathyrus hirsutus, Lathyrus annuus, Lathyrus ochrus, Lathyrus nissolia, Lathyrus aphaca, Lathyrus tingitanus, Lathyrus setifolius, Lathyrus articulatus* and *Lathyrus sativus,* are preferentially autogamous. *Lathyrus odoratus* is preferentially outcrossing (Ben Brahim et al. 2001).

3.2.6.1 *Lathyrus sativus*

The grass pea *(Lathyrus sativus)* is predominantly self-pollinating although outcrossing has been found to vary significantly between *Lathyrus sativus* genotypes ranging from 9.8 to 27 % depending on the flower colour (Rahman et al. 1995) and occurs on a larger scale in late-opening flowers in field conditions due to higher levels of insect activity (Kearney and Smartt 1995). Pollen takes part in pollinators, particularly bees, attraction (Ben Brahim et al. 2001). An even higher outcrossing rate of 36 % was found by Gutiérrez-Marcos et al. (2006) that applied isozyme analysis in a worldwide collection of *Lathyrus sativus* and studied the genetic structure of the populations.

In the experiment carried out by Rahman et al. (1995) in Bangladesh for two years, plots with different recessive genotypes (red, pink and white flowered) each, were surrounded by blue-flowered dominant genotypes. The genotype with the red flowers showed the highest intercrossing rate (27.8 %) and the one with the white flowers the lowest (9.8 %), while in the pink flowered variety an intermediate outcrossing rate (19.4 %) was recorded. The considerable intercrossing rate of *Lathyrus sativus* may have negative implications in maintaining the varietal purity in seed multiplication (Chowdhury and Slinkard 1997), in genetic contamination of improved cultivars with landraces in farmers' fields and vice versa and also in collecting and conserving grass pea genetic resources (Rahman et al. 2001; Gutiérrez-Marcos et al. 2006). On the other hand, outcrossing provides natural variability that can be exploited in grass pea-breeding programmes (Hillocks and Maruthi 2012).

The high intrapopulation variation detected with inter sequence simple repeats (ISSR) markers within Tunisian, Portuguese and Ethiopian cultivated *Lathyrus sativus* and *Lathyrus cicera* populations and within wild Tunisian *Lathyrus ochrus* could be attributed to rates of outcrossing (Belaïd et al. 2006). Analysis of the diversity of 20 Ethiopian *Lathyrus sativus* accessions (12 plants per accession) with expressed sequence tag-derived simple sequence repeats (EST-SSR) showed an increase of the distribution of the alleles among populations that also could indicate outcrossing (Shiferaw et al. 2012). The observed heterozygosity between 2 and 10 % in Italian grass pea landraces, detected with six polymorphic loci, revealed with the application of SSR markers, may be attributed to reproductive biology of grass pea and confirmed natural outcrossing (Lioi et al. 2011).

3.2.7 Lupins (*Lupinus* spp.)

There are many lupin species that are used by humans as crop plants. These include large-seeded lupins from the Mediterranean region; *Lupinus albus* (syn *Lupinus termis*), *Lupinus angustifolius* and *Lupinus luteus,* and species from the Americas *Lupinus mutabilis*. *Lupinus angustifolius* is cultivated as a grain crop in Australia and *Lupinus mutabilis* (Andean or Pearl lupin) is consumed traditionally by indigenous people in Ecuador, Bolivia and Peru (Hill 1995; Dracup and Thomson 2000; Clements et al. 2008).

The genus *Lupinus* attracts pollinators with a display of multicoloured flowers and rewards of pollen and fragrance (Kazimierska and Kazimierski 2002). The genus is composed of many species that, depending on genetic and environmental factors, present the whole range of pollination modes from strictly self-pollination and self-pollination with facultative cross-pollination to prevailing cross-pollination (Kazimierska and Kazimierski 2002). Annual species tend to be self-fertilized while perennial lupins are cross-pollinated. Even within each species, the outcrossing rates vary depending on the genotype, the location linked to the pollinator fauna species and population. Grain lupin varieties, particularly sweet albus lupin, even though self-pollinated, cross freely with the aid of bees (http://www.ogtr.gov.au/internet/ogtr/publishing.nsf/content/42D3AAD51452D5ECCA2574550015E69F/$File/biologylupin2013-2.pdf).

Most cultivated lupin species are regarded as self-pollinated although there is a degree of outcrossing. When breeding lupins, adequate pollination barriers are needed (Hill 1995; Kazimierska and Kazimierski 2002). The outcrossing rate of *Lupinus albus* that can reach as high as 9% and has caused contamination of the sweet (low-alkaloid) cultivars, used for feed and snacks in Australia, with bitter seeds that have high alkaloids and better fitness (Richards et al. 2008).

Gene flow due to outcrossing from cultivated sweet lupins *(Lupinus angustifolius)* to wild relatives that coexist in the same environment was evaluated in the agricultural zone of Western Australia (Hamblin et al. 2005). The estimation of the outcrossing level showed one cross in 3600 plants within the first 1.5 m from neighbouring crops, while no outcrossing was detected beyond 2.25 m. Gene flow from wild populations to cultivated forms of lupins has never been observed in fields and the likelihood is extremely low (Hamblin et al. 2005).

3.2.8 Pea (*Pisum sativum*)

The pea has been considered in practice a strictly self-fertilizing crop (Bouwan 1992). However, *Xylocopa* and *Megachile* do visit pea flowers and can be responsible for natural outcrossing (Cousin 1997). Polowic et al. (2002) studied the frequency of outcrossing from a transgenic line of peas into three cultivars by using two dominant traits, normal leaf form, and a highly expressed β-glucuronidase (*gus*A) gene as markers of pollen transfer. However, normal leaf form was considered a no reliable marker. 0.07 % mean of intercrossing ranking from 0.00 to 0.11 % was found. Additionally, Vershinin et al. (2003) suggested that despite the predominant inbreeding of the genus, significant outcrossing between all the *Pisum* species should have occurred.

3.2.9 Soybean (*Glycine* spp.)

3.2.9.1 *Glycine max* (L.) Merr

Soybean has been considered naturally self-fertilized. In a 2-year study with two cultivars using flower colour, cross-pollination rates varied from 0.03 % to as high as 6.2 % (Ray et al. 2003).

Chiari et al. (2005b) evaluated the effect of honeybee pollination on seed production and quality of soybean seed using caged and uncovered honeybee plots in Brazil. They found that seed production was higher in caged plots with honeybees (50.64 %) than caged plots without honeybees and even higher in uncovered areas (57.3 %). This increase in seed production was explained in a separate study by the same investigators, by examining pollination by honeybees. They found that honeybees visited 2.24 flowers on average in uncovered plots, and 1.58 flowers in caged plots (Chiari et al. 2005a). Additionally, the percentage of flower abortion was 82.91 % in caged plots without honeybees, 53.95 % in caged plots with honeybees, and 52.66 % in uncovered plots. Higher soybean yield in Brazil were achieved with wild bee pollinators (6.34 % increased). The addition of honeybee colonies further increased yield (18.09 %; de O. Milfont et al. 2013).

3.2.9.2 Wild Annual Soybean (*Glycine soja*)

The wild annual soybean, *G. soja* Sieb. et Zucc., is believed to be predominately self-pollinating; however, little effort has been made to evaluate its breeding system. Kiang et al. (1992) estimated an outcrossing rate of 2.3 %. Fujita et al. (1997) analysed the genetic structure of four *G. soja* populations in Japan by examining allozyme variation. They obtained higher within-population genetic variation and lower genetic divergence among populations than would be expected for a selfing plant species. The mean outcrossing rate estimate was 13 % ranging from 9.3 to 19 % among the four populations. This higher outcrossing rate was supported by observations of frequent visits during flowering by honeybees and carpenter bees (*Xylocopa* spp.).

3.2.9.3 Wild Perennial Soybean Species

The perennial wild relative of soybean, *G. argyrea* (Tind.), has both self-fertilized cleistogamous flowers and chasmogamous flowers on the same plant (Brown et al.

1986). The chasmogamous flowers are visited by insect pollinators and ranged from zero to complete outcrossing, with an average of about 40 %. *G. clandestina* (Wendl.) is a closely related perennial species to *G. argyrea* and has both cleistogamous and chasmogamous flowers (Schoen and Brown 1991). The floral biology of *G. clandestina* and *G. argyrea* allows chasmogamous flowers to spontaneously self-fertilize when left unpollinated; for example, in the glasshouse and in the field when conditions for insect-mediated pollination are absent or suboptimal. Schoen and Brown (1991) sampled two populations of *G. clandestina* (1500 and 750 m elevation) and one population of *G. argyrea.* In the 1500 m population of *G. clandestina,* approximately 60 % of the overall rate of self-pollination in chasmogamous flowers was attributable to whole-flower selfing. This contrasts to the zero whole-flower selfing in chasmogamous flowers recorded with the 750 m population of *G. clandestina.* The difference in cross-pollination of the chasmogamous flowers between the two *G. clandestina* populations was considered to be related to the contrasts in the environmental conditions for insect-mediated cross-pollination. The chasmogamous flowers that did not receive pollinators would self-fertilize spontaneously (Schoen and Brown 1991). In the *G. argyrea* population, only about 4 % of the chasmogamous self-pollination was attributable to whole-flower selfing (Schoen and Brown 1991).

4 General Approaches to Manage Pollination

4.1 Basic Questions

Pollination is a critical step in breeding strategies if cross-pollination is required for breeding purposes or for hybrid seed production or if reduction of cross-pollination is necessary in seed stock multiplication. Pollination services depend on both managed and unmanaged pollinator populations. Any reflection on how to manage pollination should be focused at two levels; open field including farmer's-fields and at enclosed production systems. This section targets these two systems.

The pollination phenomenon was considered important for the production of local crops to promote sustainable development (IPBES 2013). There is growing evidence that improved pollination practices can help support higher yield. Garibaldi et al. (2013) suggested that new practices for integrated management of both honeybees and diverse wild insect assemblages will enhance global crop yields. The creation of habitat diversity within a farm in the form of shelters, nest sites, water, larval food plants, or nectar plants can promote populations of suitable pollinators (Blaauw and Isaacs 2014). Christmann and Aw-Hassan (2012) suggested farming with alternative pollinators (FAP) as an integrated agro-ecological-socio-economic approach and a self-sustaining win-win-strategy for farmers, agro-ecosystems and climate change adaptation. Palmer et al. (2009) proposed the Crop-Design System in which breeders and farmers develop, by participatory plant breeding (PPB), cultivars with enhanced heterosis-mediated yield and resilience as a result of the provision of floral resources within the crop for supporting insect pollinator populations to be used as agents of crossings to increase heterozygosity. Moreover, these techniques can be much less expensive in terms of time and labour than the alternative of hand pollination (Westerkamp and Gottsberger 2000; Potts et al. 2011).

LI/O systems attract more bee pollinators compared to conventional fields (Morandin and Winston 2005; Greenleaf and Kremen 2006; Power and Stout 2011; Andersson et al. 2012) and can benefit bee biodiversity, insect–flower interaction networks and insect-mediated pollination. However, usually, in LI/O farming, bees come freely from nearby natural habitats. According to Kasina et al. (2009c), in western Kenya, pollination in small-scale farms is unmanaged and supported by nearby ecosystems; although, large horticultural farms have initiated beekeeping projects within their farms, where they keep honeybees for pollination of legumes such as beans *(Phaseolus vulgaris)* and cowpeas *(Vigna unguiculata)*. However, they do so without considering whether honeybees are effective pollinators of these crops. Interestingly, Kasina et al. (2009a) investigated the dependence of some crops on bee pollinators as a cautionary measure to inform on the need to manage pollinators. They found that crops that are assumed not to need bee pollination such as some legumes had significantly higher yields when provided with bees. Wild bees, carpenter bees (*Xylocopa* spp.) and leafcutter bees (*Megachile* spp.) were the most effective pollinators of beans and cowpeas. These bees are large, and their weight causes flower tripping. *A. mellifera* "stole" nectar through the sides of the flower without causing flower tripping and therefore did not affect pollination. Both beans and cowpeas increase seed formation after exposure to pollinators. Also, it was found that flowers under the exposure to pollinators produce seeds that have higher protein levels compared to seeds from flowers without pollinators (Kasina et al. 2009a).

4.2 Methods for Pollination Control: Enclosed Spaces

4.2.1 General Approach

The management of pollinating insects under the controlled conditions of enclosed spaces, including field cages, growth chambers and greenhouses is applied to breeding strategies such as hybrid seed production and seedstock multiplication or ex situ germplasm regeneration.

Currently, a number of bee species are managed for pollination for a range of different crops. Honeybees, *A. mellifera* L., are the major contributors to legume pollination. The use of other bees is less common. Additional examples of managed bees are *Megachile rotundata, Bombus* spp. and *Osmia* spp. (Delaplane and Mayer 2000). The issue of supporting native pollinators by designing a crop with appropriate flower traits has been proposed (Suso and Maalouf 2010) as a good practice for long-term sustainability and it is being incorporated in international schemes to improve agri-environmental practices (Suso et al. 2013).

However, Westerkamp and Gottsberger (2000) consider that in spite of the high diversity of flowers, which requires an adequate diversity of pollinators, almost all animal pollination is simplistically ascribed to the manageable but often less efficient pollinator, the European honeybee, *A. mellifera* L. It is not only the inappropriate match between honeybees and the great diversity of flowers which often makes them inefficient pollinators. Moreover, it is also especially the missing "know how" in flower handling, which results in the malfunction of the honeybee at certain flowers such as zygomorphic flowers. They proposed as the very first step of solution to try to understand floral function.

The effectiveness of honeybees and leafcutter bees in cross pollination between two cultivars of faba beans within cages was assessed by Currie et al. (1990). Honeybees were more effective than leafcutter bees as pollinators of faba beans in caged plots.

The aptitude of leafcutter bees to pollinate male-sterile soybean plants (*ms2* gene) in caged plots was evaluated in four experiments from both quantitative and qualitative points of view. The seed set on *msms* plants was satisfactory (Roumet and Magnier 1993).

4.2.2 Exemplar Case Analysis: Multiplication/Regeneration

To maintain the genetic integrity of accessions and to take special care to prevent outcrossing between accessions but facilitate crossing between individuals within accessions to avoid inbreeding depression during legume species regeneration or seedstock multiplication is needed. Ideally, plant accessions should be grown outdoors, in isolation fields, and separate from each several meters depending on the species level of outcrossing. This allows natural pollinators to pollinate within accessions but not between them (Street et al. 2008). Alternatively, it is possible to use isolation cages or screenhouses to limit the movement of pollinators between accessions.

The survey carried out by the European Cooperative Programme for Plant Genetic Resources (ECPGR; Grain Legume Management Survey 2005, (http://www.ecpgr.cgiar.org/worksgroup/grainlegumes.htm, Suso et al. 2011) demonstrated that legume breeders are aware of gene flow problems and practiced some form of pollination control or isolation procedure with seed multiplication. Results of the survey showed that spatial isolation is the most common practice in the legume community. Though, the use of isolation facilities along with suitable insects as pollen vectors is the method most recommended. Many gene banks therefore use isolation facilities along with suitable insects as pollen vectors.

Respondents of the survey agree that regeneration needs gradually move towards the promotion of protocols aiming at encouraging and exploiting the link between seed multiplication and use. Seed multiplication protocols in gene banks must be closely integrated with pre-breeding and anticipate user demands to develop custom-tailored materials, for instance, materials for LI/O farming systems. With seed multiplication, emphasis needs to be taken in the neglected aspect of maintaining the level of heterozygosity and heterogeneity in landraces and in understanding how pollinators and crops interact to shape the specific environments in which curators make their multiplications.

4.3 Spatial Isolation

In legume species, PMGF cannot be ignored in seed multiplication. Actually, due to potential release of transgenic crops and to avoid contamination, the intercrossing even in strictly self-pollinating legumes such as pea *(Pisum sativum)* has been studied (Polowic et al. 2002). Generally, physical distance is used to prevent gene flow (details for specific references at point 3.2). However, when dealing with a large number of undersized and diverse germplasm stocks, the distance needed for isolation could be very large and therefore the method is not usually practical because the land required would exceed the capacity. For insect-pollinated species, pollen flow can be further interrupted and reduced by planting barriers of the same species or of another species.

Suso et al. (2006) and (2008b) had the objective of testing the effectiveness of different isolation zones on the control of PMGF. The study was centred on *Vicia faba,* but we present these results because it is considered a model plant due to its alternatively outbreeder–inbreeder behaviour. Therefore, the analysis of the PMGF dynamic might serve as a base line for similar studies in other legumes with similar pollination mechanisms. Suso et al. (2006) and (2008b) tested three isolation strategies: (a) a barren zone, an isolation zone devoid of all vegetation; (b) the same size isolation zone sown with two different trap crops: (1) a faba bean male-sterile variety and (2) a tetraploid genotype; and (c) the same size isolation zone sown with non-pollinated crop, a *Vicia narbonensis* population. The male-sterile variety does not release viable pollen, and the tetraploid genotype does not cross with diploid faba bean genotypes. Consequently, the trap crops were used as pollen sink for bees to deposit pollen as they moved away from the test plots. *Vicia narbonensis* was used because of its similarity to *Vicia faba,* but because it is an autogamous crop, it would discourage insect pollinators from leaving the faba bean plots. Pollen-mediatated gene flow is largely dependent on floral traits. Floral advertisement seemed to be important in explaining gene flow between plots surrounded by a barren zone. With regard to plots surrounded by a *Vicia narbonensis* population, the role of a reward trait, pollen production, was established. In contrast, in plots surrounded by faba bean trap crops, ovary length played the most important and consistent role in accounting for variation in gene flow among plots.

Pollen gene flow is genotype, border and site specific, and novel information on isolation distances and good "safe" guidelines in legumes is needed specifically for commercial seed production under organic cultural practices—either by seed companies or by farmers producing seed for their own needs. Because most of the organic farmers have much more biological diversity on their farms than the conventional seed farms, they have many more insect pollinators (wild bees, wasps and syrphid flies). The current recommendations for isolation distances in the legume self-pollinated species might be inadequate for avoiding crossing between varieties.

5 Pollination in Crop Breeding

5.1 General Approach

In the first section, we emphasized the importance of understanding how flowers function in order to use pollinators as agents of outcrossing. Availability of outcrossing mechanisms could be important in the development of hybrids and population improvement breeding schemes such as recurrent selection for accelerating the rate of gene recombination. The search of these mechanisms may be more effective in legume plants with zygomorphic flowers and pollinator-dispersed pollen than in species dependent on wind for outcrossing. Plants can exploit bee behaviour by flower morphology and by using advertising signals and rewards to increase the number of seeds. Legume zygomorphic flowers promote precise pollination by restricting which pollinators can visit a flower, the direction from which they can approach, and their movement within the flower. The success of a breeding programme for insect-aided outcrossing depends on breeders mimicking flower behaviour and advantageously adopting strategies that take advantage of the complex interplay that exists between floral features and pollinators. For instance, one means of controlling pollinator bee visitation behaviour is through flower reward traits. But when using nuclear or cytoplasmic male sterility in F_1-seed production, the recipient line lacks pollen as a reward. However, it is possible to overcome a lack of pollen with the reward of nectar and other attractive traits. Breeding programmes leading to parental lines for the production of hybrid seeds must be carried out in the presence of pollinator bees to ensure that overall pollinator-related traits are adequate. Ensuring suitable floral morphology and structure for efficient cross-pollination must be part of such breeding programmes (Kobayashi et al. 2009).

Unfortunately, many vegetable and horticultural crops that depend on pollination have tended towards greater reliance on self-pollination as they have been subjected to modern breeding programmes (FAO 2008). The shift from outcrossing or facultative selfing to strict inbreeding has been described as the single most common trend in legume domestication (Rick 1988). The transition from outcrossing to high levels of self-fertilization may be accompanied by a change in the traits functionally related to pollinators (Suso and Río 2014). As floral trait values associated with autonomous selfing rates are generally the opposite of those associated with efficient outcrossing, selection for pollinator attractiveness and selection for efficient selfing are unlikely to act in parallel. Selfing favours the evolution of plants with lower pollinator attraction and altered morphology and the evolution of plants with lower and often inadequate pollination visitation (Fishman and Willis 2008). Thus, we face a situation in which the recuperation and improvement of functional floral traits that may have been diminished or lost through domestication is more than suitable in order to facilitate the conversion of selfing species to allogamy.

5.2 *Breeding Hybrids*

5.2.1 General Approach

Palmer et al. (2011) reviewed the phenomenon of heterosis in food legumes and noticed that food legumes, in general, have not benefitted from male sterility systems to produce hybrids. Hybrid pigeon pea is the only success story in pulses (Saxena et al. 2013). They concluded that there are a number of factors that are crucial for the successful development of hybrids but the lack of an efficient pollen transfer mechanism from pollen parent to pod parent is the major limitation. If this methodology or other technology becomes viable, food legumes would be a major beneficiary of this science. Also, Fu et al. (2014) stated that in order to utilize heterotic potential, there remains a need to develop high-efficient pollination control technologies on a species-specific basis.

Saxena et al. (2013) also recognized that the commercial exploitation of hybrids is directly linked to the ease with which their hybrid seeds could be produced and delivered economically to farmers. Thus, the efficiency of mass pollen transfer from male to female parent through air or insects to affect cross-fertilization plays an important role in commercializing the hybrids in different crops. In most food legumes, the absence or low natural cross-fertilization is the major bottleneck in exploiting hybrid vigour at a commercial scale.

Kobayashi et al. (2010) argued that cultivars are generally bred by the artificial selection of agronomic traits that are of commercial interest but with little regard to pollinator-related traits and preferences. Consequently, insect-pollinated cultivars may not be attractive to pollinators, resulting in low seed production. This is a problem for commercial seed production of autogamous or partially allogamous legume crops. To enhance F1-seed productivity through the efficient application of pollinators, including honeybees and native insects, we must fully understand the pollination process and the plant–pollinator interplay.

5.2.2 Case Studies

Considerations and studies carried out on specific species are now summarized.

5.2.2.1 Soybean *(G. max)*

In soybean, despite the existence of genetic male sterility and heterosis expression, no soybean hybrids are used in commercial production. Unless better or more efficient pollinator systems can be found, the genetic male sterility used to develop hybrids will not suffice for the commercial release of hybrid soybean (Cober et al. 2010). Palmer et al. (2012) considered that the exploitation of the heterosis depends on basic information available on different aspects of the interplay between plant and pollinator. The limiting step in the study of soybean heterosis is the few hybrid seeds available for the agronomic performance tests. To determine and identify heterotic combinations or associations requires that a large number of parental recombinations be evaluated in multiple years, many locations within years and adequate replications per location. The seed requirement for each parental combination is very large. The first step in soybean heterosis studies is to understand the plant–pollinator interplay. Ortiz-Perez et al. (2008) indicated that bee preference for certain parental lines was the key factor in hybrid seed production. To address these issues, Suso et al. (2010), using *M. rotundata* as main pollinator and two soybean inbred lines, reported that floral size and shape are of primary importance in guiding pollinators' foraging decisions. High and low seed-set lines differed in flower size and shape. Longer and more lobed flowers increased seed set. Increasing the attractiveness and facilitating the manipulation of the flowers, by less flattened standards, provide a useful means of improving seed set. From a follow-up experiment (Pappas et al. 2012), the proboscis extension response system (PERS) was used to determine if honeybees detected differences between the volatiles emitted by flowers from high and low seed-set lines and parental lines. Honeybees responded more favourably to the high set lines, and parents, than to the low seed-set lines. Differences between highly pollinator attractive genotypes and poorly attractive genotypes were likely the result of organic volatiles intensity, rather than one or two unique volatiles. This preference was utilized in phenotypic recurrent selection to produce large quantities of hybrid soybean seed (Pappas et al. 2012).

An open flower mutant, apetally, in soybean was also male sterile, but female fertile (Palmer et al. 2004). The apetalous mutant might have utility as a female parent in hybrid seed production. The manual cross-pollination success rate with apetalous plants as female parent was comparable to cross-pollinations made with fertile female sibling plants. However, the unprotected stigma of the apetalous mutant may be more vulnerable to desiccation under low humidity, and outcrossed seed set would be reduced.

5.2.2.2 Pigeon Pea *(Cajanus cajan)*

Saxena et al. (2013) analysed the development of seed production technology in pigeon pea. They argued that the benefits of hybrid technology cannot be realized unless sufficient quantities of genetically pure hybrid seed are commercially produced and sold at affordable prices. To harvest good hybrid seed yields, it is imperative to select suitable seed production sites with good insect pollinator activity. Thus, the hybrid seed set on the male-sterile plants is chiefly determined by the availability of bee populations in the vicinity. In pigeon pea, the main pollinating vectors are *Megachile lanata, Apis florea* and *A. mellifera*. Good hybrid yield is obtained, up to

25–30 % outcrossing. This is primarily attributed to prolonged flowering in pigeon pea as an evolutionary consequence. Pollinating insects may visit the male-sterile plants several times, and at each visit, a certain proportion of the flowers are pollinated to set the pods while the un-pollinated flowers drop. This is followed by the emergence of new flowers on the same plant, and again a proportion of them set pods through open pollination. This cycle continues, and at the end of the season, plenty of crossed pods are observed on each male-sterile plant.

5.2.2.3 Mung Bean *(Vigna radiata)*

Mung bean is a self-pollinating crop that displays significant hybrid vigour in seed yield of F1 hybrids offering the possibility to use hybrid varieties as a breakthrough to raise the yield plateau of mung bean. Sorajjapinun and Srinives (2011) considered that changing flower form in mung bean may affect pollination rate and increase outcrossing. They proposed to encourage hybrid seed set and to develop potential characters that promote higher outcrossing rate such as open flower (chasmogamy). Thus, chasmogamy controlled by a single recessive gene can be used to promote natural outcrossing rate as a step towards large-scale hybrid seed production in mung bean. However, the mechanism of outcrossing requires pollinators; they observed that major pollinators of mung bean flowers were bees.

5.2.2.4 Chickpea *(Cicer arietinum)*

In chickpea, open flower mutants also have been proposed to be used to increase the level of outcrossing (Pundir and Reddy 1998; Srinivasan and Gaur 2012). Chickpea is a highly self-pollinated crop. The outcrossing is reported to be in the range of 0.0–1.9 % (Tayyar et al. 1995; Toker et al. 2006). Thus, cleistogamy poses challenges in the development of hybrids. Flowers with all petals open (open flowers) that expose male and female reproductive organs would facilitate cross-pollination. A study conducted to estimate the multilocus outcrossing in an open flower population using microsatellite markers revealed a 5.9 % outcrossing rate (Rubio et al. 2010). The higher frequency of cross-pollination of the open flower phenotype could be used for the production of hybrid seeds, if lines producing heterotic hybrids could be identified.

5.3 Breeding Populations: Faba Bean as a Model Plant

Heterosis determines higher productivity, resilience and fertility. It is fully realized in hybrid cultivars and partly in open-pollinated varieties (OPVs) or synthetics that are obtained via inter-mating selected parental genotypes for high general combining ability (GCA)/specific combining ability (SCA). The development of these populations will allow: (1) to fully exploit the "panmitic-mid-parent heterosis" of crosses between genetically distant populations (heterotic groups) and to capitalize not only GCA but also SCA effects. In some legume crops, such as faba beans, due to instability of male-sterile systems and where it is not economically practical to obtain hybrid seed by manual crossing, managing pollinator behaviour through selection for floral traits that enhance pollinator adequate visitation rates has been proposed (Palmer et al. 2009, 2012) to increase the level of outcrossing. This strategy, selection of pre-mating floral traits, has been advocated by other researchers (Ceccarelli 1978; Abdel-Ghani et al. 2003) to generate OPVs with high levels of heterozygosity. This also avoids the dangers of genetic uniformity associated with the development of hybrid varieties and allows the beneficial effects of heterosis available to low-income farmers in a faster way (Nandety 2010). Preliminary experimentations have been carried out to understand the relationship between outcrossing and floral traits in order to determine if floral variation can be effectively utilized for the development of almost exclusively cross-pollinated varieties (with high heterozygosity levels; Suso et al. 2005; Suso and Maalouf 2010).

Suso et al. (2005) proposed that outcrossing might be enhanced by artificial selection for appropriate aspects of floral design and display. Floral traits relevant to increase the level of outcrossing include enhanced inflorescence production during the beginning of flowering but with few flowers open simultaneously per inflorescence. In addition, low nectar reward and short floral tubes should be taken into account. However, as outcrossing varies geographically (Suso and Moreno 1999; Suso et al. 2008a), the most suitable floral traits for promoting outcrossing depend on local environmental conditions (Vogler et al. 1999), particularly the composition of the pollinator fauna.

Suso et al. (2010) carried out artificial selection for outcrossing in open-pollinated faba bean. They found that whether the selection was for increased or decreased outcrossing, the selected groups shifted in the opposite direction of the selected type. The patterns of increase and decrease in outcrossing in the selected populations were due to a simultaneous multidimensional change in floral traits, thus limiting or contra-balancing the effects of artificial selection. They concluded that for the improvement of faba bean populations, direct selection on outcrossing cannot be a selection criterion. However, floral traits, in combination with pollinator behaviour, should be used as indirect selection criteria to increase the level of outcrossing. This approach, in turn, will allow the development of both pollinator-friendly varieties and enhance the environmental services of faba beans.

5.4 Concluding Remarks

Strategies for using heterosis more widely to increase yield and yield stability in legumes centre on finding ways of reducing the cost and increasing the efficiency of producing hybrid seeds. High-efficient pollination insect-aided technologies on a species-specific basis are lacking. Unfortunately the information available is insufficient for an in-depth assessment of the mating patterns in most grain legume species. However, there is experimental evidence that pollinator behaviour influences plant mating patterns and that plants can modify pollinator behaviour through changes in floral traits. To advance in the application of the understanding of the reproductive biology tool for heterosis breeding, more multipopulational studies, combining ecological and genetic factors, are required in order to know the variability and relationships among floral traits, mating systems and floral visitors under different local conditions and ecological contexts. Plant–pollinator interplay may be expected to vary among populations, generating a complex pattern of differential adaptation. Thus, studies are particularly necessary to assess the year to year and location to location dynamics of the plant–pollinator interplay.

References

Abdel-Ghani AH, Parzies HK, Ceccarelli S et al (2003) Evaluation of floral characteristics of barley in the semi-arid climate of north Syria. Plant Breed 122:273–275

Alcorn K, Whitney H, Glover B (2012) Flower movement increases pollinator preference for flowers with better grip. Funct Ecol 26:941–947

Andersson MS, Vicente MC (2010) Gene flow between crops and their wild relatives. Johns Hopkins University Press, Baltimore

Andersson GKS, Rundlöf M, Smith HG (2012) Organic farming improves pollination success in strawberries. PLoS ONE 7(2):e31599 doi:10.1371/journal.pone.0031599

Angioi SA, Rau D, Nanni L et al (2011) The genetic make-up of the European landraces of the common bean. Plant Genet Resour 9:197–201

Aouar-Sadli M, Louadi K, Doumandji SE (2008) Pollination of the broad bean (*Vicia faba* L. var. *major*) (Fabaceae) by wild bees and honey bees (Hymenoptera: Apoidea) and its impact on the seed production in the Tizi-Ouzou area (Algeria). Afr J Agr Res 3:266–272

Ashman TL, Majetic CJ (2006) Genetic constraints on floral evolution: a review and evaluation of patterns. Heredity 96:343–352

Asiwe JAN (2009) Insect mediated outcrossing and geneflow in cowpea (*Vigna unguiculata* (L.) Walp): implication for seed production and provision of containment structures for genetically transformed cowpea. Afr J Biotech 8:226–230

Barrett SCH (1998) The evolution of mating strategies in flowering plants. Trends Plant Sci 3:335–341

Barrett SCH (2008) Major evolutionary transitions in flowering plant reproduction: an overview. Int J Plant Sci 169:1–5

Barrett SCH, Harder LD (1996) Ecology and evolution of plant mating. Trends Ecol Evol 11:73–79

Bartomeus I, Potts SG, Steffan-Dewenter BF et al (2014) Contribution of insect pollinators to crop yield and quality varies with agricultural intensification. Peer J 2:e398. doi:10.7717/peerj.328

Baudoin JP, Degreef J, Hardy O, Janart F, Zoro Bi I (1998) Development of an in situ conservation strategy for wild lima bean (*Phaseolus lunatus* L.) populations in the central valley of Costa Rica. In: Owens SJ, Rudall PJ (eds) Reproduction biology. Royal Botanic Garden Press, Kew, pp 417–426

Baudoin JP, Rocha O, Degreef J et al (2004) Ecogeography, demography, diversity and conservation of *Phaseolus lunatus* L. in the central valley of Costa Rica. Systematic and ecogeographic studies of crop genepools 12. International Plant Genetic Resources Institute, Rome

Belaïd Y, Chtourou-Ghorbel N, Marrakchi M (2006) Genetic diversity within and between populations of *Lathyrus* genus (Fabaceae) revealed by ISSR markers. Genet Resour Crop Evol 53:1413–1418

Ben Brahim N, Combes D, Marrakchi M (2001) Autogamy and allogamy in genus *Lathyrus*. Lathyrus Lathyrism Newsl 2:21–26
Benachour K, Louadi K, Terzo M (2007) Role of wild and honey bees (Hymenoptera: Apoidea) in the pollination of *Vicia faba* L. var. *major* (Fabaceae) in constantine area (Algeria). Ann Soc Entomol Fr 43:213–219
Blaauw BR, Isaacs R (2014) Flower plantings increase wild bee abundance and the pollination services provided to a pollination-dependent crop. J Appl Ecol 51:890–898
Bliss FA (1980) Common bean. In: Fehr WR, Hadley HH (eds) Hybridization of crop plants. American Society of Agronomy. Crop Science Society of America, Madison, pp 273–284
Bond DA, Kirby EJM (1999) *Anthophora plumipes* (hymenoptera: Anthophoridae) as a pollinator of broad bean (*Vicia faba* major). J Apicult Res 38:199–203
Bond DA, Kirby EJM (2001) Further observations of *Anthophora plumipes* visiting autumn-sown broad bean (*Vicia faba* major) in the United Kingdom. J Apicult Res 40:113–114
Bond DA, Poulsen MH (1983) Pollination. In: Hebblethwaite PD (ed) The faba bean (*Vicia faba* L.). Butterworths, London, pp 77–101
Bouwman AJ (1992) Maintenance breeding and multiplication of pea and faba bean cultivars. Maintenance of protein peas (*Pisum sativum*) and field beans (*Vicia faba*). Euphytica 61:213–215
Brandenburg A, Dell'Olivo A, Bshary R et al (2009) The sweetest thing: advances in nectar research. Curr Opin Plant Biol 12:486–490
Brown AHD, Grant JE, Pullen R (1986) Outcrossing and paternity in *Glycine argyrea* by paired fruit analysis. Biol J Linn Soc 29:283–294
Burkle LA, Marlin JC, Knight TM (2013) Plant–pollinator interactions over 120 years: loss of species, co-occurrence, and function. Science 339:1611–1615
Ceccarelli S (1978) Single-gene inheritance of anther extrusion in barley. J Hered 69:210–211
Charlesworth D (2006) Evolution of plant breeding systems. Curr Biol 16(17):726–735
Chaturvedi SK, Gupta DS, Jain R (2011) Biology of food legumes. In: Pratap A, Kumar J (eds) Biology and breeding of food legumes. CABI, Oxford, pp 35–48
Chiari WC, de Toledo VDA, Ruvolo-Takasusuki MCC et al (2005a) Floral biology and behavior of Africanized honeybees *Apis mellifera* in soybean (*Glycine max* L. Merrill). Braz Arch Biol Techn 48:367–378
Chiari WC, de Toledo VDA, Ruvolo-Takasusuki MCC et al (2005b) Pollination of soybean by honeybees. Braz Arch Biol Techn 45:31–36
Chowdhury MA, Slinkard AE (1997) Natural outcrossing in grass pea. J Hered 88:154–156
Christmann S, Aw-Hassan AA (2012) Farming with alternative pollinators (FAP)—an overlooked win-win-strategy for climate change adaptation. Agr Ecosyst Environ 161:161–164
Çildir H, Kahraman A, Doğan M (2012) Petal and sepal epidermal micromorphology of six *Lathyrus* Taxa (Fabaceae) and their systematic value. Not Bot Hort Agr 40:35–41
Clements JC, Sweetingham MS, Smith L et al (2008) Crop improvement in *Lupinus mutabilis* for Australian agriculture—progress and prospects. In: Palta JA, Berger JB (eds) Lupins for health and wealth, Proceedings 12th international lupin conference, 14–18 September 2008, Fremantle, Australia. International Lupin Association, Canterbury
Cober ER, Cianzio SR, Pantalone VR et al (2010) Soybean. In Vollmann J, Rajcan I (eds.) Oil crops, handbook of plant breeding 4. Springer, NewYork
Cousin R (1997) Peas (*Pisum sativum* L.). Field Crop Res 3:111–130
Cunningham SA, Le Feuvre D (2013) Significant yield benefits from honeybee pollination of faba bean (*Vicia faba*) assessed at field scale. Field Crops Res 149:269–275
Currie WE, Jay SC, Wright D (1990) The effects of honeybees (*Apis mellifera* L.) and leafcutter bees (*Megachile rotundata* F.) on outcrossing between different cultivars of beans (*Vicia faba* L.) in caged plots. J Apicult Res 29:68–74
Davis AR (2001) Searching and breeding for structural features of flowers correlated with high nectar-carbohydrate production. Acta Hort 561:107–121
Davis AR, Gunning BES (1992) The modified stomata of the floral nectar of *Vicia faba* L.1. Development, anatomy and ultrastructure. Protoplasma 166:134–152

Davis AR, Peterson RL, Shuel RW (1988) Vasculature and ultrastructure of the floral and stipular nectaries of *Vicia faba* (Leguminosae). Can J Bot 66:1435–1448
De O Milfont M, Rocha EEM, Lima AON et al (2013) Higher soybean production using honeybee and wild pollinators, a sustainable alternative to pesticides and autopollination. Environ Chem Lett 11:335–341
Debouck DG, Smartt J (1995) Beans, *Phaseolus* spp. (Leguminosae-Papilionoideae). In: Smartt J, Simmonds NW (eds) Evolution of crop plants. Longman Scientific and Technical, Harlow, pp 287–294
Delaplane KS, Mayer DF (2000) Crop pollination by bees. CABI International, Wallingford
Delgado-Salinas A, Thulin M, Pasquet R et al (2011) *Vigna* (Leguminosae) sensu lato: the names and identities of the American segregate genera. Am J Bot 98:1694–1715
Dracup M, Thomson B (2000) Restricted branching narrow-leafed lupin 2. Cross-pollination. Aust J Agr Res 51:1011–1015
Erskine W, Muehlbauer FJ (1991) Allozyme and morphological variability, outcrossing rate and core collection formation in lentil germplasm. Theor Appl Gen 83:119–122
FAO (Food and Agriculture Organization). (2008) A contribution to the international initiative for the conservation and sustainable use of pollinators: rapid assessment of pollinator's status. Conference of the parties to the Convention on Biological Diversity, Rome, Italy
Faria JC, Carneiro GE, Aragão FJ (2010) Gene flow from transgenic common beans expressing the bar gene. GM Crops 1:94–98
Farré-Armengol G, Filella I, Llusia J et al (2013) Floral volatile organic compounds: between attraction and deterrence of visitors under global change. Perspect Plant Ecol Evol Syst 15:56–67
Fatokun CA, Ng Q (2007) Outcrossing in cowpea. J Food Agr Environ 5:334–338
Ferreira JJ, Alvarez E, Fueyo MA et al (2000) Determination of the outcrossing rate of *Phaseolus vulgaris* L. using seed protein markers. Euphytica 13:259–263
Ferreira JL, de Souza Carneiro JE, Teixeira AL et al (2007) Gene flow in common bean (*Phaseolus vulgaris* L.). Euphytica 153:165–170
Fishman L, Willis JH (2008) Pollen limitation and natural selection on floral characters in the yellow monkeyflower, *Mimulus guttatus*. New Phytol 177:802–810
Fu D, Xiao M, Hayward A et al (2014) Utilization of crop heterosis: a review. Euphytica 197:161–173
Fujita RM, Ohara M, Okazaki K et al (1997) The extent of natural cross-pollination in wild soybean (*Glycine soja*). J Hered 88:124–128
Garibaldi LA, Steffan-Dewenter I, Winfree R et al (2013) Wild pollinators enhance fruit set of crops regardless of honey bee abundance. Science 339:1608–1611
Gasim S, Abel S, Link W (2004) Extent, variation and breeding impact of natural cross-fertilization in German winter faba beans using hilum colour as marker. Euphytica 136:193–200
Giurfa M, Dafni A, Neal PR (1999) Floral symmetry and its role in plant-pollinator systems. Int J Plant Sci 160 (6 Suppl.):S41–S50
Greenleaf SS, Kremen C (2006) Wild bees enhance honey bees pollination of hybrid sunflower. Proc Natl Acad Sci U S A 103:13890–13895
Gutiérrez-Marcos JF, Vaqueroa F, Sáenz de Miera LE et al (2006) High genetic diversity in a world-wide collection of *Lathyrus sativus* L. revealed by isozymatic analysis. Plant Genet Resour 4:159–171
Hamblin J, Barton J, Sanders M et al (2005) Factors affecting the potential for gene flow from transgenic crops of *Lupinus angustifolius* L. in Western Australia. Austral J Agr Res 56:613–618
Harder LD (1985) Morphology as a predictor of flower choice by bumble bees. Ecology 66:198–210
Hardy O, Dubois S, Zoro Bi I et al (1997) Gene dispersal and its consequences on the genetic structure of wild populations of lima bean (*Phaseolus lunatus*) in Costa Rica. Plant Genet Resour Newsl 109:1–6
Heil M (2004) Induction of two indirect defenses benefits lima bean (*Phaseolus lunatus*, Fabaceae) in nature. J Ecol 92:527–536

Heil M, Koch T, Hilpert A et al (2001) Extrafloral nectar production of the ant-associated plant, *Macaranga tanarius*, is an induced, indirect, defensive response elicited by jasmonic acid. Proc Natl Acad Sci USA 98:1083–1088

Hernández-Cumplido J, Benrey B, Heil M (2010) Attraction of flower visitors to plants that express indirect defense can minimize ecological costs of ant–pollinator conflicts. J Trop Ecol 26:555–557

Hill GD (1995) Lupins. In: Smartt J, Simmonds NW (eds) Evolution of crop plants, 2nd edn. Longman Scientific and Technical, Harlow, pp 277–282

Hillocks RJ, Maruthi MN (2012) Grass pea (*Lathyrus sativus*): is there a case for further crop improvement? Euphytica 186:647–654

Horneburg B (2006) Outcrossing in lentil (*Lens culinaris*) depends on cultivar, location and year, and varies within cultivars. Plant Breed 125:638–640

Ibarra-Perez FJ, Ehdaie B, Waines JG (1997) Estimation of outcrossing rate in common bean. Crop Sci 37:60–65

Ibarra-Perez FJ, Barnhart D, Ehdaie B et al (1999) Effects of insect tripping on seed yield of common bean. Crop Sci 39:428–433

IPBES (Intergovernmental Platform on Biodiversity and Ecosystem Services) (2013) Report of the second session of the plenary of the intergovernmental science-policy platform on biodiversity and ecosystem services. United Nations Environment Program, Antalya

Kalisz S, Ree RH, Sargent RD (2006) Linking floral symmetry genes to breeding system evolution. Trends Plant Sci 11:568–573

Kasina M, Hagen M, Kraemer M et al (2009a) Bee pollination enhances crop yield and fruit quality in Kakamega, western Kenya. E Afr Agric For J 75:1–11

Kasina M, Kraemer M, Martius C et al (2009b) Diversity and activity density of bees visiting crop flowers in Kakamega, Western Kenya. J Apicul Res 48:134–139

Kasina JM, Mburu J, Kraemer M et al (2009c) Economic benefit of crop pollination by bees: a case of kakamega small-holder farming in Western Kenya. J Econ Entomol 102:467–473

Kazimierska EM, Kazimierski T (2002) Biology of flowering, embryological and caryological peculiarities. In: Kurlovich (ed) Lupins: geography, classification, genetic resources and breeding. OY International North Express, St. Petersburg, pp 205–239

Kearney J, Smartt J (1995) The grasspea. In: Smartt J, Simmonds NW (eds) Evolution of crop plants, 2nd edn. Longman Scientific and Technical, Harlow, pp 266–270

Kevan PG, Lane MA (1985) Flower petal microtexture is a tactile cue for bees. Proc Natl Acad Sci U S A 82:4750–4752

Kiang YT, Chiang YC, Kaizuma N (1992) Genetic diversity in natural populations of wild soybean in Iwate prefecture, Japan. J Hered 83:325–329

Kingha BMT, Fohouo FNT, Ngakou A et al (2012) Foraging and pollination activities of *Xylocopa olivacea* (Hymenoptera, Apidae) on *Phaseolus vulgaris* (Fabaceae) flowers at dang (Ngaoundere-Cameroon). J Agr Ext Rural Dev 4:330–339

Kobayashi K, Horisaki A, Niikura S et al (2009) Floral morphology affects seed productivity through pollination efficiency in radish (*Raphanus sativus* L.). Euphytica 168:263–274

Kobayashi K, Tsukamoto S, Tanaka A et al (2010) Selective flower visitation behavior by pollinators in a radish F1 seed production field. Breed Sci 60:203–211

Koltowski Z (2004) Flowering biology, nectar secretion and insect foraging of the runner bean (*Phaseolus coccineus* L.). J Apicul Sci 48:53–60

Köpke U, Nemecek T (2010) Ecological services of faba bean. Field Crops Res 115:217–233

Kouam EB, Pasquet RS, Campagne P et al (2012) Genetic structure and mating system of wild cowpea populations in West Africa. BMC Plant Biol 12:113

Lavin M, Delgado A (1990) Pollen brush of Papilionoideae (Leguminosae): morphological variation and systematic utility. Amere J Bot 77:1294–1312

Lewers KS, Palmer RG (1997) Recurrent selection in soybean. Plant Breed Rev 16:275–313

Link W (1990) Autofertility and rate of cross-fertilization: crucial characters for breeding synthetic varieties in faba (*Vicia faba* L.). Theor Appl Genet 79:713–717

Link W, Ederer W, Metz P et al (1994) Genotypic and environmental variation for degree of cross-fertilization in faba bean. Crop Sci 34:960–964
Lioi L, Sparvoli F, Sonnante G et al (2011) Characterization of Italian grasspea (*Lathyrus sativus* L.) germplasm using agronomic traits, biochemical and molecular markers. Genet Resour Crop Evol 58:425–437
Lloyd DG, Schoen DJ (1992) Self-fertilization and cross-fertilization in plants.1. Functional dimensions. Intl J Plant Sci 153:358–369
Marinho CR, Souza CD, Barros TC et al (2014) Scent glands in legume flowers. Plant Biol 16:215–226
Martínez-Castillo J, Zizumbo-Villarreal D, Gepts P et al (2007) Gene flow and genetic structure in the wild–weedy–domesticated complex of *Phaseolus lunatus* L. in its mesoamerican center of domestication and diversity. Crop Sci 47:58–66
Mayer C, Adler L, Armbruster W et al (2011) Pollination ecology in the 21st century: key questions for future research. J Poll Ecol 3:8–23
Mitchell RJ (2004) Heritability of nectar traits: why do we know so little. Ecol 85:1527–1533
Mondor EB, Addicott JF (2003) Conspicuous extra-floral nectaries are inducible in *Vicia faba*. Ecol Lett 6:495–497
Mondor EB, Tremblay MN, Messing RH (2006) Extrafloral nectary phenotypic plasticity is damage and resource dependent in *Vicia faba*. Biol Lett 2:583–585
Morandin LA, Winston ML (2005) Wild bee abundance and seed production in conventional, organic, and genetically modified canola. Ecol Appl 15:871–881
Nandety A (2010) Recurrent Selection for increased outcrossing rates of barley from semi-arid regions of Syria and Jordan. University of Hohenheim, Hohenheim
Narbona E, Dirzo R (2010) A reassessment of the function of floral nectar in *Croton suberosus* (Euphorbiaceae): a reward for plant defenders and pollinators. Am J Bot 97:672–679
Nayak GK, Roberts SPM, Garratt M et al (2015) Interactive effect of floral abundance and semi-natural habitats on pollinators in field beans (*Vicia faba*). Agr Ecosyst Environ 199:58–66
Neal PR, Anderson GJ (2005) Are 'mating systems' 'breeding systems' of inconsistent and confusing terminology in plant reproductive biology? Or is it the other way around? Plant Syst Evol 250:173–185
Nepi M, Soligo C, Nocentini D et al (2012) Amino acids and protein profile in floral nectar: much more than a simple reward. Flora 207:475–481
Ojeda I, Francisco-Ortega J, Cronk QCB (2009) Evolution of petal epidermal micromorphology in Leguminosae and its use as a marker of petal identity. Ann Bot 104:1099–1110
Ortiz-Perez E, Wiley H, Horner HT (2008) Insect-mediated cross-pollination in soybean [*Glycine max* (L.) Merrill]: II. Phenotypic recurrent selection. Euphytica 162:269–280
Özbek H (2013) New data on large Carpenter-bees of Turkey with considerations about their importance as pollinators. J Entomol Res Soc 15:79–89
Pacini E, Nepi M, Vesprini JL (2003) Nectar biodiversity: a short review. Plant Syst Evol 238:7–21
Palmer RG, Matson AL, Knap HT (2004) Registration of an apetalous male-sterile genetic stock (T368) of soybean. Crop Sci 44:2282–2283
Palmer RG, Perez P, Ortiz-Perez E et al (2009) The role of crop-pollinator relationships in breeding for pollinator-friendly legumes: from a breeding perspective. Euphytica 170:35–52
Palmer RG, Gai J, Dalvi VA et al (2011) Male sterility and hybrid production technology. In: Pratap A, Kumar J (eds) Biology and breeding of food legumes. CABI International, Oxford, pp 193–207
Palmer RG, Horner H-T, Suso MJ et al (2012) Soybean heterosis: the first steps. International conference on utilization of heterosis in crops abstracts, Xi´an, China
Pando JB, Fohouo FNT, Tamesse JL (2011) Foraging and pollination behaviour of *Xylocopa calens* Lepeletier (Hymenoptera: Apidae) on *Phaseolus coccineus* L. (Fabaceae) flowers at Yaounde (Cameroon). Entomol Res 41:185–193
Pappas AL, Koziel JA, Healy RA et al (2012) Floral and aromatic analyses of the soybean perennial Glycine syndetika for future agronomic exploitation. Paper presented at XXII International Congress on Sexual Plant Reproduction. Melbourne, Australia 1 13–17 February 2012, p 21

Park S, Michaels T, Myers J et al (1996) Outcrossing rates of common bean grown in Ontario and Idaho. Annu Rept Bean Improv Coop 39:90–91

Pierre J, Le Guen J, Delegue MHP et al (1996) Comparative study of nectar secretion and attractivity to bees of two lines of spring-type faba bean (*Vicia faba* L. var *equina* Steudel). Apidologie 27:65–75

Pierre J, Suso MJ, Moreno MT et al (1999) Diversite et efficacite de l'entomofaune pollinisatrice (Hymenoptera: Apidae) de la feverole (*Vicia faba* L.) sur deux sites, en France et en Espagne. Ann Soc Entomol Fr(NS) 35(suppl.):312–318

Polowic PL, Vandenberg A, Mahon JD (2002) Field assessment of outcrossing from transgenic pea (*Pisum Sativum* L.) plants. Transgenic Res 11:515–529

Potts SG, Biesmeijer JC, Bommarco R et al (2011) Developing European conservation and mitigation tools for pollination services: approaches of the STEP (Status and Trends of European Pollinators) project. J Apicult Res 50:152–164

Power EF, Stout JC (2011) Organic dairy farming: impacts on insect–flower interaction networks and pollination. J Appl Ecol 48:561–569

Prenner G (2013) Papilionoid inflorescences revisited (Leguminosae-Papilionoideae). Ann Bot 112:1567–1576

Pundir RPS, Reddy GV (1998) Two new traits—open flower and small leaf in chickpea (*Cicer arietinum* L.). Euphytica 102:357–361

Rahman MM, Kumar J, Rahman MA (1995) Natural outcrossing in *Lathyrus sativus* L. Indian J Genet 13:204–207

Rahman MA, Rahman MM, Sarkar MA (2001) Progress in isolation and purification of *Lathyrus sativus* breeding lines. Lathyrus Lathyrism Newsl 2:39–40

Ray JD, Kilen TC, Abel CA et al (2003) Soybean natural cross-pollination rates under field conditions. Environ Biosafety Res 2:133–138

Richards AJ (2001) Does low biodiversity resulting from modern agricultural practice affect crop pollination and yield? Ann Bot 88:165–172

Richards MF, Luckett DJ, Cowley RB et al (2008) Controlling bitter-seed contamination in the NSW *Lupinus albus* industry. In: Palta JA, Berger JB (eds) Proceedings of the 12th International Lupin Conference, 'Lupins for Health and Wealth' Proceedings of the 12th International Lupin Conference, 14–18 September 2008, Fremantle, Australia. International Lupin Association, Canterbury, pp 275–278

Rick CM (1988) Evolution of mating systems in cultivated plants. In: Gottlieb LD, Jain S (eds) Plant evolutionary biology. Chapman and Hall, London, pp 133–147

Ritland K (1990) A series of FORTRAN computer programs for estimating plant mating systems. J Hered 81:235–237

Ritland K (2002) Extensions of models for the estimation of mating systems using *n* independent loci. Heredity 88:221–228

Roumet P, Magnier I (1993) Estimation of hybrid seed production and efficient pollen flow using insect pollination of male sterile soybeans in caged plots. Euphytica 70:61–67

Rubio J., Fernandez-Romero MD, Millán T et al (2010) Outcrossing rate and genetic structure on an open-flowering population of *Cicer arietinum* based on microsatellite markers. In: Legumes for global health legume crops and products for food, feed and environmental benefits. 5th International Food Legumes Research Conference (IFLRC V) & 7th European Conference on Grain Legumes (AEP VII). Book of Abstracts, April 26–30, 2010, Antalya, Turkey, p 220

Saxena KB, Kumar RV, Tikle AN et al (2013) ICPH 2671– the world's first commercial food legume hybrid. Plant Breed 132:479–485

Schoen DJ, Brown HD (1991) Whole- and part-flower self-pollination in *Glycine clandestina* and *G. argyrea* and the evolution of autogamy. Evolution 45:1651–1664

Sexton R, Stopford AP, Moodie WT et al (2005) Aroma production from cut sweet pea flowers (*Lathyrus odoratus*): the role of ethylene. Physiol Plant 124:381–389

Shaw DV, Kahler AL, Allard RW (1981) A multilocus estimator of mating system parameters in plant populations. Proc Natl Acad Sci U S A 78:1298–1302

Shiferaw E, Pe ME, Porceddu E et al (2012) Exploring the genetic diversity of Ethiopian grass pea (*Lathyrus sativus* L.) using EST-SSR markers. Mol Breed 30:789–797

Sorajjapinun W, Srinives P (2011) Chasmogamous mutant, a novel character enabling commercial hybrid seed production in mungbean. Euphytica 181:217–222

Srinivasan S, Gaur PM (2012) Genetics and characterization of an open flower mutant in chickpea. J Hered 103:297–302

Street K., Ismail A., Rukhkyan N (2008) Regeneration guidelines: faba bean. In: Dulloo ME, Thormann I et al (eds) Crop specific regeneration guidelines [CD-ROM]. CGIAR System-wide Genetic Resource Programme, Rome

Suso MJ, Moreno MT (1999) Variation in outcrossing rate and genetic structure on six culitvars of *Vicia faba* L. as affected by geographic location and year. Plant Breed 118:347–350

Suso MJ, Maalouf F (2010) Direct and correlated responses to upward and downward selection for outcrossing in *Vicia faba*. Field Crop Res 116:116–126

Suso MJ, Río R (2014) Faba bean gene-pools development for low-input agriculture: understanding early stages of natural selection. Euphytica 196:77–93

Suso MJ, Río R (2015) A crop-pollinator inter-play approach to assessing seed production patterns in faba bean under two pollination environments. Euphytica 201:231–251

Suso MJ, Pierre J, Moreno M.T et al (2001) Variation in outcrossing levels in faba bean cultivars: role of ecological factors. J Agr Sci 136:399–405

Suso MJ, Harder LD, Moreno MT et al (2005) New strategies for increasing heterozygosity in crops: *vicia faba* mating system as a study case. Euphytica 143:51–65

Suso MJ, Gilsanz S, Duc G et al (2006) Germplasm management of faba bean (*Vicia faba* L.): monitoring intercrossing between accessions with inter-plot barriers. Genet Resour Crop Evol 53:1427–1439

Suso MJ, Hunady I, Solis I et al (2008a) Germplasm management of *Vicia faba* L.: comparative study of the mating system of local and common cultivars growing under different agro-ecological conditions. Plant Genet Resour Newsl 155:46–51

Suso MJ, Nadal S, Roman B et al (2008b) *Vicia faba* germplasm multiplication—floral traits associated with pollen-mediated gene flow under diverse between-plot isolation strategies. Ann Appl Biol 152:201–208

Suso MJ, Nadal S, Palmer RG (2010) Potential power of the plant-pollinator relationship as a tool to enhance both environmental and production services of grain legumes in the context of low-input agriculture: what do we know? In: Goldringer I, Dawson J, Rey F, Vettoretti A (eds) Breeding for resilience: a strategy for organic and low-input farming systems. Eucarpia 2nd Conference of the Organic and Low-Input Agriculture Section, Paris, 1-3 Dec, pp 23–6

Suso MJ, Vishnyakova M, Ramos A et al (2011) An international survey on state of the art of grain legume management in gene banks. J Agr Sci Tech-Iran B 1:975–981

Suso MJ, Bocci R, Chable V (2013) La diversidad, una herramienta poderosa para el desarrollo de una agricultura de bajos-insumos. Ecosistemas 22:10–15

Tayyar R, Federici CV, Waines JG (1995) Natural outcrossing in chickpea (*Cicer arietinum* L.). Crop Sci 36:203–205

Teuber LR, Barnes DK, Rincker CM (1983) Effectiveness of selection for nectar volume, receptacle diameter, and seed yield characteristics in alfalfa. Crop Sci 23:283–289

Teuber LR, Rincker CM, Barnes DK (1990) Seed yield characteristics of alfalfa populations selected for receptacle diameter and nectar volume. Crop Sci 30:579–583

Toker C, Canci H, Ceylan FO (2006) Estimation of outcrossing rate in chickpea (*Cicer arietinum* L.) sown in autumn. Euphytica 151:201–205

Tucker SC (2003) Floral development in legumes. Plant Physiol 131:911–926

Vershinin AV, Allnutt TR, Knox MR et al (2003) Transposable elements reveal the impact of introgression, rather than transposition, in *Pisum* diversity, evolution, and domestication. Mol Biol Evol 20:2067–2075

Vogler DW, Peretz S, Stephenson AG (1999) Floral plasticity in an iteroparous plant: the interactive effects of genotype, environment, and ontogeny in *Campanula rapunculoides* (Campanulaceae). Am J Bot 86:482–494

Wäckers FL, van Rijn PCJ, Bruin J (eds.) (2013) Plant-provided food for carnivorous insects: a protective mutualism and its applications. Cambridge University Press, New York

Wang Z, Luo Y, Li X et al (2008) Genetic control of floral zygomorphy in pea (*Pisum sativum* L.). Proc Natl Acad Sci U S A 105:10414–1049

Webster BD, Lynch SP, Tucker CL (1979) A morphological study of the development of reproductive structures of *Phaseolus lunatus* L. J. Am Soc Hortic Sci 104:240–243

Westerkamp C, Gottsberger G (2000) Diversity pays in crop pollination. Crop Sci 40:1209–1222

Westerkamp C, Weber A (1999) Keel flowers of the Polygalaceae and Fabaceae: a functional comparison. Bot J Linn Soc 129:207–221

Whitney HM, Bennett KMV, Dorling M et al (2011) Why do so many petals have conical epidermal cells? Ann Bot 108:609–616

Willmer P (2011) Pollination and floral ecology. Princeton University Press, Woodstock

Yuan YW, Byers KJRP, Bradshaw HD (2013) The genetic control of flower–pollinator specificity. Curr Opin Plant Biol 16:422–428

Zhang X, Mosjidis JA (1998) Rapid prediction of mating system of *Vicia* species. Crop Sci 38:872–875

Zoro Bi I, Maquet A, Baudoin JP (2005) Mating system of wild *Phaseolus lunatus* L. and its relationship to population size. Heredity 94:153–158

Chapter 13
Grain Legume Cropping Systems in Temperate Climates

Thomas F. Döring

1 Introduction

Grain legumes (or pulses) are of paramount importance as sources of protein. At the global scale, the role of grain legumes as protein providers is currently mostly fulfilled in their function as feed for livestock, but there is also substantial usage of grain legumes for direct human consumption. This means that this group of plants realizes a critical complementary function for human and animal nutrition, especially in comparison to cereals. The protein digestibility for dairy cattle and monogastrids is particularly high in grain legumes. The key role of grain legumes for feeding humans and animals is expected to increase in the future as current breeding activities are aiming at optimising protein quality (e.g. in terms of amino acid profile) and minimising the content of anti-nutritive substances such as glycosides, alkaloids and phenol derivates in grain legumes (Kolbe et al. 2002).

However, the importance of grain legumes is not restricted to their role as a supplier of high-quality protein. In particular, grain legumes exhibit a critical function from an agronomic point of view as well (Köpke and Nemecek 2010). Because of their ability to fix aerial nitrogen (N_2) in symbiosis with soil-living bacteria, grain legumes provide crops following them in the rotation with an essential plant nutrient. Indeed, N can be seen as the main driver of biomass production and a key determinant of crop yield. As pre-crops, grain legumes therefore play a central role, especially in terms of N provision for organic and low-input cropping systems. Further, apart from providing N in the crop rotation, grain legumes are known to have additional beneficial effects on the following crop, in particular by helping to break the life cycle of soil-borne diseases of cereals. Finally, grain legumes are also important from a broader agro-ecological point of view, as they provide nectar and pollen resources for pollinators.

T. F. Döring (✉)
Department of Agronomy and Crop Science, Humboldt University Berlin,
Albrecht-Thaer-Weg 5, Berlin, Germany
e-mail: thomas.doering@agrar.hu-berlin.de

A. M. De Ron (ed.), *Grain Legumes*, Handbook of Plant Breeding 10,
DOI 10.1007/978-1-4939-2797-5_13

Despite the multiple benefits of grain legumes, many countries have experienced a relatively steady decline in the area grown with grain legumes over the last few decades. A major direct driver of this development has been a dramatic change in the subsidies paid to farmers for growing grain legumes in the European Union. In addition, numerous other factors have contributed to this shift, including the replacement of the role of grain legumes as N provider in the crop rotation by the usage of mineral N fertiliser and the replacement of grain legumes in human consumption by meat as a source of protein. However, at a global scale, not all grain legume species have experienced the same trend. In fact, over the past two decades, there has been a substantial expansion of the area planted with the globally most important grain legume, soybean *(Glycine max)*. Specifically, the global area of soybean has increased by 64.4 % between 1993 and 2012, according to data by the Food and Agriculture Organisation. The combination of these trends has led to an increasing monopolisation of the grain legume spectrum by soybean. Used mainly as animal feed, soybean is a crop dominated by highly intensive production systems, increasingly based on large-scale monocultures and intercontinental transport of the harvested grain.

In order to regain protein self-sufficiency at a regional and national scale, there have therefore been repeated calls and initiatives to expand the area cropped with various grain legumes in Europe. A source of innovation and experience and a driver of research in this area has been the organic farming sector. Much of the research that this chapter draws on is therefore focussed on grain legumes in organic systems. Interestingly, the yield gap between organic and conventional management is relatively low for legumes in comparison to cereals (Seufert et al. 2012).

Further, for the past decade, there have been increasing efforts to expand the climatic range of soybean and introduce this crop in new, that is, more northern, regions including Central Europe (Fig. 13.1). Globally, main cropping areas of grain legumes are northern France, India, west Brazil, Argentina, South Australia and

Fig. 13.1 A soybean crop on a sandy soil at the field research station of Humboldt University in Dahlem, Berlin, where demonstration trials of soybean varieties have been conducted since the late 1980s. (Photo: T. Döring.)

Russia (Leff et al. 2004). Future climate change is likely to result in major shifts in the geographical distribution of grain legume growing areas. For managing and supporting sustainable change in this dynamic situation, it is of high importance to understand the role of grain legumes in cropping systems and to know the requirements for optimal grain legume management in the field.

With a focus on temperate climates, and a geographical bias towards Europe, this chapter reviews the agronomy of grain legumes and their role in cropping systems, building on previous reviews of the area (e.g. Jensen et al. 2010). The highly integrative science of agronomy generally investigates the interactions between crop plants and the biotic and abiotic environment under different crop management regimes. The agronomist's tools include the selection of crop species and varieties, the design of crop rotations, the management of seed densities, the design of intercropping systems, the integrative management of plant nutrition and plant protection measures and the choice of appropriate tillage systems. In this context, a cropping system can be understood as pattern and sequences of crops cultivated on a given piece of land over a given period of time, together with the entirety of the pertaining management measures such as crop fertilisation and tillage operations. Usually, the period of time refers to a minimum of one rotation, that is, typically between 3 and 8 years. In addition, the view of the cropping system includes the interaction of the cropping management with the farm resources and other farm enterprises such as livestock production units.

The main task that cropping system design typically focuses on is to obtain high yields and high-quality levels of the harvested product; in the case of grain legumes a major aim is to achieve high protein contents. In addition, cropping system design also targets high yield stability, and the ability of crops to perform under stress situations such as drought. Indeed, instability of grain legume yields has been identified as a major problem, impeding progress and wider adoption of grain legumes in practice. In particular, grain legumes are perceived to be associated with high risk, especially in terms of their response to stresses such as low water availability. A widely accepted view is that grain legumes are characterised by a low ability to compensate influences of the environment. Therefore, this chapter investigates some agronomic determinants of yield stability and potential solutions to the problem of "markedly fluctuating yields" in grain legumes (Kolbe et al. 2002).

The chapter concentrates on seven major species of grain legumes, namely faba beans *(Vicia faba)*, field peas *(Pisum sativum)*, soybeans *(G. max)*, lentils *(Lens culinaris)* and lupins *(Lupinus angustifolius, L. albus* and *L. luteus)*. Occasionally, it will also draw on examples from common bean *(Phaseolus vulgaris)* and chickpea *(Cicer arietinum)*. The predominantly tropical and subtropical grain legume species, such as cowpea *(Vigna unguiculata)* and pigeon pea *(Cajanus cajan)* are not covered.

2 Grain Legumes in the Crop Rotation

Crop rotations encapsulate the essentials of a given cropping system. This section therefore summarises some general rules of crop rotation design and reviews the effects of grain legumes on the soil and the following main crop. It further describes effects of the main crop preceding the grain legume, elucidates the interactions between grain legumes and subsidiary crops, such as green manures in the crop rotation, and finally discusses some examples of grain-legume-based crop rotations.

2.1 General Rules of Crop Rotation Design

The term crop rotation refers to the practice of growing a series of different crop species on the same field over a number of years. With the general aim to maintain and increase long-term soil fertility, and thereby achieve high and stable crop yields over several rotations, there are some general rules for designing successful crop rotations. These rules refer to (i) the selection of crop species, (ii) the assignment of proportions of these crop species within the rotation and (iii) the design of the particular sequence of the selected crops.

As a first step, the selection of crop species for the rotation needs to follow the known requirements of the crop species with regard to soil and climatic conditions that are found on the farm. For instance, typical crops following yellow lupins are rye and potatoes, because all of these crop species require relatively light soils (Zimmermann 1958). There are large differences among the grain legume species in relation to, for example, tolerated values of soil pH (Fig. 13.2a) or water requirements (Table 13.1) and such information forms the basis for crop species selection in the rotation.

Second, economic as well as ecological considerations need to be taken into account when determining the proportion of various crop species in the rotation. For instance, grain legumes are (often) economically outcompeted by other cash crops such as cereals in terms of economic performance per unit area land. In addition, direct support payments may influence the decision on the crop proportions as well.

Further, the proportion of a crop species in the rotation is restricted by effects of pests and diseases on the crop. In particular, soil-borne pests and pathogens can build up high population levels if the same crop species is grown on the same field over successive seasons. Similar considerations apply to infestation with weeds. In particular, because several grain legume species are relatively weak competitors, weed species adapted to the growing patterns associated with grain legumes, for example, spring sowing, may increase to intolerable levels over time if the proportion of grain legumes in the rotation is too high.

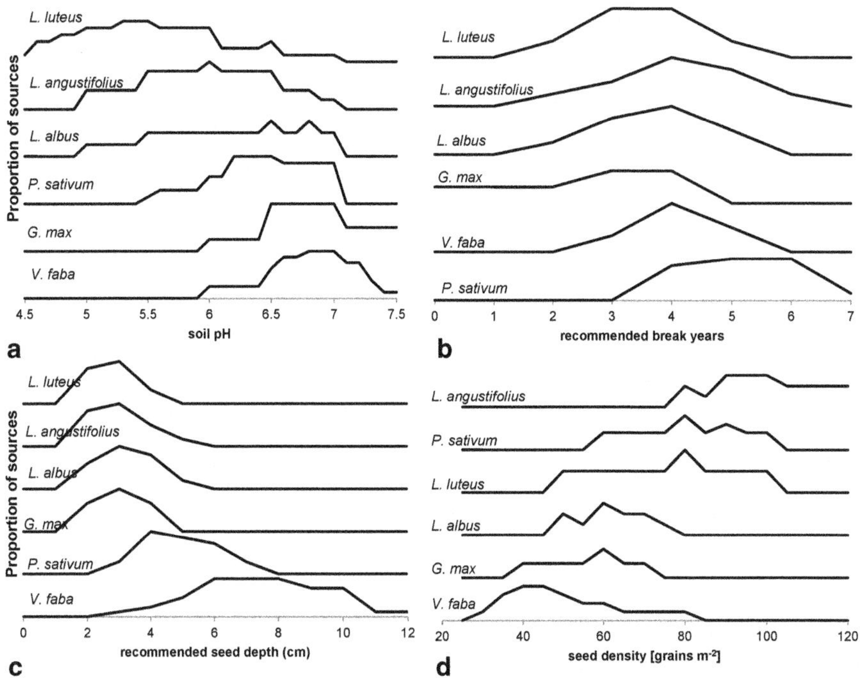

Fig. 13.2 Recommendations for grain legume growing with regard to (**a**) soil pH requirements, (**b**) break years in the rotation, (**c**) sowing depth and (**d**) seed density. The y-axis shows the proportion of sources recommending a given value. (Data sources include Schlipf 1898; Döring et al. 1956; Zimmermann 1958; Kay 1979; Mahler and McDole 1987; Sperber et al. 1988; Franzmann 1992; Kolbe et al. 2002; Kahnt 2008; Guddat and Karalus 2009; Rühl et al. 2009; Jensen et al. 2010; Alpmann and Schäfer 2014)

Because of weed, pest and disease problems, it is therefore usually recommended that not more than about a quarter of the rotation is assigned to grain legumes (Kolbe et al. 2002). Thus, a year of growing a grain legume needs to be followed by a number of break years that is required to minimise pest, disease and weed problems. The number of break years might be reduced when the grain legume is used as (the minor) partner in intercropping systems with cereals or when used for forage. Importantly, different grain legume species differ in the number of recommended break years (Fig. 13.2b). However, the available information also highlights that the recommendations on the number of required break years for grain legumes vary strongly among different sources. Because systematic investigations on the effects of the number of break years on grain legume yield and quality are rare at the moment, the information available to researchers and practitioners carries a relatively large degree of uncertainty.

A major criterion for designing the sequence of crops within the rotation is that the harvesting time of one crop and the sowing or planting time of the following crop cannot overlap but need to leave a time window for preparing the seedbed. For

Table 13.1 Soil and water requirements and agronomic management of various grain legume species. (Data sources: Water requirements: Sperber et al. 1988, Horneburg 2003, Kahnt 2008, Guddat and Karalus 2009; soil conditions: Döring et al. 1956, Franzmann 1992, Kolbe et al. 2002, Horneburg 2003, Guddat and Karalus 2009; sowing and harvest times: Schlipf 1898, Döring et al. 1956, Sperber et al. 1988, Franzmann 1992, Kolbe et al. 2002, Guddat and Karalus 2009, Hiltbrunner and Kessler 2009, Rühl et al. 2009)

Species	Optimal soil conditions	Water demand	Sowing time	Harvest time
Vicia faba (spring)	Deep medium–heavy or light with good water availability	High–very high; requires even distribution	Very early spring: III.2-II.3(-I.4)	8-III.9
Vicia faba (winter)	Similar to spring form	High–very high	Autumn: II.9–II.10	I.8
Pisum sativum (spring)	Light–medium humus-rich soil; no waterlogging	High	Early: (I.3-) II.3–II.4	II.7–II.8
Pisum sativum (winter)			Autumn: II.9–I.11	7
Lens culinaris	Quick to warm up, can tolerate stony, shallow, nutrient poor soil	Can tolerate rel. dry conditions	Spring: I4-III.4	7
Lupinus angustifolius	Medium soil conditions, between *Lupinus luteus* and *Lupinus albus*	Moderate	Early spring: (I.3-) II.3–I.4	8–9
Lupinus luteus	Light soil, sand to sandy loam	Moderate	Early spring: (II.3-)III.3–I.4	8–9
Lupinus albus	Typically on loam	Moderate	Early spring: II.3–I.4	8–9 (late)
Glycine max	Quick to warm up, rich in humus, no water logging	Low–moderate	Late spring: II.4–I.5	9

example, early sowing of a main winter crop such as oil-seed rape (e.g. in late August) requires early harvest of the preceding main crop with enough time to perform the necessary tillage operations. As a pre-crop of oil-seed rape, faba bean with its relatively long vegetation time (150–180 days in temperate climates), is therefore not suitable under these conditions; in this case, an alternative crop would be pea with its shorter vegetation time; however, plant pathogens for which both peas and oil-seed rape are hosts mean that this may not be an optimal cropping sequence either (see below).

Further, when determining the order of crops in the rotation, plant nutrition effects need to be taken into account, with particular reference to N. In particular, grain legumes usually leave some N in the soil for the following crop (see below). Therefore, a highly demanding crop species such as winter wheat should be grown in the season after the grain legume, in order to make ideal use of the soil conditions left by the grain legume. Also, rotation designers should avoid stringing crop species together that share the same pathogen or pest species. This means that in most cases grain legumes need to be followed by crop species from a different plant family.

Generally, rotation design often follows the rule to alternate species with contrasting properties. This can refer to various traits such as rooting depth, water requirements, spring or autumn sowing and effects on soil organic matter. These issues are discussed in the following two sections.

2.2 Effects of Grain Legumes on the Soil

The most important effect of growing grain legumes on the following main crop is the enrichment of the soil with N. The symbiotic bacteria enabling grain legumes to fix N_2 from the air belong to various species of rhizobia and live within nodules formed by the plant root. However, grain legumes do not exclusively rely on symbiosis for N acquisition; in particular, when the soil is high in mineral N, the proportion of N derived from the atmosphere through symbiosis decreases.

The N taken up by the grain legumes, either with or without symbiosis, follows three main routes. Part of the N is converted into grain protein and harvested with the grain, that is, removed from the field. A second part is present in the form of organically bound N in above-ground and below-ground plant residues. When these residues decay after the death of the plant, the nitrogen is mineralised and can then be taken up by the following crop. However, there is also the risk that the mineralised N is lost due to either leaching as nitrate (NO_3^-) or in gaseous form as N_2O to the air. Cropping systems therefore need to be designed in a way to minimise these potential losses (Jensen et al. 2010). A third fraction of nitrogen is released by the living roots of the grain legumes. Here, the nitrogen takes the form of low-molecular-weight chemicals (soluble root exudates, amino acids, hormones and enzymes), and high-molecular-weight substances (from mucilage, dead cells, cell lysates and decomposed root material). The N released from roots, termed N rhizodeposition, can constitute a significant proportion of the N balance in crop rotations (Mayer et al. 2003).

With regard to the N effects of grain legumes in the rotation, there are four key findings of agronomic research. First, the N supply from grain legumes is, in most cases, lower than from legume-based leys, for example, grass clover or lucerne-based pre-crops (Hossain et al. 1996). This is mainly because of the large content of N in the product of the harvested grain so that a large proportion of the N taken up by the grain legume is eventually taken off the field and is therefore not available to the next crop. Also, grain legumes are not as long in the field as the perennial forage legumes in fertility building leys.

Second, as indicated above, there is a negative correlation between mineral N in the soil and the proportion of N in the crop from biological N_2 fixation (Schwenke et al. 1998; Schmidtke and Rauber 2000). This means that N_2 fixation is not only low when mineral N is applied as fertiliser but also when large amounts of organically bound N are microbially mineralised during the growth of the grain legume plant.

Third, the N balance after grain legumes may occasionally also be neutral or even negative. The N balance depends on several factors, including the identity of the crop species. Crops that are able to achieve high levels of N_2 fixation, for example, faba bean and field pea, have been found to be more likely to lead to positive N balances, whereas grain legumes that achieve only modest levels of N_2 fixation, for example, chickpea and common bean, can be either N neutral or lead to decrease of soil N (Walley et al. 2007).

Fourth, the N-fixing ability and other parameters determining the amount of N available for the next crop are notoriously variable and difficult to estimate. This poses a serious problem for the evaluation of the pre-crop value of grain legumes. This variability, evident from individual studies (Evans et al. 1989; Walley et al. 2007; Urbatzka et al. 2011) and from compilations in handbooks for practitioners (Sperber et al. 1988; Franzmann 1992; Kolbe et al. 2002; Horneburg 2003), makes it extremely difficult to quantify expected effects of grain legumes on the following crop and give recommendations on the choice of legume species. One reason for the high variability is that the amount of N released from roots (rhizodeposition) is extremely difficult to quantify (Mayer et al. 2003); a further problem is the dynamic response of N_2 fixation by grain legumes to the mineral N fraction in the soil.

Several approaches have been developed for estimating the N fixed by grain legumes at the farm level, in order to support rotation planning. However, some models widely used in practice have been shown to be extremely poor in terms of their ability to reproduce experimentally measured values. One model based on grain yield and a species-specific N_2 fixation factor was demonstrated to be particularly poor, as it systematically overestimated N balances at low levels and underestimated true values at high levels (Kolbe 2009). More accurate estimates were obtained by using nonlinear functions and a larger range of input variables including the grain legume species, grain yield, soil N_{min} in spring and the N harvest index (Kolbe 2009).

Apart from effects on soil N dynamics, grain legumes have several other effects on the soil, thereby influencing the following crop. In particular, grain legumes improve soil tilth and soil structure by their root systems. For example, in experiments on the effects of various grain legume pre-crops of cotton, it was found that penetration resistance of the soil was lower after most grain legume pre-crops (including faba bean and field pea) than when cotton was used as a pre-crop. However, effects of soybean–cotton rotation in comparison with the cotton–cotton rotation yielded inconsistent results with regard to penetration resistance (Rochester et al. 2001).

Grain legumes may also affect soil structure indirectly through accumulation of soil organic matter. Depending on the amount of above- and below-ground plant residues left after harvest, grain legumes may contribute to the maintenance and enrichment of soil organic matter (humus). Grain legume species are generally thought to be moderately positive with respect to soil organic matter balance in the rotation. When the grain is harvested, their positive effects on soil humus are lower than of grass–clover leys. A further effect of grain legumes on the soil is the potential to mobilise nutrients from deeper layers of the soil, particularly in species with relatively deep tap roots, such as faba beans and lupins. In contrast, lentils, for example, have a relatively weak root system (Horneburg 2003).

2.3 *Effects of Grain Legumes on the Following Main Crop*

In most cases, cumulative effects of grain legume pre-crops on the yield and quality of following cereals have been reported to be positive (e.g. Kirkegaard et al. 2008). Grain legumes are usually recommended as good pre-crops for winter cereals, especially winter wheat and winter triticale, but also for maize. Early-sown crops such as winter rape and winter barley can be grown after peas. In a study from Germany, yields of winter wheat were on average 0.92 t ha^{-1} higher after grain legume pre-crop than after a cereal pre-crop (Albrecht and Guddat 2004). In relative terms, yield increases were 5.6–18.2 % in winter wheat and 8.7–28.7 % in other cereals. These values are similar to those obtained in the northern Prairies where wheat and barley grain yields increased by 12–21 % after grain legume pre-crops in comparison to cereal pre-crops (Wright 1990). However, it is probably more appropriate to compare grain legume pre-crops with other noncereals when assessing the rotation effects on cereals. In a French study, reporting the results of 17 field trials, it was found that yields of winter wheat following either peas or oil-seed rape did not respond significantly to the pre-crop species in 14 out of the 17 trials (Plancquaert and Desbureaux 1985); out of the remaining three cases, significant differences in wheat grain yields favoured peas as pre-crops in one case and oil-seed rape in two cases.

Pre-crop effects of grain legumes are highly context dependent. For example, the pre-crop benefit of grain legumes has been found to be higher in organic than conventional systems (Albrecht 2002); in line with this finding, a series of field trials in East Germany showed that the effect of the grain legume pre-crop, measured as the relative increase of yield in the cereal, decreased with increasing yield level of the cereal. That is, when the cereal yield was close to its optimum, additional gains obtained from a grain legume pre-crop were small (Albrecht and Guddat 2004). Similar results were obtained in a long-term trial in the UK, testing pre-crop effects of faba beans at different N levels (Dyke and Prew 1983). Thus, when comparisons are made between treatments with optimal fertiliser levels, pre-crop effects of grain legumes may disappear. This was observed for cotton lint yields where grain legume pre-crops had no significant effects at optimal fertiliser level (Rochester et al. 2001). The pre-crop effects of grain legumes are considered to be mainly based on N dynamics. However, there are also additional effects on soil-borne pathogens of cereals (e.g. take-all, *Gaeumannomyces graminis* var. *tritici*) and on soil structure (see above).

Unfortunately, effects of grain legumes on the stability and resilience of the following crop (e.g. cereals) have not been subject to extensive research so far. Effects of grain legume pre-crops on the yield of the following cereal could be stabilising or destabilising. On the one hand, it could be expected that a cereal following a grain legume would be more stable than when following a cereal, because, after a cereal, the unpredictable build-up of soil-borne cereal diseases may destabilise yields in comparison to a grain legume pre-crop. On the other hand, however, the variable and weather-dependent N availability after grain legumes may mean that stability of the following cereal grain yield may actually be lower than after other pre-crops.

In accordance with these contrasting predictions, a long-term trial at Broadbalk (Rothamsted, UK) found effects of a faba bean pre-crop on yield variability of wheat over 10 years to be small and inconsistent (Dyke and Prew 1983); for example, a measure of yield variability (the ratio of largest to smallest wheat yields) was higher in a wheat–wheat system than in a bean–wheat sequence when farmyard manure was applied (1.8:1 and 1.7:1, respectively), but the order was reversed when mineral fertiliser was applied (1.4:1 and 1.7:1 respectively). In light of these inconclusive results, it is clear that more research is needed to clarify the question of grain legume effects on yield stability in the rotation (see Sect. 5).

2.4 Selecting Pre-Crops for Grain Legumes

Grain legumes are generally not considered to be particularly demanding with respect to pre-crops. They can be grown after crops with high N demand, for example, recommended pre-crops before faba beans include winter wheat, winter barley and silage maize (Freyer 2003). As mentioned before, however, grain legumes are self-intolerant, that is, they should not directly be grown in successive years. Yield losses in direct succession of grain legumes have been reported to be substantial, for example, 22 % in faba beans and 29 % in field peas in comparison to nonlegume pre-crops (Bachthaler cited in Freyer 2003). Other sources report yield reduction in peas of 25–50 % (Alpmann and Schäfer 2014). In a comparison between growing peas every 2 years and every 4 years in a total of seven environments, it was found that peas in the 2-year pattern had on average a grain yield reduction of −24.5 % compared with peas in the 4-year pattern (Plancquaert and Desbureaux 1985; calculated by the author from published data). This yield reduction was mainly attributed to soil-borne pea diseases (Peronospora and Ascochyta). However, yield effects of the number of break years ranged considerably around this mean, namely between −89 and + 19 %. Also, the effects of the number of break years were significant in only three out of the seven environments.

One of the main reasons for required break years is the build-up of soil-borne fungal pathogens over time. These can dramatically impact on the establishment of the crop and therefore can have severely yield-reducing effects (Stoddard et al. 2010). For example, in faba bean, the minimum number of break years required to reduce the risk of fungal infections to tolerable levels is considered to be 4 years for Ascochyta blight, Cercospora leaf spot, chocolate spot *(Botrytis fabae)*, and faba bean rust *(Uromyces viciae-fabae)*. Substantially longer intervals suggested for the control of powdery mildew, that is, at least 10 years, are unlikely to be practical though (Stoddard et al. 2010). Peas and faba beans share the fungal pathogen *Sclerotinia sclerotiorum* with oil-seed rape (canola, *Brassica napus*) and sunflowers *(Helianthus annuus)*, and direct successions of these crops should therefore be avoided. Also, in rotations that include oil-seed rape and pea, populations of the fungal pathogen *Fusarium avenaceum* are believed to build up over time (Feng et al. 2010).

In addition, weed levels may increase as well if breaks between grain legumes are too short. Most grain legume species are prone to weed infestation because of the relatively wide row width and comparatively low plant densities in currently practiced cropping systems. Especially peas, soybeans and lentils are mentioned as species with low competitive ability against weeds. Vulnerable stages include both early development and late growth stages when seeds mature, and the vegetative parts of the crops are dying off. Therefore grain legumes should be grown after crops that have a high competitive ability against weeds or are at least complementary in their typical weed species community. When spring-sown grain legumes follow a winter cereal, it is recommended to insert a highly competitive green manure between the harvest of the cereal and the following legume in order to suppress weeds. Weed reduction is also the main motivation for selecting mechanically weeded row crops such as potatoes and maize as pre-crops before grain legumes. For example, harrowed row crops have been recommended as suitable pre-crops before lentils (Horneburg 2003). However, if both grain legumes and preceding row crops are sown in spring, this direct succession might pose a risk of accumulating weed species adapted to spring sowing. Of particular relevance for weed control in grain legumes are broomrapes (*Orobanche* sp.). However, because of long viability of the seeds of these parasitic weeds and due to the broad host range of *Orobanche,* crop rotation is of limited effect in this case (Stoddard et al. 2010).

Unfortunately, the most weed-suppressing cereal species, oats or rye, are not ideal as pre-crops of grain legumes because both are hosts for plant parasitic nematodes that also infect grain legumes (Sperber et al. 1988; Freyer 2003). This view is not unanimously shared and some authors do recommend rye and oats as pre-crops of grain legumes (Kolbe et al. 2002). In any case, however, attention should be paid to the host ranges of plant parasitic nematodes when selecting the pre-crop for grain legumes. Further, in terms of the spatial planning of the rotation, it is also necessary to avoid growing grain legumes as direct neighbours because of the risk of mobile pest insects and insect-transmitted plant viruses (Jones et al. 2008).

2.5 Grain Legumes, Cover Crops and Green Manures

Cover crops have several functions in cropping systems (Clark 2008). Primarily, they fill gaps in otherwise vegetation-free periods of the rotation, covering the soil in the intervals between main crops. They help to reduce soil erosion, improve soil structure and increase soil biological activity. They can contribute to the reduction of weeds, soil-borne plant diseases and pests. For example, some brassica crops are used for biofumigation to reduce soil-borne fungal diseases and nematodes (Larkin and Griffin 2007). However, in organically and conventionally managed field trials, it was recently found that a biofumigation crop *(Brassica juncea)* did not have any significant effect on foot diseases or establishment in faba beans and peas and effects on yields remained inconsistent (Jacob et al. 2014). Apart from this investigation, however, there is currently little experience with biofumigation in grain

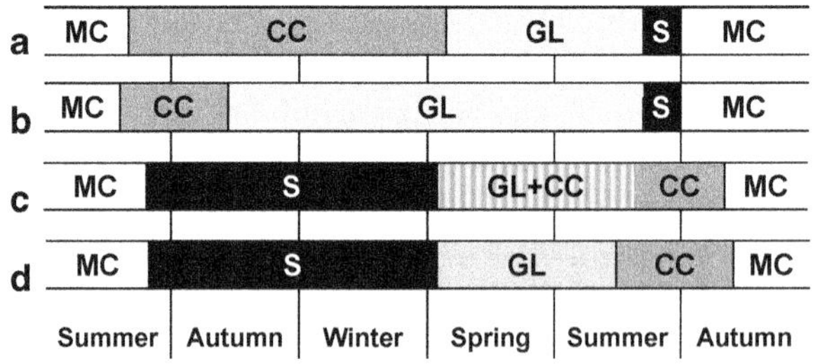

Fig. 13.3 Examples of simultaneous integration of grain legumes *(GL)* and cover crops *(CC)* into rotations with nonlegume main crops *(MC)*, for example, winter cereal and stubble *(S)*; **a** spring-sown grain legume following a cover crop. **b** Autumn-sown grain legume following a cover crop. **c** Cover crop undersown into a spring-sown grain legume. **d** Summer-sown cover crop following an early-harvested grain legume

legumes. More research is therefore necessary to explore the potential of cover cropping for disease reduction in these crops.

A further function of cover crops, in case of leguminous species, is to increase N levels in the soil for the subsequent crop, thereby acting as a green manure. Finally, nonleguminous cover crops may be used to reduce the risk of N losses from the soil. As catch crops, their main role is to retain the N and prevent it from being leached or volatilised. After the catch crop has died off, the retained N is then released again to be taken up by the next main crop.

The combination of grain legumes and cover crops can take several forms in the rotation (Fig. 13.3). The cover crop can precede the grain legume, be grown simultaneously, or follow the harvest of the grain legume. The first option of a cover crop preceding a grain legume applies mainly to spring-sown grain legumes following a main crop that is harvested in the summer of the previous year (Fig. 13.3a). It is particularly relevant in regions where autumn-sown forms of grain legumes cannot be used because of harsh winters. Here, a cover crop will be essential because of the long period that the soil is not covered by main crops. But also with autumn-sown grain legumes, cover crops can be used when the preceding main crop is harvested relatively early, that is, as in barley (Fig. 13.3b).

Further, cover crops can be undersown into grain legumes (Fig. 13.3c). For example, faba beans can be undersown with brassica crops such as mustard in order to keep soil nitrate levels low in the autumn (Köpke 1998). Thereby, losses of grain legume-derived N from the soil are reduced. An underlying mechanism of this effect is that the nonleguminous companion crop of the grain legume has a wider C to N ratio than the grain legume, so that the decomposition of plant residues in the soil is slowed down in comparison to a pure grain legume crop. Also, trials have been conducted on undersowing grass such as cocksfoot (*Dactylis glomerata*) into young faba beans; after harvesting the faba beans, the grass was then kept in the field either until winter wheat was sown in the autumn or until maize was established in the following spring. While the grass did not have any effects on the yield of the beans, there were effects on yields of the following crops. In the above-ground biomass of the grass 20–30 kg ha^{-1} N were bound, but the release of the retained N was relatively slow. This explains why the directly following crop showed reduced yields in the undersown treatment, whereas increased main crop yields were observed in year after that (Gröblinghoff et al. cited in Zerhusen-Blecher and Schäfer (2013)). For undersowing, it is recommended to use a slow-growing-partner variety with a small height, high tolerance against low light levels and high diseases resistance levels.

A different function of undersowing in grain legumes is the suppression of weeds. Experiments on subterranean clover (*Trifolium subterraneum,* at 2000 seeds m^{-2}) undersown into peas (90 seeds m^{-2}) concluded that weed suppression was an important benefit of this system. However, there was no effect on the yield of the following wheat, nor on N_2 fixation of the pea. Also, effects of undersowing on weeds already disappeared in the following wheat crop (Köpke et al. 2011).

The risk of postharvest N losses can be reduced by growing catch crops after grain legumes and before the next main crop (Fig. 13.3d). For example, because of the relatively early harvest date of peas, it is useful to plant a fast-growing summer catch crop such as mustard after the peas and before the following autumn-sown wheat. The practice of following a cover crop after grain legume harvest is particularly important when the period between harvesting the grain legume and sowing the next main crop is characterised by high precipitation as it is often the case when the main crop after a grain legume is spring sown. In this case, it is a recommended practice to plant a catch crop that is killed by frost in the winter. The dead mulch can then be incorporated into the soil in spring.

A major problem when combining grain legumes and leguminous cover crops is that there is currently insufficient knowledge about how the two components interact in the field. In particular, host ranges of fungal and viral pathogens as well as nematodes infecting both grain legumes and leguminous cover crops are currently not well enough characterised to evaluate in how far the integration of cover crops into grain legume cropping systems poses phytopathological risks. One example of a fungal pathogen infecting both grain legumes and forage legumes is *Fusarium solani* which attacks common bean and pea but also white clover (*Trifolium repens*).

3 Growing Schedule for Grain Legumes as a Sole Crop

3.1 Sowing and Inoculation

Sowing times affect grain legumes in cropping systems in a number of ways. In temperate climates, sowing times primarily depend on the frost hardiness and minimal temperature required for germination, but water availability also plays a role for choosing the right sowing time; in particular, because of their high seed weight, grain legumes have a relatively high water demand at germination. Generally, earlier sowing results in higher grain yield of the grain legume, for example, as observed for autumn-sown chickpeas (Horn et al. 1996). This effect is mainly due to the longer time period available to the plant at early sowing. For instance, in comparison to other grain legume species, faba bean as a long-day plant species has a relatively long vegetation period (150–180 days) and therefore needs to be sown particularly early (Table 13.1). Insofar as sowing times affect the amount of grain legume biomass and of the nitrogen fixed, sowing times can affect the yield and quality of a subsequent cereal crop (Heenan 1995). Early sowing can also be beneficial in terms of weed control, because of stronger competition of the crop against weeds. On the other hand, however, sowing later, that is, waiting until the soil temperature is higher often results in more uniform and stronger crop emergence. This can be of particular importance in organic systems. Also, sowing too early entails the risk of damaging the soil when soil moisture levels are too high. With vulnerable soils, it is therefore better to wait until later in spring. As lighter soils are less prone to waterlogging and warm up more quickly, it is recommended to sow earlier in lighter soils and later in heavier soils.

Sowing time can also be a potential tool for aphid and virus control. For example, it is recommended to sow early so the plants are well developed before aphids start to colonise the crop. However, in a study on faba beans in Germany, aphid infestation was not consistently reduced through early sowing; nevertheless early sowing resulted in lower virus incidence, possibly because of stronger mature plant resistance in the early- than in the late-sown beans (Saucke et al. 2009). While grain legumes grown to maturity are typically sown in the spring or autumn, summer sowing is possible if they are used as cover crops. For example, a lupin cover crop may be sown directly after the cereal harvest and can then be kept until the following main crop is sown in the autumn or the next spring.

Sowing depth mainly depends on seed size and therefore recommendations vary considerably, both among and within grain legume species (Fig. 13.2c). Generally, species with hypogaeic germination (faba bean, pea and vetch) should be sown deeper, whereas species with epigaeic germination (soybean, common bean and lupin) require shallower sowing. Deeper sowing is known to reduce bird damage, leads to lower damage from mechanical weed control, and it also reduces lodging risk. As for all crops, it is important to achieve even sowing depth.

Optimal seed density is, at least theoretically, mainly a function of plant size, and plant spacing should be as even as possible to optimise resource use and minimise competition (Weiner et al. 2010; Olsen et al. 2012). As for sowing depth, ranges for recommended seed densities vary (Fig. 13.2d). Generally, there is scope to increase plant density above currently recommended levels in order to achieve high weed suppression. However, the success of these "high-density" cropping systems depends on high spatial uniformity, and appropriate genotypes to avoid that plant resources are allocated towards individual competitiveness (shade avoidance) and away from grain yield (Weiner et al. 2010). In practice, however, seed densities are also bound by row width, which in turn depends on the sowing and weeding technology available on the farm. Row widths recommended for grain legumes are wider for soybeans (45 cm) and faba bean (30–40 cm) and narrower for lentils (15–25 cm Horneburg 2003), peas (10–15 cm Sperber et al. 1988) and lupins (12–24 cm). If the crop is too dense, it is more sensitive to drought and produces more vegetative growth and a lower number of pods. Plants in high seed densities are taller, which can increase the risk of lodging, and there is also a higher risk of plant diseases (Sperber et al. 1988) because of increased humidity and smaller distance between plants. However, if densities are too low, yield potential is lowered because the number of pods produced during the growing season cannot compensate the low plant density. In addition, there is a higher risk of early weeds competing with the young plants; finally, low seed density can lead to a longer flowering time and uneven maturity (Sperber et al. 1988).

Because of the specific combinations of grain legume species and rhizobial species, inoculation of grain legume seed with rhizobia is recommended when the naturally occurring rhizobia populations are likely to be low or absent from the target site. In fact, it was already recommended in the nineteenth century to inoculate soil where legumes do not grow by sprinkling some soil onto it from a site where legumes do grow (Schlipf 1898). However, experimental evidence from Canada shows that inoculation with rhizobia increased seed yield of pea only in the minority of field trials (9 out of 22). At the same time, though, the study also showed that the inoculation benefit was more than three times greater on fields with no history of legumes than on fields with previous legume cropping (McKenzie et al. 2001). From a practical point of view, it is possible for farmers to test the presence of suitable bacteria by growing grain legumes in pots; investigating the presence of root nodules can then be used as an easy diagnostic tool. More recently, it has been shown that co-inoculation of legumes with rhizobia and mycorrhiza can be beneficial for crop growth under low P and/or low N conditions (Wang et al. 2011). However, more research is needed to elucidate the complex interactions between these symbionts and the crops, as well as the applicability in practice.

3.2 Nutrient Management and Irrigation

Nutrient management in grain-legume-based cropping systems comprises four key aspects. These are (1) nutrient requirements of the different grain legume species, (2) effects of grain legumes on the nutrition of other crops, of either intercropping partners (see Sect. 4.1) or the following crops in the rotation (see Sect. 2.3), (3) effects of other crops on the nutrition of grain legumes and (4) the interplay between grain legumes and organic manure. In discussing these issues, this section mainly concentrates on the key nutrients N and P. In addition, the dependency of grain legumes on soil pH is briefly conferred.

Nitrogen fertilisation of grain legumes is usually not economic (Freyer 2003) partly because N_2 fixation is impeded by high levels of mineral N in the soil, though, in a study with faba bean, it has been shown that N_2 fixation is not strongly suppressed at high N_{min} levels (Hardarson et al. 1991). When applying organic fertilisers to grain legumes, it is recommended to use materials with low availability of N, for example, well-rotted farmyard manure or compost, whereas the application of slurry is not advisable. Conversely, grain legumes provide a source of N which can be brought back to the field as organic manure in the nonlegume parts of the rotation. Usually, the organic manure is derived from livestock production, that is, the grain legumes are fed to the animals. However, grain legume seeds can also be processed into a grist to be used directly as organic manure in horticultural (Heuberger et al. 2005) or agricultural (Heinze et al. 2011) crops. Faba bean grist usually contains about 3–5 % N (Heuberger et al. 2005; Raupp 2005; Heinze et al. 2011). Grain legume grist has been suggested as a replacement of animal-derived manure on organic stockless farms. Although long-term effects of this vegetal fertiliser are yet to be determined, available evidence suggests that crop yields are not significantly different between farmyard manure application and grain legume grist fertilisation when N levels of both are comparable (Oltmanns and Raupp 2006; Heinze et al. 2011).

Unfortunately, the N autarky of grain legumes comes at a cost. In particular, symbiotic N_2 fixation requires relatively high levels of phosphorus for nodule development and function (Cassman et al. 1981). In fact, N_2 fixation is sensitive to P deficiency (Tang et al. 2001; Olivera et al. 2004). Although grain legumes are generally known for their relatively good nutrient scavenging ability, P can be limiting in young grain legume plants, and the P mobilizing efficiency can be low until relatively late in the season (Kolbe et al. 2002). P acquisition and use in grain legumes is based on several different mechanisms. These include acidification of the rhizosphere through the release of organic acids; exudation of phosphatase, changes of root architecture at low P levels, for example, the formation of proteoid roots in lupins, improved transport of P and use efficiency, and symbioses with mycorrhiza (Graham and Vance 2003). The specificities of mycorrhizal symbiosis can also inform the design of the crop rotation. For instance, it needs to be taken into account that brassica crops do not form symbiotic relationships with mycorrhiza, and some brassica cover crops are known to reduce mycorrhiza populations in the soil (Larkin et al. 2011). Furthermore, different legume species differ in which fractions of P they can access (Rose et al. 2010).

Nutrient management of both N and P are closely linked to the soil pH. Grain legume species are known to differ in their requirements regarding soil pH (Fig 13.2a). Optimal ranges of soil pH in grain legumes are partly a result of the symbiotic rhizobia, which are sensitive to low pH. At high pH, grain legumes can suffer from Ca chlorosis, which can lead to plant death in extreme cases. Through the process of N_2 fixation, grain legumes normally reduce the pH in the rhizosphere (Nyatsanga and Pierre 1973; Hinsinger et al. 2003), but alkalization of the rhizosphere has also be observed in grain legumes (Betencourt et al. 2012b). Changes in the soil pH strongly affect the availability of various macro- and micronutrients. Soil acidification caused by legumes can be partly compensated for by returning crop residues to the soil (Yan et al. 1996).

Grain legumes, and especially faba beans with their long vegetation time, are known to be drought sensitive during and after flowering. Therefore, irrigation directly before and during flowering can have positive effects, in particular, when the dry period is before flowering. However, in temperate climates, irrigation in peas before flowering has been shown not to affect grain yield (Sperber et al. 1988), even in dry years. On the other hand, irrigation can also have negative effects on grain legumes. In particular, late irrigation can lead to lodging, and this may reduce grain yields.

3.3 Tillage and Weed Control

Cropping systems are intricately linked with issues of tillage. Broadly, three forms of tillage can be distinguished that are relevant for cropping systems. In conventional tillage (CT) systems, ploughing is used prior to the preparation of the seedbed. With reduced tillage (RT), mechanical disturbance of the soil is not as strong as with CT, either by reducing the depth of tillage or by leaving strips untilled. In no-till systems (NT), the seed is directly drilled into the untilled soil. Three major aspects of tillage are of particular relevance in grain legume cropping, namely the mineralisation of soil organic nitrogen, water supply for the crop and weed control.

In principle, it can be expected that decreased tillage intensity leads to lower amount of available N, because of lower rates of mineralisation and nitrification, and because of increased N immobilisation. Reducing tillage intensity can also increase available soil water. It is therefore reasonable to assume that both effects together will stimulate N_2 fixation in grain legume under RT. However, in a long-term experiment conducted in southern Spain, the effect of no tillage versus CT on N_2 fixation in faba bean was not significant. In particular, there were no significant effects of the tillage system on the percentage of nitrogen derived from the atmosphere or on the total amount of N_2 fixed (López-Bellido et al. 2006).

Weed infestation in grain legumes causes several problems. Most importantly, they compete with the crop for water, light and nutrients, thereby leading to reductions of grain yield. In organic cultivation, especially late weed infestation is considered to be a major problem in grain legumes. In addition to direct effects of competition, weeds also indirectly affect grain legumes through uneven and late maturation of crops and lower harvestability. Further, cleaning of the harvested seeds is necessary at high weed infestation levels, and higher moisture of grain from weedy fields requires longer drying after harvest. While the competitive ability of grain legumes early in the season is thought to be relatively strong, weeds such as *Chenopodium album* can thrive later in the season when the crop is senescing. However, also early weed infestation can be severe, especially when soybeans are grown in colder climates where the crop's early development is slow. In Central Europe, lentils are considered to be particularly weak competitors (Horneburg 2003). Problem weeds in lentils include wild oats *(Avena fatua)* and cleavers (*Galium aparine,* on more fertile soils). However, in some regions, lentils, as a late closing spring crop on nutrient-poor soils, also act as habitat for weed species with nature conservation value, such as *Adonis aestivalis, Caucalis platycarpos* and *Misopates orontium* (Horneburg 2003).

Because of the general vulnerability of grain legumes to weed infestation, prophylactic mechanical weed control is considered to be necessary in many grain-legume-based cropping systems. For weed control reasons, ploughing before grain legumes is usually preferred over RT. If grain legumes are sown in spring, ploughing in autumn is recommended, because spring ploughing delays the sowing date, leading to higher water losses in the soil than autumn ploughing, and is also thought to be inferior in terms of weed control. NT systems are widespread in soybean growing but rely heavily on the application of nonselective herbicides that have been criticised because of developing resistance in some weed species (Waltz 2010) and toxicity (Gasnier et al. 2010; Romano et al. 2010).

In order to increase crop competitiveness of grain legumes, a uniform seedbed is needed, and some time should be left for the soil to settle after ploughing. Mechanical weed control during the growing season can be done with various techniques, including tine harrowing and ridging (Kolbe et al. 2002). Specific techniques mainly depend on the growth stage of the grain legume and the row width. However, it should be observed that mechanical weeding might contribute to the spread of fungal diseases in the crop, for example, anthracnose in lupins (Kolbe et al. 2002). In addition to mechanical weeding, rotational means to keep weed levels at bay in grain legumes, for example, by growing a tall competitive cereal such as oats before or after the grain legume. Further, variety selection can also contribute to reduce weed problems. Generally, it is advised to select tall early maturing grain legume varieties with low lodging risk (Kolbe et al. 2002). In peas, semi-leafless types have been shown to be worse competitors than full-leaf types (Urbatzka et al. 2013), though semi-leafless peas are better for mutual support and reduced risk of lodging (Sperber et al. 1988).

3.4 Pest and Disease Control

As in virtually all crops, pests and diseases pose great challenges to the production of grain legumes (Emden et al. 1988). Pests in grain legumes are manifold. Insect pests with high economic importance include the pea moth (*Cydia nigricana* F.) in peas, bruchid beetles such as *Bruchus rufimanus,* and *Acanthoscelides obtectus,* and the weevil *Sitona lineatus.* Several aphid species infest grain legumes, including the black bean aphid *(Aphis fabae)* and the pea aphid *(Acyrthosiphon pisum)*. Further, the soybean aphid *(Aphis glycines)* has caused substantial plant protection and pest monitoring costs after it was accidentally introduced to North America (Ragsdale et al. 2011). Among soil-borne pests, nematodes play a key role in limiting grain legume production. For instance, in peas, the nematodes *Ditylenchus dipsaci* and *Heterodera göttingiana* can cause severe damage. Finally, grain legumes are also vulnerable to bird damage. Specifically, birds such as pigeons (*Columba palumbus* L.), carrion crows (*Corvus corone* L.) and jackdaws (*Corvus monedula* L.) may often damage the germinating seed, breaking the young plant. Periurban areas are especially vulnerable to bird damage (Kolbe et al. 2002).

Grain legumes are also affected by a large number of plant diseases including plant pathogenic viruses such as the *Bean leaf roll virus* and the *Pea enation mosaic virus*. Fungal diseases of high importance include *Fusarium* species, *Pythium*, *Ascochyta*, as well as the leaf diseases false mildew *(Peronospora viciae)* in peas and faba beans, *Botrytis cinerea* in peas and chocolate spot *(Botrytis fabae)* in faba bean. The seed-borne fungal disease anthracnose, caused by *Colletotrichum acutatum* has had devastating effects on lupin farming. In Germany, it affected the previously preferred lupin species (white and yellow lupin) more than the narrow-leafed lupin *(L. angustifolius)*. Accordingly, the disease led to a complete change of lupin species grown in the country.

Major efforts are being made to develop grain legume varieties with resistance or tolerance to pests and diseases. However, both pests and diseases of grain legumes can also be reduced by adjusted management of the cropping system. Such indirect control measures include (1) crop breaks in the rotation (see Sect. 2.1); (2) keeping large distances between fields where grain legume crops are grown, or had been grown in the previous year; (3) keeping distance to forage legumes such as lucerne and clover both in time and space; (4) deep incorporation of plant residues before sowing to reduce fungal infection risk; (5) aiming for uniform crop development, uniform flowering and maturation by diligent seedbed preparation, and moderate row width; (6) using certified seed against seed-borne diseases; (7) sowing early against pest infestation and virus transmission but avoiding very early sowing into cold soils to reduce the risk of fungal infections and (8) using moderate plant densities.

4 Agronomy of Grain Legume Intercropping

4.1 Benefits of Intercropping Grain Legumes

From various systems, it is known that increased plant diversity in the field has multiple benefits (Cardinale et al. 2011; Döring et al. 2012; Costanzo and Bàrberi 2014), including increased productivity (Tilman et al. 2001), reduction of pests and diseases (Finckh and Wolfe 2006), better resource use, and higher yield stability (Tilman et al. 2006). As described in this section, the scientific and applied literature has confirmed that these advantages can also be observed when increasing the diversity in grain-legume-based cropping systems, either by mixing them with other species (intercropping; Hauggaard-Nielsen et al. 2008) or by using intraspecific diversity in the field (cultivar mixtures and populations; Pyndji and Trutmann 1992; Atik et al. 2012).

Cultivar mixtures in grain legumes have received relatively little attention from research so far. In contrast, numerous intercropping combinations involving grain legumes have been tried in research and practice. Combinations of grain legumes and cereals include spring-sown faba bean with spring oats (Helenius and Jokinen 1994; Kahnt 2008), spring barley (Schlipf 1898; Agegnehu et al. 2006; Kahnt 2008) and spring wheat (Bulson et al. 1997; Wolfe et al. 2013); field pea with spring oats (Schlipf 1898; Zimmermann 1958; Rauber et al. 2001; Kolbe et al. 2002; Kahnt 2008; Urbatzka et al. 2011) or spring barley (Schlipf 1898; Jensen 1996; Hauggaard-Nielsen et al. 2001; Kolbe et al. 2002; Kahnt 2008); lentil with rye, spelt, oats or barley (Schlipf 1898; Horneburg 2003); lupins with oats or rye (Zimmermann 1958), winter-sown pea with winter rye (Urbatzka et al. 2011); summer vetch with spring oats (Kahnt 2008; Böhm 2013), or chickpea with durum wheat (Betencourt et al. 2012a). Grain legumes have also been intercropped with oil crops, for example, in combinations of faba bean and oil-seed rape (Jamont et al. 2013) or field pea with false flax (Saucke and Ackermann 2006), and with grasses (Franzmann 1992; Kolbe et al. 2002). Finally, combinations of two grain legume species, such as faba bean and field pea (Zimmermann 1958; Kolbe et al. 2002; Kahnt 2008), winter faba bean with winter vetch (Kahnt 2008), or summer vetch with white or blue lupin (Kahnt 2008) have been trialled.

Already in the nineteenth century, it was suspected that intercropping grain legumes with nonlegumes leads to more efficient resource use through complementation (Schlipf 1898). This view has largely been confirmed by research, for example, for nitrogen use in various intercropping systems. In a mixture of peas and oats, it was found that a higher proportion of N was derived from the atmosphere by the intercropped pea than by the sole cropped pea. When intercropped with pea, the oat plants took up more soil N from deeper layers. Thus, the N leaching risk was lower after the intercrop than after the monocropped pea (Neumann et al. 2007). This is supported by other studies reporting that N use is more efficient in the grain legume–cereal mixtures and the N balance in the soil is closer to zero, that is, there is less over- or undersupply (Hauggaard-Nielsen et al. 2008).

Although the intercropping partners also compete for resources, there is less niche overlap than in monocultures. In addition to niche separation among the intercropping partners, resource use of one partner can also be facilitated by the other. The most significant mechanism of facilitation in intercropping grain legumes is that nitrogen fixed by the legume may be transferred to a nonlegume intercropping partner. Evidence for such N transfer has been found in several intercropping including grain legumes and cereals (Aufhammer 1999). For example, N transfer was observed from soybean to sorghum, in particular, when the planting pattern was such that the distance between the partners was low (Fujita et al. 1990); However, there have also been cases where N transfer was not significant, for example, in an experiment studying N-transfer from pea to barley (Jensen 1996) or in an intercropping system with soybean to maize (Hamel et al. 1991). Surprisingly, N transfer can also take place in the opposite direction. In a study testing intercropping rapeseed and faba bean in rhizotrons, N was transferred in both ways, from faba bean to rapeseed and vice versa (Jamont et al. 2013). Generally, the percentage of N derived from fixing is greater in intercropping than in monoculture grain legumes (Jensen 1996).

Facilitation in intercropping has also been observed for phosphorus. For example, in a mixture of white lupins and spring wheat grown in pots with and without root contact between the partners, it was found (Horst and Waschkies 1987) that the lupin increased availability of soil phosphorus through exudation of organic acids.

In particular, the lupin made three times more P available than it needed; as a result, there was an increased yield of the mixture on P-deficient soil (dry matter of wheat doubled). Another more recent example is an experiment on a chickpea–durum wheat intercrop which highlighted the nutrient mobilizing ability of grain legumes and their ability to make P better available for other crops (Betencourt et al. 2012a); in this experiment, the intercropping of chickpea and durum wheat in a P-deficient soil resulted in higher durum wheat biomass per plant compared to the monocrop, whereas chickpea was not affected significantly. In the P-deficient soil, intercropping also led to significantly higher levels of P (water extracts and Olsen extracts) in the rhizosphere than in the sole crops.

A further important benefit of intercropping grain legumes is improved weed control. For instance, better weed suppression was observed when autumn-sown faba beans were intercropped than when wheat or beans were sown in monoculture (Wolfe et al. 2013). Similar observations of weed suppression in intercrops compared to monocrops were made for lentil intercropping (Horneburg 2003) and peas and false flax (*Camelina sativa;* Saucke and Ackermann 2006). In a field experiment, weed cover was strongly reduced in the intercrop in comparison to the monocrop of peas, but only at one of two sites in comparison to the other partner, false flax (*Camelina sativa*; Paulsen et al. 2006). A further consequence of better weed suppression in the intercrop is that soil inorganic N is used for grain production of the nonlegume intercropping partner instead of weed biomass (Hauggaard-Nielsen et al. 2001).

In many cases, grain legumes are grown in intercropping systems with cereals to reduce the risk of lodging; for example, a cereal can physically support a pea or lentil crop which can use its tendrils to climb. This also facilitates the harvesting process as the lentils climb higher when grown in a mixed stand with cereals than when grown in monoculture (Horneburg 2003). Reduction of the lodging risk can also be achieved by intercropping two legume species, for example, when faba bean acts as the supporting crop for pea (Schlipf 1898). Further, because intercropping increases soil cover in comparison to monocrops, it reduces the risk of soil erosion. This risk is particularly high where rows are spaced relatively widely in the (monocrop) grain legumes.

Intercropping has also been shown to reduce the incidence of pests and diseases due to effects of diluting hosts, non-hosts acting as physical barriers, induced resistance, modification of microclimate and manipulation of host-finding behaviour of pests. For example, in comparison to monocropped barley, net blotch *(Pyrenophora teres)* infestation was significantly reduced when barley was intercropped with lupins, peas, faba beans, or any combination of the three legume species (Hauggaard-Nielsen et al. 2008). In an intercropping field trial in Nigeria, pest damage in soybean was consistently lower over 2 years when soybean was intercropped with millet than in the soybean monocrop (Sastawa et al. 2004).

However, intercropping effects are not always reliable (Trenbath 1993) or can even be counterproductive (Helenius 1990). For instance, while several practical guides (Schlipf 1898; Zimmermann 1958; Franzmann 1992; Aufhammer 1999) recommend intercropping for the control of the black bean aphid in faba beans, other sources reported that no effect of intercropping oats with faba beans on the black bean aphid could be observed (Kolbe et al. 2002). Similarly, mixed results were found for effects of intercropping cowpea on insect pests (Jackai and Daoust 1986). Also, intercropping wheat with faba beans can lead to higher disease infection with powdery mildew *(Blumeria graminis)* in wheat (Chen et al. 2007), possibly because of higher levels of foliar N in wheat leaves, and also because of a moister climate in the intercrop than in the wheat monocrop.

In terms of yield, effects of intercropping are usually quantified by the land equivalent ratio (LER), which is the area of monocultures required to achieve the same yield as obtained in the mixture. For example, in a two-partner intercrop, an LER of 1.2 means that 1.2 ha of land would be needed for growing the two partners separately (in monocultures) to obtain the same yield as harvested from 1 ha of the intercrop. In intercrops of grain legumes and cereals, the LER has usually been found to be above 1, with a range between 0.91 and 1.51 (Hauggaard-Nielsen et al. 2008). However, the LER is also dependent on N input, with the LER being lower and the proportion of the legume partner being smaller at high N levels (Jensen 1996). This has also been confirmed by a study on intercropping lentil and naked barley where the yield advantage of the intercrop over both sole crops was only apparent at low levels of mineral soil N (Schmidtke et al. 2004).

Finally, intercropping grain legumes has also been shown to result in higher yield stability of combined yields in comparison to sole crops (Jensen 1996). In lentil–barley mixtures, it has been observed that yield stability is achieved through complementarity; in dry years, there are more lentils, whereas the balance in wet years is more towards barley (Horneburg 2003). While the view that intercropping grain legumes leads to high yield stability is widespread (Schlipf 1898; Zimmermann 1958; Aufhammer 1999), also the opposite effect has been found (Hauggaard-Nielsen et al. 2008). Further, details about effects of intercropping on yield stability in grain legumes are presented in Sect. 5.3.

4.2 *Managing Intercropped Grain Legumes*

Agronomic management of intercrops is more complex than sole cropping of the partners, partly because requirements of both (or all) partners need to be balanced. One of the most important criteria when the intercrop is grown for threshing is that the components of the mixture have similar times in the season at which they mature. For example, for intercropping peas with cereals, it is recommended to select pea varieties with suitable maturity so that the maturation is matched in time with the cereals (Franzmann 1992). An alternative solution to the problem of different maturation times is to harvest the intercrops before maturity, and to use either the green crop for silage or with slightly later harvesting to thresh before the grains are completely dry and conserve the combined intercrop in a process called crimping. Such early harvests are particular relevant for cropping system design as they affect the entire planning of the rotation.

Also, if intercropping partners are sown at the same time, the required seed depths of the partners cannot be too different, unless specialised sowing machinery for separate sowing depths of the different partners is available. In terms of seed densities, experience shows that yield benefits are largest when the added densities of each partner are above 100 % (better than substitutive mixtures), but not completely additive (i.e. below 200 %). A further issue in managing intercrops is weeding. This refers both to the design of mechanical weeding in row crops such as maize and, in conventional agriculture, to the selection of herbicides that can be used in both crops simultaneously (Pekrun et al. 2013).

Generally, the design of intercropping systems with grain legumes is currently a tedious case-by-case work. Although many general principles have been established, transferability of experience from one system to the other is still limited. Despite these limitations, however, research on intercropping is well advanced, in that it provides detailed accounts of optimal agronomic management for many crop species combinations. Still, adoption rates of intercropping are well below the potential. The main underlying problem for the lack of implementation in practice appears to lie in the supply chain, which, in the case of feed for farm animals is structurally not well set up to deal with mixed grain products. Future work should therefore aim to remove socioeconomic constraints to intercropping.

5 Yield Stability in Grain Legumes

5.1 Concepts of Stability

For virtually all crops, achieving high levels of yield stability is an important goal. This goal is shared by farmers, plant breeders, cropping systems designers and governments concerned about food security. However, yield stability is not a simple concept. Instead, it comprises numerous different statistical approaches (e.g. Becker and Léon 1988). Making progress towards greater yield stability therefore requires specification of the kind of stability that is to be promoted. For example, when Deghani et al. investigated yield stability in lentils in Iran (Dehghani et al. 2008), they calculated a total of 19 univariate yield stability measures. Some of the stability indices used in the study gave strongly differing rankings for yield stability among the tested genotypes. This finding indicates that the results of stability analyses strongly depend on which specific statistical approach is employed.

Parameters of stability can refer to temporal variation (across years), variation across locations, or both. When looking at temporal yield stability, it is crucial to define the spatial area over which yields are aggregated. For example, farmers' interests are likely to be more directed towards temporal yield stability at the farm scale than at the field scale, because within-field fluctuations over time can be compensated within the farm (Porter et al. 1998). In addition, however, temporal yield stability at the regional and national scale is also of importance as the farm's economic performance will be affected for example, through price fluctuations on the market.

Some relatively simple measures of yield stability that have been used in the context of evaluating yield stability in grain legumes include the ratio between highest and lowest yield (Dyke and Prew 1983), the coefficient of variation (CV), that is, standard deviation expressed as percentage of the mean (Smith et al. 2007), or the frequency of years in which yields are below a given minimum.

Many of the other stability parameters currently in use belong to one of two broad groups, being based either on variance components or on (linear) regression (Annicchiarico 2002). Yield stability parameters based on variance components calculate how much the yield of a genotype fluctuates in individual year-by-location combinations (environments) around the mean yield of that element, that is, the yield of the element averaged across all tested environments. Large fluctuations indicate low stability.

Regression-based approaches to stability first calculate for each environment E the average yield x_E achieved across all genotypes G, and then determine the linear regression of yields y_G of individual elements G in all environments against x. One suggested definition of yield stability is that a genotype shows maximal stability if its linear function $y=a+bx$ has a slope of $b=1$, that is, slopes of $b<1$ and $b>1$ indicate lower stability. For $b=1$, there are no significant interactions between genotype and environment.

Finally, a novel measure of yield stability is based on Taylor's power law (Cohen 2013; Döring et al. 2014). This stability index is able to operate over a large range of yield levels as it takes the frequently observed dependence of means and variances into account. Therefore, it appears to be particularly suitable for comparisons of different crop species. Here, means m and variances s^2 of yields over environments are calculated for each genotype (or species). Then, deviations from the linear regression of $\log(s^2)$ against $\log(m)$ can be interpreted as an index of yield variability, that is, the smaller these power-law residuals (POLAR), the higher the yield stability.

5.2 Are Grain Legume Yields Unstable?

Already in 1840, the German agricultural writer Alexander von Lengerke stated that unstable yields are one of reasons why grain legumes are not being grown more by farmers (von Lengerke 1840). Similarly, at the end of the nineteenth century, Johann Schlipf mentioned in a handbook of general agricultural practice that grain legumes show unstable yields because of their low ability of compensation (Schlipf 1898). More recently, this view of grain legumes as being characteristically unstable was reiterated (Sperber et al. 1988; Duc 1997; Horneburg 2003), and reasons given included vulnerability to pests and diseases (Schlipf 1898) or the general response to environmental and weather factors (Franzmann 1992). The view that grain legumes are yield unstable is also held by farmers. In a European survey, von Richthofen and colleagues asked farmers in Switzerland, Spain, Belgium and Germany to rate reasons for not growing grain legumes. Out of 21 reasons given in total, high variability of yield was among the four most important reasons in all four countries, the most important being the low (economic) performance in comparison to row crops and cereals (von Richthofen et al. 2006).

Unfortunately, however, quantitative evidence of low yield stability in grain legumes is surprisingly scarce. One reason for the lack of robust data is that analyses of yield stability have predominantly been the domain of plant geneticists and breeders, whereas agronomists have not engaged in this area with the same enthusiasm. Most available data of yield stability in grain legumes refer to intraspecific differences in yield stability (Bond 1987; El-Moneim and Cocks 1992; Link et al. 1994; Dehghani et al. 2008). In contrast, comparisons among grain legume species, and between grain legumes and other crops, are relatively rare. Also, comparisons between different cropping systems in terms of their effects on yield stability of grain legumes (Smith et al. 2007) have so far not been well covered by research.

At least at larger levels of spatial aggregation, and for Central Europe, yield stability in grain legumes may possibly not be as bad as its reputation. Data from the German official record over the years 1993–2012 indicate that temporal yield stability in faba beans and field peas is comparable with that of cereals when yields are aggregated at the national level (Fig 13.4). In the two grain legume species, the national average yield was below 90 % of the average yield in only 3 out of 20 years.

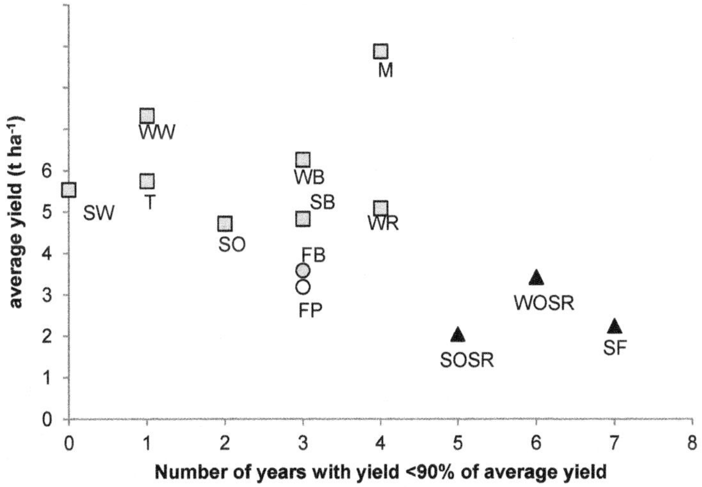

Fig. 13.4 A simple measure of temporal yield stability for grain legumes *(circles)*, oil crops *(triangles)* and cereals *(squares)*, calculated for national average yield data from Germany, 1993–2012. The *x*-axis shows the number of years out of 20 in which the yield of a given crop was below 90 % of the 20-year yield average of that crop. The *y*-axis shows the mean yield over the entire 20-year period. *FB* field beans, *FP* field peas, *M* maize, *SB* spring barley, *SF* sunflower, *SO* spring oats, *SOSR* spring oil-seed rape, *SW* spring wheat, *T* triticale, *WB* winter barley, *WOSR* winter oil-seed rape, *WR* winter rye, *WW* winter wheat

This was the same result as observed for winter barley and spring barley. Also, it was considerably better than for the oil crops (sunflower and oil-seed rape), where yields were < 90 % of the mean yield in 5 or more years. An analysis of regression-type stability of the same data confirms this picture, showing that field peas and faba beans have similar regression slopes as barley and winter rye (data not shown). Unfortunately, for other legume species such as lupins, continuous yield data was not available for the investigated period.

According to a further analysis using regional yield data from Germany, temporal yield stability of grain legumes is not consistently lower in comparison to other crops (Fig. 13.5). While the CV over years was significantly higher in grain legumes than in cereals ($p < 0.01$), the difference in CV between grain legumes and oil crops was not significant (Fig. 13.5a). The CVs of field peas and faba beans did not differ significantly. However, as a measure of stability, the CV has the disadvantage that CV values negatively correlate with mean yields, that is, greater means tend to lead to lower CVs. This bias is not observed when stability is calculated as the residuals from the power-law regression line (Fig. 13.5b, c). For this measure of stability, differences between grain legumes and other crop groups were not significant. Data from experimental field trials show a similar picture (Fig. 13.5d).

It is clear from these analyses that the view of particularly unstable yields in grain legumes may not be universally valid at all spatial levels. However, it is possible that at smaller spatial scales, grain legume yields do fluctuate significantly more than yields of other crops (Reckling et al. 2015). In fact, as far as small-scale environmental fluctuations may cancel each other out at higher spatial scales, yield stability is expected to be decreasing with the level of spatial aggregation. This is especially plausible where yields respond strongly to variations in precipitation across years. While this has been suggested for some grain legumes, in particular faba bean with its long vegetation time, further research is necessary to clarify the situation.

5.3 Factors Affecting Yield Stability in Grain Legumes

There is a multitude of potential factors that can have an effect on yield stability. Again, however, while suggestions on the nature of destabilizing factors abound, quantitative evidence is as yet relatively scarce. Generally, it is believed that grain legume yields are strongly dependent on weather, especially drought and heat. Further, their compensatory ability through plasticity of yield components is considered to be low. Genetic variation in yield stability among grain legumes does exist but stability is not strongly heritable. For faba beans, it was concluded that "due to low heritability, yield stability of faba bean inbred lines is a recalcitrant trait for practical breeding purposes" (Link et al. 1994). However, it was also found that late maturation in faba beans, though associated with high yield potential, implied lower yield stability due to the risk of lodging.

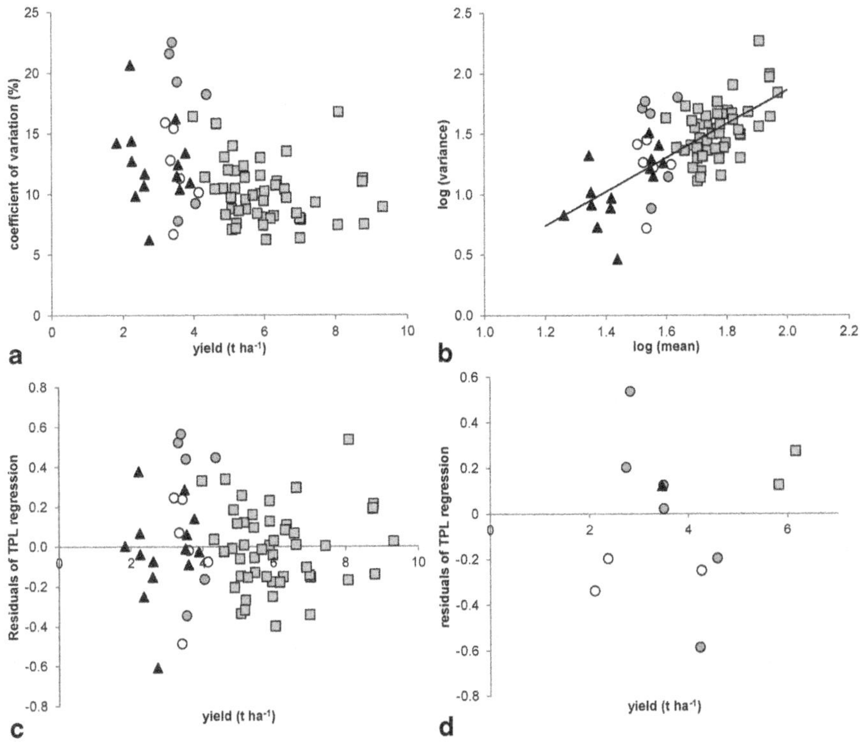

Fig. 13.5 Yield stability of grain legumes at regional and field trial level in faba beans *(grey circles)*, field peas *(white circles)*, oil crops *(black triangles)* and cereals *(grey squares)*. **a** Mean yields (in t ha^{-1}) versus coefficient of variation (across years, in %), data from seven official regional yield surveys in Germany (Baden-Württemberg, Niedersachsen, Oberbayern, Rheinland-Pfalz, Sachsen-Anhalt, Schleswig-Holstein, Thüringen) where at least 10 years of data were available; median number of years = 14; differences between groups of crops were tested with Tukey's HSD, using the Programme *R*, v. 3.0. **b** logarithm of the mean yield log(m) versus logarithm of the variance of the yield log(s^2) (calculated across years, on data in dt ha^{-1}), same dataset as in panel (**a**); the regression line follows $\log(s^2) = a \log(m) + b$, with $a = 1.349 \pm 0.171$ (s.e.), $b = -0.847 \pm 0.288$, Adj. $R^2 = 0.450$, $p < 0.001$. **c** residuals of individual data points from the regression line displayed in panel (**b**) (power-law residuals, POLAR). **d** POLAR calculated for experimental data of various crops. (Sperber et al. 1988; Link et al. 1994; Jensen 1996; Duc 1997; Kolbe et al. 2002; Smith et al. 2007; Fikere et al. 2008) *HSD* honest significant difference, *TPL* Taylor's power law

In terms of the effects of cropping systems, rotation will have a positive effect on yield stability in comparison to monocultures (see Sects. 2.1 and 3.5). Also, the variability of yields pooled from intercropping partners is generally lower than in the constituent monocrops, because of compensatory and complementary effects. Similarly, cultivar mixtures are associated with higher stability than single varieties, for example, as recently shown for cowpeas in Uganda (Okonya and Maass 2014). However, the yield variability of the grain legume partner in an intercropping mixture can be greater than in the sole crop (Böhm 2013). This could be because of the weather sensitivity of the competitive balance between the intercropping partners as the asymmetric competition amplifies weather effects between years. Similarly, weeds can constitute a positive feedback mechanism, magnifying other destabilizing factors. If grain legumes show weak growth, weeds will fill the gap left by the crop and reduce grain legume yield even more.

With a view to compare different cropping systems, Smith et al. (2007) measured temporal stability of soybean yields under different management conditions using CV as the measure of yield stability. In a 3-year rotation, four management systems were compared, namely CT, NT, Low Input (LI) and Organic (ORG). While mean yields of soybean were similar across the cropping systems, temporal yield stability differed between the systems. The interannual variation of soybean yields was lowest in the NT system and highest in the ORG system, with the other systems in between. Yield variability of soybean was comparable to that of corn, but higher than in wheat. However, the latter comparison is confounded by the fact that there was a greater variation of precipitation in the wheat phase than in the soybean phase of the trial. Thus, the study adds further, albeit indirect, evidence that the view of grain legumes being inherently less yield stable than other crops may not be universally true.

6 Conclusions and Outlook

Designing cropping systems with grain legumes is complex and requires integrative consideration of multiple issues, including aspects of soil physical, chemical and biological properties, competition among crop plants and between crops and weeds, as well as pest and disease epidemiology. Crop diversity in time, as in crop rotational design, and in space, such as in intercropping, provides ample opportunities to balance the needs from all these areas. In addition, currently underexplored genetic diversity at the species level offers new chances for further developing grain-legume-based cropping systems (Bell et al. 2011). This idea of diversification also extends to more complex systems such as agroforestry, both in tropical (Redhead et al. 1983) and temperate (Isaac et al. 2013) climates.

Unfortunately, the advance of molecular biology and the huge progress in the understanding of genetic and physiological mechanisms in grain legumes have not been paralleled by comparable advances in the agronomy of grain legumes and associated cropping system design. One area with particularly large gaps is the issue of how stability and resilience of grain legume yields are influenced by agronomic management. Thus, more research is clearly needed to optimise and innovate cropping systems for better performance and stability of grain legumes.

However, at the global level, there is a massive and increasing bias of current grain legume cropping towards a single species—soybean, as well as the widespread use of simplified and de-diversified cropping systems associated with this species. Research alone will not be able to turn this situation around. More is needed than just developing recommendations for optimised grain legume cropping. In fact, most of the key issues for successful cropping of grain legumes are already well established. Instead, barriers to greater implementation and adoption of a diverse range of grain legume cropping systems need to be identified. In addition, the demand for grain-legume-based products needs to increase.

Further, experience over the past few decades shows that political support can dramatically increase acreage grown with grain legumes, but that waning support can equally lead to a downward spiral of decreasing interest and an erosion of the socioeconomic framework that is needed for growing grain legumes. To make use of the considerable benefits of grain legumes both in agro-ecosystems and for human nutrition, increased efforts in integrative agronomic research need to be matched by critical analyses of whole food chains and enthusiastic and sustained public support.

Acknowledgments I would like to thank Frank Ellmer, Wolfgang Köhn, Peer Urbatzka and Martin Wolfe for helpful discussions on grain legume agronomy.

References

Agegnehu G, Ghizaw A, Sinebo W (2006) Yield performance and land-use efficiency of barley and faba bean mixed cropping in Ethiopian highlands. Europ J Agron 25:202–207

Albrecht R (2002) Vorfruchtwert von Körnerleguminosen in getreidebetonten Fruchtfolgen. Abschlussbericht Themenblatt-Nr. 41.04.420. Thüringer Landesanstalt für Landwirtschaft, Jena

Albrecht R, Guddat C (2004) Welchen Wert haben Körnerleguminosen in der Fruchtfolge? Thüringer Landesanstalt für Landwirtschaft. http://www.tll.de/ainfo/pdf/kleg0104.pdf. Accessed 20 Oct 2014

Alpmann D, Schäfer BC (2014) Der Wert von Körnerleguminosen im Betriebssystem. Union zur Förderung von Oel- und Proteinpflanzen e. V., Berlin

Annicchiarico P (2002) Genotype x environment interactions: challenges and opportunities for plant breeding and cultivar recommendations, vol 174. Food and Agriculture Organization, Rome

Atik O, El-Ahmed A, Baum M et al (2012) Effects of mixed pathotypes of *Didymella rabiei* on the development of *Ascochyta* blight on chickpea. Arab J Plant Prot 30(2):273–266

Aufhammer W (1999) Mischanbau von Getreide- und anderen Körnerfruchtarten—Ein Beitrag zur Nutzung von Biodiversität im Pflanzenbau. Ulmer, Stuttgart

Becker HC, Léon J (1988) Stability analysis in plant breeding. Plant Breed 101:1–23

Bell LW, Bennett RG, Ryan MH et al (2011) The potential of herbaceous native Australian legumes as grain crops: a review. Renew Agr Food Syst 26:72–91

Betencourt E, Duputel M, Colomb B et al (2012) Intercropping promotes the ability of durum wheat and chickpea to increase rhizosphere phosphorus availability in a low P soil. Soil Biol Biochem 46:181–190

Böhm H (2013) Anbau der Saatwicke (*Vicia sativa* L.) im Gemenge mit Hafer—eine Alternative in der Körnerleguminosenproduktion? Mitt Ges Pflanzenbauwiss 25:80–81

Bond D (1987) Recent developments in breeding field beans (*Vicia faba* L.). Plant Breed 99:1–26

Bulson HAJ, Snaydon RW, Stopes CE (1997) Effects of plant density on intercropped wheat and field beans in an organic farming system. J Agr Sci 128:59–71

Cardinale BJ, Matulich KL, Hooper DU et al (2011) The functional role of producer diversity in ecosystems. Am J Bot 98:572–592

Cassman KG, Whitney AS, Fox RL (1981) Phosphorus requirements of soybean and cowpea as affected by mode of N nutrition. Agron J 73:17–22

Chen Y, Zhang F, Tang L et al (2007) Wheat powdery mildew and foliar N concentrations as influenced by N fertilization and belowground interactions with intercropped faba bean. Plant Soil 291:1–13
Clark A (2008) Managing cover crops profitably. Sustainable Agriculture Network, Beltsville
Cohen JE (2013) Taylor's power law of fluctuation scaling and the growth-rate theorem. Theor Popul Biol 88:94–100
Costanzo A, Bàrberi P (2014) Functional agrobiodiversity and agroecosystem services in sustainable wheat production. A review. Agron Sust Dev 34:327–348
Dehghani H, Sabaghpour SH, Sabaghnia N (2008) Genotype × environment interaction for grain yield of some lentil genotypes and relationship among univariate stability statistics. Span J Agr Res 6:385–394
Döring H, Knust W, Mehnert M et al (1956) Richtzahlen und Tabellen für die Landwirtschaft. Deutsche Akademie der Landwirtschaftswissenschaften zu Berlin Institut für Agrarökonomik. Deutscher Bauernverlag, Berlin
Döring TF, Storkey J, Baddeley JA et al (2012) Legume based plant mixtures for delivery of multiple ecosystem services: weed diversity and weed control. In: McCracken K (ed) SAC-SEPA biennial conference: valuing ecosystems: policy, economic and management interactions. Edinburgh, United Kingdom, pp 163–168
Döring TF, Schweitzer K, Baumecker M et al (2014) Yield stability of wheat is lowest at high levels of mineral nitrogen input13th ESA Congress, 25–29 August 2014. Debrecen, Hungary
Duc G (1997) Faba bean (*Vicia faba* L.). Field Crops Res 53:99–109
Dyke GV, Prew RD (1983) Beans in crop rotations. In: Hebblethwaite PD (ed) The faba bean (*Vicia faba* L). Butterworths, London, pp 263–269
El-Moneim AA, Cocks P (1992) Adaptation and yield stability of selected lines of *Lathyrus* spp. under rainfed conditions in West Asia. Euphytica 66:89–97
Emden HF, Ball SL, Rao MR (1988) Pest, disease and weed problems in pea, lentil, faba bean and chickpea. In: Summerfield RJ (ed) World crops: cool season food legumes. (Current Plant Science and Biotechnology in Agriculture) Springer, Amsterdam, pp 519–534
Evans J, O'Connor GE, Turner GL et al (1989) N_2 fixation and its value to soil N increase in lupin, field pea and other legumes in south-eastern Australia. Austral J Agr Res 40:791–805
Feng J, Hwang R, Chang KF et al (2010) Genetic variation in *Fusarium avenaceum* causing root rot on field pea. Plant Pathol 59:845–852
Fikere M, Tadesse T, Letta T (2008) Genotype x environment interactions and stability parameters for grain yield of faba bean (*Vicia faba* L.) genotypes grown in southeastern Ethiopia. Int J Sust Crop Prod 3:80–87
Finckh MR, Wolfe MS (2006) Diversification strategies. In: Cooke BM, Jones G Kaye B (eds) The epidemiology of plant disease, 2nd edn. Springer, Amsterdam
Franzmann A (1992) Körnerleguminosen. In: Neuerburg W, Padel S (eds) Organisch-biologischer landbau in der praxis. Verlagsunion Agrar, München
Freyer B (2003) Fruchtfolgen konventionell, integriert, biologisch. Ulmer, Stuttgart
Fujita K, Ogata S, Matsumoto K et al (1990) Nitrogen transfer and dry matter production in soybean and sorghum mixed cropping system at different population densities. Soil Sci Plant Nutr 36:233–241
Gasnier C, Benachour N, Clair E et al (2010) Dig1 protects against cell death provoked by glyphosate-based herbicides in human liver cell lines. J Occup Med Toxic 5:29
Graham PH, Vance CP (2003) Legumes: importance and constraints to greater use. Plant Physiol 131:872–877
Guddat C, Karalus W (2009) Anbau und Ertragspotenzial von Körnerleguminosen in den Anbaugebieten der ostdeutschen Bundesländer. Thüringer Landesanstalt für Landwirtschaft, Erfurt
Hamel C, Furlan V, Smith D (1991) N_2-fixation and transfer in a field grown mycorrhizal corn and soybean intercrop. Plant Soil 133:177–185

Hardarson G, Danso S, Zapata F et al (1991) Measurements of nitrogen fixation in faba bean at different N fertilizer rates using the ^{15}N isotope dilution and 'A-value' methods. Plant Soil 131:161–168

Hauggaard-Nielsen H, Ambus P, Jensen ES (2001) Interspecific competition, N use and interference with weeds in pea–barley intercropping. Field Crops Res 70:101–109

Hauggaard-Nielsen H, Jørnsgaard B, Kinane J et al (2008) Grain legume–cereal intercropping: the practical application of diversity, competition and facilitation in arable and organic cropping systems. Renew Agr Food Syst 23:3–12

Heenan DP (1995) Effects of broad-leaf crops and their sowing time on subsequent wheat production. Field Crops Res 43:19–29

Heinze S, Oltmanns M, Joergensen R et al (2011) Changes in microbial biomass indices after 10 years of farmyard manure and vegetal fertilizer application to a sandy soil under organic management. Plant Soil 343:221–234

Helenius J (1990) Effect of epigeal predators on infestation by the aphid *Rhopalosiphum padi* and on grain yield of oats in monocrops and mixed intercrops. Entom Exp Appl 54:225–236

Helenius J, Jokinen K (1994) Yield advantage and competition in intercropped oats (*Avena sativa* L.) and faba bean (*Vicia faba* L.). Application of the hyperbolic yield-density model. Field Crops Res 37:85–94

Heuberger H, Kreuzmair A, Weh F et al (2005) Vegetabile Dünger als Stickstoffquelle für Topfbasilikum–Freisetzung und Aufnahme von Stickstoff aus vegetabilen Düngern bei Basilikum (*Ocimum basilicum* L.) in Topfkultur. Z Arzn Gewürzpfl 3:140–143

Hiltbrunner J, Kessler W (2009) Bioackerbohnen Merkblatt. Forschungsinstitut für biologischen Landbau, Frick

Hinsinger P, Plassard C, Tang C et al (2003) Origins of root-mediated pH changes in the rhizosphere and their responses to environmental constraints: a review. Plant Soil 248:43–59

Horn C, Birch C, Dalal R et al (1996) Sowing time and tillage practice affect chickpea yield and nitrogen fixation. 1. Dry matter accumulation and grain yield. Anim Prod Sci 36:695–700

Horneburg B (2003) Frischer Wind für eine alte Kulturpflanze—Linsen im ökologischen Anbau, ihre Geschichte und Verwendung. Dreschflegel eV, Göttingen

Horst WJ, Waschkies C (1987) Phosphatversorgung von Sommerweizen (*Triticum aestivum* L.) in Mischkultur mit Weißer Lupine (*Lupinus albus* L.). Agroecol Sust Food Syst 150:1–8

Hossain S, Strong W, Waring S et al (1996) Comparison of legume-based cropping systems at Warra, Queensland. 2. Mineral nitrogen accumulation and availability to the subsequent wheat crop. Soil Res 34:289–297

Isaac ME, Carlsson G, Ghoulam C et al (2013) Legume performance and nitrogen acquisition strategies in a tree-based agroecosystem. Agroecol Sust Food Syst 38:686–703

Jackai L, Daoust R (1986) Insect pests of cowpeas. Annu Rev Entom 31:95–119

Jacob I, Salzeder G, Aigner A et al (2014) Auswirkung einer Biofumigation durch Senf auf Ackerbohne und Futtererbse. In: Wiesinger K, Cais K, Obermaier S (eds) Angewandte Forschung und Beratung für den ökologischen Landbau in Bayern—Öko-Landbau-Tag 2014. Bayerische Landesanstalt für Landwirtschaft, Freising-Weihenstephan, pp 111–117

Jamont M, Piva G, Fustec J (2013) Sharing N resources in the early growth of rapeseed intercropped with faba bean: does N transfer matter? Plant Soil (in press) doi:10.1007/s11104-11013-11712-11102

Jensen ES (1996) Grain yield, symbiotic N_2 fixation and interspecific competition for inorganic N in pea-barley intercrops. Plant Soil 182:25–38

Jensen ES, Peoples MB, Hauggaard-Nielsen H (2010) Faba bean in cropping systems. Field Crops Res 115:203–216

Jones RAC, Coutts BA, Latham LJ et al (2008) Cucumber mosaic virus infection of chickpea stands: temporal and spatial patterns of spread and yield-limiting potential. Plant Pathol 57:842–853

Kahnt G (2008) Leguminosen im konventionellen und ökologischen Landbau. DLG Verlag, Frankfurt

Kay DE (1979) Food legumes—crop and product digest No 3. Tropical Products Institute, London
Kirkegaard J, Christen O, Krupinsky J et al (2008) Break crop benefits in temperate wheat production. Field Crops Res 107:185–195
Kolbe H (2009) Vergleich von Methoden zur Berechnung der biologischen N_2-Fixierung von Leguminosen zum Einsatz in der landwirtschaftlichen Praxis. Pflanzenbauwissenschaften 13:23–36
Kolbe H, Karalus W, Hänsel M et al (2002) Körnerleguminosen im Ökologischen Landbau—Informationen für Praxis und Beratung. Sächsische Landesanstalt für Landwirtschaft, Dresden
Köpke U (1998) Körnerleguminosen: N_2-Fixierung, Vorfruchtwirkung und Fruchtfolgegestaltung—Auswirkungen auf die Belastung von Agrarökosystemen. Schriftenreihe des Bundesministeriums für Ernährung, Landwirtschaft und Forsten, Reihe A: Angewandte Wissenschaft 367:52–63
Köpke U, Nemecek T (2010) Ecological services of faba bean. Field Crops Res 115:217–233
Köpke U, Rauber R, Schmidtke K et al (2011) Entwicklung neuer Strategien zur Mehrung und optimierten Nutzung der Boden-fruchtbarkeit. Gemeinsamer Abschlußbericht zu den Projekten BLE-08OE020, BLE-08OE145, BLE-08OE146, BLE-08OE147BÖL-Berichte-ID20737. Bonn, Göttingen, Dresden, Germany
Larkin RP, Griffin TS (2007) Control of soilborne potato diseases using *Brassica* green manures. Crop Prot 26:1067–1077
Larkin RP, Honeycutt CW, Olanya OM (2011) Management of *Verticillium* wilt of potato with disease-suppressive green manures and as affected by previous cropping history. Plant Dis 95:568–576
Leff B, Ramankutty N, Foley JA (2004) Geographic distribution of major crops across the world. Glob Biogeochem Cycl 18:GB1009. doi:1010.1029/2003GB002108
Link W, Stelling D, Ebmeyer E (1994) Yield stability in faba bean, *Vicia faba* L. 1. Variation among inbred lines. Plant Breed 112:24–29
López-Bellido L, López-Bellido RJ, Redondo R et al (2006) Faba bean nitrogen fixation in a wheat-based rotation under rainfed mediterranean conditions: effect of tillage system. Field Crops Res 98:253–260
Mahler RL, McDole RE (1987) Effect of soil pH on crop yield in Northern Idaho. Agron J 79:751–755
Mayer J, Buegger F, Jensen ES et al (2003) Estimating N rhizodeposition of grain legumes using a ^{15}N in situ stem labelling method. Soil Biol Biochem 35:21–28
McKenzie R, Middleton A, Solberg E et al (2001) Response of pea to rhizobia inoculation and starter nitrogen in Alberta. Can J Plant Sci 81:637–643
Neumann A, Schmidtke K, Rauber R (2007) Effects of crop density and tillage system on grain yield and N uptake from soil and atmosphere of sole and intercropped pea and oat. Field Crops Res 100:285–293
Nyatsanga T, Pierre WH (1973) Effect of nitrogen fixation by legumes on soil acidity. Agron J 65:936–940
Okonya JS, Maass BL (2014) Potential of cowpea variety mixtures to increase yield stability in subsistence agriculture: preliminary results. Int J Agron 2014:1–7
Olivera M, Tejera N, Iribarne C et al (2004) Growth, nitrogen fixation and ammonium assimilation in common bean (*Phaseolus vulgaris*): effect of phosphorus. Physiol Plant 121:498–505
Olsen JM, Griepentrog H-W, Nielsen J et al (2012) How important are crop spatial pattern and density for weed suppression by spring wheat? Weed Sci 60:501–509
Oltmanns M, Raupp J (2006) Ackerbohnenschrot im Vergleich zu Rottemist: Effekte auf Ertrag und Qualitätsparameter von Kartoffeln. Mitt Ges Pflanzenbauwiss 18:80–81
Paulsen HM, Schochow M, Ulber B et al (2006) Mixed cropping systems for biological control of weeds and pests in organic oilseed crops. Asp Appl Biol 79:215–220
Pekrun C, Hubert S, Schmidt W (2013) Unkrautkontrolle in Mais-Stangenbohnen-Gemengen. Mitt Ges Pflanzenbauwiss 25:245–246
Plancquaert P, Desbureaux J (1985) The agronomic effects of peas in rotation with winter wheat and oil seed rape—a progress report. In: Hebblethwaite PD, Heath MC, Dawkins TCK (eds) The pea crop. Butterworths, London, pp 193–202

Porter PM, Lauer JG, Huggins DR et al (1998) Assessing spatial and temporal variability of corn and soybean yields. J Prod Agr 11:359–363

Pyndji MM, Trutmann P (1992) Managing angular leaf spot on common bean in Africa by supplementing farmer mixtures with resistant varieties. Plant Dis 76:1144–1147

Ragsdale DW, Landis DA, Brodeur J et al (2011) Ecology and management of the soybean aphid in North America. Annu Rev Entom 56:375–399

Rauber R, Schmidtke K, Kimpel-Freund H (2001) The performance of pea (*Pisum sativum* L.) and its role in determining yield advantages in mixed stands of pea and oat (*Avena sativa* L.). J Agron Crop Sci 187:137–144

Raupp J (2005) Stickstoffmineralisation von Stallmist, Ackerbohnenschrot, Luzernegrünmehl und Rizinusschrot unter kontrollierten Bedingungen im Brutversuch. In: Heß J, Rahmann G (eds) Ende der Nische, Beiträge zur 8 Wissenschaftstagung Ökologischer Landbau. Kassel University Press, Kassel

Reckling M, Döring TF, Bloch R, Stein-Bachinger K, Bachinger J. (2015) 20 Jahre Modellbetrieb für Ökolandbau in Müncheberg. Teil 2: Wie ertragslabil sind Körnerleguminosen? In: Häring AM, Hörning B, Hoffmann-Bahnsen R, Luley H, Luthardt V, Pape J Trei G (eds) 13. Wissenschaftstagung Ökologischer Landbau Eberswalde 17.-20. 3. 2015, Verlag Dr. Köster, Berlin: pp 301–302

Redhead J, Maghembe J, Ndunguru B (1983) The intercropping of grain legumes in agroforestry systems. Plant research and agroforestry Proceedings of a consultative meeting held in Nairobi, Kenya, 8–15 April 1981. International Council for Research in Agroforestry, pp 117–124

Rochester IJ, Peoples MB, Hulugalle NR et al (2001) Using legumes to enhance nitrogen fertility and improve soil condition in cotton cropping systems. Field Crops Res 70:27–41

Romano RM, Romano MA, Bernardi MM et al (2010) Prepubertal exposure to commercial formulation of the herbicide glyphosate alters testosterone levels and testicular morphology. Arch Toxicol 84:309–317

Rose T, Hardiputra B, Rengel Z (2010) Wheat, canola and grain legume access to soil phosphorus fractions differs in soils with contrasting phosphorus dynamics. Plant Soil 326:159–170

Rühl G, Bramm A, Greef J (2009) Aspekte des Anbaus von Körnerleguminosen. J Kulturpfl 61:312–317

Sastawa B, Lawan M, Maina Y (2004) Management of insect pests of soybean: effects of sowing date and intercropping on damage and grain yield in the Nigerian Sudan savanna. Crop Prot 23:155–161

Saucke H, Ackermann K (2006) Weed suppression in mixed cropped grain peas and false flax (*Camelina sativa*). Weed Res 46:453–461

Saucke H, Juergens M, Döring TF et al (2009) Effect of sowing date and straw mulch on virus incidence and aphid infestation in organic faba-beans (*Vicia faba* L.). Ann Appl Biol 154:239–250

Schlipf JA (1898) Handbuch der Landwirtschaft, 13th edn. Paul Parey, Berlin

Schmidtke K, Rauber R (2000) Stickstoffeffizienz von Leguminosen im Ackerbau. In: Möllers C (ed) Stickstoffeffizienz landwirtschaftlicher Kulturpflanzen Initiativen zum Umweltschutz 21. Erich Schmidt Verlag, Berlin, pp 48–69

Schmidtke K, Neumann A, Hof C et al (2004) Soil and atmospheric nitrogen uptake by lentil (*Lens culinaris* Medik.) and barley (*Hordeum vulgare* ssp. *nudum* L.) as monocrops and intercrops. Field Crops Res 87:245–256

Schwenke GD, Peoples MB, Turner GL et al (1998) Does nitrogen fixation of commercial, dryland chickpea and faba bean crops in north-west new south wales maintain or enhance soil nitrogen? Austral J Expt Agr 38:61–70

Seufert V, Ramankutty N, Foley JA (2012) Comparing the yields of organic and conventional agriculture. Nature 485:229–232

Smith RG, Menalled FD, Robertson GP (2007) Temporal yield variability under conventional and alternative management systems. Agron J 99:1629–1634

Sperber J, Barisich R, Edinger E et al (1988) Öl- und Eiweißpflanzen: Anbau, Kultur, Ernte. Österreichischer Agrarverlag, Vienna

Stoddard FL, Nicholas AH, Rubiales D et al (2010) Integrated pest management in faba bean. Field Crops Res 115:308–318

Tang C, Hinsinger P, Jaillard B et al (2001) Effect of phosphorus deficiency on the growth, symbiotic N. Agron 21:683–689

Tilman D, Reich PB, Knops J et al (2001) Diversity and productivity in a long-term grassland experiment. Science 294:843–845

Tilman D, Reich PB, Knops JMH (2006) Biodiversity and ecosystem stability in a decade-long grassland experiment. Nature 441:629–632

Trenbath BR (1993) Intercropping for the management of pests and diseases. Field Crops Res 34:381–405

Urbatzka P, Graß R, Haase T et al (2011) The level of N_2-fixation of different genotypes of winter pea in comparison to spring pea in pure and mixed stands. J Kulturpfl 63:374–386

Urbatzka P, Salzeder G, Rehm A (2013) Vergleich verschiedener Wuchstypen von Sommererbsen unter den Bedingungen des ökologischen Landbaus. Mitt Ges Pflanzenbauwiss 25:193–194

von Lengerke A (1840) Landwirthschaftliche Statistik der deutschen Bundesstaaten. Zweiter Band, erste Abteilung. George Westermann, Braunschweig

von Richthofen J-S, Pahl H, Nemecek T (2006) Was Körnerleguminosen in Anbausystemen leisten. Raps 24:35–39

Walley FL, Clayton GW, Miller PR et al (2007) Nitrogen economy of pulse crop production in the northern great plains. Agron J 99:1710–1718

Waltz E (2010) Glyphosate resistance threatens Roundup hegemony. Nat Biotech 28:537–538

Wang X, Pan Q, Chen F et al (2011) Effects of co-inoculation with arbuscular mycorrhizal fungi and rhizobia on soybean growth as related to root architecture and availability of N and P. Mycorrhiza 21:173–181

Weiner J, Andersen SB, Wille WKM et al (2010) Evolutionary agroecology: the potential for cooperative, high density, weed-suppressing cereals. Evol Appl 3:473–479

Wolfe M, Fradgley N, Winkler L et al (2013) Beans and wheat intercropping: a new look at an overlooked benefit. Org Res Cent Bull 112:8–9

Wright AT (1990) Yield effect of pulses on subsequent cereal crops in the northern Prairies. Can J Plant Sci 70:1023–1032

Yan F, Schubert S, Mengel K (1996) Soil pH changes during legume growth and application of plant material. Biol Fert Soils 23:236–242

Zerhusen-Blecher P, Schäfer BC (2013) Stand des Wissens und Ableitung des Forschungsbedarfes für eine nachhaltige Produktion und Verwertung von Ackerbohne und Erbse. Fachhochschule Südwestfalen, Agrarwirtschaft Fachbereich, http://download.ble.de/12NA118/Abschlussbericht%2812NA118_12NA077%29.pdf. Accessed 20 Oct 2014

Zimmermann M (1958) Schlipf Praktisches Handbuch der Landwirtschaft, 32nd edn. Paul Parey, Hamburg

Index

A. M. De Ron (ed.), *Grain Legumes,* Handbook of Plant Breeding 10,
DOI 10.1007/978-1-4939-2797-5

Printed by Printforce, the Netherlands